MECHANISMS OF PHYSICAL AND EMOTIONAL STRESS

ADVANCES IN EXPERIMENTAL MEDICINE AND BIOLOGY

Recent Volumes in this Series

Volume 238
BIOTECHNOLOGICAL APPLICATIONS OF LIPID MICROSTRUCTURES
Edited by Bruce Paul Gaber, Joel M. Schnur, and Dennis Chapman

Volume 239
HOST DEFENSES AND IMMUNOMODULATION TO INTRACELLULAR PATHOGENS
Edited by Toby K. Eisenstein, Ward E. Bullock, and Nabil Hanna

Volume 240
PROTEASES: Potential Role in Health and Disease II
Edited by Walter H. Hörl and August Heidland

Volume 241
MOLECULAR BIOLOGY OF HEMOPOIESIS
Edited by Mehdi Tavassoli, Esmail D. Zanjani, Joao L. Ascensao, Nader G. Abraham, and Alan S. Levine

Volume 242
VASCULAR ENDOTHELIUM IN HEALTH AND DISEASE
Edited by Shu Chien

Volume 243
EICOSANOIDS, APOLIPOPROTEINS, LIPOPROTEIN PARTICLES, AND ATHEROSCLEROSIS
Edited by Claude L. Malmendier and Petar Alaupovic

Volume 244
THE EXPANDING ROLE OF FOLATES AND FLUOROPYRIMIDINES IN CANCER CHEMOTHERAPY
Edited by Youcef Rustum and John J. McGuire

Volume 245
MECHANISMS OF PHYSICAL AND EMOTIONAL STRESS
Edited by George P. Chrousos, D. Lynn Loriaux, and Philip W. Gold

---

MECHANISMS OF PHYSICAL AND EMOTIONAL STRESS

Edited by

George P. Chrousos,
D. Lynn Loriaux,
and Philip W. Gold

National Institute of Health
Bethesda, Maryland

PLENUM PRESS • NEW YORK AND LONDON

Library of Congress Cataloging in Publication Data

Mechanisms of physical and emotional stress / edited by George P. Chrousos, D. Lynn Loriaux, and Philip W. Gold.
(Advances in experimental medicine and biology; v. 245)
p. cm.
Based on the proceedings of a symposium held Nov. 6–8, 1986, at the National Institutes of Health Clinical Center, Bethesda, Md.
Includes bibliographies and index.
ISBN 0-306-43017-7
1. Stress (Physiology)—Congresses. 2. Stress (Psychology)—Congresses. I. Chrousos, George P. II. Loriaux, D. Lynn. III. Gold, Philip, W., 1944–
[DNLM: 1. Stress—congresses. 2. Stress, Psychological—congresses. W3 AD215 / QZ 160 M486 1986]
QP82.2.S8M39 1988
612—dc19
DNLM/DLC 88-28384
for Library of Congress CIP

Based on the Proceedings of an NIH Symposium on Mechanisms of Physical and Emotional Stress, held November 6–8, 1986, in Bethesda, Maryland

A Division of Plenum Publishing Corporation
233 Spring Street, New York, N.Y. 10013

Printed in the United States of America

We would like to dedicate this book to our wives and children.

—George P. Chrousos

—D. Lynn Loriaux

—Philip W. Gold

PREFACE

It has been over 50 years since Hans Selye formulated his concept of stress. This came after the isolation of epinephrine and norepinephrine and after the sympathetic system was associated with Walter Cannon's "fight or flight" response. The intervening years have witnessed a number of discoveries that have furthered our understanding of the mechanisms of the stress response. The isolation, identification and manufacture of glucocorticoids, the identification and synthesis of ACTH and vasopressin, and the demonstration of hypothalamic regulation of ACTH secretion were pivotal discoveries. The recent identification and synthesis of CRH by Willie Vale and his colleagues gave new impetus to stress research.

Several new concepts of stress have developed as a result of advances in bench research. These include the concept of an integrated "stress system", the realization that there are bi-directional effects between stress and the immune system, the suggestion that a number of common psychiatric disorders represent dysregulation of systems responding to stress, and the epidemiologic association of stress with the major scourges of humanity.

The idea that there is an integrated set of anatomical structures that function to produce the behavioral, physiological and biochemical changes directed toward maintaining homeostasis is supported by new findings in neurobiology demonstrating anatomical and functional connections between the hypothalamus, the arousal center in the pons, and several sympathetic nuclei in the hindbrain. Additional integrative interactions between the adrenal and sympathetic divisions of the stress system are present in the periphery, as glucocorticoids and catecholamines are complementary and permissive to each other's action in regulating and maintaining metabolic and cardiovascular homeostasis.

A number of new and interesting associations between the "stress system" and the immune system have been revealed over the years supporting the long suspected link between endocrinology and immunology. Glucocorticoids at "pharmacologic" doses have been known to suppress the immune system since the early 50's. These doses, however, can be within the daily production rate of cortisol during stress. Recently, Allan Munck has suggested that one of the roles of glucocorticoids is to hold the immune system in check following tissue damage. This idea suggests an explanation for the unexplainable-from a teleologic point of view - immunosuppressive actions of cortisol. On the other hand, it appears that several products of the immune system, including interleukin I, cachexin and thymosin, are potent activators of the pituitary-adrenal axis. These findings, together with the immunosuppressive effects of cortisol reveal a regulatory feedback loop by which the immune system keeps its own activity under control by using the endocrine system as an intermediate.

In the past decade it has become apparent that a number of psychiatric states, including depression, anxiety, anorexia nervosa, and opiate or alcohol withdrawal are associated with hyperactivity of the hypothalamic-pituitary-adrenal axis. It has also been shown that CRH itself exerts a number of effects reminiscent of both the stress response and of the pathophysiologic manifestations of the above psychiatric disorders. The idea that these disorders represent consequences of a hyperactive or hyperresponsive stress system is a very appealing one and suggests new approaches to therapy.

Finally, one needs to mention the major consequences that chronic stress may have on the health of many segments of the human population. The "diseases of adaptation" should include, among others, accelerated atherosclerosis, tumorigenesis, several gastrointestinal and allergic disorders, alcoholism and narcotic addiction.

This volume which represents the updated proceedings of a symposium on the "Mechanisms of Physical and Emotional Stress" that took place at the National Institutes of Health Clinical Center in November 1986, will attempt to put together a number of reports from diverse areas with stress as a common focus. The editors would like to thank the participants for a stimulating and enlightening conference and the authors for their valuable contributions.

The Editors

CONTENTS

INTRODUCTION

BASIC MECHANISMS OF THE STRESS RESPONSE

STRESS AND IMMUNITY

PHYSICAL STRESS

STRESS, DEVELOPMENT AND REPRODUCTION

EMOTIONAL STRESS AND PSYCHIATRIC ILLNESS

INTRODUCTION

INTRODUCTION

THE CONCEPT OF STRESS AND ITS HISTORICAL DEVELOPMENT

"He who thus considers things in their first growth and origin, whether a state or anything else, will obtain the clearest view of them."

Aristotle, circa 450 BC

The term "stress" has been used, and occasionally abused, by scientists and by the lay public, in almost every single language of the civilized world. Many definitions and meanings have been ascribed either consciously or unconsciously to the word. Nevertheless, despite a lack of general agreement about its meaning, the term has prevailed because it attempts to address a basic principle of Nature, that of maintenance of balance, equilibrium, or harmony in the face of disturbing forces on the one hand and counteracting reestablishing forces on the other. One potential reason for the confusion surrounding the term "stress" is that it has been variously used to describe the disturbing forces, the disturbed balance or disequilibrium, and/or the results of the counteracting, reestablishing forces.

The term "stress" is, according to dictionaries, synonymous with the term "strain." The roots of these words go far back into the origins of Indo-European language. In ancient Greek, it was the root for "strangale," which means a halter, and the verb "strangaleuin," which means to strangle. In Latin, it gave the root for the verb "stringere," which means to draw tight, to press. Thus, the roots of stress have always meant the exertion of primarily adverse forces. Modern words like "distress" or "strenuous" imply similar connotations.

Hans Selye, in the late 1930's, was the first man to use the term "stress" more or less as we know it today. He borrowed the term from physics and set it to mean the mutual actions of forces that take place across any section of the body. The selection of the term was thus conceptually, etymologically, and phonetically sound.

The brief historical approach to stress that will be attempted will be divided into three parts. First, we will try to answer the question, "What is stress?" or at least to propose a current, conventionally accepted definition. This will not only allow us to look back at its historical development but also will allow scientists from diverse disciplines, to communicate with each other. Second, we will try to answer the question, "How was the concept of stress developed to mean what it means today," and, third, we will try to answer the question, "Which are the developmental milestones that led to our current understanding of stress mechanisms?" This, of necessity, will be a very brief overview of an extraordinarily complex and multifaceted subject.

What is Stress?

Stress is the state of threatened balance, equilibrium, or harmony. The "balanced" state directly translated from the Greek "homeostasis" means "steady state." The disturbing forces or threats to homeostasis we call "stressors", and the counteracting reestablishing forces we call "adaptive responses". Homeostasis pertains not only to a steady milieu of chemical or physiological parameters but also to states of mind. Stressors are not only physical, like cold, hot, pain or loss of blood, but also emotional, like psychological loss or uncertainty. Finally, the adaptive responses of the body can be specific to a given stressor or, in other settings, general, and independent of the type of stressor.

Excessive or chronic activation of adaptive responses to stressors can have their own noxious effects, themselves acting as threats to homeostasis. They produce a syndrome that was first noted by Hans Selye and gave him the impetus for formulating his concept of a nonspecific generalized response against any major threat to homeostasis.

Homeostasis, in summary, is the steady, harmonious state of mind and body; 'stress' is the recognition by the body of a stressor and therefore the state of threatened homeostasis; stressors are threats against homeostasis; and adaptive responses are the body's attempt to counteract the stressor and reestablish homeostasis. These responses can be specific to the stressor, or generalized and nonspecific, probably depending on the magnitude rather than the type of the stressor.

How was the Concept of Stress Developed?

The first written reference to homeostasis was by the natural philosopher Empedocles about 450 B.C. He considered all matter as a mixture of essential elements existing together in a state of harmonious balance (Table 1). Hippocrates, a contemporary, extended the concept in living beings considering the state of harmonious balance as health and the state of dysharmony as disease.

The next scientist to address the concept of homeostasis, was Claude Bernard. He talked about the stability of the internal environment and described the "milieu interieur formed by the circulating organic liquid which surrounds and bathes all the tissue elements." Walter Cannon coined the term homeostasis, at the turn of the century and provided a clear description of this state. He wrote:

"The living being is stable. It must be so in order not to be destroyed, dissolved or disintegrated by the colossal forces, often adverse, which surround it. In a sense it is stable because it is modifiable, the slight instability is the necessary condition for the true stability of the organism".

The first reference to disturbed homeostasis and stressors was by Hippocrates in 400 B.C. He considered diseases in general as states of disequilibrium and imbalance. He ascribed natural causes to diseases, defining them as stressors. An excerpt of his words on effects of external natural forces on epilepsy is given below:

"As for this disease called divine, surely it has its nature and causes, as have other diseases. It arises - like them - from things which enter and quit the body, such as cold, the sun and the winds, things everchanging and never at rest. Such things are divine or not as you will, for the distinction matters not and there is no need to make such division anywhere in nature, for all are alike divine or all are alike human. All have their antecedent causes which can be found by those who seek them."

Hippocrates was also the first to talk about the adaptive responses of the body. He introduced the concept of the healing forces of Nature, later called by the Romans "vis medicatrix naturae". These seem to be the same specific responses and nonspecific adaptive reactions of the body to stress that were later suggested by Selye.

Epicurus, in 300 BC spoke of emotional stressors and suggested that "ataraxia" (impertubability of mind) is a superior pleasure. He preached modification of responses to emotional stimuli as a way of improving the quality of life.

TABLE 1. THE CHRONOLOGY OF STRESS AS A CONCEPT

Empedocles (500-430 BC)	Matter consists of essential elements and qualities in opposition or alliance to one another.
Hippocrates (460-375 BC)	A harmonious balance of the elements and qualities of life is health -- Dysharmony is disease. Vis medicatrix naturae
Epicurus (341-270BC)	"Ataraxia" (imperturbability of mind) as a superior pleasure.
Sydenham (AD 1624-1689)	Symptoms and signs of a disease arise also from the reaction of the patient's system.
Claude Bernard (1813-1878)	The "milieu interieur"
Walter Cannon (1871-1945)	"Homeostasis" Bodily responses to emotions "Fight or flight reaction"
Hans Selye (1907-1982)	"The general adaptation syndrome" (the "stress" syndrome); "Diseases of adaptation"

Selye, in defining the "General Adaptation Syndrome" or "Stress Syndrome", picked up where Hippocrates had left off. He realized that patients with severe illnesses often developed a number of common symptoms and signs, primarily gastrointestinal, weight loss, and depression. He also showed this experimentally in the rat, where many different types of noxious influences led to enlarged adrenal cortices, shrinkage of the thymus and lymph nodes, and development of gastrointestinal symptoms. On the basis of these observations, he suggested that the body has nonspecific mechanisms to adapt to a variety of noxious influences and called this the "General Adaptation Syndrome" or "Stress Syndrome". Selye also talked about "diseases of adaptation", a concept originally suggested by Sydenham in the seventeenth century.

Milestones Leading to Our Current Understanding of Stress Mechanisms

The general adaptive responses that occur during stress can be divided into adaptive redirection of behavior or "behavioral adaptation" and into adaptive redirection of energy or "peripheral adaptation" (Table 2).

"Behavioral adaptation" can be explained as facilitation of adaptive and inhibition of nonadaptive neural pathways subserving various behaviors. The adaptive changes in behavior include an improved alertness and attention span level, a decrease of reflex time, and suppression of feeding and sexual behaviors. "Peripheral adaptation" includes redirection of O_2 and nutrients to both the CNS and the stressed body site, increased detoxification from toxic products produced in excess during stress, and containment of the stress reaction.

The coordination of behavioral and peripheral adaptive responses to stressors occurs in the brain. Neural circuits, neurotransmitters, and neuropeptides are the central effectors of these changes. Some of these central effectors will produce the right behaviors, others will stimulate the peripheral effectors, such as the sympathetic system and the pituitary--adrenal axis.

The behavioral and peripheral components of the stress response are not isolated from each other but interact at various levels. Common central effectors like the catecholamines, vasopressin, or CRH may be shared, and biochemical and neural connections may be present. Interactions at other levels also exist. A peripheral effector substance, the glucocorticoids, has profound effects on brain function. Behaviors also can alter the response of the peripheral effector systems. For instance, during the "fight or flight" response, exercise can act as a stressor *per se* to activate the sympathetic system and the pituitary-adrenal axis.

It is of interest that certain diseases which are responsible for a disproportionate share of human misery are associated with disinhibition or increased sensitivity of the central effectors of the stress response. Thus, depressive illness and panic anxiety disorder appear to be associated with mild chronic hyperactivation of the adaptive responses to stress and hypersensitivity of these responses to emotional/social stressors. Study of the normal and abnormal stress response may lead to an enhanced understanding of "emotional" illness and provide us with better tools to combat it.

TABLE 2. ADAPTIVE RESPONSES DURING STRESS

BEHAVIORAL ADAPTATION
Facilitation of adaptive and inhibition of nonadaptive neural pathways

- Alertness level
- Attention span level
- Decrease of reflex time
- Suppression of feeding behavior
- Suppression of sexual behavior

PERIPHERAL ADAPTATION
Adaptive redirection of energy

- O_2 & nutrients directed to the CNS and the stressed body site
- Detoxification from toxic products
- Containment of the stress reaction

We will conclude by briefly recounting the milestones that led to our current concept of stress. Advances in our understanding of the brain, the sympathetic system, the hypothalamic-pituitary unit and the adrenal cortex are prominent in our hierarchy.

Herophilus called the brain "the seat of intelligence and the center of the nervous system," in 300 B.C. The concepts of synapse, levels of control and integration, neurotransmission, and neurosecretion were established over 2000 years later. The presence and function of neuropeptides, in addition to neurotransmitters, and advances in functional neuroanatomy, are the recent products of this long train of investigations.

The description of the adrenal medulla, the discovery of the catecholamines, and the involvement of catecholamines in the stress response have been the products of the past 150 years.

Harris proposed the humoral control of the pituitary by the hypothalamus in 1948. CRF was demonstrated to be an active "principle" in 1955, and CRH was isolated by Vale, and colleagues, in 1981.

Finally, the adrenal glands were only described in 1563 by Eustachius. Their function was suggested by Addison and Brown-Sequard in the last century. The isolation of glucocorticoids, their involvement in the stress response, and their important therapeutic properties are recent developments of the past fifty years.

We owe much to the succession of men and women that, step by step, provided the body of truths that constitute our modern concept of stress. It is impossible to give all these people the credit they deserve. "The whole succession of men through the ages should be considered as one man, ever living and always learning" (Pascal - A.D. 1633-1662).

George P. Chrousos, M.D.
D. Lynn Loriaux, M.D., Ph.D.
Philip W. Gold, M.D.

Sources

1. F. Adams, "The genuine works of Hippocrates", The Williams and Wilkins Company, Baltimore (1939).
2. I. N. Dambasis, "The Hippocratic Philosophy", Parnassos (Athens) 13:34-48 (1971).
3. F. H. Garrison, "An Introduction to the History of Medicine", W. B. Saunders Company, Philadelphia (1929).
4. B. L. Gordon, "Medicine Throughout Antiquity". F. A. Davis Company, Philadelphia, 1949.
5. V. C. Medvei, "A History of Endocrinology", MTP Press Ltd, Boston (1982).
6. H. Selye, "Stress", Acta Inc. Medical Publisher, Montreal (1950).
7. C. Singer, "A Short History of Science", University Press, Oxford (1941).
8. H. O. Taylor, "Greek Biology and Medicine", Marshall Jones Company, Boston (1922).
9. R. F. Witzmann, "Steroids: Keys to Life", Van Norstrand Reinhold Company, New York (1981).
10. T. Zvingerus, "Commentarii Hippocratis", Episcopiorum Opera, Basileae, (1569).

BASIC MECHANISMS OF THE STRESS RESPONSE

SYMPATHOADRENAL MEDULLARY SYSTEM AND STRESS

I. J. Kopin, G. Eisenhofer, and D. Goldstein*

National Institutes of Health
National Institute of Neurological and Communicative Disorders and Stroke
*National Heart, Lung, and Blood Institute
Bethesda, Maryland 20892

HISTORICAL INTRODUCTION

Claude Bernard, in 1879, introduced the concept that as organisms evolved to become more independent of their external environment, they developed more complex means of preserving their internal environment, the *milieu interieur*. He wrote that "all vital mechanisms, however varied they may be, have only one object, that of preserving the conditions of life in the internal environment." Modern concepts of stress stem from this tenet.

The neuroendocrine determinants of stress responses were first indicated by Oliver and Shaffer, who, in 1895, described the remarkable physiological changes produced after injection of adrenal extracts. These observations led to the isolation and characterization of epinephrine (or adrenaline) by Abel in 1899. The similarity of effects of epinephrine to those of sympathetic nerve stimulation suggested to Elliott in 1905 that an epinephrine-like substance might be a chemical mediator released from sympathetic nerve endings. Almost a half-century later, von Euler discovered that the transmitter was norepinephrine.

Walter Cannon, in 1929, summarized a generation of work which centered on the theory that the sympathoadrenomedullary system reacts in various emergency situations, such as pain, bleeding, exposure to cold, and rage, by secretion of epinephrine into the bloodstream, and that sympathoadrenomedullary activation plays an important role in preserving the *milieu interieur*. Cannon coined the term "homeostasis" to describe "the coordinated physiological reactions which maintain the steady states of the body" by integrated cooperation of wide range of organs.

Selye proposed that stress is a specific response pattern elicited regardless of the stimulus and allowing the organism to adapt and to re-establish normalcy (see Selye, 1983). He did not view stress as necessarily damaging or unpleasant--he used these characteristics to define distress. Selye defined stages of the stress reaction: an initial "alarm reaction", characterized by an immediate sympathoadrenomedullary discharge; a subsequent "stage of resistance", characterized by activation of the hypothalamic-pituitary-adrenocortical axis; and a syndrome of adrenal hypertrophy, gastrointestinal ulceration, and thymic and lymphoid shrinkage, which he called the "General Adaptation Syndrome", which could proceed to the last stage, exhaustion and death. During the stage of resistance, derangements in hormonal responses and abnormal tissue changes were proposed to result in "diseases of adaptation". Although Selye's early concepts were expressed in terms of biological processes, they were extended to include psychological, interpersonal, and even sociocultural processes (see e.g., Jenkins, 1979).

HOMEOSTASIS AND DISTRESS

Stress defies easy definition. Cannon would have defined stress in terms of stimuli--in particular, noxious stimuli or stressors--which are countered by activation of physiological systems including the sympathoadrenomedullary system. Selye defined stress in terms of a specific reaction pattern, with emphasis on the pituitary-adrenocortical system. In this presentation, stress is defined neither by the stimuli nor the responses, but as a state in which expectations--whether genetically programmed or acquired--do not match current or anticipated perceptions of the internal or external environment. Stress responses evolved because they were advantageous in natural selection, providing means to anticipate and react rapidly to threats to the survival or well-being of the organism, preserving the internal milieu by short-term adjustments in activities of several systems. When these responses do not maintain homeostasis and resistance fails due to inadequate, inappropriate, or excessive activation of the compensatory systems, physical and psychological damage results. This condition, often associated with emotional experiences such as anxiety, fear, or anger, is distress; avoidance of distress is one of the main objectives motivating learning. In contrast, successful homeostatic responses usually are unconscious.

Homeostatic and distress responses also differ in that homeostatic responses are stimulus-specific, producing changes which are appropriate for adaptation to the disturbance, whereas distress responses usually are more generalized and can be precipitated when the homeostatic mechanisms are overwhelmed. Homeostatic responses are relatively predictable within a relatively narrow range of individual variation, whereas distress responses can differ markedly among individuals, depending on constitutional factors as well as the conditioning history of the organism. Both homeostatic mechanisms and distress responses include activation of the sympathoadrenomedullary system.

Homeostatic responses generally are mediated by nerve networks, such as the sympathoneural and parasympathetic systems (Table 1), whereas distress responses often involve hormones such as epinephrine, adrenal corticosteroids, and vasopressin. Challenges which throughout evolution have posed direct threats to well-being, such as cardiovascular collapse, hypoglycemia,

Table 1. **Stimuli Which Evoke Autonomic Responses**

Stimulus-Specific Responses (Autonomic Nervous System)	*Generalized Responses* (Sympathoadrenomedullary)
Orthostasis, Exercise (C.V.S.*, Lung)	**Exhaustion, C.V.S.* Collapse**
Temperature (C.V.S.*, Lung, Sweat)	**Pain**
Food Intake, Waste Discharge (GI Motility, Secretion)	**Hypoglycemia**
Fluid Balance (G.U.**, Renal, Bladder)	**Hypovolemia, Hemorrhage**
Alertness (C.V.S.*)	**"Fight or Flight"** (Fear, Rage Anxiety, Depression)

*C.V.S. = Cardiovascular **G.U. = Genitourinary

asphyxiation, hemorrhage, and fight/flight situations, require and are countered by global, metabolic responses determined mainly by elaboration of hormones, even at low intensities of stimulation. The organism invariably appears to be distressed. As the intensity increases, sympathoneural responses also can occur. Stimuli such as orthostasis, altered environmental temperature, and food ingestion tend to elicit more regionally specific responses of nerve networks, causing complex adjustments of glandular secretion and blood volume distribution. The organism often is unaware of these responses, but if the intensity of stimulation overwhelms the homeostatic mechanisms--e.g., with forced prolonged standing or exercise to exhaustion--more generalized responses are elicited and the organism experiences distress.

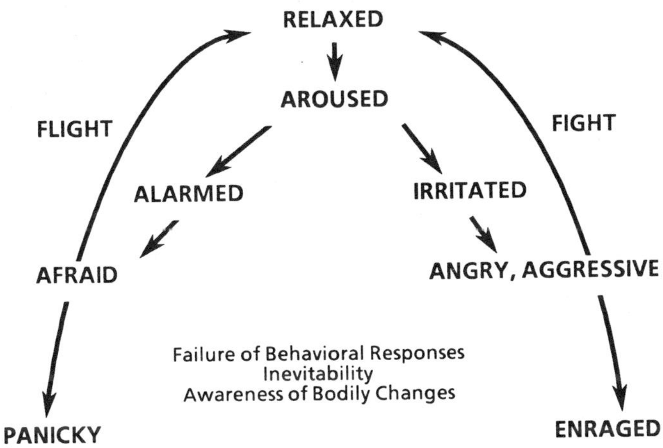

Fig. 1 Arousal States in Response to Environmental Stimuli or Internal Needs

Arousal states resulting from environmental stimuli or internal needs may be conceptualized in terms of continua (Figure 1). The organism strives to attain a safe, sated state. Alerting reactions can evolve to alarm or irritation and then panicky fear or aggressive rage, the physiological and biochemical concomitants of distress becoming progressively more intense. Failure to escape, avoid, or reverse the disturbance can lead to extreme states where behaviors and physiological responses no longer achieve the desired goals. Persistent or chronically repeated distress responses such as these can lead to a variety of psychosomatic syndromes.

DETERMINANTS OF SYMPATHOADRENOMEDULLARY FUNCTION

The neuronal modulation of sympathoadrenomedullary outflow began to be unfolded when von Euler discovered in 1948 that although epinephrine is the hormone released from the adrenal medulla, the precursor of epinephrine, norepinephrine, is the neurotransmitter released at peripheral sympathetic nerve endings. Axelrod and co-workers defined the metabolism and disposition of catecholamines (Axelrod, 1960), demonstrated the importance of reuptake as a mode of terminating the actions of the neurotransmitter (Hertting and Axelrod, 1961), and showed that a wide variety of psychoactive drugs act by altering catecholamine disposition or metabolism.

Release of epinephrine from the adrenal medulla and of norepinephrine from sympathetic nerve terminals depends on nerve impulse outflow from the spinal cord via preganglionic cholinergic neurons whose cell bodies are located in the intermediolateral gray matter of the thoracic and upper lumbar segments. At each level, myelinated axons from the preganglionic neurons emerge in an anterior root of a spinal nerve, form the white ramus communicans, and enter the corresponding ganglion of the sympathetic chain. In the ganglia the fibres may form synapses (usually many) with postganglionic (noradrenergic) neurons or traverse the ganglia without forming synapses, continuing to merge into the greater or lesser splanchnic nerves which supply the adrenal medulla or synapse with neurons located in ganglia clustered around the branches of the abdominal aorta (coelic or aorta-renal ganglia). Preganglionic fibres from the upper thoracic segments ascend along the sympathetic chain to terminals in the stellate or cervical sympathetic ganglia; preganglionic fibres from the lower thoracic or lumbar segments descend to sympathetic ganglia of the lower lumbar or sacral spinal nerves. The spinal sympathetic ganglia send postganglionic unmyelinated fibres, via a gray ramus communicans, to the spinal nerves and thereby reach various areas of the peripheral tissues. Unmyelinated fibres from the cervical ganglia are distributed via branches of the vagus to reach the thoracic viscera and via plexuses on the carotid artery and its branches to reach the eye and blood vessels, glands, and muscles of the head and neck.

Immediate (within seconds or less) changes in the rates of catecholamine release from sympathetic terminals or the adrenal medulla depend mainly on altered rates of release of acetylcholine from the preganglionic neurones, but regulatory processes have been implicated at various levels. The release of NE at postganglionic nerve terminals and of epinephrine from the adrenal medulla are subject to modulatory effects mediated by presynaptic receptors for angiotensin, dopamine, prostaglandins, and other substances. The density of such receptors may be modified slowly (requiring hours or days). This may be determined by genetic factors or be responsive to physiological "resetting". Release of acetylcholine and its efficacy in depolarizing postganglionic neurons can also be modulated physiologically.

The regional localization and pharmacologically-induced alterations of norepinephrine in the hypothalamus and mesencephalon led Martha Vogt (1954) to suggest that catecholamines are also central nervous system neurotransmitters. Carlsson (1959) first suggested that dopamine, the precursor of norepinephrine, was also a distinct brain neurotransmitter. Subsequent development of methods for assay of catecholamines and the metabolites, for determining localization of catecholamine-synthesizing enzymes using immunohistochemical methods, and for fluorescence histochemical demonstration of catecholamines in tissues (Falck, 1962) allowed definitive localization of catecholaminergic neurons and their projections in brain as well as studies of the effects of exposure to various stressors. These techniques, and newer methods for tracking neuronal pathways by axonal transport of fluorescent dyes or horseradish peroxidase, as well as for immunohistochemical demonstration of peptides and other neuromodulators, promise to enable more complete descriptions of the central neural regulation of sympathetic outflow. It already is clear that regionally specific responses determining physiological adaptations and more generalized activation are under complex neurological regulation involving an array of neurotransmitters (see below).

There appear to be several intraspinal and supraspinal mechanisms for influencing the outflow of nerve impulses to the adrenal medulla and to various segments of the sympathetic innervation fields. These are modulated via spinal reflexes, reflexes from baroreceptors, and by descending pathways from various brain areas. Many of the neurotransmitter substances thought to be involved are listed in Table 2. The roles of growth factors, genetic determinants and neuronal plasticity in long term modifications of responsivity are poorly defined.

Table 2 **Neurotransmitters in fibres terminating in region of spinal preganglionic sympathetic neurones**

Neurotransmitter (Descending from)	***Local Spinal Neurotransmitter***
Serotonin (Raphe)	**Substance P**
Norepinephrine (A_1, A_5)	**Enkephalin**
Epinephrine (C_1)	**Somatostatin**
Oxytocin-neurophysin (Paraventricular N)	**Neurotensin**
TRH (? Paraventricular N)	**GABA**
Substance P (Medulla)	**Glycine**
Enkephalin (Medulla)	**Glutamate**
	Aspartate

Table 3 **Physiological Responses to Sympathoadrenomedullary Activation**

Tachycardia	**Hyperventilation, Bronchodilation**
Vasoconstriction GI Tract Kidney Spleen Cutaneous	**Sweat** **Piloerection** **Hyperglycemia, Hyperlipemia,**
Inhibition of GI Tract Motility Secretion	**Proptosis, Pupillary Dilation** **Renin-angiotensin-aldosterone System Activation**
Vasodilation Skeletal muscle	**Platelet Activation**

PHYSIOLOGICAL INDICES OF SYMPATHO-ADRENOMEDULLARY RESPONSES TO STRESS

Physiological components of the generalized response to stressors include cardiovascular, renal visceral, cutaneous, and metabolic responses, each of which may by itself be a homeostatic mechanism. These are listed in Table 3. The cardiovascular responses to assuming an erect posture include vasoconstriction and increases in heart rate. These sympathetically mediated responses are imperceptible but are required to counter hydrostatic pooling of blood in the legs and maintain cardiac output, blood pressure and cerebral perfusion (Bannister, 1983; Goldstein et al., 1982). Adrenomedullary secretion, however, is stimulated only slightly if at all during orthostasis (Burke et al., 1977; Robertson et al., 1979) whereas sympathetic nerve activity to leg skeletal muscle is increased (Wallin and Sudlof, 1982). Similarly, in mild exercise sympathetic neuronal activation, rather than adrenal medullary secretion, predominates. When homeostatic mechanisms fail, e.g., with orthostatic hypotension attending drug-induced blockade of adrenergic receptors, more generalized responses are observed, with adrenal medullary secretion, marked tachycardia, peripheral vasoconstriction, sweating, and hyperglycemia.

Gastrointestinal responses also may be homeostatic or a component of generalized stress responses. Postprandial increases of visceral blood flow at the expense of coronary and cerebral perfusion may account for episodes of angina pectoris or orthostatic hypotension in susceptible elderly individuals. Splanchnic vasoconstriction during exercise or during exposure to high temperature are homeostatic responses. Inhibition of gastrointestinal activity and sphincter relaxation, however, are characteristic responses in acute distress.

Cutaneous responses to extremes of environmental temperature promote body cooling (vasodilation and sweating) or conserve heat (cutaneous vasoconstriction and piloerection). Metabolic responses to stimuli such as lowered blood glucose, exposure to cold or exercise, involve both sympathetic and adrenal medullary responses, but the latter predominate when extremes (hypoglycemia, hypothermia, exhaustion, etc.) cause insufficiency of the homeostatic mechanism.

The physiological changes attending the generalized stress response may be viewed as an integrated adaptive response necessary to preserve life, preparing the organism for "fight or flight" (Cannon, 1929). Thus, cutaneous and visceral vasoconstriction with associated decreases in skeletal muscular vascular resistance results in shunting of blood flow to support extraordinary muscular efforts. Similarly, increments in blood glucose and free fatty acids, as well as enhancement of oxygen delivery by bronchodilation and increases in cardiac output (sustained by tachycardia and splenic contraction), support increased energy needs. Other stress responses have also received teleological explanations: piloerection provides insulation from cold or disguises small size; proptosis and pupillary dilation attend increased vigilance; platelet activation and cutaneous vasoconstriction limit blood loss. Cannon (1929) emphasized the advantages of anticipatory sympatho-adrenomedullary responses in the evolution of emotional behaviors: "But fear and aggressive feeling, as anticipatory responses to critical situation, make ready for action and thereby have had great survival values. And the remarkable system of internal adjustments which attend these emotions and which mobilize the forces of the body are such as to have had great survival values. Thus the bodily changes may be reasonably interpreted". Although sympatho-adrenomedullary stress responses have survival value, repeated or severe stress may result in inappropriate persistance of increased sympathetic nerve activity (e.g., Johnson et al., 1983). Such persistent responses have provided the basis for several hypotheses relating stress to subsequent psychosomatic disorders.

MEASUREMENTS OF URINARY AND PLASMA CATECHOLAMINES AS INDICES OF ACUTE RESPONSES TO STRESSORS

Cannon's observations on the physiological responses that prepare the organism for "fight or flight" were attributed to sympatho-adrenal medullary discharge of catecholamines, but it was not until 1954, after new, sensitive fluorimetric methods were introduced, that stress responses were demonstrated to produce increases in urinary catecholamine excretion. It was shown that during anxiety of airplane flight pilots and passengers excreted increased amounts of catecholamines (Euler and Lundberg, 1954). Elmadjian (1963) reviewed the various early studies that showed normal subjects excrete increased amounts of catecholamines during a variety of situations associated with anxiety or aggression. In general, in these situations increases in epinephrine were more striking than those of norepinephrine. Often there has been poor agreement among physiological responses, subjective feelings, and changes in catecholamine excretion. Frankenhauser (1971) noted that epinephrine excretion was more closely related to the intensity than the quality of an emotion. Increased tension, expectancy (pleasant or unpleasant), and demand for concentration also were associated with higher epinephrine excretion rates.

During the interval between 1955 and 1970, there were several important discoveries regarding the disposition and metabolism of catecholamines. Accurate methods for measurements of catecholamine metabolites in urine became available, and later, sensitive and specific radioenzymatic assays for catecholamines in plasma were introduced. The main urinary metabolites of both epinephrine and norepinephrine were found to be 3-methoxy-4-hydroxymandelic acid (vanillylmandelic acid, VMA) and 3-methoxy-4-hydroxyphenylglycol (MHPG). Assay of these metabolites could not descriminate adrenal medullary from sympathetic responses. The hope that MHPG excretion might be an index of brain norepinephrine metabolism fostered many studies in psychiatric patients, but the subsequent discovery that VMA largely is formed from MHPG, made interpretation of such studies tenuous

Table 4. **Stress Research**

ANIMAL MODELS

- Slight Disturbances (turn on light, open cage, handle, transfer) (alerting reaction)
- Major Stressors
  - Physical (temp. extremes, elec. shock, immobilization, etc.)
  - Psychological (threat, natural or experimental, alarm)
  - Pharmacological (ether, 2-deoxyglucose, insulin, etc.)
  - Tissue Damage (fracture, hemorrhage, chemical injury, etc.)

HUMAN VOLUNTEERS

- Laboratory Experiments (mental activity, cold, centrifugation, etc.)
- On-the-job (workers, aviators, air traffic controllers, etc.)
- Opportunistic
  - Patients (trauma, surgical operations, acute or chronic illness)
  - Psychological Stressors (patient relatives, near disasters, etc.)

(see review by Kopin, 1985). The ability to measure precisely both epinephrine and norepinephrine by radioenzymatic assays and, more recently, by high performance liquid chromatography with electrochemical detection (HPLC-ED) stimulated hundreds of studies in which a variety of stimuli were used to evoke changes in plasma catecholamines in experimental animals and in humans (see Table 4).

Increases in catecholamines in plasma were found with relatively mild stimuli associated with usual daily activities (Figure 2). Thus standing up doubles plasma norepinephrine without greatly affecting epinephrine. Drinking coffee, smoking, or mental activity elevates both norepinephrine and epinephrine, whereas public speaking (medical residents presenting grand rounds) produces proportionately larger changes in epinephrine than norepinephrine.

The effects of some experimental procedures used in humans to evoke catecholamine responses are shown in Figure 3. Considering the discomfort and cardiovascular effects of immersion of the hand and forearm into ice-cold water, the plasma catecholamine responses to the cold pressor test seem surprisingly small. Mild exercise, in contrast, evokes larger increases in both norepinephrine and epinephrine. Mental challenge requiring increased alertness is attended by a sympathetically mediated cardiovascular response,

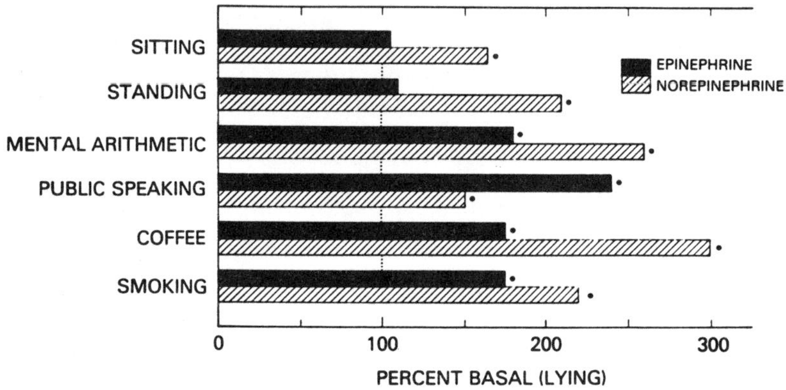

Fig. 2 Effects of Usual Activities on Plasma Catecholamines

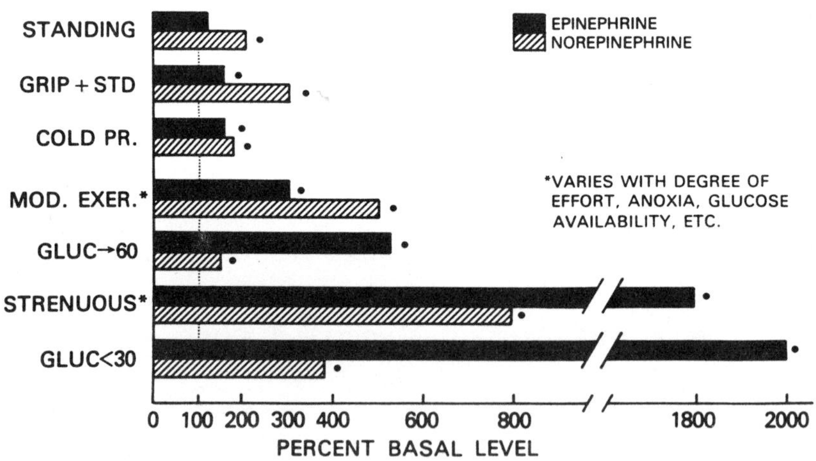

Fig. 3 Evoked Increases in Plasma Catecholamines

but performance of a non-distressing mental task does not appear to produce elevated plasma epinephrine levels (Eisenhofer et al., 1985; Goldstein et al., 1987).

With decreases in blood glucose, or strenuous exercise the relative increases in epinephrine exceed those of norepinephrine, as is the case for public speaking. A similar difference in the sympathoadrenal versus adrenomedullary responses is seen with diabetic ketosis, acute myocardial infarction, and hemorrhagic shock (Fig. 4). Such severe stressors evoke proportionately greaterresponses in epinephrine than in norepinephrine, whereas during extensive surgery or in the post-operative state, the norepinephrine response is often relatively greater. Sympathetic neuronal responses appear to regulate blood volume and distribution of the blood to various tissues, whereas epinephrine release from the adrenal medulla is a hormonal response to metabolic disturbances, preparation for emergencies, situations requiring vigilance or lacking predictability, or failure of sympathoneural homeostatic mechanisms. The hormonal response especially influences cardiac output, both by positive inotropic and chronotropic effects and by skeletal muscle vasodilation.

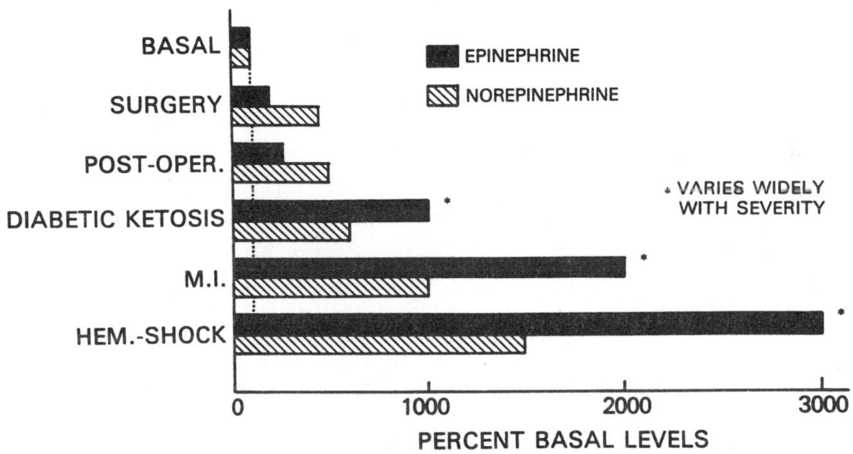

Fig. 4 Plasma Catecholamine Levels In Catastrophic Illness

The validity of venous plasma norepinephrine as an index of sympathetic neuronal discharge has been questioned (Brown et al., 1981). Differences in overall sympathetic activation or in sympathetic activity in other regions may not be reflected adequately by changes in catecholamine levels in plasma obtained from blood in the antecubital vein, the usual site of sampling. The concentration of norepinephrine in the venous blood of any region is determined by the concentration of the catecholamine in arterial blood, extraction of the catecholamine from the blood during its passage through the tissues, and the overflow of the neurotransmitter released from the sympathetic nerve terminals in the tissue. The arterial concentration of norepinephrine is the result of the levels in mixed venous blood after removal of a portion of the catecholamine during passage of blood through the lungs. Thus, the norepinephrine concentration in venous blood of any particular tissue would not necessarily reflect the overflow of norepinephrine into blood throughout the body.

In a recent study (Goldstein et al., 1987) of plasma catecholamines during a mental challenge of playing a video electronic game, increases in plasma norepinephrine in antecubital venous blood bore no relationship to increments in cardiac output or systolic blood pressure. Arterial norepinephrine responses correlated weakly ($r = 0.48$) with increments in cardiac output, and were not related to responses of systolic blood pressure. There were, however, significant correlations between changes in total body norepinephrine spillover (determined by isotope dilution methods during constant infusion of ^{3}H-ℓ-norepinephrine) and changes in cardiac output ($r = 0.68$) or systolic blood pressure ($r = 0.60$). Furthermore, while there was a linear relationship between forearm clearance of norepinephrine and blood flow, the slope and intercept of the regression line relating the forearm clearance and blood flow were such that at low flow rates as much as 70-80% of the arterial norepinephrine was extracted, whereas at faster flow rates only 30-40% was extracted. Thus regional hemodynamics, activity of norepinephrine removal systems, and regional differences in sympathetic activity interact in determining changes in antecubital venous norepinephrine levels and diminish reliability of such changes as an index of sympathetic activity.

CONSEQUENCES OF CHRONIC INCREASES IN SYMPATHOADRENAL MEDULLARY ACTIVITY

Stimulation-evoked increases in neurohumoral secretion occur almost instantaneously and may be maintained for short intervals, during which the stores of catecholamine provide a reserve for continued responses. With frequently repeated or prolonged exposure to a stressful stimulus, mechanisms are brought into play to adapt to the stressor. Increased demand for the production of catecholamines results in increased levels of tyrosine hydroxylase (TH), the rate limiting enzyme in catecholamine synthesis, as well as of dopamine-β-hydroxylase (DBH) and phenylethanolamine-N-methyl transferase (PNMT)) (see Kvetnansky, 1980)). In the adrenal medulla, TH and DBH are mainly a result of neuronally involved catecholamine release. Increments in TH levels subsequent to repeated immobilization, hypoglycemia, or cold exposure, are prevented by denervation (which also blocks release of epinephrine). Adrenal DBH decreases during exposure to stressful stimuli, because it is released by exocytosis, along with epinephrine and other vesicular contents,. Restoration of DBH becomes increasingly rapid with repeated stress responses and levels of the enzyme become greater than normal during a subsequent interval without exposure to stress. PNMT, the enzyme that converts norepinephrine to epinephrine, mainly is under hormonal control. Glucocorticoids in high concentration from the adrenal cortex bathe the medullary cells and maintain PNMT levels. This enzyme decreases after hypophysectomy and increases slightly when adrenal cortical secretory activity is enhanced. The increments in enzyme levels in the adrenal medulla during chronic or repeated stress serve to maintain the ability of the gland to secret increased quantities of epinephrine. Similar increments in TH and DBH have been found in sympathetically innervated tissue. Basal levels of the enzymes and presumably their regulatory mechanisms (synthesis and degradation) are under genetic control (Ciaranello, 1977). Responses to stimuli are likely to vary among individuals, as they do among inbred strains of rats (McCarty & Kopin, 1978), on the basis of such genetically controlled factors as enzyme synthesis and degradation and responsivity of the neuronal systems modulating catecholamine release.

The physiological responses to released catecholamines are dependent also on the receptors and receptor activated mechanisms. With repeated exposure, receptors are frequently down-regulated so that enhanced catecholamine release over a prolonged interval might elicit responses of lesser magnitudes. Variability in responsiveness to adrenoceptor stimulation by administered catecholamines has been cited as a possible explanation for differences among individuals or groups in adrenoceptor-mediated responses to some stressors (Eisenhofer et al., 1985). The health consequences of prolonged of chronically repeated distress are being explored actively. The availability of biochemical indices of the magnitude of distress should aid these studies.

SUMMARY

The sympatho-adrenomedullary system plays an important role in stages of stress responses and in turn is subject to regulatory processes which are modulated by the neuronal and hormonal responses to exposure to stressors. Homeostatic mechanisms which prevent marked physiological changes in the course of daily life and psychological alerting are mild stimuli. As described by Cannon the homeostatic mechanisms enlisted for rapid alterations of bodily functions in preparation for "fight or flight" are mediated mostly by sympathoadrenomedullary discharge of catecholamines. Such initial neurohumorally mediated bodily changes are included in the early alarm reaction in stress, as defined by Selye. Norepinephrine is released from sympathetic nerve terminals into the neuroeffector junction and then into blood; epinephrine is discharged directly into the circulation from the adrenal medulla.

When homeostatic mechanisms fail to correct the effects of a disturbance, or when psychological alerting produces emotional reactions, an alarm reaction begins with general activation of the sympathoadrenomedullary system and recruitment of additional neurohumoral systems (e.g., hypothalamic - pituitary - adrenal cortical, vasopressin, renin-angiotensin). Catecholamine neurotransmitters in the central nervous system as well as peripheral catecholamines are likely to have an important role in regulating these stress responses but exactly how is unknown. With repeated or prolonged exposure to stressors, a phase ensues where there are changes in catecholamine biosynthetic enzymes and receptor responsivity; other biochemical responses may result in a variety of pathological changes . Changes in norepinephrine release during homeostasis (e.g., local changes in blood flow, glandular secretion, fat pad metabolic activation) may not be reflected in altered plasma levels of the catecholamine in antecubital venous blood. Increases in arterial concentrations of catecholamines and in their urinary excretion occur when sympatho-adrenomedullary activity is sufficient to produce significant increases in overflow of norepinephrine from the neuroeffector junctions into the circulation or when epinephrine is discharged from the adrenal medulla. Massive increases in catecholamine discharge are readily detected in plasma even from antecubital vein blood, but less generalized responses may be masked unless regional spillover rates are assessed using isotope dilution methods.

References

Abel, J.J. 1899, Ueber den blutdruckerregenden Bestandtheil der Nebenniere, das Epinephrin. Hoppe-Selyer's Z. Physiol. Chem., 28:318.

Axelrod, J., 1950, Metabolism of epinephrine and other sympathomimetic amines. Physiol. Rev. 39:751.

Axelrod, J., 1965, The metabolism, storage, and release of catecholamines. Recent Progress Hormone Res., 21:597.

Bannister, R., 1983, Autonomic Failure. New York: Oxford.

Bernard, C., 1879, Les Phénomènes de la Vie, Paris.

Brown, M.J., Jenner, D.A., Allison, D.J., and Dollery, C.T., 1981, Variat ons in individual organ release of noradrenaline measured by an improved radioenzymatic technique: limitations of peripheral venous measurements in the assessement of sympathetic nervous activity. Clin. Sci., 61:585-590.

Burke, D., Sundlof, G., Eriksson B-M, Dominiak, P., Grobecker, H., and Lindblad, L.E., 1977, Postural effects on muscle sympathetic activity in man. J. Physiol., 272:399-414.

Cannon, W.B., 1929, Bodily changes in pain, hunger, fear and rage. Boston.

Carlsson, A., 1959, The occurrence, distribution, and physiological role of catecholamines in the central nervous system. Pharmacol. Rev., 11:490.

Ciaranello, R.D., 1972, Catecholamine biochemical genetics: at the crossroads of biochemistry and behavior. in Usdin, E., et al., ed. "Neuroregulators and Psychiatric Disorders", New York, Oxford University Press, pp. 497-508.

Eisenhofer, G., Lambie, D.G., and Johnson, R.A., 1985, Beta-adrenoceptor responsiveness and plasma catecholamine as determinents of cardiovascular reactivity to mental stress. Clin Sci., 69:483-492.

Elliott, T.R., 1905, The action of adrenalin. J. Physiol. (London), 32:401.

Elmadjian, F., 1963, Excretion and metabolism of epinephrine and norepinephrine in various emotional states, in Proceedings of the Fifth Pan American Congress of Endocrinology, pp. 341-369.

Esler, M., Jenings, G., Korner, P., Blomberg, P., Burke, F., Willett, I. and Leonard, P., 1984, Total and organ specific noradrenaline kinetics in essential hypertension. Clin. Exper. Hyperts., Part A., Theory Pract., 6:507-621.

Euler, U.S., von, and Lundberg, U., 1954, Effect of flying on the epinephrine excretion in Air Force personnel J. Appl. Physiol., 6:551-555.

Falck, B., Hillarp, N.O., Thieme, G., and Torp, A., 1962, Fluorescence of catecholamines and related compounds condensed with formaldehyde. J. Histochem. Cytochem., 10:348.

Frankenhouse, M., 1971, Behavior and circulating catecholamines, Brain Res., 31:241-262.

Goldstein, D.S., Spanarkel, M., Pitterman, A., Toltzis, R., Gratz, E., Epstein, S., Keiser, H.R., 1982, Circulatory control mechanisms in vasodepressor syncope. Am. Heart J., 104:1071-1075.

Goldstein, D.S., Eisenhofer, G., Sax, F.L., Keiser, H.R., and Kopin, I.J., 1987, Plasma norepinephrine pharmacokinetics during mental challenge, Psychosam. Med., in press.

Hertting, G. and Axelrod, J., 1961, Fate of tritiated noradrenaline at sympathetic nerve endings, Nature, 192:172.

Johnson, T.S., Young, J.B., and Landsberg, L., 1982, Sympathoadrenal responses to acute and chronic hypoxia in the rat. J. Clin. Invest., 71:1263-1272.

Kvetnansky, R., 1980, Recent progress in catecholamines under stress in (ed) Usdin, E., Kvetnansky, R., and Kopin, I."Catecholamines and Stress: Recent Advances", Elsevier, New York, pp. 7-20.

McCarty, R. and Kopin, I.J.,1978, Sympatho-adrenal medullary activity and behavior during exposure to footshock stress: A comparison of seven rat strains. Physiol. Behav., 21:567-572.

Oliver, G., and Shaffer, E.A., 1895, The physiological effects of extracts from the suprarenal capsules. J. Physiol., 18:230.Robertson, D.A., Johnson, G.A., Robertson, R.M., Nies, A.S., Shand, D.G., and Oates, J.A., 1979, Comparative assessment of stimuli that release neuronal and adrenomedullary catecholamines in man. Circulation , 59:637-643.

Selye, H. 1973, The evolution of the stress concept. Amer. Scientist 1:692.Jenkins, C.D., 1979, Psychosocial modifiers of response to stress. J. Human Stress, 5:3-125.

Vogt, M., 1954, Norepinephrine and epinephrine in the central nervous system. Pharm. Rev., 6:31.

von Euler, U.S., 1948, Identification of the sympathomimetic ergone in adrenergic nerves of cattle (sympathin H) with laevo-noradrenaline. Acta. Physiol. Scand., 16:63.

BEHAVIORAL PHARMACOLOGY OF STRESS:

FOCUS ON CNS CORTICOTROPIN-RELEASING FACTOR

George F. Koob[1], Karen Thatcher-Briton[2], Abdelouahhab Tazi[3], and Michel Le Moal[3]

1. Dept of Basic and Clinical Research, Scripps Clinic and Research Foundation, 10666 North Torrey Pines Road, La Jolla, CA 92037
2. Psychobiologie des Comportements Adaptatifs, INSERM Unite-254, Rue Camille St. Saens, 33078 Bordeaux, France
3. Dept of Psychiatry, San Diego Veterans, Administration Medical Center & UCSD, Sch. of Med., La Jolla, CA 92161

INTRODUCTION

Stress is a hypothetical construct which has been extensively studied for the past three decades at different levels of analysis. It is a ubiquitous concept in physiology, psychology and medicine that often eludes precise definition. Hans Selye conceptualized stress as "a nonspecific response to any demand upon the body (usually, but not always, noxious), or anything which causes an alteration of homeostatic processes (1)". A more modern version of this definition, taken from Burchfield (2) has emphasized the concept of psychological homeostasis or the maintenance of a normal mood state at rest. All emotions then are changes from this state. Indeed, as Dr. John Mason showed so elegantly, without emotional changes accompanying a stressor, the stress response is minimal (3). Thus commonly accepted physiological "stressors"(heat, excercise, hunger) do not elicit a "stress response" when they are presented in a way that eliminates their psychological (emotional) effects (fear, conflict, uncertainty, frustration).

The importance of psychological variables in the stress response emphasizes the need for a neurobiological substrate to process the interaction of sensory stimuli with the ultimate psychoendocrine stress response. It is presumably through central nervous system pathways of the limbic system to the hypothalamus that the stress response is triggered by psychological factors. There is, however an alternate means by which behavioral or physiological responses to stress or anxiety might be mediated by neurohormones classically involved in stress. That is via a direct neurotropic action in the central nervous system.

The sequencing and subsequent synthesis of corticotropin releasing factor (CRF) (4) has provided a direct opportunity to study the central control of the HPA axis, and numerous studies support the hypothesis that

CRF is the critical mediator of stress-induced changes in the pituitary adrenocortical response to stress. Although CRF was originally isolated and characterized on the basis of its ability to induce ACTH release, it is now known to exert a surprisingly broad spectrum of action in the central nervous system. CRF appears to act within the brain as a mediator of certain autonomic nervous system, visceral and behavioral actions independent of its action on the pituitary. These effects are not abolished by hypophysectomy or adrenalectomy. These actions, in concert with the wide anatomical distribution of CRF in brain regions outside the areas hypothalamic-pituitary-adrenal axis (5), provide a tenable basis for hypothesizing that CRF may simultaneously activate and coordinate neuroendocrine, autonomic circulatory, metabolic and behavioral responses to stressful stimuli.

CRF and Central Nervous System Arousal

Corticotropin releasing factor CRF has central nervous system activating properties when administered directly into the central nervous system. Intraventricular (ICV) injection of CRF produces elevation of plasma epinephrine, norepinephrine and glucose (6,7). These effects are reproducable in hypophysectomized animals but are abolished by ganglionic blocking agents (8) suggesting an involvement of the sympathetic autonomic system. CRF injected ICV also produces a profound dose-dependent activation of the electroencephalogram (EEG) (9). Doses of 0.015 to 0.15 nmol produced a long lasting activation of EEG, and after a 2-hour delay some interictal spikes occurred in the amygdala and hippocampus. Higher doses (1.5-3.75 nmol) were characterized by consistent amygdala interictal spikes and after discharges and, after a delay of 4 to 7 hours, some motor seizures. These seizures developed over time in a manner not unlike those produced by amgdala "kindling" paradigms. At the cellular level, CRF produces increases in the firing frequencies of cells within the locus coeruleus (10) a system thought to be of importance in the mechanisms by which the brain is able to attend selectively to certain novel external events.

The autonomic and electrophysiological activation produced by central administration of CRF is paralleled by a dose-dependent locomotor activation (11,12). These effects appear to be independent of direct mediation by the pituitary-adrenal system since they were observed in hypophysectomized and dexamethasone-treated rats (see figure 1) (13,14,15). Taken together with the fact that this activation is seen only with central administration, these observations suggested that CRF exerted its effects within the central nervous system. Although other peptides such as the endorphins and ACTH have been shown to produce increases in spontaneous behavioral activity. the locomotor activation caused by CRF is not antagonized by the opiate antagonist naloxone, or by low doses of the dopamine receptor antagonist alpha flupenthixol (12). Nor is this activation reversed by 6-hydroxydopamine lesions of the region of the nucleus accumbens, lesions that reverse the locomotor-stimulated effects of indirect sympathomimetics (16). ACTH, itself, injected centrally failed to increase locomotor activity but instead produced an increase in grooming behavior as has been observed by others (17).

There is a significant amount of empirical and theoretical evidence which suggests that performance of learned behavior can vary with the level of activation or "arousal" of the animal. Correlations have been obtained between independent measures of "arousal" such as exploratory behavior, electrencephalogram and galvanic skin response and the speed of acquisition of a response (18). In addition, Berlyne (18) has proposed that both animals and man strive to maintain intermediate amounts of arousal potential and that stimuli that evoke such intermediate levels of arousal will be pleasant and stimuli that evoke zero or too much arousal will be unpleasant. It follows that optimal learning will correspond to these optimal levels of arousal, and more recent treatments of this theoretical issue have centered on "arousal" as the intervening variable

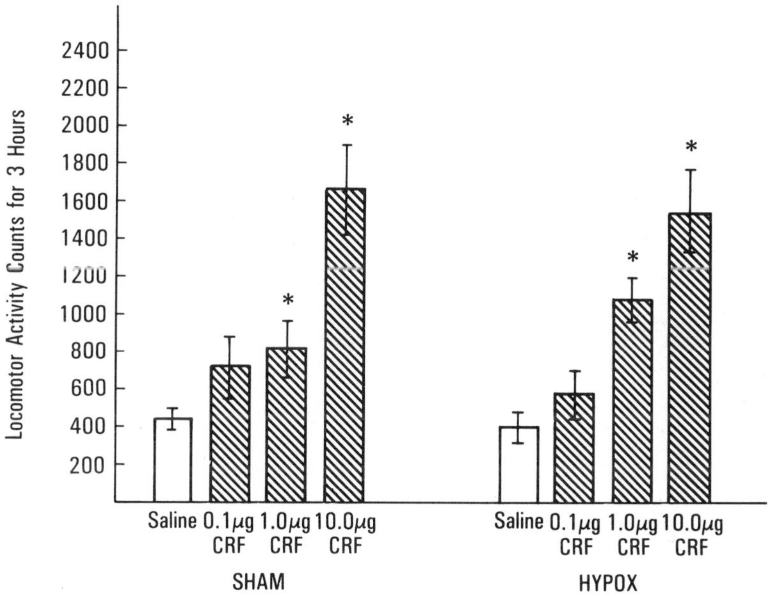

Fig. 1. Effects of CRF injected ICV on locomotor activity in photocell cages in hypophysectomized rats treated chronically with rat growth hormone and in sham operated animals. Results represent the total activity counts over 3 hours (mean$\pm$ S.E.M.) n=9 in each group. Significantly different from saline injection. $p<0.05$ main effects ANOVA. (Taken with permission from 13).

in the incentive-motivation required for all learning (19). Others have argued that both stress and arousal can be considered as representing "state" phenomena reflecting diffuse central nervous system connections and that activity in the pituitary-adrenal system reflects the aroused state of an animal (20). In this conceptual framework activity in the pituitary adrenal system was considered a reliable, and sensitive measure of arousal.

A central nervous system (CNS) component stress-arousal continuum is proposed that may be a CNS extension of the pituitary adrenal system. The hypothesis for the present treatise is that central nervous system corticotropin releasing factor forms a basic arousal system that mediates behavioral responses to stressors. Thus the previously described homology of stress and arousal (20) may be reflected in a specific neurochemical substrate in the CNS.

Performance in a variety of test situations varies with the level of activation or "arousal" of the animal. Given that CRF has "arousal" properties it follows that a reflection of this activation would be the performance of an animal during acquisition of a learned response. Preliminary results provided support for this hypothesis. CRF administered prior to the daily acquisition of an appetitively motivated visual discrimination task significantly improved acquisition (21). CRF injected rats, made significantly fewer errors and much more rapidly acquired the visual discrimination reaching 70%-80% correct by days 3 and 4 whereas the control animals only reached this level by days 5 and 6 (figure 2).

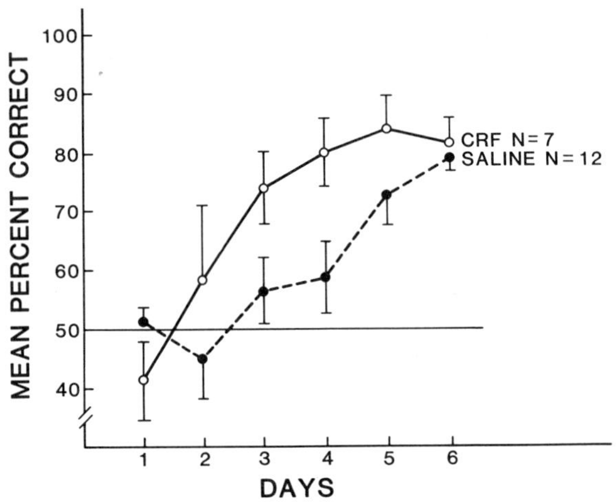

Fig. 2. Rats were implanted with intraventricular cannula and were tested in a 10 day maze adaptation and visual discrimination task using a Y maze. CRF injections (150 pmoles ICV), N=7, or saline ICV, n=12, were given 1 hour before each of the six actual discrimination learning days, the four adaptation days were injection-free. Both groups of rats acquired the visual discrimination during the six days of acquisition training and the rate of learning of the control rats was similar to that reported by others using a similar adaptation procedure in a T-maze (22).

CRF, Stress and Performance

A classical test of behavioral responsiveness to stress involves exposing a rat to a novel, open field environment and observing its behavior. When mildly food deprived rats were exposed to an open field with food in the center, rats injected ICV with CRF showed a marked decrease in exploratory and ingestive behavior and remained close to the corners of the field (23), in contrast to an untreated animal which made increasingly frequent forays into the center of the field to consume food (24). Similar results were observed with non deprived rats in an open field without food (11). These behavioral changes were consistent with heightened "emotionality" or increased sensitivity to the, presumably, stressful aspects of the test.

These findings led to the suggestion that CRF might be related in some way to the expression of human anxiety or fear. Since it is not possible to identify animals with anxiety, CRF was tested in a series of tests that are sensitive to anti-anxiety (anxiolytic) compounds such as the benzodiazepines.

One of the most consistent behavioral effects of antianxiety agents is their ability to "release" behavior previously suppressed by punishment, and one of the most sensitive animal models to measure this effect is the conflict test. The conflict test is an operant conditioning paradigm in which a reinforcing stiumlus (food) is accompanied by an aversive stimulus (shock). Antianxiety agents consistently "release" or increase behaviors previously suppressed by punishment in the conflict test with a rank order of potency similar to that observed clinically (25).

Rats treated with CRF display a pattern of responding in the conflict test that is the opposite to that observed in rats treated with benzodiazepines. CRF (0.5 and 1.0 μg. i.c.v.) produced a dose-dependent attenuation of both punished and non-punished responding in the conflict test, decreasing non-punished responding to 40% of baseline (26). As expected, chlordiazepoxide (5 mg/kg i.p.) significantly increased the number of footshocks taken during the punishment component (Figure 3). In addition, chlordiazepoxide reversed the response-attenuating effects of CRF, restoring non-punished responding to 75% of baseline and punished responding to nearly 100% of baseline (Figure 3). These results suggest that exogenously administered CRF can produce changes in the rat consistent with an exaggeration of the stressfulness of the environment and that this behavioral state can be reversed by classical 'antianxiety' drugs.

Another paradigm differentially sensitive to anxiogenic and anxiolytic compounds is the acoustic startle test. The acoustic startle reflex is an easily quantified muscular contraction in response to an intense acoustic stimulus. CRF (1 μg ICV) significantly potentiated acoustic startle amplitude (27). Pretreatment with the benzodiazepine chlordiazepoxide in doses that did not by themselves lower startle baseline attenuated this effect. These results also supported the hypothesis that CRF might act to potentiate behavioral responses normally expressed during states of enhanced fear or anxiety.

Although the increases in performance discussed above with CRF were consistent with an improvement in learning due to an arousing action, one would presumably observe an inverted U dose-response curve where lower doses should have weak effects and higher doses should interfere with performance. Alternatively, if the unconditioned stimulus for learning is aversive such as in an avoidance situation, lower doses of CRF might become effective and moderate doses might disrupt performance. Preliminary results support this hypothesis. CRF at doses as low 0.01 μg ICV facilitated acquisition of one way and two way active avoidance and higher doses 1.0 μg ICV completely disrupted performance (28). Perhaps even more interesting was the observation that the rats receiving 0.1 and 1.0 μg of CRF and trained in these avoidance tasks showed extreme "emotionality" for long periods after testing. These rats assumed boxing positions, vocalized, and fought with other rats for hours following testing. This unusual behavior was never observed in rats receiving these doses without avoidance experience.

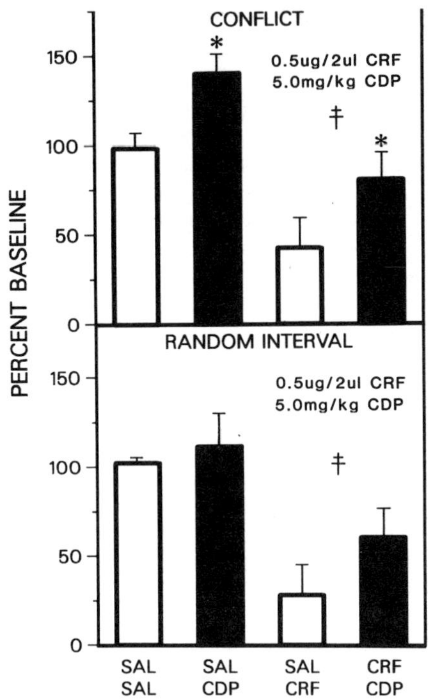

Fig. 3. The interaction of 0.5 μg corticotropin releasing hormone injected ICV on responding during the random interval and conflict components of an operant conflict test. Results are expressed as percent of baseline from the previous day. *denotes significant difference from saline, main effect CDP; double cross significant difference from saline, main effect CRF, $P<0.05$, two-way analysis of variance (Taken with permission from 26).

This unusual interaction of CRF with the stress produced by aversive events provided the basis for investigating whether using a stress-induced fighting paradigm might exaggerate this phenomenon. This fighting procedure has been shown to be sensitive to environmental parameters such as stress level (29) and to be modulated by peptides (30).

Rats previously implanted with ICV cannula, were exposed in weight-matched pairs to mild inescapable electric footshock during daily consecutive sessions. In such a paradigm, upright postures (boxing positions) or fighting (upright postures followed by physical contacts), develop within a few sessions. These behaviors depend on many environmental parameters, including the size of the test chamber, and mainly the shock intensity (30). Thus, the frequency of fighting responses is increased gradually with the shock intensity, probably as a consequence of the gradual increase in physiological changes accompanying the increased stress.

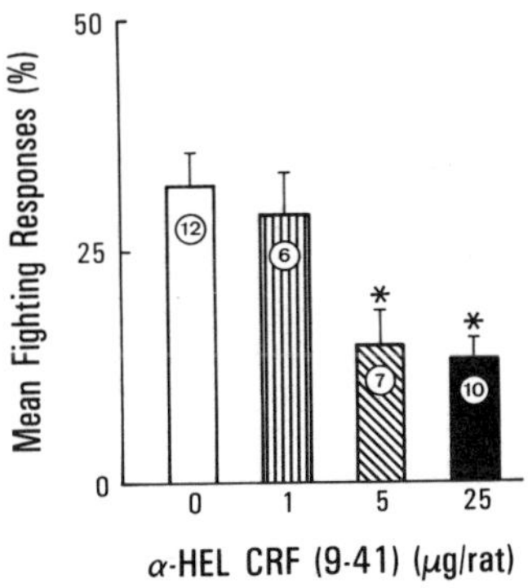

Fig. 4. Effects of 0, 1, 5 and 25.0 μg/rat of α-Hel CRF (9-41) injected ICV on fighting responses. Pairs of weight-matched rats were tested at 0.6mA shock intensity. The CRF antagonist was injected 5 minutes before testing. No significant change was observed in the number of pairs showing the "boxing position" at any dose (data not shown). The data are the mean over 3 consecutive daily 10 minute sessions. Numbers inside the columns represent the number of pairs in each group. *$p < 0.05$ in comparison to saline-treated rats, Newman-Keuls test following a significant analysis of variance. (Taken with permission from 28).

Rat CRF at doses of 0, 0.01, 0.1 and 1.0 μg/rat was injected ICV increased the frequency of these responses in a dose-dependent, and shock dependent manner (28). At the lower shock intensities (0.3 and 0.4 mA),there was no fighting between the control rats. However, "boxing positions" were observed at these shock levels in control rats and "boxing" was increased significantly by 0.1 and 1.0 μg of CRF. CRF doses of 0.01 and 0.1 μg significantly facilitated fighting at the 0.5 mAshock level. CRF at the highest dose and highest shock level totally disrupted the behavior of the animals.

More importantly, however, a slightly higher shock level (0.6 mA) produced a higher fighting frequency in control animals and this fighting was reversed by 5 and 25 μg/rat of α helical (9-41), a CRF antagonist, (see Figure 4). These rats had not received exogenously administered CRF. This CRF-antagonist has been shown to decrease both in vitro and in vivo baseline release of ACTH, as well as ether-induced ACTH secretion (31). At the behavioral level, α helical CRF(9-41) has been shown to block the activating and "anxiogenic" actions of exogenously administered CRF in the rat (32). These observations and the present results, suggest that under certain conditions of high arousal and stress endogenous CRF systems may play a role in mediating behavioral responses.

CONCLUSIONS

These results describing neuronal activation, sympathetic activation, EEG arousal, general behavioral activation, and stress enhancing actions of CRF all suggest a possible role for CRF as a fundamental activating system. The functional significance of this system may have developed as a means for an organism to mobilize not only the pituitary adrenal system, but also to activate the central nervous system in response to environmental challenge. Indeed, results from our laboratory suggest that treatment with CRF can alter learning performance and this is dose and task related. Aversive situations appear much more sensitive to exogenous CRF and preliminary results suggest that these aversive situations may be sensitive to administration of a weak CRF antagonist. Clearly, a hypothetical central nervous system activation system somehow linked to the pituitary adrenal system, that can improve behavioral performance at low levels of output but attenuate behavioral performance at high levels of output would be of certain survival value. It is also not difficult to imagine a possible role for such a system in clinical disorders such as anxiety, affective disorders and other psychopathology.

For decades there has been speculation that the elusive CRF molecule might serve as the "first mediator of stress-responses" (33). Yet many studies have emphasized the non-specificity of the pituitary-adrenal stress response in that it is activated by a variety of stimuli and is relatively slow to develop. This observation led Mason (3) to hypothesize that the "primary mediator" underlying the pituitary adrenal response to various stressors may be the neural substrate for emotional arousal.

One might hypothesize then that CRF is not only the first mediator of the pituitary stress response but also may be involved in mobilizing the neural substrates for emotional arousal. Studies demonstrating that CRF elicits neuronal, sympathetic and behavioral activation, coupled with stress-enhancing actions that are responsive to anxiolytic compounds, suggest a possible role for corticotropin releasing factor in a fundamental activating system that mobilizes an organism to meet an environmental challenge. Questions such as what pathways project to CRF-secreting cells, what neurochemical changes are concerned with the regulation of CRF release and what are the interactions between central and pituitary-adrenal CRF systems remain to be elucidated.

REFERENCES

1. H. Selye, "Selye's guide to stress research." Van Nostrand Reinhold Co., New York V-VIII (1980).
2. S. Burchfield, The stress responses: a new perspective. Psychosomatic Med. 41:661-672 (1979).

3. J.W. Mason, A re-evaluation of the concept of "non-specificity" in stress specificity in stress theory. J. Psychiatric Res. 8:323-333 (1971).
4. W. Vale, J. Spiess, C. Rivier, and J. Rivier, Characterization of a 41-residue ovine hypothalamic peptide that stimulates the secretion of corticotropin and β-endorphin. Science 213:1394-1397 (1981).
5. L.W. Swanson, P.E. Sawchenko, J. Rivier, and W. Vale, Organization of ovine corticotropin releasing factor immunoreactive cells and fibres in the rat brain: An immunohistochemical study. Neuroendocrinology 36:165-186 (1983).
6. M.R. Brown, L.A. Fisher, J. Rivier, J. Speiss, C. Rivier, and W. Vale, Corticotropin releasing factor: Effects on the sympathetic nervous system and oxygen consumption. Life Sci. 30:207-210 (1982).
7. M.R. Brown, L.A. Fisher, J. Spiess, J. Rivier, C. Rivier, and W. Vale, Comparison of the biologic actions of corticotropin releasing factor and sauvagine. Regulatory Peptides 4:107-114 (1982).
8. M.R. Brown, and L.A. Fisher, Central nervous system effects of corticotropin releasing factor in the dog. Brain Res. 280:75-79 (1983).
9. C.L. Ehlers, F.E. Bloom, S.J. Henriksen, J. Rivier, and W.W. Vale, Corticotropin releasing factor produces increases in brain excitability and convulsive seizures in rats. Brain Res. 278:332-336 (1983).
10. R.J. Valentino, S.L. Foote, and G. Aston-Jones, Corticotropin releasing factor activates noradrenergic neurons of the locus coeruleus. Brain Res. 270:363-367 (1983).
11. R. Sutton, G.F. Koob, M. Le Moal, J. Rivier, and W. Vale, Corticotropin releasing factor (CRF) produces behavioral activation in rats. Nature 299:331-333 (1982).
12. G.F. Koob, N. Swerdlow, M. Seeligson, M. Eaves, R. Sutton, J. Rivier, and W. Vale, CRF-induced locomotor activation is antagonized by alpha-flupenthixol but not by naloxone. Neuroendocrinol. 39:359-464 (1984).
13. M. Eaves, K. Thatcher-Britton, J. Rivier, W. Vale, and G.F. Koob, Effects of corticotropin releasing factor on locomotor activity in hypophysectomized rats. Peptides 6:923-926 (1985).
14. D.R. Britton, M. Varela, A. Garcia, and J. Rivier, Dexamethasone suppresses pituitary-adrenal but not behavioral effects of centrally administered CRF. Life Sci. 38:211-216 (1986).
15. K.T. Britton, G. Lee, R. Dana, S.C. Risch, and G.F. Koob, Activating and "anxiogenic" effects of CRF are not inhibited by blockade of the pituitary-adrenal system with dexamethasone. Life Sci. 39:1281-1286 (1986).
16. N.R. Swerdlow and G.F. Koob, Separate neural substrates of the locomotor-activating properties of amphetamine, heroin, caffeine, and corticotropin releasing factor (CRF) in the rat. Pharmacology, Biochemistry and Behav. 23:303-307 (1985).
17. W.H. Gispen, V.M. Weigant, H.H. Greven, and D. De Wied, The induction of excessive grooming in the rat by intraventricular application of peptides derived from ACTH: structure activity. Life Sci. 17:645-655 (1975).
18. D.E. Berlyne, "Conflict Arousal and Curiosity," McGraw Hill Book Co. Inc., New York (1960).
19. P.R. Killeen, S.J. Hanson, and S.R. Osbourne, Arousal: its genesis and manifestation as response rate. Psychological Rev. 85:571-581 (1978).

20. J.W. Hennessy, and S. Levine, A psychoendocrine hypothesis. in: "Progress in Psychobiology and Physiological Psychology", J.M. Sprague, A.N. Epstein, eds., Academic Press 8:133-178 (1979).

21. G.F. Koob, and F.E. Bloom, Corticotropin-releasing factor and behavior. Fed. Proc. 44:259-263 (1985).

22. G.N.O., Brito, G.J. Thomas, S.J. Gingold, and D.M. Gash, Behavioral characteristics of vasopressin-deficit rats (Brattleboro Strain). Brain Res. Bull. 6:71-75 (1981).

23. D.R. Britton, G.F. Koob, J. Rivier, and W. Vale, Intraventricular corticotropin releasing factor enhances behavioral effects of novelty. Life Sci. 31:363-367 (1982).

24. D.R. Britton and K.T. Britton, A sensitive open field measure of anxiolytic drug activity. Pharmacology, Biochemistry and Behavior 15:577-582 (1981).

25. L. Cook and J. Sepinwall, Behavioral analysis of the effects and mechanism of action of benzodiazepines. in: "Mechanisms of action of benzodiazepines," E. Costa, P. Greengard, eds., Raven Press, New York (1975).

26. K.T. Britton, J. Morgan, J. Rivier, W. Vale, and G.F. Koob, Chlordiazepoxide attenuates CRF-induced responses suppression in the conflict test. Psychopharmacology 86:170-174 (1985).

27. N. Swerdlow, M. Geyer, W.W. Vale, and G.F. Koob, Cortocotropin releasing factor potentiates acoustic startle in the rats: Blockade by chlordiazepoxide. Psychopharmacol. 88:142-152 (1986).

28. A. Tazi, R. Dantzer, M. Le Moal, J. Rivier, W. Vale, and Koob, G.F. Corticotropin-releasing factor antagonist blocks stress-induced fighting in rats. Regulatory Peptides, in press (1987).

29. A. Tazi, R. Dantzer, R. Mormede, and M. Le Moal, Effects of post-trial injection of β-endorphin on shock-induced fighting are dependent on baseline of fighting. Behav. and Neural Biol. 43:322-326 (1985).

30. A. Tazi, R. Dantzer, R. Mormede, and M. Le Moal, Effects of post-trial administration of naloxone and β-endorphin on shock-induced fighting in rats. Behav. and Neural Biol. 39:192-202 (1983).

31. J. Rivier, C. Rivier, and W. Vale, Synthetic competitive antagonists of corticotropin-releasing factor: Effect on ACTH secretion. Science 224:889-891 (1984).

32. K.T. Britton, G. Lee, W. Vale, J. Rivier, and G.F. Koob, Corticotropin releasing factor antagonists blocks activating and "anxiogenic" actions of CRF in the rat. Brain Res. 369:303-306 (1986).

33. H. Selye, and G. Henser, "Fifth annual report on stress," M.D. Publications, New York (1956).

GLUCOCORTICOID RECEPTORS AND BEHAVIOR:

IMPLICATIONS FOR THE STRESS RESPONSE

Bruce S. McEwen and Roberta E. Brinton
Laboratory of Neuroendocrinology
Rockefeller University
1230 York Avenue, New York, N.Y. 10021

Robert M. Sapolsky
Department of Biological Sciences
Stanford University, Stanford, CA 94305

INTRODUCTION

Stress-induced release of glucocorticoids is one of two modes of operation of the pituitary-adrenal axis, the other being the diurnal variation of glucocorticoid levels. Both operating modes are, of course, driven by CNS release of the releasing factor, CRH, in combination with other modulators such as vasopressin and oxytocin (Vale, this volume; Plotsky, this volume; (1)). However, stress-induced release is due to environmental and experiential factors such as heat, cold, anxiety, trauma and exertion, and it results in high levels of glucocorticoids in the blood. On the other hand, diurnal variation in pituitary-adrenal function is based upon an endogenous oscillator (2), although both the light-dark cycle and the anticipation of food entrain the rhythm (3, 4).

When glucocorticoids are elevated in the blood by diurnal variations or by stress, the hormone level is sensed by the brain and pituitary. In addition to regulating further hormone output, glucocorticoid actions on the brain also modify brain function and change behavior. Sensing of glucocorticoids in the blood is accomplished by the uptake of hormone by brain tissue and its interaction with receptors within the cell and on the cell surface. Membrane receptors for glucocorticoids are poorly characterized, with only one published binding study (5); and the existence of these receptors is largely inferred from rapid feedback actions of glucocorticoids on hormone release and on other membrane events (6).

Intracellular glucocorticoid receptors are well-characterized, and they are known to mediate changes in gene transcription (7). Two types of intracellular glucocorticoid receptors are now recognized in brain tissue (8), and the existence and properties of these receptors forms the basis of this chapter. We shall discuss these two receptor types in relation to the feedback effects of glucocorticoids secreted in response to the diurnal mode and the stress mode of operation of the pituitary-adrenal axis, and we shall subdivide these actions into three categories, referred to as "the three A's": "activational" effects associated with the diurnal cycle; "adaptational" effects resulting from stress-induced glucocorticoid

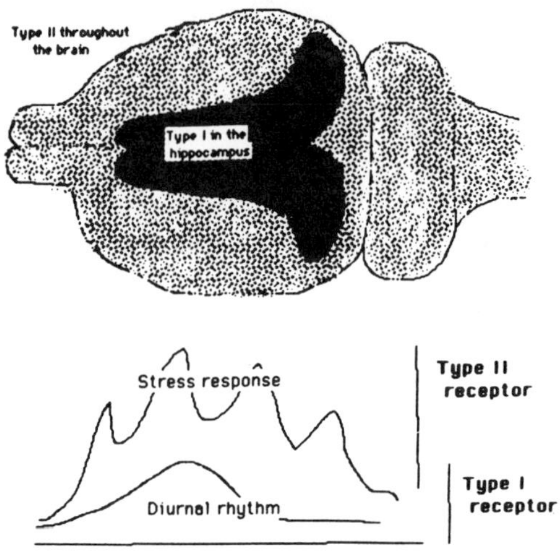

Figure 1. Schematic representation of distribution of Type I and Type II glucocorticoid receptors in rat brain. Note that the Type II receptor is found in all brain regions including the hippocampus, whereas the Type I receptor is highly concentrated in the hippocampal formation.

secretion; and "atrophy" or damage, induced by prolonged glucocorticoid exposure.

TWO RECEPTOR SUBTYPES

It has been recognized for some time that two types of receptors exist for adrenal steroids, and they are frequently referred to as Type I and Type II. In the original designation by Funder et.al. (9). Type I referred to mineralocorticoid receptors and Type II to glucocorticoid receptors. We made the observation in 1968, that the principal adrenal steroid of the rat, corticosterone, is taken up and concentrated by the hippocampus, a sub-cortical limbic system structure involved in both cognitive and vegetative processes (10). Although 3H aldosterone (ALDO) is also highly concentrated by the hippocampus in adrenalectomized (ADX) rats (11), it is not selectively retained in the presence of corticosterone (CORT) levels which mimic the normal excess of CORT over ALDO (12, 13). Thus the hippocampal Type I receptor, which has a proteolytic fragmentation pattern identical to that of the kidney (14), operates principally as a receptor for CORT, whereas that in the kidney operates as a receptor for aldosterone (15).

Recently, with the aid of the synthetic Type II-specific glucocorticoid, Ru26988, Reul and DeKloet (16) demonstrated the existence of both Type I and Type II receptors in hippocampus, as well as the presence of Type II receptors outside of hippocampus, where Type I receptors are much less plentiful. See Figure 1. The brain-wide distribution of Type II receptors has also been noted by immunocytochemistry (17, 18). Because of the lower affinity of the Type II receptor for CORT, compared to the Type I receptor (19), it is likely that Type II receptors are only extensively occupied by stress levels of CORT and by CORT at the diurnal peak, whereas Type I receptors are extensively occupied by low levels of CORT such as occur during the diurnal cycle (16; 20; 21). See Figure 1. This "two-level" recognition system for CORT would explain the selective hippocampal uptake of CORT when its levels are low in the blood and the more widespread uptake of CORT at stress blood levels (22).

FUNCTIONAL IMPLICATIONS OF TWO RECEPTORS

In order to understand the functional role of two glucocorticoid receptors, we return to the two modes of adrenocortical operation, the diurnal rhythm and stress-induced secretion. Based upon the relative affinities for CORT, it appears likely that Type I receptors in hippocampus are occupied principally during the diurnal cycle, whereas Type II receptors in hippocampus and elsewhere in the brain are occupied as CORT levels reach their diurnal peak and when CORT levels rise even higher with stress (16). This is an interesting possibility because the two modes of adrenocortical function are associated with different behaviors in which glucocorticoids play a role.

Diurnal variations of adrenocortical function are analogous to the rhythms of gonadal steroids in the estrous cycle. Cyclicity of hormone secretion results in cyclicity of behavioral states caused by the hormones, and we call the induction of these behaviors by hormones "activation". The peak of glucocorticoid secretion occurs in all mammals just prior to waking, whether or not activity is a nocturnal or diurnal event. In rats we know that accompanying waking is an increase in locomotor and exploratory activity (23) and that glucocorticoids increase this activity (24) and may do so via the hippocampus (23-26). Waking is also accompanied by an increased appetite and food-seeking behavior, which are both increased by glucocorticoids (27). In fact, presenting food at only one time of day entrains the diurnal glucocorticoid rhythm and results in an anticipatory rise of glucocorticoids prior to food presentation (3, 4). We infer a role for the hippocampus in this process, because withholding food reward results in an even larger rise in CORT; this rise is blocked by fornix ablation which eliminates a major output from hippocampal neurons (28). (It should be noted that glucocorticoid actions in paraventricular nucleus to promote feeding are also important (29)).

A third aspect of diurnally-varying neural activity is what can be called "synaptic efficacy", and this can be measured in terms of the magnitude of evoked potentials in the CA1 field of Ammon's horn after stimulation by known current intensities in dentate gyrus (30). Rats with adrenals in place, show peak synaptic efficacy in the active or waking phase, whereas ADX rats show peak synaptic efficacy in the sleeping phase, ie., 12h out of phase (30). If this effect of ADX is mediated by adrenal steroids, then we might have an explanation for cognitive deficits in "jet lag" in terms of synaptic efficacy. That adrenal steroids do shift the period of sleep when given intermittently supports the notion that periodic elevations of adrenal steroids in the diurnal rhythm coordinates neural events associated with sleep and waking states (31, 32).

Stress-induced elevations in glucocorticoids result in a different set of neural and behavioral responses which are consistent with the involvement of another receptor type having a different distribution within the brain. One type of evidence for this is in steroid specificity of effects. Stress-induced glucocorticoid secretion mediates suppression of swimming behavior in the Porsolt swimming test (33), and this effect can be blocked by ADX and reinstated by dexamethasone (DEX), a steroid with high affinity for Type II receptors in vivo (34). In contrast, glucocorticoid effects on exploratory activity are produced by CORT, and not by dexamethasone (DEX), and may involve Type I receptors (24).

Results showing glucocorticoid mediation of adaptation to the Porsolt test raise a key issue in the role of pituitary-adrenal function in stress. Is the role of adrenal steroids to assist in the adaptation of the individual to a stressful situation? Behavioral response to severe stress, eg., 1-2h restraint for a rat, includes depressed exploratory behavior, reduced food intake and increased anxiety (35). With time, and even with repeated stress, behavior returns to normal, indicating adaptation. Stone who measured changes in cerebral cortical beta receptors and noradrenergic-stimulated cAMP formation with repeated restraint stress, has argued that the time course of change in the neurochemistry parallels the adaptation process. (36-38). Indeed, it appears that adrenocortical secretions brought about by stress are the mediators of the reduction in noradrenaline-stimulated cAMP formation, and that they do so by suppressing the efficacy of the alpha receptor mediated potentiation of beta receptor-stimulated cAMP formation (38). Glucocorticoids also regulate cAMP formation stimulated by several neuroactive substances in hippocampus, with reductions in VIP- and NA-stimulated cAMP accumulation and increases in histamine-stimulated cAMP formation occurring after DEX treatment (39, 40).

What is interesting about the hippocampal response to DEX is that it occurs in ADX male rats but not in ADX females (41). Does this mean that adaptation mediated by glucocorticoids has a different character in males and females? We do not know for sure, but it is evident in behavioral studies of adaptation to restraint stress that females do not adapt as readily as males, even though they appear to be less severely affected by the stress (35). Further studies will have to determine the endocrine basis of the sex differences, whether it represents an effect of early brain sexual differentiation or a more immediate effect of circulating gonadal steroids. Also remaining to be determined is the exact mechanism for the glucocorticoid-mediated suppression. Steroid effects on cAMP accumulation are widespread in tissues of the body (39), and the mechanism may involve genomic regulation of coupling proteins involved in cAMP generation or of the receptors which are coupled to adenylate cyclase. With regard to the female response to DEX, this is not due to reduced Type II receptor levels, although we have recently found that the negative regulation of these receptors by endogenous glucocorticoids is less pronounced in females, another indication of a possible sex difference in glucocorticoid sensitivity (42).

We have only begun to touch the surface regarding the changes in the brain which are induced by single or repeated stress or by single or repeated glucocorticoid administration. Among other changes which must be explored in greater detail are changes in GABA-benzodiazepine receptors (43) high affinity GABA transport into synaptic endings (44) increased levels of Synapsin I (45) and increased central serotonin sensitivity (46-48). With respect to the latter, although serotonin sensitivity appears to be increased by stress, glucocorticoids appear to counteract this increase (48), and they do so, at least in part, by decreasing 5HT-1 receptor levels in hippocampus and several other brain regions (8). Thus although glucocorticoids also potentiate increased serotonin turnover in acute stress (49-51) they appear to dampen the serotonin sensitivity with prolonged stress. A similar biphasic response may also occur for noradrenaline (52).

Although it is premature to formulate comprehensive theories about such a complex disorder as endogenous depressive illness, it is tempting to note that the effects of glucocorticoids to suppress NA-stimulated cAMP accumulation are same direction as the tricyclic anti-depressants. Thus glucocorticoids are, in this one sense, an endogenous anti-depressant, and it is noteworthy that in endogenous depression, the glucocorticoid levels are frequently elevated with accompanying resistance to DEX suppression, as if the glucocorticoid response system has gone awry and become desensitized. One can only wonder whether glucocorticoid receptor levels are reduced in endogenous depression to the point that responses to endogenous and exogenous glucocorticoids are also significantly reduced.

PROLONGED GLUCOCORTICOID EXPOSURE: DOWN REGULATION AND NEURONAL LOSS

With repeated, intense stress and also after repeated administration of CORT, glucocorticoid receptors are down-regulated (53). The duration of stress required for this down-regulation is somewhere between 1 and 3 weeks, where 3-4 stress sessions are given during daylight hours to male rats (McEwen, Spencer and Brinton, unpublished; (53)). Although the

CONDITIONS WHICH EXACERBATE NEURONAL LOSS

1. **ELEVATED GLUCOCORTICOID LEVEL**
2. **ELEVATED LEVEL OF EXCITATORY AMINO ACIDS OR EXCITOTOXINS**

CONDITIONS WHICH REDUCE NEURONAL LOSS

1. **REDUCED GLUCOCORTICOID LEVEL**
2. **INCREASED LEVELS OF BRAIN FUELS**

Figure 2. Summary of conditions which promote neuronal loss and which protect against it.

initial studies examined total binding of 3H DEX and found down-regulation, principally in hippocampus, we now know that both Type I and Type II receptors are susceptible to down-regulation in hippocampus (McEwen, Brinton and Spencer, unpublished). However, the Type I receptors, with their higher affinity for CORT, appear to be down-regulated to a greater extent than Type II receptors (42).

One way of looking at receptor down-regulation is that it represents a means for the target tissue to reduce its sensitivity to hormone. On the one hand, this may be deleterious, if the consequence is a failure of the tissue or organism to adapt to stress (eg., as in depressive illness). Furthermore, negative feedback will also be impaired, and CORT levels will be elevated. On the other hand, reduced target sensitivity might at least partially protect against the inhibitory effects which glucocorticoids exert on immune function that may be deleterious in terms of lesser resistance to pathogens and to tumors.

There is another aspect to glucocorticoid sensitivity which involves another deleterious effect of chronic glucocorticoid exposure. With age in the male rat, neurons are lost in subfields of the hippocampal formation, and this loss is due, in part, to glucocorticoids (54-59). Evidence for this conclusion is that ADX at mid-age protects the rat from severe neuronal loss (55). Moreover, 12 weeks of CORT administration to younger rats produces a pattern of neuronal loss which resembles that in aging hippocampus (57, 58).

The ability of CORT to promote neuronal loss does not stop with the examples noted above, for glucocorticoids present in the circulation at the time of experimental hypoxia-ischemia exacerbate neuronal loss not only in hippocampus but also in other brain regions (60). Moreover, damage induced by excitotoxins is also exacerbated by high CORT levels and reduced by ADX (61, 62). See Figure 2. This has led to the proposal that glucocorticoids impair neuronal energy generating capacity, possibly by reducing transport of glucose and other metabolizable fuels; this idea is supported by the demonstration that damage is reduced by giving metabolizable sugars or their metabolites (63). See Figure 2. Excitotoxin damage in cultured hippocampal neurons is also exacerbated by elevated CORT in the culture medium (Sapolsky, submitted). Ironically, perhaps, an inhibitor of Type II receptors, Ru38486, blocks the CORT effect whereas a Type I receptor blocker, Ru26318, does not (Sapolsky and Vale, unpublished). Thus the receptor which is less readily down-regulated by CORT appears to be the one to mediate the damage induced by glucocorticoids.

It should be noted, however, that glucocorticoids are not operating by themselves to bring about neuronal loss. Rather, the neurochemistry of excitatory amino acids may play an important role as also does the level of circulating sugars in the blood. Conceivably, individual animals which are more prone to neuronal loss may be at one end of the normal distribution with regard to neuronal excitability and energy metabolism. Nutrition and stressful life experiences must also be considered, as well as the individual differences in the regulation of pituitary-adrenal function.

An important factor in shaping individual patterns of performance and stress-susceptibility is neonatal experience. Meaney and coworkers (64, 65) have established that neonatal handling of rats increases hippocampal glucocorticoid receptor levels as well as the ability to shut off the pituitary-adrenal stress response. Recently they have produced initial evidence supporting the notion that neonatal handling leads to slower development of severe "dementia" with age (Meaney, personal communication).

CONCLUSIONS

We have shown that glucocorticoids play an important role in 3 phases of nervous system function: the "activation" of neural activity during the diurnal cycle of pituitary adrenal function in which we postulate a principal role of the Type I glucocorticoid receptor localized in the hippocampus; the "adaptation" to stress in which we postulate a principal role for the Type II receptor found throughout the brain; and the loss of neurons with age and after prolonged exposure to elevated glucocorticoids ("atrophy") in which initial evidence indicates a role for Type II receptors. The latter events occur only after prolonged periods and lead us to believe that there are many back-up systems which protect the brain against self-destruction, including the down-regulation of receptors by elevated glucocorticoids. Nevertheless, self-destruction does apparently become a factor with increasing age in individuals whose genetic make-up and early life experiences have made them more susceptible to stress.

In spite of these fascinating and somewhat disturbing consequences of long-term glucocorticoid exposure, we prefer to place greatest emphasis on the protective effects of glucocorticoids, especially as they relate to the ability of the organism to adapt to stress. Changes in the brain brought about by stress-induced elevation of glucocorticoids appear to counteract some of the more direct neural effects of stress, such as elevated central 5HT sensitivity and increased central noradrenergic sensitivity; and they bring about restoration of behavior to normal after a stressful experience.

HYPOTHESIS: Glucocorticoids prevent the nervous system from over-reacting to "stressful" stimuli - in this respect, they act as endogenous "antidepressants"

NEURAL RESPONSE TO STRESS	GLUCOCORTICOID RESPONSE
increases -noradrenaline -serotonin turnover	decreases -noradrenaline -serotonin sensitivity

Figure 3. Summary of hypothesis that glucocorticoids prevent brain from overreacting to stressful stimuli.

How does this view of stress and brain function coincide with view about the peripheral actions of glucocorticoids in stress?

In their recent review of glucocorticoids and stress, (66) propose that "...the physiological function of stress-induced increases in glucocorticoid levels is to protect not against the source of stress itself, but against the normal defense reactions that are activated by stress" and that they "...accomplish this function by turning off those defense reactions, thus preventing them from overshooting themselves and threatening homeostasis." See Figure 3. We therefore conclude that Munck's view of glucocorticoid action in stress relative to processes such as the inflammatory process and the immune response is entirely consistent with our views of glucocorticoid actions on brain function.

ACKNOWLEDGMENTS

This work was supported by NIMH Grant MH41256 to BMc and by an Individual USPHS Research Fellowship MH09224 to RB. RMS was supported by a pre-doctoral grant from the National Institute on Aging and by a post-doctoral grant from the Life Sciences Research Foundation.

REFERENCES

1. D. Gibbs, Vasopressin and oxytocin: hypothalamic modulators of the stress response: a review, Psychoneuroendocrinology, 11:131-140 (1986).
2. C. Richter, "Biological Clocks in Medicine and Psychiatry", Charles C. Thomas Publisher, Springfield, Ill. (1965).
3. K. Honma, S. Honma and T. Hiroshige, Feeding-associated corticosterone peak in rats under various feeding cycles, J. Physiol, 246:R721 -R726 (1984).
4. D. Krieger, Food and water restriction shifts corticosterone, temperature, activity and brain amine periodicity, Endocrinology, 95:1195-1201 (1974).
5. A. Towle and P. Sze, Steroid binding to synaptic plasma membrane: differential binding of glucocorticoids and gonadal steroids, J. Steroid Biochem., 18:135-143 (1983).
6. M. Keller-Wood and M. Dallman, Corticosteroid inhibition of ACTH secretion, Endocr Rev, 5:1-24 (1984).
7. K. Yamamoto, Steroid receptor regulated transcription of specific genes and gene networks, Ann Rev Gen, 19:209-252 (1985).

8. B.S. McEwen, E.R. DeKloet and Rostene, W., Adrenal steroid receptors and actions in brain, Physiol Rev, 66:1121-1188 (1986).
9. J. Funder, D. Feldman and I. Edelman, The roles of plasma binding and receptor specificity in the mineralocorticid action of aldosterone, Endocrinology, 92:994-1004 (1973).
10. B. S. McEwen, J. Weiss and L. Schwartz, Selective retention of corticosterone by limbic structures in rat brain, Nature, 220:911-912 (1968).
11. A. Ermisch and H. Ruhle, Autoradiographic demonstration of aldosterone-concentrating neuron populations in rat brain, Brain Res, 147:154-158 (1978).
12. B.S. McEwen, L. Lambdin, T. Rainbow and A. DeNicola, Aldosterone effects on salt appetite in adrenalectomized rats, Neuroendo, 43:38-43 (1986).
13. B. Yongue and E. Roy, Radioimmunoassay of hippocampal cell nuclear aldosterone and corticosterone, Soc Neurosci Abstr, 10:140.8 (1984).
14. O. Wrange and Z-Y. Yu, Mineralocorticoid receptor in rat kidney and hippocampus characterization and quantitation by isoelectric focusing, Endocrinology, 113:243-250 (1983).
15. J.W. Funder and K. Sheppard, Adrenocortical steroids and the brain, Ann. Rev. Physiol., 49:397-411 (1987).
16. J. Reul and E.R. DeKloet, Two receptor systems for corticosterone in rat brain microdistribution and differential occupation, Endocrinology, 117:2505-2511 (1985).
17. K. Fuxe, A. Harfstrand, L. Agnati, Z-Y Yu, A. Cintra, A. Wikstrom, S. Okret and E. Cantoni, J. Gustafsson, Immunocytochemical studies on the localization of glucocorticoid receptor immunoreactive nerve cells in the lower brain stem and spinal cord of the male rat using a monoclonal antibody against rat liver glucocorticoid receptor, Neurosci Letts, 60:1-6 (1985b).
18. K. Fuxe, A. Wikstrom, S. Okret, L. Agnati, A. Harfstrand, Z. Yu, L. Granholm, M. Zoli, W. Vale and J. Gustafsson, Mapping of glucocorticoid receptor immunoreactive neurons in the rat tel- and diencephalon using a monoclonal antibody against rat liver glucocorticoid receptor, Endocrinology, 117:1803-1812 (1985).
19. H. Veldhuis, C. Van Koppen, M. Van Ittersum and E.R. DeKloet, Specificity of the adrenal steroid receptor system in rat hippocampus, Endocrinology, 110:2044-2051 (1982).
20. B.S. McEwen, G. Wallach and C. Magnus, Corticosterone binding to hippocampus. Immediate and delayed influence of the absence of adrenal secretion, Brain Res, 70:321-334 (1974).
21. W. Rotsztejn, M. Normand, J. Lalonde and C. Fortier, Relationship between ACTH release and corticosterone binding by the receptor sites of the adenohypophysic and dorsal hippocampus following infusion of corticosterone at a constant rate in the adrenalectomized rat, Endocrinology, 97:223-230 (1975).
22. B.S. McEwen, B. Stephenson and L.C. Krey, Radioimmunoassay of brain tissue and cell nuclear corticosterone, J Neurosci Meth, 3:57-65 (1980).
23. C. Kim, H. Choi, J. Kim, H. Chang, R. Park and I. Kang, General behavioral activity and its component patterns in hippocampectomized rats, Brain Res, 19:379-394 (1970).
24. H. Veldhuis, E.R. DeKloet, I. Van Zoest and B. Bohus, Adrenalectomy reduces exploratory behavior activity in rat a specific role of corticosterone, Horm Behav, 16:191-198 (1982).

25. P. Iuvone and C. Van Hartesveldt, C., Locomotor activity and plasma corticosterone in rats with hippocampal lesions, Behav Biol, 16:515-520 (1976).
26. P. Iuvone and C. Van Hartesveldt, Diurnal locomotor activity in rats: effects of hippocampal ablation and adrenalectomy, Behav Biol, 19:228-237 (1977).
27. M. Jhanwar-Uniyal, R. Roland and S. Leibowitz, Diurnal rhythm of a2-noradrenergic receptors in the paraventricular nucleus and other brain areas: relation to circulating corticosterone and feeding behavior, Life Sci, 38:473-482 (1986).
28. B. Osborne, Behavioral and adrenal responses and meal expectancy in rats with fornix transection, Physiol Behav, 37:499-502 (1986).
29. C. Roland, P. Bhakthavatsalam and S. Leibowitz, Interaction between corticosterone and alpha-1-noradrenergic system of the paraventricular nucleus in relation to feeding behavior, Neuroendo, 42:296-305 (1986).
30. R. Dana and J. Martinez, J., Effect of adrenalectomy on the circadian rhythm of LTP, Brain Res, 308:392-395 (1984).
31. J. Gillin, L. Jacobs, F. Snyder and R. Henkin, Effects of ACTH on the sleep of normal subjects and patients with Addison's disease, Neuroendo, 15:21-31 (1974).
32. J. Johnson and C. Sawyer, Adrenal steroids and the maintenance of a circadian distribution of paradoxical sleep in rats, Endocrinology, 89:507-512 (1971).
33. R. Porsolt, M. Le Pichon and M. Jalfre, Depression: a new animal model sensitive to anti-depressant treatments, Nature, 266:730-732 (1977).
34. D. Jefferys, D. Copolov, D. Irby and J. Funder, Behavioral effect of adrenalectomy: reversal by glucocorticoids or [D-ala2 -met5] enkephalinamide, Eur J Pharm, 92:99-103 (1983).
35. G. Kennett, F. Chaouloff, M. Marcou and G. Curzon, Female rats are more vulnerable than males in an animal model of depression the possible role of serotonin, Brain Res, 382:416-421 (1986).
36. E. Stone and J. Platt, Brain adrenergic receptors and resistance to stress, Brain Res, 237:405-414 (1982).
37. E. Stone, A. Slucky, J. Platt and R. Trullas, Reduction of the cyclic adenosine 3'5'-monophosphate response to catecholamines in rat brain slices after repeated stress, J Pharm Exp Ther, 233:382-388 (1985).
38. E. Stone, J. Platt, A. Herrera and K. Kirk, Effect of repeated restraint stress, desmethylimipramine or adrenocorticotropin on the alpha and beta adrenergic components of the cyclic AMP response to norepinephrine in rat brain slices, J Pharm Exp Ther, 237:702-707 (1986).
39. A. Harrelson, W. Rostene and B.S. McEwen, Adrenocortical steroids modify neurotransmitter-stimulated cAMP accumulation in the hippocampus and limbic brain of the rat, J Neurochem, 48:1648-1655 (1987).
40. V. Roberts, R. Singhal and D. Roberts, Corticosterone prevents the increase in noradrenaline-stimulated adenyl cyclase activity in rat hippocampus following adrenalectomy or metopirone, Eur J Pharm, 103:235-240 (1984).
41. Harrelson, A. and McEwen, B.S., Gonadal steroid modulation of neurotransmitter-stimulated cAMP accumulation in the hippocampus of the rat, Brain Res, 404:89-94 (1987).
42. B.S. McEwen, R.E. Brinton and L. Snyder, Regulation of receptors for adrenal steroids in rat hippocampus, Abstr Soc Neurosci, 12 #290.2 (1986).

43. E. Souza, N. Goeders and Kuhar, M., Benzodiazepine receptors in rat brain are altered by adrenalectomy, Brain Res, 381:176-181 (1986).
44. A. Miller, C. Chaptal, B.S. McEwen and E. Peck, Modulation of high affinity GABA uptake into hippocampal synaptosomes by glucocorticoids, Psychoneuroendorinol, 3:155-164 (1978).
45. E. Nestler, T. Rainbow, B.S. McEwen and P. Greengard, Corticosterone increases the level of protein I, a neuron-specific protein, in rat hippocampus, Science, 212:1162-1164 (1981).
46. G. Kennett, S. Dickinson and G. Curzon, Central serotonergic responses and behavioral adaptation to repeated immobilization: the effect of the corticosterone synthesis inhibitor metyrapone, Eur J Pharm. 119:143-152 (1985).
47. G. Kennett, S. Dickinson and G. Curzon, Central serotonergic responses and behavioral adaptation to repeated immobilization: the effect of the corticosterone synthesis inhibitor metyrapone, Eur J Pharm, 119:143-152 (1985).
48. S. Dickinson, G. Kenneth and G. Curzon, Reduced 5-hydroxytryptamine -dependent behavior in rats following chronic corticosterone treatment, Brain Res, 345:10-18 (1985).
49. E. Azmitia and B.S. McEwen, Corticosterone regulation of tryptophan hydroxylase in midbrain of the rat, Science, 166:1274-1276 (1969).
50. E. Azmitia and B.S. McEwen, Adrenocortical influence on rat tryptophan hydroxylase activity, Brain Res, 78:291-302 (1974).
51. P. Sze, Glucocorticoid regulation of the serotonergic system of the brain, Advances in Biochemical Psychopharmacol, 15:251-265 (1976).
52. B.S. McEwen, Glucocorticoid-biogenic amine interactions in relation to mood and behavior, Biochem Pharm, 36:1755-1763 (1987).
53. R. Sapolsky, L.C. Krey and B.S. McEwen, Stress down-regulates corticosterone receptors in a site-specific manner in the brain, Endocrinology, 114:287-292 (1984).
54. P. Landfield, G. Rose, L. Sandles, T. Wohlstadter and G. Lynch, Patterns of astroglial hypertrophy and neuronal degeneration in the hippocampus of aged, memory- deficient rats, J Geron, 32:3-12 (1977).
55. P. Landfield, J. Waymire and G. Lynch, Hippocampal aging and adrenocorticoids: quantitative correlation, Science, 202:1098-1101 (1978).
56. P. Landfield, R. Baskin and T. Pitler, Brain-aging correlates retardation by hormonal-pharmacological treatments, Science, 214:581-584 (1981).
57. R. Sapolsky, L.C. Krey and B.S. McEwen, Prolonged glucocorticoid exposure reduces hippocampal neuron number: implications for aging, J Neurosci, 5:1222-1227 (1985).
58. R. Sapolsky, L.C. Krey and B.S. McEwen, The neuroendocrinology of stress and aging: the glucocorticoid cascade hypothesis, Endocr Rev, 7:284-301 (1986).
59. R. Sapolsky, L.C. Krey, B.S. McEwen and T.C. Rainbow, Do vasopressin -related peptides induce hippocampal corticosterone receptors? Implications for aging, J Neurosci, 6:1479-1485 (1984).
60. R. Sapolsky and W. Pulsinelli, Glucocorticoids potentiate ischemic injury to neurons: therapeutic implications, Science, 229:1397 -1399 (1985).
61. R. Sapolsky, A mechanism for glucocorticoid toxicity in the hippocampus: increased neuronal vulnerability to metabolic insults, J Neurosci, 5:1228-1232 (1985).
62. R. Sapolsky, Glucocorticoid toxicity in the hippocampus temporal aspects of synergy with kainic acid, Neuroendo, 43:440-444 (1986).
63. R. Sapolsky, Glucocorticoid toxicity in the hippocampus: reversal by supplementation with brain fuels, J Neurosci, 6:2240-2244 (1986).

64. M. Meaney and D. Aitken, The effects of early, postnatal handling on hippocampal glucocorticoid receptor concentrations: temporal parameters, Brain Res, 22:30-304 (1985).
65. M. Meaney, D. Aitken, S. Bodnoff, L. Iny, J. Tatarevicz and R. Sapolsky, Early, postnatal handling alters glucocorticoid receptor concentrations in selected brain regions, Behav Neurosci, 99:765 -770 (1985).
66. A. Munck, P. Guyre and N. Holbrook, N., Physiological functions of glucocorticoids in stress and their relation to pharmacological actions, Endocr Res, 5:25-44 (1984).

CRH EFFECTS ON CENTRAL NORADRENERGIC NEURONS: RELATIONSHIP TO STRESS

Rita J. Valentino

Department of Pharmacology
George Washington University Medical Ctr.
Washington, DC 20037

CRH AS A STRESS NEUROTRANSMITTER

Criteria

The stress response is generally considered to be composed of endocrine, behavioral, and autonomic components, coordinated in such a way that it mobilizes the organism to best respond to a stressful stimulus and ultimately return to a homeostatic state. It is now well established that corticotropin-releasing hormone (CRH) is the primary neurohormone responsible for initiating the endocrine component of the stress response, i.e., release of ACTH from anterior pituitary cells[1,2,3]. Evidence supporting this role is summarized in other chapters of this series. An additional, broader, role for CRH may also be hypothesized; as a neurotransmitter acting in extrahypothalamic circuits to initiate an integrated CNS response to stress that includes the behavioral and autonomic components. To support such a role, several criteria should be met. For example, CRH should be localized in CNS regions whose potential function is to modify behavior or autonomic acitivity. It should alter activity of neurons in these target regions, and a mechanism should exist to translate CRH electrophysiological effects to appropriate behaviors or autonomic responses that are normally observed in the stress response. Most importantly, it should be demonstrated that stressful stimuli elicit release of CRH in these extrahypothalamic circuits resulting in subsequent changes in neuronal activity and the appropriate behaviors or autonomic responses.

This chapter cites some of the findings that are supportive of a role for CRH as a stress neurotransmitter. Specifically, the focus will be on CRH as a neurotransmitter in the noradrenergic nucleus, locus coeruleus (LC). This is based on the proposed role of the LC in stress, the localization CRH within the LC, and as discussed in detail below, the electrophysiological effects of CRH on LC neurons. The hypothesis that CRH acts as a neurotransmitter within the LC is an attractive one because it offers a mechanism by which the CNS can integrate and coordinate a multicomponent response to stress. The implications of this model in stress and psychiatric disease are discussed.

CRH Localization

Results of anatomic studies demonstrate that CRH meets at least one of the criteria for a stress neurotransmitter. CRH-like immunoreactivity has been visualized in the rat brain in several extrahypothalamic regions, particularly in circuits throughout the basal forebrain and brainstem[4-8]. Cell bodies containing CRH have been visualized in cortex, bed nucleus of the stria terminalis, central nucleus of the amygdala, parabrachial nucleus, and lateral dorsal tegmental nucleus. CRH-containing fibers occur throughout the brainstem associated with nuclei concerned with autonomic function. CRH localization in nonhuman primate CNS is similar to that reported in the rat[9]. This localization provides an anatomic substrate for CRH as a stress neurotransmitter. The demonstration of specific binding sites for CRH in parallel regions by autoradiographic studies is consistent with the concept that CRH released from fibers can interact with specific receptors on neurons in terminal regions and alter cellular activity[10,11,12].

CRH Electrophysiological Effects

The effects of CRH on neuronal activity after central administration or local application have been described in several brain regions which are target areas for CRH-containing neurons. Increases in spontaneous discharge of cells in the cortex[13], hypothalamus[13], hippocampus[14] and LC[15,16], and decreases in activity of neurons in the lateral septum[13] and thalamus[13] have been reported. Effects on LC neurons will be described in detail below. Long-lasting EEG activation, after-discharges, and seizure-like activity have also been reported after central administration of relatively low doses of CRH[17]. Although preliminary, the demonstration of these electrophysiological effects on neurons which are innervated by fibers containing CRH further support a neurotransmitter role for CRH.

Behavioral and Physiological Effects of CRH

The behavioral and physiological effects associated with administration of a "stress neurotransmitter" should resemble those observed in the generalized stress response. Unfortunately, these effects are not always consistent between individual animals and may vary with the type of stressor. However, it would be adaptive for sympathetic tone to be enhanced and parasympathetic tone to be suppressed. These are physiological changes that mobilize the organism to respond in a fight or flight manner. Behaviors that are often observed as part of a stress response are those associated with arousal and increased vigilance, while vegetative behaviors (i.e., eating) are decreased. Central administration of CRH results in many of these expected effects. Increases in sympathetic tone produced by CRH administration are reflected in increased heart rate, blood pressure, oxygen consumption, glucose utilization, and plasma concentrations of epinephrine and norepinephrine[18-20]. Decreases in parasympathetic tone are reflected as decreased gastric acid secretion[21,22].

Behavioral effects associated with CRH administration are also consistent with a role as a stress neurotransmitter. Locomotor activity is enhanced in rats that are placed in activity chambers, and decreased in open field paradigms[23,24]. In this latter model crossovers are fewer, and grooming behavior is increased. If food is placed in the center of the open field, CRH-injected rats make fewer approaches towards the food than do saline-treated rats, and rearing is suppressed. It has been suggested that these behaviors reflect neophobia and are an indication of stress associated with novel stimuli in the external environment[25]. Finally, food-reinforced responding and punished responding are also suppressed by CRH[26,27]. Similar environment-specific effects of CRH have been reported in primates[28,29]. When restrained monkeys are injected with CRH, struggling is prominent. However, similar doses of CRH will inhibit motor activity in unrestrained monkeys and result in huddling and lying down in the home cage.

The reported physiological and behavioral effects of centrally administered CRH occur at similar doses, are not mimicked by peripheral administration, and can be produced in hypophysectomized, adrenalectomized animals. These findings indicate a site of action at central extrahypothalamic sites.

It is tempting to speculate that stressful stimuli elicit release of CRH from peptide-containing fibers, subsequently altering activity of neurons in target areas involved in behavioral arousal and autonomic activation. Such a mechanism would allow the CNS to simultaneously respond to stressors by several modes, i.e. endocrine, autonomic and behavioral. To test this hypothesis it is necessary to demonstrate, 1) that CRH alters activity of neurons in the CNS that have been implicated in some component of the stress response, 2) that this change in electrical activity is translated to a behavioral or autonomic effect and 3) that these actions are mimicked by the presentation of stressful stimuli and may be prevented by antagonists of CRH.

THE LC AS A FUNCTIONAL SITE OF ACTION FOR CRH

CRH Localization in the LC

CRH effects on LC cells are a reasonable model in which to study CRH neurotransmission since CRH immunoreactive fibers have been visualized both in rat and monkey LC[4-9]. In the monkey, a particularly high density of fibers is observed in the ventrolateral region[9]. In both species, these fibers are of unknown origin. Recent studies of afferents to the LC demonstrate that major inputs originate in the prepositus hypoglossal nucleus and the paragigantocellularis nucleus with a minor input originating from the paraventricular hypothalamus[30]. As the latter two regions contain CRH immunoreactivity, these are potential sources for CRH visualized within the LC. Interestingly, CRH levels in the LC increase dramatically in both acute and chronically stressed rats and similar changes do not occur in other brain regions[31]. It is not clear however, if this increased CRH reflects decreased release from fibers or increased rates of peptide processing due to high turnover.

LC Role in Stress

The LC has long been implicated in the stress response. This is based on its anatomical and physiological characteristics as well as reported changes in norepinephrine turnover in stressed animals. For example, the projections arising from this small cluster of cells innervate all regions of the CNS, and the LC is the sole source of norepinephrine in regions such as the cortex and hippocampus[32-35]. This widespread projection system can serve as an amplification mechanism for drugs or hormones acting directly on LC cells.

Activity of LC neurons in unanesthetized animals is correlated to levels of arousal. Thus, spontaneous discharge is highest in the active waking state when the animal is orienting to the environment, is decreased in the quiet awake state when the animal is attending to internal behaviors such as eating or grooming., is further decreased during slow wave sleep, and is inactive when the animal is in REM sleep[36,37]. Moreover, a wide range of external stimuli activate LC cells producing a discharge pattern characterized by a burst of activity followed by a period of relative inhibition[37,38,39,40]. Target cells of the LC projection system respond to LC stimulation or norepinephrine with an enhancement of evoked activity relative to spontaneous activity, and the response to both inhibitory and excitatory neurotransmitters is enhanced[41-49]. Based upon these physiological characteristics and the divergent anatomy of the LC projection system, it has been speculated that increased LC activity helps to mobilize the organism to respond towards stimuli in the external environment. Thus, activation of the LC would be predicted in stress. This is supported by studies demonstrating increased norepinephrine turnover in cortical regions in stressed rats[50-54]. Increases in norepinephrine turnover have been observed in rats exposed to footshock[50,51], rats presented with a previously neutral stimulus that becomes conditioned to footshock[52,53], and also in rats that are given experimental subarachnoid hemorrhage[54]. Because these increases in norepinephrine turnover occur primarily in CNS regions that receive their sole norepinephrine input from the LC, these results imply activation of this system by the stressors, electric shock and hemorrhage.

Recordings from LC cells of anesthetized animals are consistent with the concept that the LC is activated by stressors. Footshock increases spontaneous discharge rates of LC neurons of anesthetized rats. Likewise, hypovolemia[55,56] and hypoxia[57] also have been reported to increase spontaneous LC discharge of anesthetized animals. Finally, in unanesthetized cats, stimuli paired with noxious events appear to be more effective in activating the LC system than nonconditioned non-noxious stimuli[40]. All of these findings support a role for the LC in response to stress. However, the mechanism for LC activation by stressful stimuli has not been elucidated. The possibility that these stimuli elicit CRH release from fibers innervating the LC and thus activate this nucleus was examined. The first steps in testing this model were to determine the effects of CRH on spontaneous and sensory-evoked activity in both anesthetized and unanesthetized rats.

CRH EFFECTS ON LC SPONTANEOUS DISCHARGE

Anesthetized Rats

Intracerebroventricular (i.c.v.) administration of CRH produces a dose-dependent increase in spontaneous discharge rates of single LC neurons in halothane anesthetized rats[15,16] (figure 1). This activation peaks within 6 - 10 min and is still apparent 16 min after injection. The doses of CRH that result in activation (1-3 µg) are similar to those that produce the reported behavioral and autonomic stress-like effects [18-20,23,24,26,27]. Several analogues of CRH with varying potency

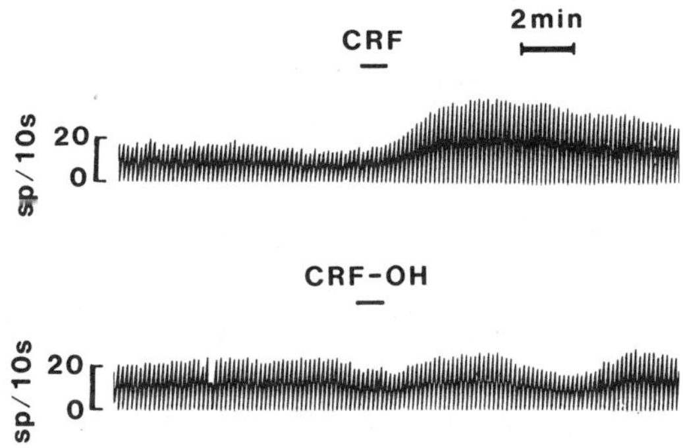

Fig. 1 Effect of i.c.v. administration of CRH or CRH-OH on single LC cells in separate rats. The solid bars above the individual ratemeter records indicate administration of CRH or CRH-OH (3 µg in 6 µl). After recording pre-peptide discharge rates for at least 5 min spontaneous firing rates were then recorded for at least 10 min after the injections. While CRH increased LC spontaneous discharge rate, CRH-OH was ineffective. (With permission, ref. 15).

and efficacy in releasing ACTH have been synthesized[1]. Two analogues that have a low efficacy in releasing ACTH, were also ineffective in increasing LC spontaneous discharge rates when administered at doses of CRH which were effective. These were the free acid analogue of CRH, CRH-OH, and Ala[14]-CRH. The lack of efficacy of these analogues in releasing ACTH and in activating LC neurons suggests that the structural requirements at both pituitary and LC binding sites are similar.

The activation of LC neurons by CRH probably occurs by a direct action on these cells since exposure of LC neurons to CRH by pressure application to a CRH-containing micropipette also increases tonic discharge of LC cells[15]. Interestingly, direct application of CRH to a neuron that was subsequently identified as being within the parabrachial nucleus was also activated by CRH. CRH activation of neurons in the parabrachial nucleus are of interest since CRH fibers have been visualized here and this nucleus is thought to play a role in certain autonomic functions such as respiration. However, CRH does not increase activity of all cells to which it is applied. Spontaneous discharge rates of cerebellar purkinje neurons and cells of the mesencephalic nucleus of the trigeminus are not altered after direct CRH application. Thus, CRH directly increases activity of LC neurons. This effect is dose-dependent, structurally specific, and regionally specific. Moreover, it occurs in the same dose-range and with the same timecourse as behavioral and autonomic effects associated with CRH administration.

Unanesthetized Rats

The physiological characteristics of LC neurons in unanesthetized rats differ from those observed in anesthetized animals in that discharge rates are more labile rather than regular and are associated with states of arousal and behavioral vigilance[36-40]. Because of this association between LC activity and state of arousal it is more meaningful to study the effects of CRH on LC neurons in unanesthetized animals. CRH was more potent in increasing LC discharge rates of unanesthetized rats as was demonstrated by a shift to the left in the dose-response curve for CRH[58] (figure 2). Interestingly, CRH efficacy appeared to increase with time in the unanesthetized preparation. Spontaneous activity recorded 40 min after CRH injection was greater than that determined 9 min after injection. This enhanced efficacy with time was not observed in anesthetized rats where LC spontaneous discharge was similar 3 min and 16 min after CRH injection.

An advantage of electrophysiological recording in unanesthetized animals is the opportunity to view gross behavioral effects and electrophysiological effects simultaneously. In many rats, increases in LC tonic activity were accompanied by increased exploratory movements in the cage for freely moving animals, and by struggling for animals that were partially restrained in a cloth harness during recording sessions. Both groups of rats were previously acclimated to the recording chamber so that all stimuli were familiar. This "behavioral activation" in familiar environments is probably similar to that previously reported[23]. It is of interest however, that this is temporally associated with increased activity of LC neurons and supports a link between CRH elicited LC activation and behavioral components of stress responses.

CRH EFFECTS ON LC SENSORY RESPONSES

Anesthetized Rats

In addition to being spontaneously active, LC discharge can be evoked by sensory stimuli[37-40]. In the anesthetized

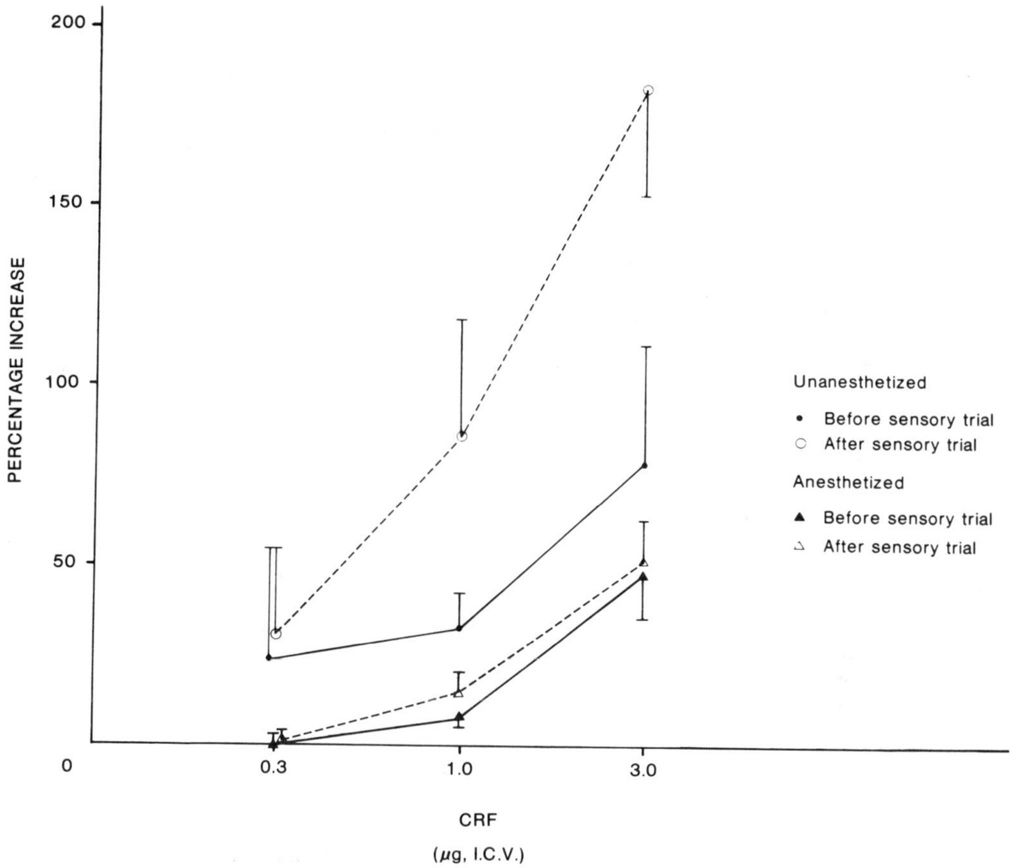

Fig. 2 Dose-response relationships for CRF effects on LC spontaneous discharge rate in anesthetized and unanesthetized rats. CRH was administered i.c.v., one dose per rat. For anesthetized rats, each point represents the postinjection rate expressed as a percentage of the mean prepeptide rate, recorded 3 min after CRH administration (solid triangles) or immediately after a block of sciatic nerve stimulation (16 min after CRH, open triangles). For unanesthetized rats, each point represents postinjection spontaneous activity expressed as a percentage of the mean prepeptide rate, recorded 9 min after injection (solid circles) or immediately after a block of auditory stimulation (45 min after CRH; open circles). Each point is the mean of 4-21 rats. Vertical lines indicate ± 1 S.E.M. CRH causes larger increases in LC tonic activity in unanesthetized rats, and these effects are even greater 45 min after injection. In contrast, CRH effects on LC tonic activity, in anesthetized rats appear to peak 3 min after injection.

state the stimulus required to evoke activity is potentially noxious, i.e., footshock produced by sciatic nerve stimulation[39]. Nonnoxious somatosensory stimuli are not effective in eliciting LC firing in anesthetized rats. In contrast, LC neurons of unanesthetized rats are responsive to all modes of sensory stimuli including auditory, visual, olfactory and somatosensory[37-40]. The pattern of discharge of LC neurons to both noxious and nonnoxious stimuli is the same and does not differ in the anesthetized or unanesthetized state. Examples of this pattern are shown in Figure 3. Shortly following the presentation of phasic stimuli, LC discharge is increased for 50-100 msec and this is followed by a relatively quiescent

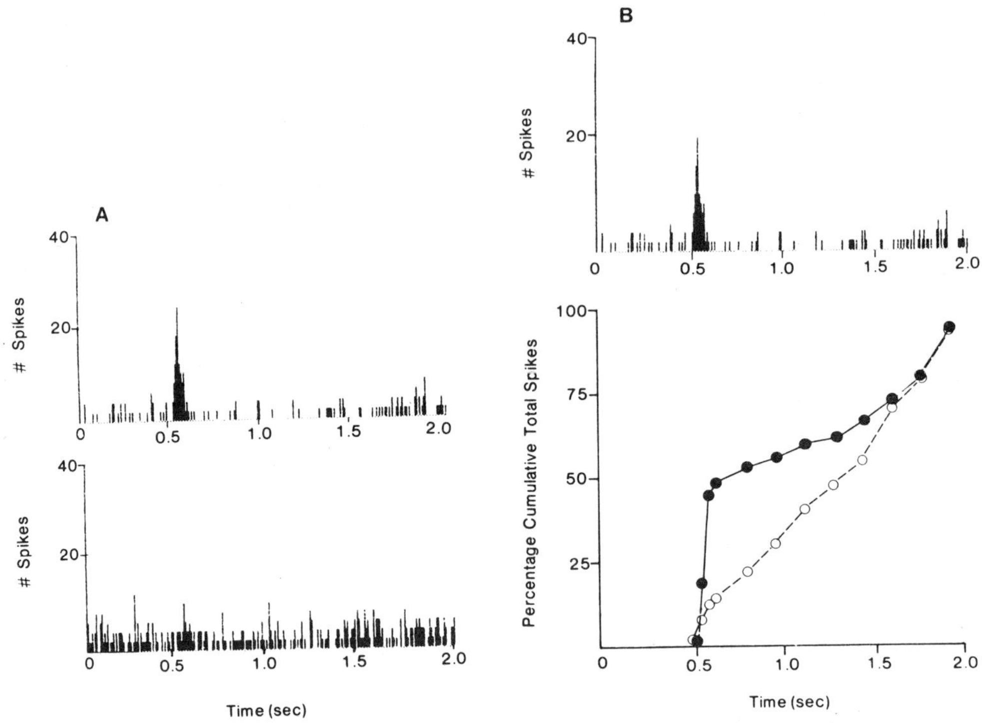

Fig. 3 Effect of CRH on LC responses to sciatic nerve stimulation in a halothane-anesthetized rat. A) Example post-stimulus time histograms (PISHs) generated before (top) and after (bottom) CRH administration (1.0 µg, i.c.v.) Repeated (80 trials) sciatic nerve stimulation were presented once every 10 sec. The PISH shows cumulative activity for 500 msec prior to stimulation and the next 1500 msec. Thus, discharge during the first 500 msec represents spontaneous discharge. Shortly following stimulation (0.5 sec) the LC neuron is activated and this is followed by a postactivation pause and a return to basal discharge rates between 1.5 and 2.0 sec. CRH disrupted this pattern such that fewer discharges occurred during the excitatory component of the response, more discharges occurred during the postactivation pause and basal rate of discharge was increased. This was quantified in B. B) The top PISH is the same control as shown in A. the bottom panel shows both control ( ) and CRH (O) histograms replotted as the normalized cumulative distribution of spikes during the period 20 msec before to 1500 msec after sciatic nerve stimulation. The curve was more linear (i.e., counts were more equally distributed among bins) following CRH. Differences between control and CRH distributions were quantified by comparing correlation coefficients of the 2 curves. This was greater in the presence of CRH since the curve became more linear. Alternatively, the maximum percentage difference between the 2 curves was calculated and if this exceeded the Kolmorgorov-Smirnoff statistic, the curves were considered significantly different. For this example, the K-S value for $p<0.05$ is 0.15 and the difference between the 2 curves at times 0.58-1.1 sec ranges between 0.19 and 0.35 (With permission, ref 16).

period. Intracellular recordings from LC neurons in slice preparations suggest that this postactivation pause is due to an afterhyperpolarization which follows a burst of action potentials[59,60].

The responsiveness of LC cells to various stimuli is one characteristic that has implicated the LC in behavioral arousal and stress. The interaction between CRH and LC sensory-evoked activity may be more important than CRH effects on LC tonic activity since it reflects the relationship between LC neurons and the environment. Since CRH appears to enhance behavioral responses to novel environmental stimuli in open field paradigms[24] it was also expected to alter LC responses to sensory stimuli.

The most consistent effect of CRH was to decrease the "signal-to-noise" in the pattern of discharge to the sensory stimulus[16]. This can be demonstrated best in anesthetized rats where tonic LC activity which may confound the LC sensory response is not as sensitive to CRH activation as in unanesthetized rats. LC cells of anesthetized rats respond to sciatic nerve stimulation with a period of excitation followed by a postactivation pause. CRH disrupts this pattern such that the number of discharges in the excitatory component of the response are decreased, more discharges occur during the postactivation pause, and tonic activity measured prior to the stimulus is often increased. The net effect then, is to disrupt the temporal pattern of LC discharge to the stimulus such that discharges are not time-locked to the stimulus (Figure 3).

After low doses of CRH (1.0 μg, i.c.v.) the effect on background activity is small and an absolute decrement in the excitatory response to the stimulus is observed. At higher doses (3.0 μg) this decrement is not significant and may be confounded by the more dramatic increase in spontaneous discharge. The signal-to-noise in the LC sensory response has been quantified by determining the ratio of discharge rate in the excitatory component of the response to the rate determined in the prestimulus interval. This ratio is significantly decreased to almost 50% by both 1.0 and 3.0 μg CRH.

Attenuation of the LC sensory response could be due to increases in LC tonic activity by CRH. For example, if the cell is firing rapidly it may not respond to phasic stimuli efficiently. This has been demonstrated with the muscarinic agonist, carbachol[61]. Like CRH, I.C.V. carbachol also increases LC spontaneous discharge and appears to decrease signal-to-noise in the LC sensory response. The degreee of disruption of the LC sensory response produced by carbachol is correlated (r=0.78) to increases in spontaneous LC discharge (figure 4). A similar plot of the relationship between CRH increases in LC spontaneous discharge and a measure of CRH attenuation of the sensory response showed that the two effects were hot as highly correlated (r=0.44) suggesting that these are two distinct actions of CRH. Taken together, the results of these studies suggest that drugs or manipulations that increase tonic discharge rates of LC neurons will attenuate the signal to-noise of the LC sensory response because of a ceiling effect on LC activity. However, while this may be partly responsible for CRH disruption of the LC sensory response

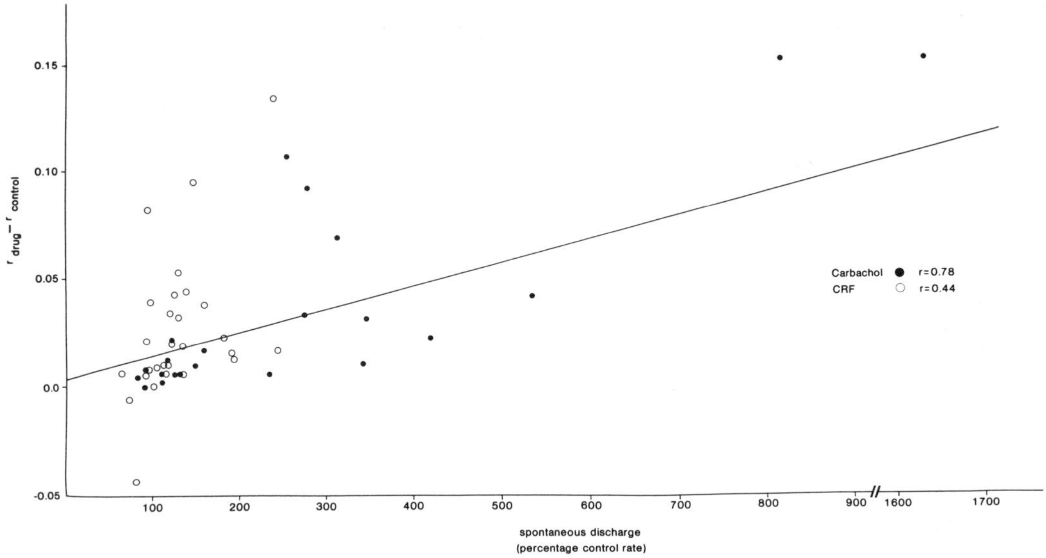

Fig. 4 Comparison of CRH and carbachol effects on the relationship between LC spontaneous activity and LC activity evoked by sciatic nerve stimulation. Single unit LC activity of halothane-anesthetized rats was recorded during repeated sciatic nerve stimulation. The abscissa indicates drug effects on spontaneous discharge which was recorded during the 500 msec prior to stimulation. The ordinate is a measure of drug-induced changes in the LC sensory response pattern (for details of quantification, see ref 16). Each point represents effects on an individual cell tested with i.c.v. carbachol (0.1 or 0.5 µg) or CRH (0.3, 1.0 or 3.0 µg). One dose was tested per cell and only one cell per rat. Both CRH and carbachol increased LC spontaneous discharge and disrupted LC sensory responses. However, for CRH, the attenuation of LC sensory responses was not systematically related to increases in LC tonic discharge (r = 0.44). In contrast, carbachol effects on LC spontaneous discharge and LC sensory responses were highly correlated (r = 0.78) suggesting that disruption of the sensory response by carbachol is the result of increased tonic activity. The results suggest that increasing LC tonic activity will result in disruption of the pattern of LC sensory responses. However, LC sensory responses can be attenuated by some agents (CRH) in the absence of increases in tonic activity.

pattern, particularly at higher doses, a true decrement in the discharge rate during the excitatory component of the LC sensory response occurs after CRH administration. Thus, in the anesthetized rat CRH both increases tonic LC discharge and attenuates discharge evoked by a potentially noxious stimulus and these may be distinct effects of CRH.

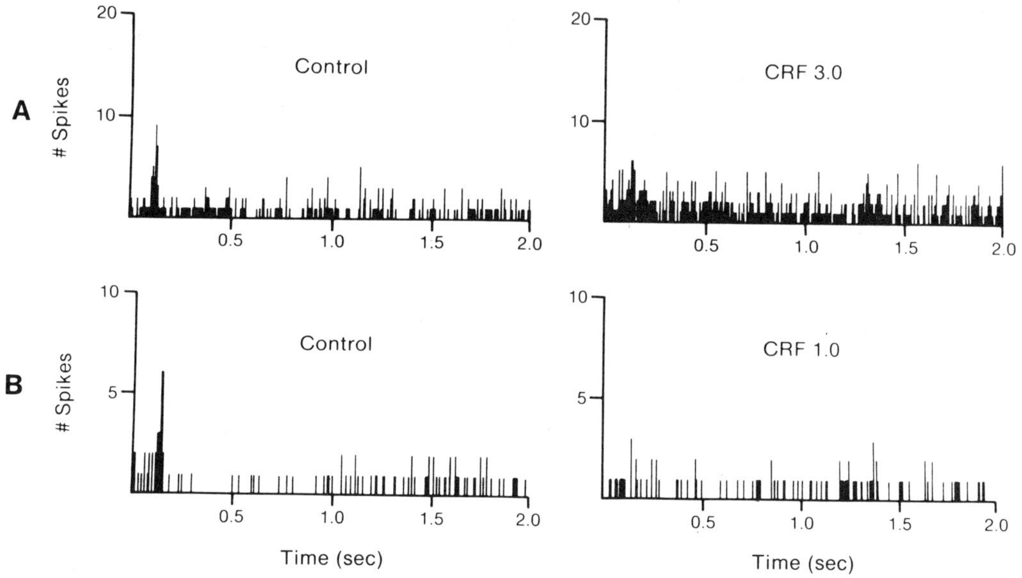

Fig. 5 Effect of CRH on LC auditory responses in unanesthetized rats. A and B show example PISHs computed during a block of auditory stimulation 15 min before and 9 min after i.c.v. injection of CRH 3.0 µg (top) or CRH, 1.0 µg (bottom). A tone was presented every 30 sec. A pulse triggered a 2 sec sweep on the computer 100 msec prior to tone presentation. The ordinate indicates the cumulative number of discharges occurring in each 8 msec bin during the block of stimulation. The top and bottom pairs of PISHs represent cumulative activity of 80 and 50 tone presentations, respectively. As shown in preCRH PISHs, tone elicits a brief activation followed by a post-activation pause. CRH increased the total number of discharges in each PISH and usually attenuated the signal-to-noise ratio in the pattern of responses such that background activity was increased more than activity during the excitatory component of the response.

Unanesthetized Rats

In the unanesthetized state LC discharge can be evoked by many modes of nonnoxious stimuli and the discharge pattern is similar to that observed when LC activity of anesthetized rats is evoked by footshock. CRH causes a disruption of the pattern of LC discharge to an auditory stimulus that is similar to the disruption of the response to sciatic nerve stimulation recorded in anesthetized rats[58] (figure 5). Thus, signal-to-noise in the LC sensory response is decreased by both 1.0 and 3.0 µg CRH. However, these doses of CRH also cause simultaneous increases in spontaneous LC discharge. Unlike its effects in anesthetized rats, CRH tended to increase the number of discharges in the excitatory component of the LC auditory response, although this effect was never significant. The tendency to increase discharges in this component was highly correlated to CRH increases in tonic LC discharge rates measured in the same rats (r=0.84). Since LC spontaneous activity of unanesthetized rats is more sensitive to CRH than in anesthetized rats it appeared that the most predominant effect in the unanesthetized animal is an increase in spontaneous discharge rate. Rates during the excitatory component of the sensory response are also increased but not proportionately to background activity, perhaps because of a ceiling effect on LC discharges. Thus, the net effect of CRH on the pattern of LC discharge to auditory stimuli is a decreased signal-to-noise. Because of confounding effects on LC tonic activity it is not clear whether the mechanism for CRH disruption of the LC auditory response in unanesthetized rats is similar to that for CRH disruption of the LC sensory response in anesthetized rats.

CELLULAR MECHANISM FOR CRH ACTIONS ON LC NEURONS

To determine the cellular mechanism by which CRH activates LC neurons intracellular studies are required. While these are not available, certain mechanisms may be hypothesized from intracellular studies of CRH effects on other neurons. To date the most extensive investigations have been done on pyramidal cells of the hippocampal slice preparation[14]. Exposure of this preparation to CRH (0.25 μM) increases firing rates of pyramidal neurons in the CA1 and CA3 region and this is accompanied by depolarization and increased input resistance. Lower concentrations reduce the magnitude and duration of the afterhyperpolarization seen after a burst of action potentials. Since this afterhyperpolarization is due to a calcium-dependent potassium conductance, CRH may activate neurons blocking influx of calcium responsible for the potassium conductance or by blocking the potassium conductance directly. The observation that CRH had no effect on calcium spikes points to a direct effect on the calcium-dependent potassium conductance. A block of this channel would result in decreased afterhyperpolarization and slower repolarization. This could allow more calcium to enter the cell through voltage sensitive channels. By reducing afterhyperpolarization and repolarization an inhibition of the calcium-dependent potassium conductance could account for activation of certain CNS neurons by CRH. Indeed, a similar afterhyperpolarization produced by a calcium-dependent potassium conductance is present in LC neurons following a burst of action potentials and this is thought to be responsible for the postactivation pause observed in the LC sensory response[59,60]. A reduction of this conductance by CRH could be partly responsible for both the decreased signal-to-noise in the LC sensory response and the increased spontaneous activity of LC cells. The reported cellular mechanisms of CRH action on hippocampal pyramidal neurons provide a model which can be tested using intracellular studies on LC neurons. This is necessary in understanding the translation between CRH binding and in vivo neuronal effects.

IMPLICATIONS OF CRH EFFECTS ON LC NEURONS

The localization of CRH fibers within the LC and its potent effects on these cells implicate the LC as a target where CRH can act in the capacity of a neurotransmitter.

Through its widespread projection system and effects on target cells the LC may act to integrate and amplify information regarding stress. An important question that arises is how CRH effects on LC neuronal activity may be translated into a generalized response to stress or even a component of the stress response.

The activation of LC tonic discharge by CRH is consistent with a role for the LC in behavioral arousal associated with the stress response. As higher rates of LC discharge in unanesthetized animals have been correlated to enhanced behavioral vigilance and orientation towards the external environment, LC activation by CRH would be an expected adaptive response[36-38,40]. Presumably, the pattern of LC discharge in response to phasic stimuli is an important signal to LC target areas providing information about the environment or internal homeostasis. A decreased signal-to-noise in this response could then have important consequences regarding behavior or autonomic function. Alternatively, the observed changes in LC neuronal activity may also represent a pathologic consequence of stress. For example, in a highly stressful situation an animal may become so aroused that it does not respond to phasic external stimuli efficiently. The electrophysiological representation of this may be very high levels of tonic LC activity that are not further increased by these phasic external stimuli. This interpretation requires further experimentation to assess the determinants of CRH effects on LC sensory responses. The type of stimulus may be an important determinant of CRH effects. Thus, CRH effects on LC responses to noxious stimuli may be different than effects on LC activity elicited by non-noxious stimulation when tested in the same animals. The informative quality of the stimulus may be important, such that CRH effects on LC responses to conditioned stimuli may differ from those effects on LC activity evoked by nonconditioned stimuli. Additionally, the pattern of stimulus presentation may also be an important variable.

Finally, the demonstrated effects of CRH on LC cells may be pharmacologic rather than physiological. Although these effects occur in the same dose range as the reported autonomic and behavioral effects of CRH it is not clear that these concentrations mimic those that are released at the synapse. Most importantly, it has not been demonstrated that CRH is released from fibers within the LC by stressful stimuli. While the dramatic increases in CRH levels in the LC measured after acute and chronic stress may reflect enhanced processing due to increased turnover, this has not been demonstrated[31]. Thus, it is important to demonstrate that stressful stimuli cause similar changes in LC activity as does CRH and that this is prevented by a CRH antagonist.

Abnormal function of the hypothalamic-pituitary axis has been demonstrated in several psychiatric conditions including depression and anorexia nervosa[62-64]. In some patients with these disorders, CRH levels in CSF are increased and most evidence indicates that this is a result of enhanced release of CRH in these disorders. However, it is not clear whether enhanced CRH release is part of the etiology or a subsequent result of the disease. In either case the consequences of a chronic CRH excess on CNS neurons must be considered. Exposure of LC neurons to excesses of CRF would be expected to result in continuous activation of these cells and altered responses to sensory stimuli. LC cells in these patients may therefore have consistently higher rates of discharge and attenuated responses to sensory stimuli. These patients may not respond appropriately to external sensory stimuli and this is often a clinical feature of certain depressed patients.

REFERENCES

1. W. Vale, C. Rivier, M.R. Brown, J. Spiess, G. Koob, L. Swanson, L. Bilezikjian, F. Bloom, and J. Rivier, Chemical and biological characterization of corticotropin releasing factor, Rec. Prog. Horm. Res. 39:245-270 (1983).
2. W. Vale, J. Spiess, C. Rivier, and J. Rivier, Characterization of a 41-residue ovine hypothalamic peptide that stimulates secretion of corticotropin and β-endorphin. Science 213:1394-1397 (1981).
3. C. Rivier, J. Rivier, and W. Vale, Inhibition of adrenocorticotropic hormone secretion in the rat by immunoneutralization of corticotropin-releasing factor, Science 218:377-378 (1982).
4. S. Cummings, R. Elde, J. Ells, and A. Lendall, Corticotropin-releasing factor immunoreactivity is widely distributed within the central nervous system of the rat: An immunohistochemical study. J. Neurosci. 3:1355-1368 (1983).
5. A.J. Fischman, and R.L. Moldow, Extrahypothalamic distribution of CRF-like immunoreactivity in the rat brain. Peptides (Fayetteville) 3:149-153 (1982).
6. I. Merchenthaler, M.A. Hynes, S. Vigh, A.V. Schally, and P. Petrusz, Immunocytochemical localization of corticotropin-releasing factor (CRF) in rat brain. Am. J. Anat. 165:383-396 (1982).
7. J.A. Olschowka, T.L. O'Donahue, G.P. Mueller, and D.M. Jacobowitz, The distribution of corticotropin-releasing factor-like immunoreactive neuron in rat brain. Peptides 3:995-1015 (1982).
8. L.W. Swanson, P.E. Sawchenko, J. Rivier, and W. Vale, Organization of ovine corticotropin-releasing factor immunoreactive cells and fibers in the rat brain: An immunohistochemical study. Neuroendocrinology 36:165-186 (1983).
9. R.J. Valentino, C.I. Cha, and S.L. Foote, Anatomic and physiologic evidence for innervation of noradrenergic locus coeruleus by neuronal corticotropin-releasing factor. Soc. Neurosci. Abstr. 12:1003 (1986).
10. E.B. De Souza, M.H. Perrin, T.R. Insel, J. Rivier, W.W. Vale, and M.J. Kuhar, Corticotropin-releasing factor receptors in rat forebrain: autoradiographic identification, Science 224:1449-1451 (1984).
11. E.B. De Souza, T.R. Insel, M.H. Perrin, J. Rivier, W.W. Vale, and M.J. Kuhar, Corticotropin-releasing factor receptors are widely distributed within the rat central nervous system: An autoradiographic study, J. Neurosci. 5:3189-3203 (1985).
12. P.C. Wynn, R.L. Hauger, M.C. Holmes, M.A. Millan, K.J. Catt, and G. Auilera, Brain and pituitary receptors for corticotropin-releasing factor: Localization and differential regulation after adrenalectomy,

Peptides 5: 1077-1084 (1984).

13. L.B. Eberly, C.A. Dudley, and R.L. Moss, Iontophoretic mapping of corticotropin-releasing factor (CRF) sensitive neurons in the rat forebrain, *Peptides* 4:837-841 (1983).
14. J.B. Aldenhoff, D.L. Gruol, J. Rivier, W. Vale, and G.R. Siggins, Corticotropin releasing factor decreases postburst hyperpolarizations and excites hippocampal neurons, *Science* 221:875-877 (1983).
15. R.J. Valentino, S.L. Foote, and G. Aston-Jones, Corticotropin-releasing factor activates noradrenergic neurons of the locus coeruleus, *Brain Res*. 270: 363-367 (1983).
16. R.J. Valentino, and S.L. Foote, Corticotropin-releasing factor disrupts sensory responses of brain noradrenergic neurons, *Neuroendocrinol*. *45*:28-36 (1987).
17. C.L. Ehlers, S.J. Henricksen, M. Wang, J. Rivier, W.W. Vale, and F.E. Bloom, Corticotropin-releasing factor produces increases in brain excitability and convulsive seizures in rats, *Brain Res.* 278:332-336 (1983).
18. M.R. Brown, L.A. Fisher, J. Rivier, J. Spiess, C. Rivier, and W. Vale, Corticotropin-releasing factor: Effects on the sympathetic nervous system and oxygen consumption, *Life Sci.* 30:207-210 (1982).
19. M.R. Brown, L.A. Fisher, J. Spiess, C. Rivier, J. Rivier, and W. Vale, Corticotropin-releasing factor: actions on the sympathetic nervous system and metabolism, *Endocrinol.* 111: 928-931 (1982).
20. L.A. Fisher, J. Rivier, C. Rivier, J. Spiess, W. Vale, and M.R. Brown, Corticotropin-releasing factor (CRF): Central effects on mean arterial pressure and heart rate in rats. *Endocrinology* 110:2222-2224 (1982).
21. Y. Tache, Y. Goto, M. Gunion, W. Vale, J. Rivier, and M. Brown, Inhibition of gastric acid secretion in rats by intracerebral injection of corticotropin-releasing factor (rCRF), *Science* 222:935-937 (1983).
22. Y. Tache, D. Hamel, and M. Gunion, Inhibition of gastric acid secretion in rats by intracisternal or intrathecal injection of rat corticotropin-releasing factor (rCRF), *Dig. Dis. Sci.* 29:86S (1984).
23. R.E. Sutton, G.F. Koob, M. LeMoal, J. Rivier, and W. Vale, Corticotropin-releasing factor produces behavioral activation in rats, *Nature* 297:331-333 (1982).
24. D.R. Britton, G.F. Koob, J. Rivier, and W. Vale, Intraventricular corticotropin releasing factor enhances behavioral effects of novelty, *Life Sci.* 31:363-367 (1982).
25. D.R. Britton, and K.K. Britton, A sensitive open field measure of anxiolytic drug activity, *Pharmacol. Biochem. Behav.* 15:577-582 (1981).
26. K. Britton, J. Morgan, J. Rivier, W. Vale, and G. Koob, Chlordiazepoxide attenuates CRF-induced response suppression in the conflict test, *Psychopharmacology* 86:170-174 (1985).
27. K.T. Britton, G. Lee, R. Dana, S.C. Risch, and G.F. Koob, Activating and "anxiogenic" effects of corticotropin-releasing factor are not inhibited by blockade of the pituitary-adrenal system with dexamethasone, *Life Sci.* 39:1281-1286 (1986).

28. N.H. Kalin, S.E. Shelton, G.W. Karemer, and W.T. McKenney, Corticotropin-releasing factor administered intraventricularly to rhesus monkeys, Peptides (Fayetteville) 4:217-220 (1983).
29. N.H. Kalin, Behavioral effects of ovine corticotropin-releasing factor administered to rhesus monkeys, Fed. Proc., 44:249-254 (1985).
30. G. Aston-Jones, M. Ennis, V.A. Pieribone, W.T. Nickell, and M.T. Shipley, The brain nucleus locus coeruleus: restricted afferent control of a broad efferent network, Science 234:734-737 (1986).
31. P.B. Chappell, M.A. Smith, C.D. Kilts, G. Bissette, J. Ritchie, C. Anderson, and C.B. Nemeroff, Alterations in corticotropin-releasing factor-like immunoreactivity in discrete rat brain regions after acute and chronic stress. J. Neurosci. 6:2908-2914 (1986).
32. R. Grzanna, and M.E. Molliver, The locus coeruleus in the rat: an immunohistochemical delineation, Neuroscience 5:21-40 (1980).
33. L.W. Swanson, and B.K. Hartman, The central adrenergic system: An immunofluorescence study of the location of cell bodies and their efferent connections in the rat utilizing dopamine-β-hydroxylase as a marker. J. Comp. Neurol. 163:467-507 (1976).
34. B.E. Jones, A.E. Halaris, and O.X. Freeman, Innervation of forebrain regions by medullary noradrenaline neurons, a biochemical study in cats with central tegmental tract lesions, Neurosci. Lett. 10:251-258 (1978).
35. R.M. Kobayashi, M. Palkovits, I.J. Kopin, and D.M. Jacobowitz, Biochemical mapping of noradrenergic nerves arising from the rat locus coeruleus, Brain Res. 77:269-279 (1974).
36. G. Aston-Jones, and F.E. Bloom, Activity of norepinephrine containing locus coeruleus neurons in behaving rats anticipates fluctuations in the sleep-waking cycle, J. Neurosci. 1:876-886 (1981).
37. G. Aston-Jones, and F.E. Bloom, Norepinephrine-containing neurons in behaving rats exhibit pronounced responses to non-noxious environmental stimuli, J. Neurosci. 1:887-900 (1981).
38. S.L. Foote, G. Aston-Jones, and F.E. Bloom, Impulse activity of locus coeruleus neurons in awake rats and monkeys is a function of sensory stimulation and arousal. Proc. Nat. Acad. Sci. (U.S.A.) 77:3033-3037 (1980).
39. J.M. Cedarbaum, and G.K. Aghajanian, Activation of locus coeruleus neurons by peripheral stimuli: Modulation by a collateral inhibitory mechanism, Life Sci. 23:1383-1392 (1978).
40. K. Rasmussen, D.A. Morilak, and B.L. Jacobs, Single unit activity of locus coeruleus neurons in the freely moving cat. I. During naturalistic behaviors and in response to simple and complex stimuli, Brain Res. 371:324-334 (1986).
41. S.L. Foote, F.E. Bloom, and G. Aston-Jones, Nucleus locus coeruleus: new evidence of anatomical and physiological specificity, Physiol. Rev. 63:844-914 (1983).
42. B.D. Waterhouse, H.C. Moises, and D.J. Woodward, Nor-

adrenergic modulation of somatosensory cortical neuronal responses to iontophoretically applied putative neurotransmitters, Exp. Neurol. 69:30-49 (1980).
43. B.D. Waterhouse, H.C. Moises, H.H. Yeh, and D.J. Woodward, 1982, Norepinephrine enhancement of inhibitory synaptic mechanisms in cerebellum and cerebral cortex: mediation by β-adrenergic receptors, J. Pharmacol. Exp. Ther. 221:495-506 (1982).
44. M. Segal, and F.E. Bloom, The action of norepinephrine in the rat hippocampus. I. Iontophretic studies, Brain Res. 72:79-97 (1974).
45. M. Segal, and F.E. Bloom, The action of norepinephrine in the rat hippocampus. II. Activation of the input pathway, Brain Res. 72:99-114 (1974).
46. M. Segal, and F.E. Bloom, The action of norepinephrine in the rat hippocampus. III. Hippocampal cellular responses to locus coeruleus stimulation in the awake rat, Brain Res. 107:499-511 (1976).
47. R. Freedman, B.J. Hoffer, D.J. Woodward, and D. Riro, Interaction of norepinephrine with cerebellar activity evoked by mossy and climbing fibers, Exp. Neurol. 55:269-288 (1977).
48. H.C. Moises, B.D. Waterhouse, and B.J. Woodward, Locus coeruleus stimulation potentiates Purkinje cell responses to afferent input: the climbing fiber system, Brain Res. 222:43-64 (1981).
49. D.J. Woodward, H.C. Moises, B.D. Waterhouse, B.J. Hoffer, and R. Freedman, Modulatory actions of norepinephrine in the central nervous system, Fed. Proc. 38:2109-2116 (1979).
50. A.M. Thierry, F. Javoy, J. Glowinski, and S.S. Kety, Effects of stress on the metabolism of norepinephrine, dopamine and serotonin in the central nervous system of the rat: modification of norepinephrine turnover, J. Pharmacol. Exp. Ther. 163:163-171 (1968).
51. J. Korf, G.K. Aghjanian, and R.H. Roth, Increased turnover of norepinephrine in the rat cerebral cortex during stress: Role of the locus coeruleus, Neuropharmacology 12:933-938 (1973).
52. G. Cassens, M. Roffman, A. Kuruc, P.J. Orsulak, and J.J. Schildkraut, Alterations in brain norepinephrine metabolism induced by environmental stimuli previously paired with inescapable shock, Science 209:1138-1139 (1980).
53. G. Cassens, A. Kuruc, M. Roffman, P.J. Orsulak, and J.J. Schildkraut, Alterations in brain norepinephrine metabolism and behavior induced by environmental stimuli previously paired with inescapable shock, Behav. Brain Res. 2:387-407 (1981).
54. R.A. Solomon, B.M. McCormack, R.N. Lovitz, D.M. Swift, and M.T. Hegemann, Elevation of brain norepinephrine concentration after experimental subarachnoid hemorrhage, Neurosurg. 19:363-366 (1986).
55. T.H. Svensson, and P. Thoren, Brain noradrenergic neurons in the locus coeruleus: inhibition by blood volume load through vagal afferents, Brain Res. 172:174-178 (1979).
56. M. Elam, T. Yao, T.H. Svensson, and P. Thoren, Regulation of locus coeruleus neurons and splanchnic, sympathetic nerves by cardiovascular afferents, Brain Res. 290:281-287 (1984).

57. M. Elam, T. Yao, P. Thoren, and T.H. Svensson, Hypercapnia and hypoxia: Chemoreceptor-mediated control of locus coeruleus neurons and splanchnic, sympathetic nerves, Brain Res. 222:373-381 (1981).

58. R.J. Valentino, and S.L. Foote, Corticotropin-releasing factor increases tonic activity of locus coeruleus neurons and disrupts their responses to auditory stimuli: Studies in unanesthetized rats (Submitted).

59. R. Andrade, and G.K. Aghajanian, Locus coeruleus activity in vitro: intrinsic regulation by a calcium-dependent potassium conductance but not α_2-adrenoreceptors. J. Neurosci. 4:161-170 (1984).

60. J.T. Williams, R.A. North, S.A. Shefner, S. Nishi, and T.M. Egan, Membrane properties of rat locus coeruleus neurons, Neuroscience 13:137-156 (1984).

61. R.J. Valentino, and E.F. Aulisi, Carbachol-induced increases in locus coeruleus activity are associated with an altered pattern of response to sensory stimuli. Neurosci. Lett (in press).

62. P.W. Gold, D.L. Loriaux, A. Roy, M.A. Kling, J.R. Calabrese, C.H. Kellner, L.K. Nieman, R.M. Post, D. Pickar, W. Gallucci, P. Avgerinos, S. Paul, E.H. Oldfield, G.B. Cutler, Jr., and G.P. Chrousos, Responses to corticotropin-releasing hormone in the hypercortisolism of depression and Cushings disease, New Eng. J. Med. 314:1329-1334 (1986).

63. P.W. Gold, H. Gwirtsman, P.C. Averginos, L.K. Nieman, W.T. Gallucci, W. Kaye, D. Jimerson, M. Ebert, R. Rittmaster, D.L. Loriaux, and G.P. Chrousos, Abnormal hypothalamic-pituitary-adrenal function in anorexia nervosa: pathophysiological mechanisms in underweight and weight-corrected patients, New Eng. J. Med. 314:1335-1342 (1986).

64. C.B. Nemeroff, E. Widerlov, G. Bissette, H. Wallerus, I. Karlsson, K. Eklund, C.D. Kilts, P.T. Loosen, and W. Vale, Elevated concentrations of CSF corticotropin-releasing factor-like immunoreactivity in depressed patients, Science 226:1342-1344 (1984).

HYPOPHYSIOTROPIC REGULATION OF STRESS-INDUCED ACTH SECRETION

Paul M. Plotsky

The Clayton Foundation Laboratories for Peptide Biology
The Salk Institute
10010 North Torrey Pines Road
La Jolla, C.A. 92037

INTRODUCTION

Actual and perceived deviations from the homeostatic state evoke numerous physiological adjustments, including activation of the hypothalamic-pituitary-adrenal axis (HPA). The primary endproducts of this cascade (adenohypophysial opioid and adrenocortical glucocorticoid secretion) result in mobilization of energy substrates, suppression of nonessential systems such as digestion and reproduction, inhibition of inflamatory responses, and alteration of pain and sensory perception (1,2). An overview of the information flow through the HPA is presented in Fig. 1. Most stimuli which elicit ACTH secretion must first be encoded as neurochemical messages within the central nervous system, undergo processing and must then be transformed into a hypophysiotropic signal recognized by corticotropes within the adenohypophysis. By now, it is appreciated that regulation of ACTH secretion is mediated by multiple factors of hypothalamic, neurohypophysial and peripheral origins. Corticotropin releasing factor (CRF; 3,4), arginine vasopressin (AVP; 5,6) and epinephrine (EPI; 7,8) are secreted from nerve endings within the zona externa of the median eminence into the hypophysial-portal circulation. Other putative regulatory factors of hypothalamic origin include oxytocin (OT; 9,10) and angiotensin II (AII; 11,12). In addition, circulating neurohypophysial AVP and OT, and catecholamines of sympathetic nervous system and adrenomedullary origins may participate in regulation of ACTH secretion under certain circumstances. Superimposed upon this regulatory network are the multiple actions of the glucocorticoids acting at both the pituitary and central nervous system levels to modulate basal and stimulus-induced secretion of ACTH (13).

The multifactorial nature of ACTH regulation and potential interactions among hypothalamic regulatory factors make it difficult to infer features of central regulatory mechanisms from measurements of systemic ACTH and glucocorticoid levels alone. Therefore our efforts have focused on direct determination of the concentration profiles of hypothalamic peptides in the hypophysial-portal circulation. These studies have been performed in urethane-anesthetized male rats and α-chloralose/urethane anesthetized male cats. The infundibular region of rats was exposed via a modified

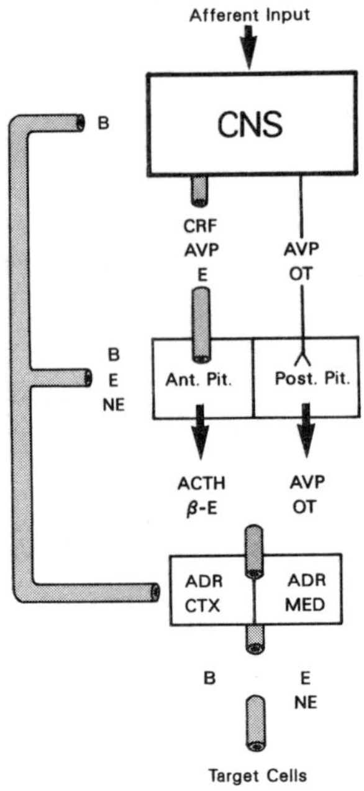

Fig. 1. Schematic of information flow through the hypothalamic-pituitary-adrenal axis. Afferent input is encoded as a neurochemical message, transduced in the CNS and then transformed into a hypophysiotropic signal by neurosecretory cells in the hypothalamus. This signal is recognized by receptive cells within the adenohypophysis and results in secretion of ACTH which then activates adrenocortical synthesis and secretion of corticosteroids. ACTH, adrenocorticotropin; ADR CTX, adrenal cortex; ADR MED, adrenal medulla; Ant Pit, adenohypophysis; AVP, arginine vasopressin; B, corticosterone; β-E, beta endorphin; CNS, central nervous system; CRF, corticotropin releasing factor; E, epinephrine; NE, norepinephrine; OT, oxytocin; Post Pit, neurohypophysis.

transpharyngeal microsurgical approach (4), while in cats a modified transoral, transphenoidal approach was utilized (14). Hypophysial-portal blood was collected from the cephalid stump of the infundibular stalk following its transection at the level of the anterior pituitary gland. While it is recognized that this acute preparation does not represent the basal-state animal, a number of neuroendocrine studies suggest that central regulatory mechanisms remain qualitatively intact in such experimental preparations (15,16).

Initially, the following primary questions were addressed: (i) Which of the putative ACTH secretagogues are present in hypophysial-portal blood?, (ii) Are any of these substances present at physiologically active concentrations?, (iii) Which of these factors are involved in encoding the final signal for ACTH secretion in response to a variety of stimuli?, (iv) What are the actions of central neurotransmitters upon release of individual hypophysiotropic agents?, and (v) Are any of these hypophysiotropic factors involved in mediation of glucocorticoid negative feedback?

HYPOPHYSIOTROPIC FACTORS IN HYPOPHYSIAL-PORTAL BLOOD

Initial assay of hypophysial-portal plasma samples revealed the presence of rather substantial portal-to-peripheral concentration gradients of immunoreactive (ir) CRF, irAVP and irOT (16). A smaller, but significant, gradient of EPI (7,16) has also been observed, while no evidence for a gradient of AII exists (17). Limited chromatographic characterization of immunologically identified portal plasma peptides supports the authenticity of these measurements. Comparison of peptide concentrations measured in portal plasma obtained from anesthetized rats with doses of synthetic peptides eliciting half-maximal ACTH secretion from cultured anterior pituitary cells provides an indirect method of estimating the potential contribution of endogenous molecules to corticotrope secretory activity. Portal plasma levels of irCRF (30-100 pM) appear to be within a concentration range capable of eliciting ACTH

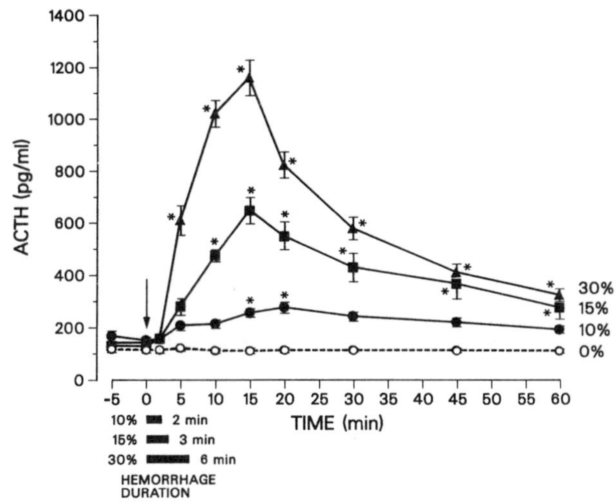

Fig. 2. ACTH secretory response to graded hemorrhage in anesthetized cats. Hemorrhage magnitude was based on total blood volume estimated from body weight. Arterial blood was withdrawn at 5%/min, as indicated by the bars under the abscissa, and reinfused at 15 min. Data expressed as mean±sem of 6 cats per hemorrhage group. *, $p<0.01$ from mean of -5 and 0 time points.

secretion from cultured pituitary cells (18). irAVP and EPI, which are intrinsically weak ACTH secretagogues capable of synergizing with irCRF to produce additivity or potentiation of the ensuing ACTH response (5,8,18), are also present in portal plasma samples (16) at concentrations (irAVP, 460-900 pM; EPI, 1.1-3.8 nM) demonstrated to possess activity *in vitro* . However, these levels are undoubtedly elevated due to the combined effects of anesthesia and surgery. It is postulated that portal irAVP and EPI concentrations are actually substantially lower in intact, unanesthetized rats. Under extreme circumstances it is conceivable that peripheral catecholamine and/or neurohypophysial peptide concentrations might reach levels permitting participation in regulation of ACTH secretion at the pituitary level by potentiation of the actions of irCRF. Another method of assessing the physiological importance of putative ACTH secretagogues to resting and stimulated ACTH secretion involves blockade of their action by administration of receptor antagonists or by passive immunoneutralization. Rivier *et al* (19,20) and Plotsky *et al* (16) have reported strong attenuation of ACTH secretion in response to a variety of stressors following these interventions with respect to irCRF, and have demonstrated lesser attenuation after blockade of catecholamine release or AVP receptors. Other laboratories have confirmed these findings under a variety of circumstances (21,22), and have postulated circadian variation in central secretory drive (23).

HYPOPHYSIOTROPIC CODING OF STIMULUS-INDUCED ACTH SECRETION

Once the presence of potent ACTH secretagogues in the hypophysial-portal circulation was established, it became incumbent to determine whether or not the concentrations of these factors changed in a predictable fashion during periods of increased or decreased ACTH secretion

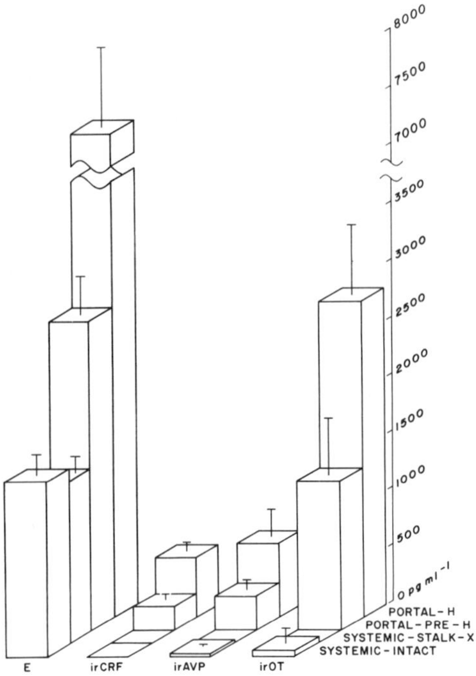

Fig. 3. Changes in concentration profile of hypophysiotropic factors in response to 15% hemorrhage in rats. Peripheral samples were collected from urethanized rats prior to transection of the infundibular stalk (SYSTEMIC-INTACT) and 30 min following transection (SYSTEMIC-STALK-X). The initial hypophysial-portal sample was obtained over a 30 min period prior to hemorrhage (PORTAL-PRE-H; third row), while the second hypophysial-portal sample was collected during the 30 min period of hemorrhage (PORTAL-H; last row). Data expressed as mean±sem in pg/ml (n=8). *, $p<0.05$; **, $p<0.01$. (From ref. 16).

Experimental paradigms were developed with which reliable changes in ACTH secretion could be evoked from anesthetized animals. An obvious first paradigm was controlled hemorrhage (hypovolemia and hypotension). As illustrated in Fig. 2, hemorrhage of varying magnitude of the estimated blood volume via an arterial line at 5%/min resulted in graded increases of circulating ACTH in anesthetized cats (Plotsky, unpublished observations). In both rats and cats, this response could be attenuated by prior administration of an AVP pressor receptor antagonist or by ganglionic blockade, suggesting the participation of other factors in mediation of the ACTH secretory response (16). Furthermore, passive immunoneutralization with antiserum against CRF completely blocked the ACTH secretory response to hemorrhage. Collection and assay of hypophysial-portal blood during this hemorrhage paradigm revealed elevations in portal concentrations of irCRF, irAVP, irOT and EPI during the period of hypovolemia, as illustrated in Fig. 3. Reversible lesion of the paraventricular nuclei by stereotaxic microinjection of procaine blocked secretion of irCRF into the portal circulation and of ACTH into the systemic circulation (4), indicating that this region was the major source of CRF destined for the portal system in agreement with immunohistochemical mapping studies (24,25). Hypotension induced by systemic infusion of nitroprusside also evoked ACTH secretion in the rat, which was associated with 1.5- to 2.0-fold elevations in portal plasma irCRF concentration unaccompanied by changes in the portal plasma levels of other factors (26).

Studies of ACTH and hypothalamic peptide secretion in response to variable hemorrhage magnitude in cats (Plotsky, unpublished observations) were consistent with previous studies in rats and with the studies of Bereiter et al (27) in demonstrating a strong correlation between hemorrhage magnitude and net ACTH secretion. In addition, our experiments provided several new observations: (i) hypophysial-portal plasma concentrations of irCRF, irAVP and EPI, but not of irOT, were correlated with hemorrhage magnitude, (ii) a regression model including only irCRF, shown in Fig. 4, accounted for 84% of the variability indicating the predominant role of irCRF in mediating the ACTH response to hemorrhage in this species, and (iii) the physiologically significant ACTH response to 10% hemorrhage appeared to be mediated by the combined effects of small, statistically non-significant increases in the portal concentrations of each of these hypothalamic factors (or the stimulatory action of an as yet unidentified factor).

Mechanisms mediating the ACTH response to insulin-induced hypoglycemia remain controversial. The response is inversely correlated with plasma glucose concentration, and may be blocked by infusion of glucose simultaneous with insulin administration (28). Passive immunoneutralization by peripheral administration of antiserum raised against CRF prevented insulin hypoglycemia-induced ACTH secretion (28). These observations suggested that CRF acted as a primary dynamic mediator of insulin hypoglycemia-induced ACTH secretion. Participation of additional factors was suggested by the observations that administration of

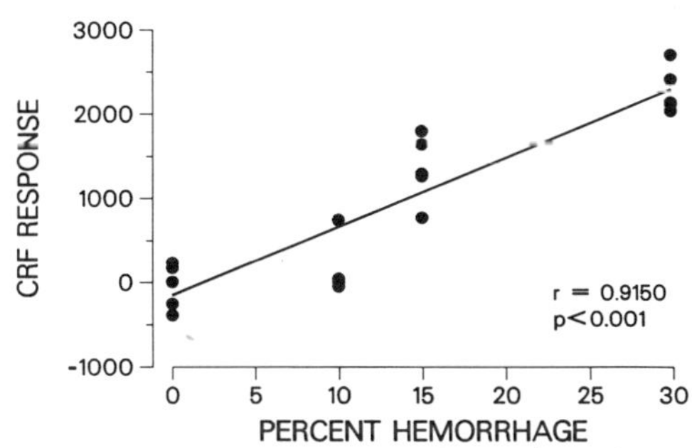

Fig. 4. Relationship between hemorrhage magnitude and irCRF secretion into the hypophysial-portal plasma of anesthetized cats.

an AVP V_1 receptor antagonist, the ganglionic blocker chlorisondamine, or aminergic receptor antagonists attenuated the response (28). As presented in Table 1, measurement of hypophysial-portal plasma peptide concentration profiles revealed an increase in irAVP concentration during the hypoglycemic episode which was unaccompanied by alterations in portal irCRF levels. Subsequent studies (Plotsky, unpublished observations) also demonstrated increased secretion of EPI into the portal circulation during hypoglycemia. Increased EPI was observed even in acutely adrenalectomized rats treated with the ganglionic blocker chlorisondamine, suggesting a central origin of the EPI. These observations are at variance with the report of Mezey et al (29) suggesting mediation of the ACTH response to hypoglycemia by peripheral catecholamines, but are supported by the recent studies of Jezova et al (30). These observations led us to postulate that the presence of irCRF was obligatory for the ACTH secretory response, but that it was acting in a permissive rather than a dynamic role. A similar conclusion was reached by Gibbs (31) using an experimental paradigm of hypothermia.

Table 1. Portal Plasma Peptide Profiles in Response to Hypoglycemia

Factor[a]	Collection Time (min) -20 to 0	0 to 20	21 to 40	41 to 60
irCRF	228± 24 (6)	247± 24 (6)	216± 23 (6)	189± 23 (5)
irAVP	769± 83 (6)	811±107 (6)	1601±258*(6)	1454±193*(5)
irOT	1344±149 (6)	1278±171 (6)	1231±153 (6)	1097±154 (5)
EPI[b]	158± 27 (5)	183± 25 (6)	243± 39*(6)	346±*51 (5)

[a] Values as mean±SEM pg/ml (number of animals).
[b] Bilaterally adrenalectomized 12 h prior to portal collection and given basal corticosterone replacement via drinking water (plasma corticosterone, 3.4±0.2 μg/dl).
* $p<0.01$ vs initial collection period (-20 to 0 min).

Table 2. Stimulus-Induced Changes in Portal Plasma Concentration Profiles of Hypophysiotropic Factors

Paradigm	ACTH	irCRF	irAVP	irOT	EPI
Prestimulus Levels[a]	236±32	269±39	622±118	1539±277	2049±246
Nitroprusside Hypoten.[b]	+	+	nc*	nc	?
15% Hemorrhage[c]	+	+	+	+	+
Atrial Pulsation[d]	-	nc	-	nc	nc
Insulin Hypoglycemia[e]	+	nc	+	nc	+

[a] Values presented as pg/ml; mean±sem.
[b] Reduction of MAP by 50% for 15 min by nitroprusside infusion (10μg/min, iv).
[c] Removal of 15% estimated blood volume at 5%/min.
[d] Sinusoidal infusion-withdrawal of 0.2 ml (2 Hz) heparinized saline from right atrium for 30 min.
[e] Injection of insulin (300 IU/kgBW, iv) following overnight fast.
* nc = no change
? = not determined

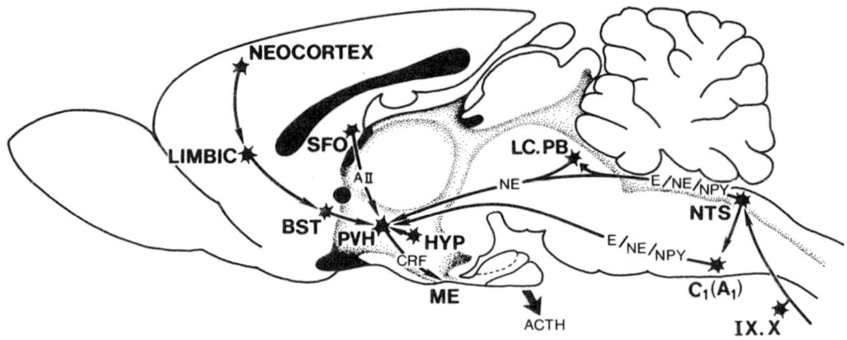

Fig. 5. Major afferent projections to the parvocellular CRF-secreting cells of the paraventricular nucleus. A_1, ventrolateral medullary noradrenergic cell group; ACTH, adrenocorticotropin; AII, angiotensin II; BST, bed nucleus of the striae terminalis; C_1, ventrolateral medullary adrenergic cell group; CRF, corticotropin releasing factor; E, epinephrine; HYP, hypothalamus; IX, nineth cranial nerve; LC, locus coeruleus; ME, median eminence; NE, norepinephrine; NPY, neuropeptide Y; NTS, nucleus tractus solitarius; PB, parabrachial region; PVN, paraventricular nucleus; SFO, subfornical organ; X, tenth cranial nerve.

Finally, inhibition of ACTH secretion by atrial pulsation (simulated volume loading) was associated with reduction of portal plasma irAVP concentration without concomitant changes in portal irCRF. The direction of hypophysial-portal plasma peptide concentration changes in response to the various experimental paradigms examined are summarized in Table 2.

CENTRAL NEUROTRANSMITTER REGULATION OF HYPOTHALAMIC PEPTIDE RELEASE

The specificity of neural regulation of individual hypothalamic peptide release into the hypophysial-portal circulation remains to be adequately characterized. Four major classes of afferent inputs to the CRF-rich parvocellular paraventricular nucleus (pPVN) have been identified on the basis of immunohistochemical and morphological studies, as schematically illustrated in Fig. 5. Cell groups in the caudal ventrolateral medulla (A_1, C_1), nucleus of the solitary tract (A_2, C_2), and the region encompassing the locus coeruleus and parabrachial area send mono- and polysynaptic projections to the pPVN (32,33). These pathways probably relay visceral sensory information (34) to the pPVN via adrenergic (35), noradrenergic (32,33), and neuropeptide Y-positive (36) fibers. A second projection to pPVN arises from the subfornical organ and contains AII-positive fibers (37). This pathway may mediate the central limb of the peripheral renin-angiotensin system by amplification and relay to the PVN. A third major pathway carries cortical and limbic input to the pPVN primarily by funneling information through a projection originating in the bed nucleus of the striae terminalis (38). Finally, many intrahypothalamic projection pathways to pPVN exist, although the function and transmitter specificity of these afferents are presently unknown (38).

Table 3. Modulation of irCRF Release into the Portal Circulation by Intracerebroventricular Neurotransmitter Administration

Transmitter	Dose (nmole)	irCRF	Receptor Specificity
Acetylcholine	0.1 - 10.0	+	Muscarinic/Nicotinic
Angiotensin II	0.1 - 1.0	+	Block with saralasin
Dynorphin 1-13	0.1 - 10.0	-	κ/μ
β-Endorphin	0.1 - 10.0	-	μ
GABA	0.1 - 1.0	-	Block with bicuculline
Norepinephrine	0.1 - 10.0	+	α1-adrenergic
Norepinephrine	5.0 - 50.0	-	β-adrenergic
Serotonin	0.1 - 10.0	+	?

Many previous reviews (39-41) present the current state of our knowledge with regard to neurotransmitter regulation of hypothalamic peptide secretion and highlight the existing controversies. The results of our initial attempts to screen transmitter actions on irCRF secretion into the portal circulation are presented in Table 3 (41a,41b). Substances were introduced into the third ventricle in a total volume of 2 μl (0.5 μl/min) in phosphate-buffered saline or artificial cerebrospinal fluid. Injection of acetylcholine (ACh), AII, or NE ($\leq$5.0 nmole) enhanced portal plasma irCRF concentration in a dose-dependent fashion, and their actions were attenuated, respectively, by pretreatment with the specific receptor antagonists pempidine (nicotinic) and atropine (muscarinic) for ACh, saralasin for AII, and coryanthine (α1) for NE. Conversely, intraventricular administration of gamma aminobutyric acid (GABA), dynorphin (DYN), β-endorphin (β-E), or NE (>5.0 nmole) attenuated portal plasma irCRF concentration in a dose-related manner. Pretreatment with bicuculline (GABA-antagonist), naltrexone (opioid receptor antagonist), or propranolol (β-adrenergic receptor antagonist) prevented inhibition mediated by subsequent injection of the related neurotransmitter. With the exception of NE, these observations are in reasonable agreement with previously published reports (39-42).

Norepinephrine has commonly been considered as an inhibitory influence on the central portion of the HPA (41,42). However, accumulating evidence suggests that NE may act in a stimulatory capacity at the central level as well (43,44). Electrical stimulation of the ventral noradrenergic ascending bundle (VNAB) has been demonstrated to evoke an increase in portal plasma irCRF levels (45) and peripheral ACTH concentration (46). Pretreatment with an α1-, but not a β-adrenergic, receptor antagonist

Table 4. **Alterations of Hypophysial-Portal Plasma irCRF and irAVP Concentration Following Pharmacological Adrenalectomy**

Treatment Time*(h)	irCRF#	irAVP#
0	230±41	1312±287
12	121±30	2361±605
24	50± 6	458± 76
72	426±79	5016±117

* Hours from initial glucocorticoid synthesis blockade.
# Data expressed as pg/ml, mean±sem; n=6 per time point.

prevented the stimulatory effects of electrical stimulation on this axis. These studies suggest the potential of aminergic input to exert either stimulatory or inhibitory actions on irCRF secretion.

More comprehensive studies of AII (17) have revealed that electrical stimulation of the subfornical organ is associated with elevation of portal plasma irCRF concentration. This response can be blocked by prior treatment with saralasin. These observations, coupled with the apparent lack of a portal-to-peripheral AII concentration gradient, suggest that AII may participate in modulation of ACTH secretion via central facilitation of irCRF secretion with little AII facilitatory activity at the pituitary level.

In addition to their role as hypophysiotropic factors, CRF and AVP exhibit many attributes of putative neurotransmitters (47,48). Immunopositive perikarya and fibers of the CRF and AVP systems are found in unique distributions throughout the nervous system, along with high affinity, specific receptors (49,50). Evidence from a variety of experiments suggested: (i) that CRF might modulate its own secretion via an ultrashort loop feedback action (51), and (ii) that AVP facilitated CRF secretion (52). These hypotheses were directly tested by measurement of irCRF and irAVP before and after intraventricular administration of either ovine CRF (which has full biological activity in the rat but does not cross-react with the rat CRF antiserum used for radioimmunoassay) or AVP. As reported by Plotsky _et al_ (53), ovine CRF (0, 1, 10, 100 pmole) was without effect on hypothalamic irCRF secretion but significantly attenuated secretion of irAVP into the hypophysial-portal circulation in a dose-related fashion. However, passive immunoneutralization of endogenous irCRF was without effect on portal plasma irAVP concentration. Intracerebroventricular injection of synthetic AVP (0, 1, 10, 100 pmole) strongly attenuated portal plasma irCRF concentration in a dose-dependent manner (54), with complete blockade of irCRF secretion at an AVP dose of 100 pmole. Conversely, intraventricular administration of an AVP V_1 receptor antagonist or AVP antiserum was associated with elevation of portal plasma irCRF concentration. These experiments suggest the possibility of mutual inhibitory interactions between these factors at the hypothalamic level which may act to damp secretion in response to stimulatory input to conserve future response capacity or to limit the duration of any given ACTH secretory episode.

Most stressors evoke the secretion of ACTH and adrenal corticoids. In addition to their diversity of actions throughout the organism (1,2), the corticoids generally act to damp the responsivity of the HPA to subsequent stressors via actions at the corticotrope and central nervous system levels (23). Removal of endogenous corticoids by adrenalectomy is associated with alterations in hypothalamic immunostaining and content of both irCRF and irAVP (24,55). Pharmacological adrenalectomy induced by administration of the metryapone (11-β-hydroxylase inhibitor) and aminoglutethimide (20-α-hydroxylase inhibitor) evoked hypothalamic irCRF and irAVP staining and content changes resembling those observed following surgical removal of the adrenal glands (56). As shown in Table 4, secretion of irCRF and irAVP into the hypophysial-portal circulation exhibited a decline followed by hypersecretion over the first 72 h after blockade of glucocorticoid synthesis, with the ratio of irAVP to irCRF in portal plasma increasing from 6:1 in untreated rats to 20:1 in 12 h blocked rats. The number of CRF- and AVP-immunopositive perikarya increased approximately 3.0- and 2.6-fold by 72 h, respectively, while the percentage of CRF-positive pPVN cells costaining for AVP rose from undetectable (intact) to 64% (72 h post-blockade). These effects were prevented by corticosterone replacement, implying specificity of the effect to removal of corticosterone.

In earlier studies, we demonstrated the ability of dexamethasone administration to prevent hemorrhage-induced secretion of ACTH and irCRF in anesthetized rats (4). Of particular note was the apparent lack of effect of dexamethasone to suppress resting ACTH or irCRF secretion. One interpretation of these observations is that: (i) the primary site of corticoid feedback was at the central level (since resting ACTH was unaffected), and (ii) dexamethasone acted at a site or sites proximal to the irCRF-secreting neurons or altered their processing of afferent input (since prehemorrhage

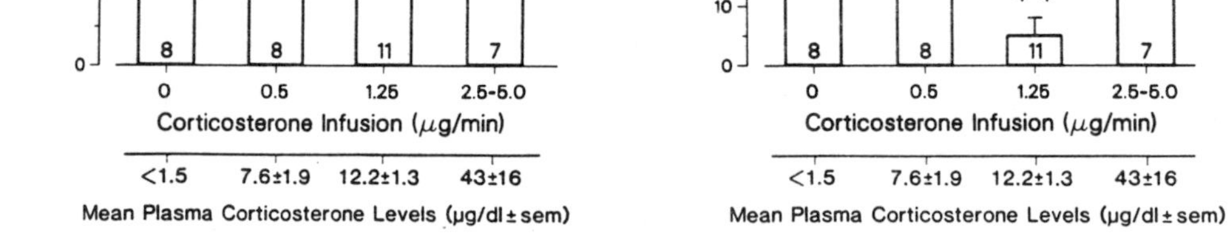

Fig. 6. Effects of delayed corticosterone feedback on irCRF secretion. Endogenous steroid synthesis was inhibited by treatment with metyrapone and aminoglutethimide. Corticosterone was infused systemically for 180 min prior to portal collection. Nitroprusside-induced hypotension (50% reduction of mean arterial pressure for 12 min) was used as the stimulus to irCRF secretion. **A.** Prestimulus irCRF levels were unaffected by all but the highest corticosterone feedback regimen. **B.** Stimulus-induced irCRF secretion was attenuated when plasma corticosterone levels $\geq$ 12.2 μg/dl. Data expressed as mean±sem; n shown by number embedded in each bar.

irCRF levels were unaffected). Utilizing pharmacological adrenalectomy, Plotsky, Otto and Sapolsky (26) recently demonstrated the ability of a delayed corticosterone feedback signal to prevent ACTH secretion in response to nitroprusside-induced hypotension. As show in Fig. 6 A and B, the primary site of action appeared to be central, as hypotension-induced irCRF release into the portal circulation was blocked at circulating corticosterone levels >8 μg/dl. Corticosterone levels >25 μg/dl not only suppressed stimulated irCRF release, but attenuated pre-hypotension irCRF concentration. These results are in accord with the report of Akana et al (57) in which corticosteroid levels of up to 7.4 μg/dl optimally restored normal body and thymic weight and normalized resting and stress-induced ACTH levels in adrenalectomized rats. Alternatively, corticosterone levels exceeding 8 μg/dl inhibited ACTH secretion and produced catabolic effects. The central sites of corticosterone feedback action remain to be elucidated. Autoradiographic and immunohistochemical maps of glucocorticoid receptor distribution certainly suggest a number of potential loci including pPVN CRF-positive perikarya as well as hippocampal, amygdaloid and brainstem regions (58,59).

CONCLUSIONS AND FUTURE DIRECTIONS

Despite the constraints inherent in the hypophysial-portal blood collection paradigm, a wealth of otherwise unavailable information has been derived from such investigations as conveyed in this brief treatise. The characteristics of hypophysiotropic coding of stimulus-induced ACTH secretion inferred from the foregoing observations may be summarized as follows: (i) coding is multifactorial involving at minimum the participation of irCRF, irAVP and EPI, (ii) it is stimulus-specific utilizing subsets of these factors to modulate the ACTH secretory response in accordance with the input stimulus qualities, (iii) release of individual hypothalamic factors into the portal circulation occurs in a graded fashion in response to stimulus intensity, and (iv) the presence of irCRF is obligatory for ACTH secretion; however, irCRF may function in either a permissive role to set pituitary gain and thus permit expression of the releasing activity of the intrinsically weaker factors or it may function in a dynamic role to actively facilitate ACTH secretion. The origin of hypophysial-portal irAVP, irOT and EPI remains to be definitively established. Present models utilizing lesion or knife-cut methodology are subject to misinterpretation due to the potential for extreme plasticity of innervation (60) during the period from lesion placement to final assessment of effect. The apparent multiplicity of ACTH secretagogues may permit finer modulation of net ACTH secretion in response to a given stimulus. In addition, it is possible that individual hypophysiotropic factors affect different aspects of the ACTH synthetic and release machinery. Preliminary studies (Boyle, Vale and Plotsky, unpublished observations), utilizing perifused cultured pituitary cell columns and secretagogue concentrations approximating those measured in portal plasma, have suggested that rate of ACTH export, maximal ACTH secretory amplitude, and relaxation time of an ACTH secretory event may be separately modulated by specific factors. Step increases (50 pM; 5 min duration) of CRF concentration from a basal perifusion concentration of 50 pM were found to augment the maximal ACTH concentration achieved without significant alteration of the rate of rise or relaxation time of the response. Alternatively, a step to 1 nM EPI enhanced the rate of rise and maximal amplitude in ACTH concentration observed in response to CRF alone. Finally, a step to 0.5 nM AVP increased the maximal amplitude and significantly prolonged the relaxation period of the ACTH response without altering the rate of rise. The intracellular mechanisms mediating these differential actions are presently under investigation in many laboratories.

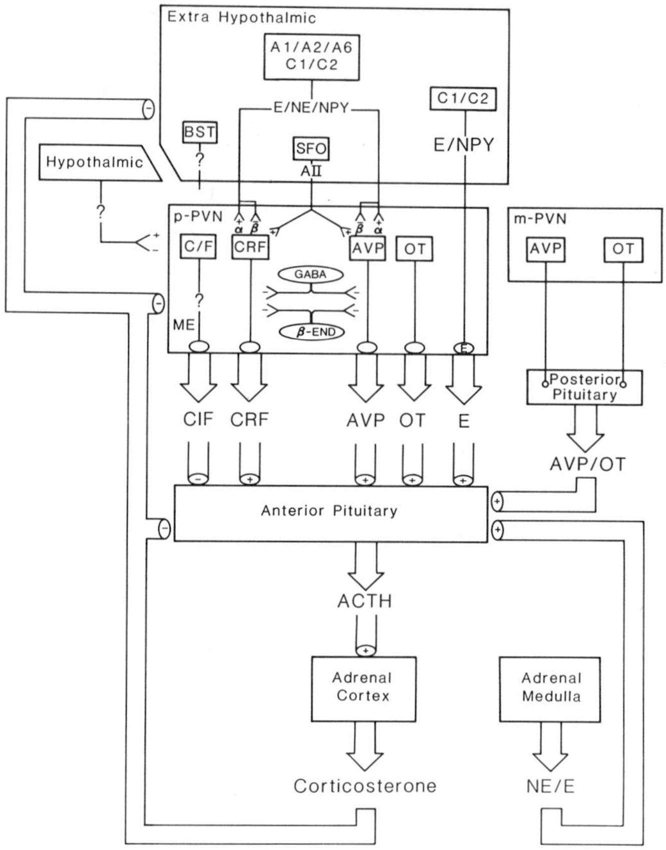

Fig. 7. Schematic diagram of the hypothalamic-pituitary-adrenal axis based on currently available literature (as reviewed in the text). A_1, ventrolateral medullary noradrenergic cell group; A_2, noradrenergic cell group of the nucleus tractus solitarius; A_6, noradrenergic group of the locus coeruleus; BST, bed nucleus of the striae terminalis; C_1, ventrolateral medullary adrenergic cell group; C_2, adrenergic cell group of the nucleus tractus solitarius; CIF, proposed corticotropin inhibiting factor; ME, median eminence; m-PVN, magnocellular region of the paraventricular nucleus; p-PVN, parvocellular division of the paraventricular nucleus.

Our present knowledge of the neural regulation of hypophysiotropic factor secretion remains but a sketchy outline. The available data supports a stimulatory role for cholinergic, serotinergic, and catecholaminergic innervation, while GABAergic and opioid systems appear to provide inhibitory tone. As our general knowledge of the range of neurotransmitter postsynaptic actions increases, it becomes clear that individual transmitters may exert multiple postsynaptic effects, some of which are dependent upon the activity of other afferents (61) or are dependent upon the steroid milieu (59). Many aspects of HPA axis function are modulated by feedback activity of corticosteroids. Participation of central pathways in mediation of these feedback processes is apparent;

however, the specific loci of steroid feedback action within the brain remains an open question. Glucocorticoid receptors have now been identified at all levels of this axis including the spinal cord, brainstem sensory input and projection regions, the limbic system and, finally, on parvocellular CRF-immunopositive perikarya. Whether mediation of the effects of steroid feedback requires activation of a single locus or cooperative activation of multiple loci remains to be determined.

In conclusion, I offer Fig. 7 in keeping with the tendency of anatomists and bioengineers to produce all-encompassing circuit diagrams. While incomplete in reflection of our present state of knowledge, perhaps it will provide a starting point for the arduous task which lies ahead. Progress in mapping the functional characteristics and mechanisms utilized by elements of the HPA axis will require a multidisciplinary approach encompassing electrophysiological, immunohistochemical, molecular biological and physiological techniques.

ACKNOWLEDGEMENTS

Research support has been provided by NIH grants AM33093 and AM26741, and The Clayton Foundation for Research-California Division. P.M.P. is a Clayton Foundation Investigator. This work would have been impossible without the expert technical assistance of Stefan Otto, Steve Sutton, and Mary Tam, the scientific collaborations of Drs. Thomas Bruhn, Robert Sapolsky, Jean Rivier, and Wylie Vale, and encouragement and constructive criticisms unstintingly offered through the years by Drs. Don Gann and Mary Dallman.

REFERENCES

1. Munck A, Guyre PM, Holbrook NJ 1984 Physiological functions of glucocorticoids in stress and their relation to pharmacological actions. Endocrine Rev 5:25

2. Yates FE, Marsh DJ, Maran JW 1980 The adrenal cortex. In: Mountcastle VB (ed) Medical Physiology, Vol 2, C.V. Mosby Co., St. Louis, MO, p. 1558

3. Vale W, Spies J, Rivier C, Rivier J 1981 Characterization of a 41-residue ovine hypothalamic peptide that stimulates secretion of corticotropin and beta-endorphin. Science 213:1394

4. Plotsky PM, Vale W 1984 Hemorrhage-induced secretion of corticotropin-releasing factor-like immunoreactivity into the rat hypophysial portal circulation and its inhibition by glucocorticoids. Endocrinology 114:164

5. Gillies GE, Lowry PJ 1982 Corticotropin-releasing hormone and its vasopressin component. In: Ganong WF, Martini L (eds) Frontiers in Neuroendocrinology, Vol 7, Raven Press, New York, p. 45

6. Rivier C, Vale W 1983 Interaction of corticotropin-releasing factor (CRF) and arginine vasopressin (AVP) on ACTH secretion in vivo. Endocrinology 113:939

7. Johnston CA, Gibbs DM, Negro-Villar A 1983 High concentrations of epinephrine derived from a central source and of 5-hydroxyindole-3-acetic acid in hypophysial portal plasma. Endocrinology 113:819

8. Giguere V, Labrie F 1983 Additive effects of epinephrine and corticotropin-releasing factor (CRF) on adrenocorticotropin release in rat anterior pituitary cells. Biochem Biophys Res Commun 110:456

9. Antoni FA, Holmes MC, Jones MT 1983 Oxytocin as well as vasopressin potentiate ovine CRF in vitro. Peptide 4:411

10. Gibbs DM 1984 High concentrations of oxytocin in hypophysial portal plasma. Endocrinology 114:1216

11. Rivier C, Vale W 1983 Effect of angiotensin II on ACTH release in vivo: role of corticotropin-releasing factor (CRF). Reg Pept 7:253

12. Sobel D, Vagnucci A 1982 Angiotensin II mediated ACTH release in rat pituitary cell cultures. Life Sci 30:1281

13. Keller-Wood ME, Dallman MF 1984 Corticoid inhibition of ACTH secretion. Endocrine Rev 5:1

14. Snyckers FD 1975 Transphenoidal selective anterior hypophysectomy in cats for microsurgical training. J Neurosurg 43:774

15. Fink G, Aiyer M, Chiappa S, Henderson S, Jamieson M et al 1983 Gonadotropin-releasing hormone release into hypophyseal portal blood and mechanism of action. In: McKerns KW, Pantic V (eds) Hormonally Active Brain Peptides, Structure and Function. Plenum Press, New York, p. 397

16. Plotsky PM, Bruhn TO, Vale W 1985 Evidence for multifactor regulation of the adrenocorticotropin secretory response to hemodynamic stimuli. Endocrinology 116:633

17. Plotsky PM, Bruhn TO, Ferguson A 1987 Evidence supporting a role for the subfornical organ in regulation of the hypothalamic-pituitary-adrenal axis in the rat. Am J Physiol, submitted

18. Vale W, Vaughan J, Smith M, Yamamoto G, Rivier J, Rivier C 1983 Effects of synthetic ovine corticotropin-releasing factor, glucocorticoids, catecholamines, neurohypophysial peptides, and other substances on cultured corticotropic cells. Endocrinology 113:1121

19. Rivier C, Vale WW 1983 Modulation of stress-induced ACTH release by corticotropin-releasing factor, catecholamines and vasopressin. Nature 305:325

20. Rivier J, Rivier C, Vale W 1984 Synthetic competitive antagonists of corticotropin releasing factor: Effect on ACTH secretion in the rat. Science 224:889

21. Linton EA, Tilders FJH, Hodgkinson S, Berkenbosch F, Vermes I, Lowry PJ 1985 Stress-induced secretion of adrenocorticotropin in rats is inhibited by administration of antisera to ovine corticotropin-releasing factor and vasopressin. Endocrinology 116:966

22. Nakane T, Audhya T, Kanie N, Hollander CS 1985 Evidence for a role of endogenous corticotropin-releasing factor in cold, ether, immobilization, and a traumatic stress. Proc Natl Acad Sci USA 82:1247

23. Dallman MF, Makara GB, Roberts JL, Levin N, Blum M 1985 Corticotrope response to removal of releasing factors and corticosteroids in vivo. Endocrinology 117:2190

24. Merchenthaler I, Vigh S, Petrusz P, Schally AV 1983 The paraventricular-infundibular corticotropin releasing factor (CRF)-pathway as revealed by immunocytochemistry in long-term hypophysectomized or adrenalectomized rats. Reg Pept 5:295

25. Swanson LW, Sawchenko PE, Rivier J, Vale WW 1983 Organization of ovine corticotropin-releasing factor immunoreactive cells and fibers in the rat brain: an immunohistochemical study. Neuroendocrinol 36:165

26. Plotsky PM, Otto S, Sapolsky RM 1986 Inhibition of immunoreactive corticotropin-releasing factor secretion into the hypophysial-portal circulation by delayed glucocorticoid feedback. Endocrinology 119:1126

27. Bereiter DA, Zaid AM, Gann DS 1986 Effect of rate of hemorrhage on release of ACTH in cats. Am J Physiol 250:E76

28. Plotsky PM, Bruhn TO, Vale W 1985 Hypophysiotropic regulation of adrenocorticotropin secretion in response to insulin-induced hypoglycemia. Endocrinology 117:323

29. Mezey E, Reisine TD, Brownstein MJ, Palkovits M, Axelrod J 1984 β-adrenergic mechanism of insulin-induced adrenocorticotropin release from the anterior pituitary. Science 226:1085

30. Jezova D, Kvetnansky R, Kovacs K, Oprsalova Z, Vigas M, Makara GB 1987 Insulin-induced hypoglycemia activates the release of adrenocorticotropin predominantly via central and propanolol insensitive mechanisms. Endocrinology 120:409

31. Gibbs DM 1985 Measurement of hypothalamic corticotropin-releasing factors in hypophyseal portal blood. Fed Proc 44:203

32. Sawchenko PE, Swanson LW 1981 Central noradrenergic pathways for the integration of hypothalamic neuroendocrine and autonomic responses. Science 214:685

33. Sawchenko PE, Swanson LW 1982 The organization of noradrenergic pathways from the brainstem to the paraventricular and supraoptic nuclei in the rat. Brain Res Rev 4:275

34. Sumal KK, Blessing WW, Joh TH, Reis DJ, Pickel VM 1983 Synaptic interactions of vagal afferents and catecholaminergic neurons in the rat nucleus solitarius. Brain Res 277:31

35. Hokfelt T, Fuxe K, Goldstein M, Johansson O 1974 Immunohistochemical evidence for the existence of adrenaline neurons in the rat brain. Brain Res 66:235

36. Sawchenko PE, Swanson R, Grzanna R, Howe PRC, Bloom SR, Polak JM 1985 Colocalization of neuropeptide Y immunoreactivity in brainstem catecholaminergic neurons that project to the paraventricular and supraoptic nuclei in the rat. J Comp Neurol 241:138

37. Lind RW, Swanson LW, Ganten D 1984 Angiotensin II immunoreactivity in the neural afferents and efferents of the subfornical organ of the rat. Brain Res 321:207

38. Swanson LW 1986 Organization of the mammalian neuroendocrine system. In: Bloom FE (ed) Handbook of Physiology, American Physiological Society, Washington DC, p. 317

39. Buckingham JC, Hodges JR 1979 Hypothalamic receptors influencing the secretion of corticotropin releasing hormone in the rat. J Physiol 290:421

40. Hillhouse EW, Burden J, Jones MT 1975 The effect of various putative neurotransmitters on the release of corticotropin releasing hormone from the hypothalamus of the rat in vitro. I. The effect of acetylcholine and norepinephrine. Neuroendocrinol 17:1

41. Weiner RI, Ganong WF 1978 Role of brain monoamines and histamine in regulation of anterior pituitary secretion. Physiol Rev 58:905

41a. Plotsky PM, Otto S, Sutton S 1987 Neurotransmitter modulation of corticotropin releasing factor secretion into the hypophysial-portal circulation. Life Sci, submitted

41b. Plotsky PM 1986 Opioid inhibition of immunoreactive corticotropin-releasing factor secretion into the hypophysial-portal circulation of rats. Reg Pept 16:235

42. Tuomisto J, Mannisto P 1985 Neurotransmitter regulation of anterior pituitary hormones. Pharmacol Rev 37:249

43. Fehm HL, Voigt KH, Lang RE, Pfeiffer EF 1980 Effects of neurotransmitters on the release of corticotropin releasing hormone (CRH) by rat hypothalamic tissue in vitro. Exp Brain Res 39:229

44. Szafarczyk A, Alonso G, Ixart G, Malaval F, Assenmacher I 1985 Diurnal-stimulated and stress-induced ACTH release in rats is mediated by ventral noradrenergic bundle. Am J Physiol 249:E219

45. Plotsky PM 1987 Facilitation of irCRF secretion into the hypophysial-portal circulation following activation of catecholaminergic pathways or central norepinephrine injection. Endocrinology, submitted

46. Szafarczyk A, Malaval F Laurent A, Gibaud R, Assenmacher I 1987 Further evidence for central stimulatory action of catecholamines on adrenocorticotropin release in the rat. Endocrinology, in press

47. Siggins GR, Gruol D, Aldenhoff J, Pittman Q 1985 Electrophysiological actions of corticotropin-releasing factor in the central nervous system. Fed Proc 44:237

48. Riphagen CL, Pittman QJ 1986 Arginine vasopressin as a central neurotransmitter. Fed Proc 45:2318

49. DeSouza EB, Insel TR, Perrin MH, Rivier J, Vale WW 1985 Corticotropin-releasing factor receptors are widely distributed within the rat central nervous system: An autoradiographic study. J Neurosci 5:3189

50. Biegon A, Terlou M, Voorhuis TD, DeKloet ER 1984 Arginine vasopressin binding sites in rat brain: a quantitative autoradiographic study. Neurosci Lett 44:229

51. Ono N, Bedran de Castro JC, McCann SM 1985 Ultrashort-loop positive feedback of corticotropin (ACTH)-releasing factor to enhance ACTH release in stress. Proc Natl Acad Sci USA 82:3528

52. Hedge GA, Yates MB, Marcus R, Yates FE 1966 Site of action of vasopressin in causing corticotropin release. Endocrinology 79:328

53. Plotsky PM, Bruhn TO, Otto S 1985 Central modulation of immunoreactive arginine vasopressin and oxytocin secretion into the hypophysial-portal circulation by corticotropin-releasing factor. Endocrinology 116:1669

54. Plotsky PM, Bruhn TO, Otto S 1984 Central modulation of immunoreactive corticotropin-releasing factor secretion by arginine vasopressin. Endocrinology 115:1639

55. Rivier CL, Plotsky PM 1986 Mediation by corticotropin releasing factor (CRF) of adenohypophysial hormone secretion. Ann Rev Physiol 48:475

56. Plotsky PM, Sawchenko PE 1987 Hypophysial-portal plasma levels, median eminence content, and immunohistochemical staining of corticotropin-releasing factor, arginine vasopressin and oxytocin after pharmacological adrenalectomy. Endocrinology 120: in press

57. Akana SF, Cascio CS, Shinsako J, Dallman MF 1985 Corticosterone: narrow range required for normal body and thymus weight and ACTH. Am J Physiol 249:R527

58. Agnati LF, Fuxe K, Yu ZY, Hafastrand A, Okret S, Wikstrom AC, Goldstein M, Zoli H, Vale W, Gustafsson JA 1985 Morphometrical analysis of the distribution of corticotropin releasing factor, glucocorticoid receptor and phenylethanolamine-N-methyl-transferase immunoreactive structures in the paraventricular hypothalamic nucleus of the rat. Neurosci Lett 54:147

59. McEwen BS, De Kloet ER, Rostene W 1986 Adrenal steroid receptors and actions in the nervous system. Physiological Rev 66:1121

60. Horn AM, Robinson IC, Fink G 1985 Oxytocin and vasopressin in rat hypophysial portal blood: experimental studies in normal and Brattleboro rats. J Endocrinol 104:211

61. Siggins G, Gruol D 1986 Mechanisms of transmitter action in the vertebrate central nervous system. In: Bloom FE (ed) Handbook of Physiology, American Physiological Society, Washington D.C., p. 1

CORTICOTROPIN RELEASING FACTOR RECEPTORS: CHARACTERIZATION AND ACTIONS IN THE ANTERIOR PITUITARY GLAND

Greti Aguilera, Abdul-Badi Abou Samra, James P. Harwood and Kevin J. Catt

Endocrinology and Reproduction Research Branch
NICHD, NIH, Bethesda, MD 20892

INTRODUCTION

Corticotropin releasing factor (CRF), which was isolated from hypothalamic extracts and sequenced in 1981, has been shown to participate in visceral and behavioral responses to stress, as well as the control of ACTH secretion (1,2). The initial event in the action of CRF in the pituitary gland is its binding to specific plasma membrane receptors, which trigger the formation of intracellular messengers responsible for the activation of ACTH release. Such receptors were first identified in rat pituitary membranes by binding studies with radioiodinated CRF (3), and were subsequently analyzed by radioassays in membranes (4) and by autoradiographic (5-8) and cytochemical techniques using biotinylated or fluorescein-conjugated CRF analogues (9,10). The use of autoradiography permitted the identification and characterization of CRF receptors in the central and peripheral nervous systems, and has greatly aided our understanding of the physiological actions of CRF in these tissues. The most common ligand used for CRF receptor studies is the radioiodinated ovine CRF derivative, Tyr-oCRF; similar receptor properties have been described using radioiodinated [$NLeu^{21}$,Tyr^{32}]oCRF (5). Analogues of rat/human CRF have given tracers with reduced biological action and lower binding activity due to peptide damage during the iodination procedure. Since ovine and human CRF bind to the CRF receptor in different species with equal affinities, oCRF can be used for studies in both rat and primates.

In addition to CRF, the normal corticotroph has receptors for several other regulatory ligands, including vasopressin (VP), angiotensin II (AII) and catecholamines. While the latter hormones are less effective than CRF as stimuli of ACTH release, they show synergism with CRF and may be of physiological importance as regulators of ACTH secretion in the intact animal (11,12). The presence of cell-surface receptors for the several putative regulators of ACTH release in the corticotroph provides the structural basis for a multifactorial control system in which the actions of cyclic adenosine monophosphate (cAMP) dependent regulators are integrated with those of calcium/phospholipid-dependent stimuli such as vasopressin, angiotensin II and catecholamines.

This review will discuss the CRF receptor and its regulation, and the mechanisms by which CRF and other ACTH secretogagues determine the functional state of the corticotroph during the physiological control of ACTH secretion.

CHARACTERIZATION OF THE PITUITARY CRF RECEPTOR

Receptors for CRF have been identified and characterized in the pituitary gland of the rat (3,5) and several primates (5,7). In all species, autoradiographic analysis of the binding of radiolabeled CRF has shown the receptors to be localized in the anterior and intermediate lobes, with no staining in the neural lobe. In the rat pituitary, CRF binding is associated with dense clusters homogeneously distributed throughout the anterior lobe (Figure 1). In primates, there are marked differences in the distribution of receptors among the various species. In the cynomolgus monkey, confluent clusters of high binding density are homogeneously distributed throughout the anterior lobe, while in the human pituitary such clusters are found preferentially in the anteromedial portion of the anterior pituitary. In contrast to human and cynomolgus monkey, the pituitary gland of the marmoset, a new world primate, exhibits very high binding density in the intermediate lobe and sparsely scattered clusters in the anterior lobe.

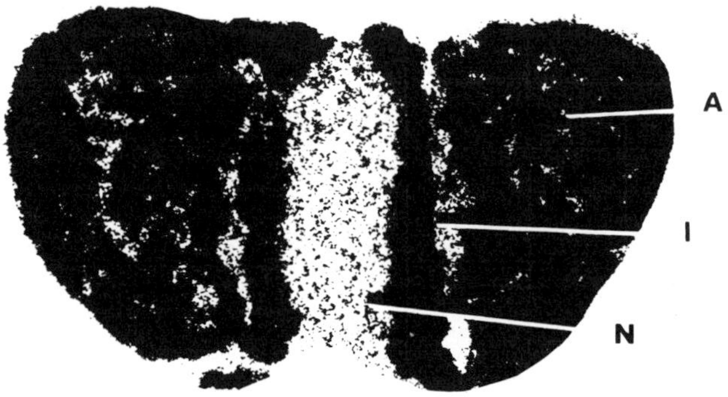

Figure 1. Autoradiograph of ^{125}I-Tyr-oCRF binding to rat pituitary. A, anterior lobe; I, intermediate lobe; N, neural lobe.

In spite of differences in distribution, the binding properties of CRF receptors are similar in pituitary membranes from rat, marmoset and cynomolgus monkey. The specific binding of ^{125}I-Tyr-CRF to anterior pituitary membrane-rich fractions is time- and temperature-dependent, and the binding reaches a steady-state within 30 to 40 min at 22°C. Scatchard analysis of equilibrium binding data show a single class of CRF receptors with Kd of 10^{-9} M (Figure 2). The number of sites, after correction for the binding activity of the tracer, is about 500 fmol/mg protein in the rat and cynomolgus monkey and is considerably lower in the marmoset (less than 50%) consistent with the autoradiographic data (7). ^{125}I-Tyr-oCRF binding is enhanced by the divalent cations, magnesium, calcium and manganese, in the 2 to 5 mM range (13,14), but is unaffected by sodium up to 140 mM. Addition of dithiothreitol (up to 1 mM), aprotinin (100 KIU/ml), and EGTA (2 mM) also increases CRF binding, by inhibiting tracer and/or receptor degradation. Binding is markedly decreased by freezing and thawing of pituitaries, or by keeping tissue or membranes at 4°C for more than 2 hr.

The binding of CRF to pituitary membranes is markedly influenced by guanyl nucleotides. Addition of the non-degradable GTP analog, Gpp(NH)p, inhibits the binding of ^{125}I-Tyr-oCRF by 60% with ID_{50} of 3 μM. This effect of Gpp(NH)p on CRF binding is also manifested during kinetic studies, when addition of the guanyl nucleotide causes rapid dissociation of the bound peptide (4,13,14). Such guanyl nucleotide action is characteristic of receptors that are coupled to adenylate cyclase by a guanyl nucleotide regulatory protein, in this case Ns.

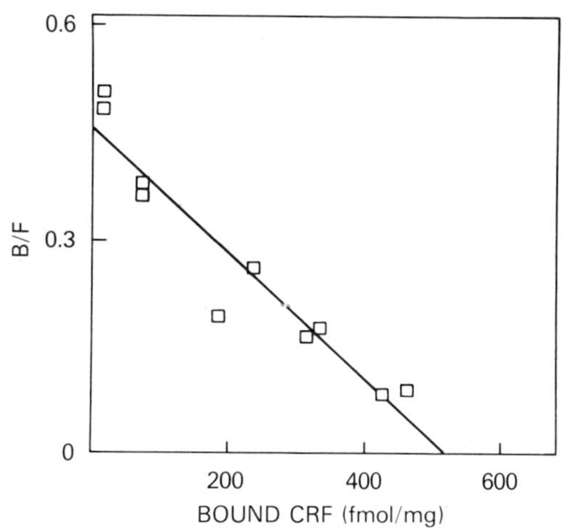

Figure 2. Scatchard analysis of the binding of [^{125}I]Tyr-oCRF to anterior pituitary membranes from cynomolgus monkey. B/F, bound to free ratio.

CORRELATION OF CRF BINDING AND ACTH SECRETION

Validation of a peptide hormone binding site as a receptor which mediates hormone action requires that occupancy results in the characteristic biological response of the cell. In the corticotroph, such a relationship was sought by comparison of the binding activities of a series of CRF peptides with their ACTH releasing potencies (3). In cultured pituitary cells, both CRF and Tyr-CRF stimulate ACTH release 8-fold, with half-maximum effective concentrations (ED_{50}) of 3.2 and 2.8 x 10^{-10} M, respectively. The CRF (15-41) fragment is a weak agonist with full intrinsic activity (ED_{50} 5 x 10^{-7} M), whereas the CRF 21-41 fragment is inactive. Also, in cultured pituitary cells from the cynomolgus monkey, CRF stimulates ACTH release with an ED_{50} in the nanomolar range (7). The maximal ACTH response to CRF occurs with about 50% receptor occupancy, consistent with the presence of spare receptors.

The excess CRF receptors in the corticotroph could be relevant to the effects of adrenalectomy, which is followed by high rates of ACTH secretion in the presence of marked down-regulation of CRF receptors (15).

The concentrations of CRF reported in portal blood (2) are in the range of the pituitary CRF receptor affinity, supporting the view that the binding sites detected by radioligand assay represent the functional receptors through which CRF regulates ACTH secretion.

REGULATION OF PITUITARY CRF RECEPTORS

The interaction of peptide hormones with the target cell is frequently accompanied by regulatory changes in their receptors (16). In many cases, changes in receptor number parallel changes in responsiveness of the cell to the hormone, and receptor regulation modulates hormonal responses. The extent to which CRF receptors undergo such changes was studied by analysis of their levels in pituitaries of adrenalectomized rats. Following adrenalectomy, when ACTH secretion and responsiveness to CRF are increased (15), pituitary CRF binding is reduced by 48, 80, 78 and 79% after 1, 2, 3 or 4 days, respectively (Figure 3). This marked fall in binding was shown by Scatchard analysis to be due to a decrease in receptor number with no change in binding affinity.

The reduction of CRF receptors after adrenalectomy is accompanied by a 60% decrease in CRF-stimulated adenylate cyclase activity at 24 hr with no further change at later times. The decrease in CRF receptors, and the fall in adenylate cyclase responsiveness following adrenalectomy, are both prevented by dexamethasone treatment. Consistent with the inhibition of CRF-activated adenylate cyclase, CRF-stimulated cyclic AMP production is also reduced in cultured pituitary cells from adrenalectomized rats. However, in contrast to the decrease in CRF receptors and cyclic AMP production, CRF-stimulated ACTH release is increased, with no change in sensitivity to CRF. Such enhanced ACTH release from corticotrophs with reduced CRF receptors and impaired activation of adenylate cyclase indicates that elevated ACTH secretion can be maintained by occupancy of relatively few receptors, and with generation of only small quantities of cyclic AMP. These findings also suggest that synergistic interactions between CRF and other ACTH secretagogues contribute to the sustained increase in ACTH secretion that follows adrenalectomy.

Since increased exposure to peptide hormones is commonly associated with down-regulation of homologous receptors and target-cell desensitization (16), elevated CRF levels in the portal blood after adrenalectomy (17-19) are a likely cause for the loss of CRF receptors. Receptor occupancy may also contribute to the decrease in receptors, since infusion of high doses of CRF (500 μg/min) for 30 min prior to preparation of pituitary membranes reduces CRF binding sites by about 30% (15). However, even in this situation, the decrease in receptors may reflect rapid internalization of hormone-receptor complexes rather than simple occupancy (9,20). Studies directed to define the role of CRF on the regulation of its pituitary receptor have shown that elevated secretion of the peptide alone cannot account for the effect of adrenalectomy, and that other factors are likely to be involved in this process. Thus, prolonged CRF infusion in normal rats produce decreases in pituitary CRF receptors of only 30 to 40%, compared with the 70 to 80% loss observed after adrenalectomy (21) (Figure 4).

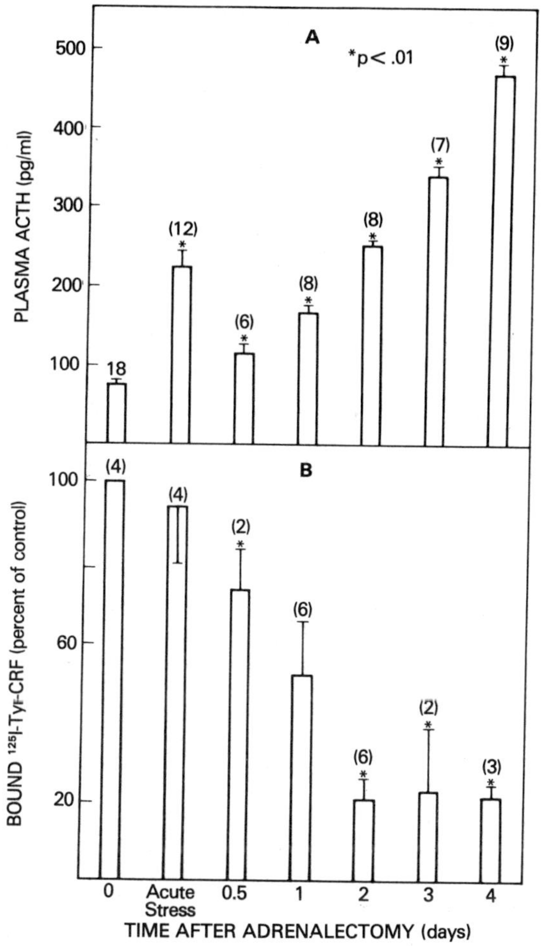

Figure 3. Effect of adrenalectomy on plasma ACTH levels (A) and pituitary CRF receptors (B) in the rat. CRF receptor concentration was calculated by Scatchard analysis from the number of cxperiments indicated in parenthesis.

CRF and vasopressin (VP) are known to interact in the regulation of ACTH secretion, and it is possible that VP influences the regulation of CRF receptors in the anterior pituitary. VP potentiates the stimulatory effect of CRF on the corticotroph (2,11) and facilitates the desensitization of cultured pituitary cells to CRF after repeated stimulation with CRF (22). There is indirect and direct evidence that the secretion of VP into the portal circulation is increased after adrenalectomy. The VP content of the hypothalamus (23), and the ratio of VP/CRF released from the median eminence _in vitro_ (17), are greatly increased after adrenalectomy, and direct measurement of VP in portal blood has shown increased peptide levels (24). Recent studies have provided further evidence for a role of VP in down-regulation of pituitary CRF receptors (25). In di/di Brattleboro rats, which lack endogenous brain VP, adrenalectomy causes only a slight decrease in CRF receptors, in contrast with the 80% loss observed in Sprague Dawley or Long Evans rats. However, the major loss of CRF receptors is reproduced in the di/di rat by infusion of VP during the postadrenalectomy period (Figure 5). Furthermore, in Sprague Dawley rats the simultaneous infusion of CRF and VP mimicked the effect of adrenalectomy on pituitary CRF receptors (25), in contrast to the minor receptor loss obtained with CRF infusion alone.

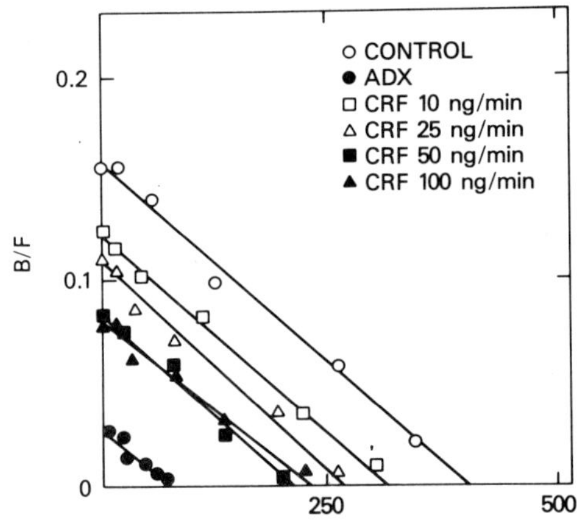

Figure 4. Scatchard analysis of the binding of ^{125}I-Tyr-CRF to anterior pituitary membranes from control, 48 hr adrenalectomized rats, and rats receiving the indicated dose of CRF by subcutaneous infusion with osmotic minipumps for 48 hr.

The possibility that the absence of glucocorticoids could contribute to the CRF receptor loss following adrenalectomy has been virtually excluded by two lines of evidence. First, hypothalamic deafferentation prior to adrenalectomy, which suppresses the release of CRF and VP from the paraventricular nucleus in the absence of glucocorticoids, prevents the decrease in CRF receptors (Figure 6). Second, low levels of glucocorticoids have been shown to decrease rather than increase pituitary CRF receptors (9,10,26). Corticosterone administration (0.5 to 150 mg/day) for 1 to 4 days to adult male rats causes a dose-dependent decrease in the number of CRF receptors in the anterior pituitary, in parallel with the reduction in ACTH secretion (Table 1). Low doses of corticosterone, which produce only transient increases in plasma steroid levels, cause significant reductions in both plasma ACTH and pituitary CRF receptors (26). Such effects of low doses of the naturally occurring glucocorticoid suggest that receptor down-regulation in the anterior pituitary may be of physiological importance and could participate in the inhibitory action of glucocorticoids on ACTH release. However, glucocorticoids inhibit not only CRF-stimulated ACTH release, but also the response to other hormones including VP, norepinephrine and angiotensin II, as well as to post-receptor stimulants such as 8-Br-cyclic AMP and phorbol esters (27). Therefore, any decrease in cell-surface CRF receptors can only partially account for the inhibitory effects of glucocorticoids on ACTH release.

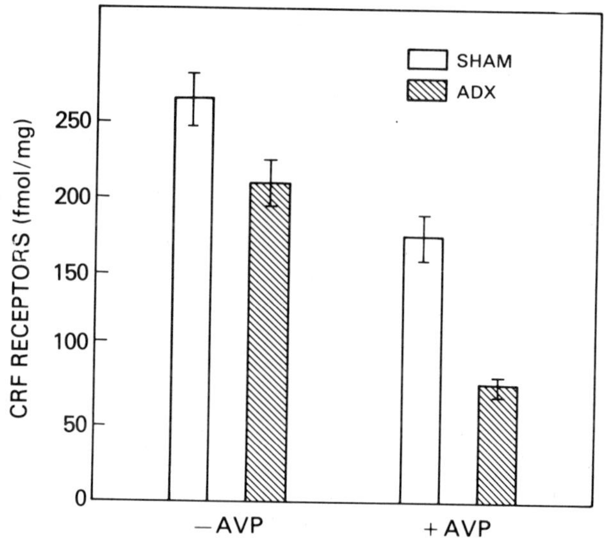

Figure 5. Effect of vasopressin infusion (50 ng/min) on anterior pituitary CRF receptor concentration in sham operated and 48-hr adrenalectomized di/di Brattleboro rats. Osmotic minipumps containing VP were implanted at the time of adrenalectomy. Bars represent the mean and SE of the values obtained by Scatchard analysis of the binding data in 4 experiments.

The decrease in CRF receptors may reflect a general inhibitory effect of glucocorticoids on the corticotroph. Glucocorticoids cause a rapid decrease in POMC gene transcription (28), and it is possible that the synthesis of other cellular proteins, including the CRF receptor, could also be reduced. The direct down-regulatory effect of glucocorticoids on pituitary CRF receptors differs from the effects of glucocorticoid replacement in adrenalectomized rats. In the latter situation, glucocorticoids are likely to block ligand-induced CRF receptor down-regulation by inhibiting the increased release of CRF and VP into the portal circulation (17,29,30).

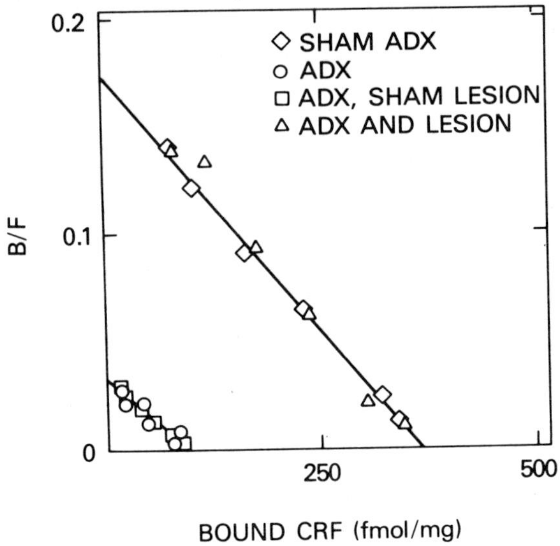

Figure 6. Hypothalamic deafferentation prevents pituitary CRF receptor downregulation 48-hr after adrenalectomy. Rats were adrenalectomized (ADX) or sham ADX 7 days after hypothalamic deafferentation.

Table 1. Effect of corticosterone administration on anterior pituitary CRF receptors in the rat

Corticosterone dose (mg/day)	No. of experiments	CRF receptors	
		Kd (nM)	Concentration (fmol/mg)
0	5	1.1 ± 0.4	395 ± 12.6
0.5	3	1.2 ± 0.8	334 ± 16.7[a]
1	2	1.7 ± 1.5	198 ± 10.6[b,c]
10	3	1.2 ± 0.6	133 ± 6.2[b]
150	5	1.3 ± 0.8	143 ± 18.5[b]

Corticosterone was administered as a single daily injection at 0900 h (100 µl) in corn oil for 4 days. [a]P<0.02, [b]P<0.01, [c]±SD.

MECHANISM OF ACTION OF CRF

Several studies in rat and monkey pituitary cells have shown that stimulation of ACTH release by CRF is accompanied by parallel increases in cyclic AMP production (2,12,31). As in many cAMP-dependent endocrine systems, the ED_{50} for CRF stimulation of cyclic AMP production is higher than that for ACTH release. However, in all studies a minor increase in cyclic AMP is observed with concentrations of CRF at the threshold for stimulation of ACTH release (Figure 7). The role of cAMP as the second messenger for CRF action is also indicated by the ability of isobutyl-methylxanthine to potentiate cAMP production and ACTH release, and the lack of additivity between the stimulatory effects of CRF and those of 8-bromo-cAMP or cholera toxin on ACTH release (27). The stimulation of cAMP production by CRF is due to increased synthesis of the nucleotide, since CRF directly stimulates adenylate cyclase in pituitary homogenates and membrane-rich fractions (12). CRF stimulation of adenylate cyclase is enhanced by guanyl nucleotides, suggesting that the CRF receptor is coupled to the enzyme by the guanyl nucleotide binding protein, N_s (13,14). The role of cAMP in the stimulation of ACTH release by CRF is also indicated by the ability of the hypothalamic peptide to stimulate cAMP-dependent protein kinase, at doses similar to those required to activate adenylate cyclase in pituitary homogenates and ACTH release in cultured pituitary cells (12).

There is evidence that cAMP-independent mechanisms such as calcium and phospholipid metabolism are also involved in post-receptor mediation of the action of CRF. Protein carboxymethylation and phospholipid methylation occur during CRF stimulation of ACTH release in AtT-20 cells (32,33). Phospholipid methylation is frequently followed by calcium-dependent activation of a membrane-bound phospholipase A_2 with liberation of arachidonic acid (34).

Calcium is not only important as a messenger system for cAMP-independent hormones, but also has a role in CRF-stimulated ACTH release. In the absence of extracellular calcium, the effect of CRF on ACTH release from normal rat anterior pituitary cells is partially inhibited (35-37). During incubation of cultured pituitary cells in normal calcium with CRF, VP, AII, or NE, the rate of ACTH release increases for about 45 min and then becomes constant for up to 3 h. In the absence of extracellular calcium, the initial rate of stimulated secretion is unaffected, but after 45 min the secretion rate decreases by 40% for CRF, and to a greater extent for the other stimuli (Figure 8). Addition of calcium after 90 min incubation in calcium-free medium restores the CRF-stimulated ACTH release rate to its initial value. The absence of extracellular calcium has no effect on CRF-stimulated cAMP accumulation, indicating that adenylate cyclase activity can be sustained by intracellular calcium levels. On the other hand, intracellular calcium depletion by incubation of the cells with EGTA completely inhibits CRF-stimulated cAMP production, and this effect is reversed by addition of calcium to permeabilized cells, consistent with the recognized calcium/calmodulin dependence of adenylate cyclase (38,39).

Such kinetic studies have defined two phases in the stimulation of ACTH release by CRF and cAMP-independent stimuli in cultured pituitary cells. The early phase is characterized by a rapid increase in ACTH release rate which is independent of extracellular calcium; the late phase is one of constant secretion rate, and shows partial extracellular calcium dependence for the stimulation by CRF, and complete calcium dependence for stimulation by non cAMP-mediated stimuli (37). The

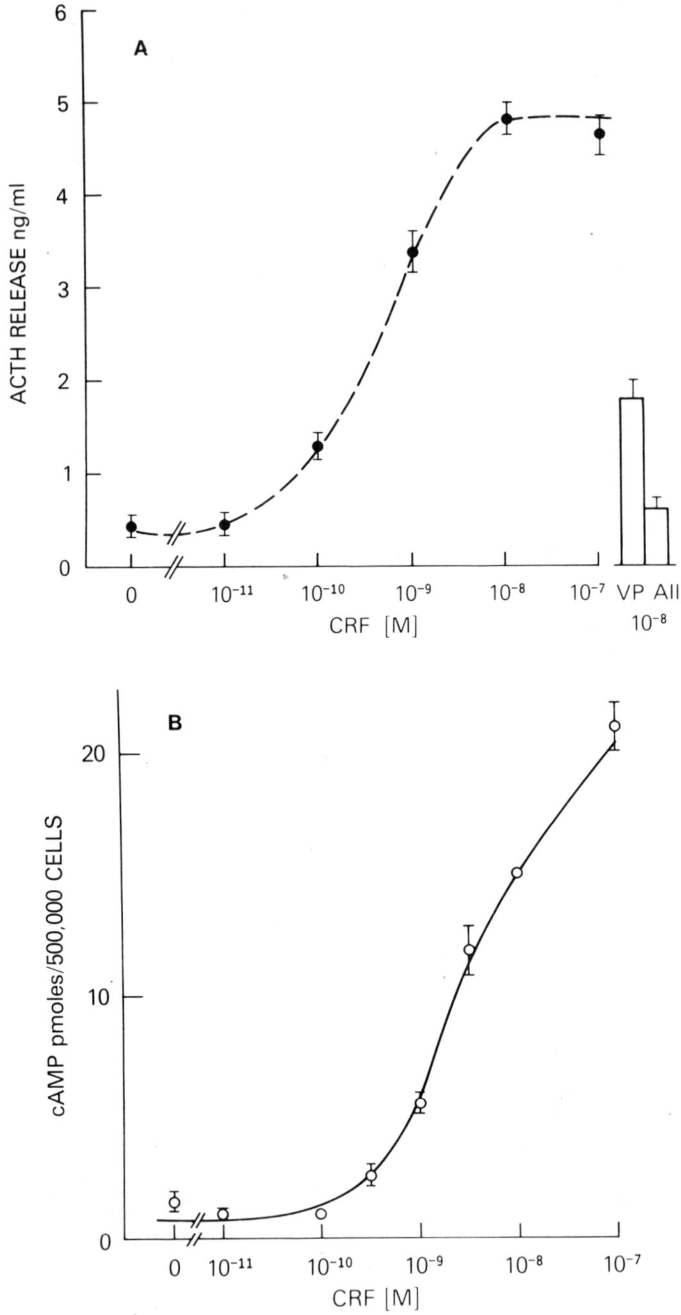

Figure 7. Dose response curves for the stimulation of ACTH (A) and cAMP production (B) by CRF in cultured cynomolgus monkey anterior pituitary cells.

decline in ACTH secretion rate after the first hour of incubation without calcium, despite the preservation of the cAMP response, and the reduced ACTH responses to 8-Br-cAMP in the absence of extracellular calcium, indicate that calcium is important at a post-cAMP level, in particular during the sustained phase of CRF action. It should be noted that the early phase of ACTH secretion stimulated by the recognized calcium-dependent stimuli, VP, NE and AII, is also unaffected by the absence of extracellular calcium. This is consistent with the notion that the early phase of hormone secretion depends on rapid rises in cytosolic calcium mediated by inositol trisphosphate, while the sustained phase depends on protein kinase C activation which requires the maintenance of intracellular calcium levels through calcium influx (40). Although the stimulatory effects of VP, AII and NE are more markedly dependent on voltage sensitive calcium channels (37), the

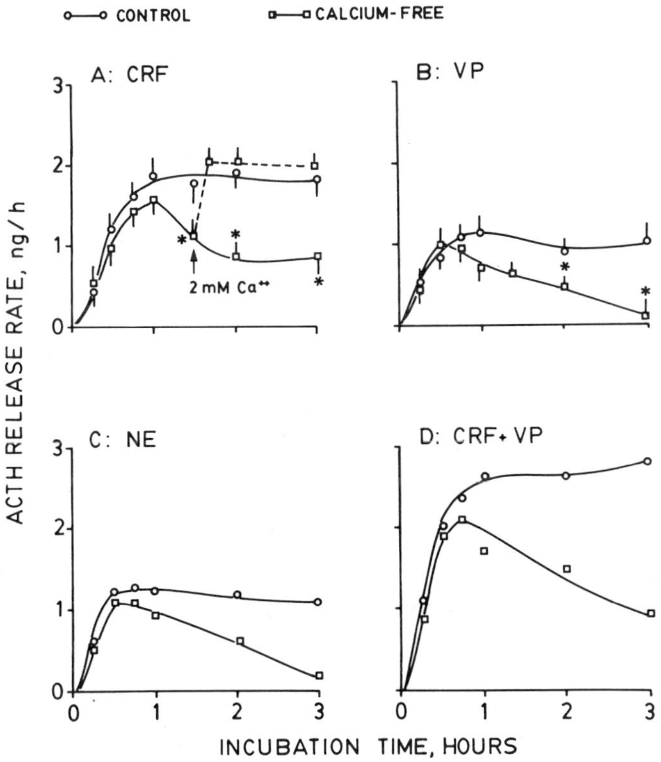

Figure 8. Effect of calcium-free medium on ACTH release rate. Cultures were washed with PBS and incubated with the secretagogue (10 nM) in the absence and presence of calcium. At each time point, 10 µl of medium was removed, diluted 50-fold with PBS and rapidly frozen until measurement of ACTH by RIA. ACTH release rate was derived by calculating the amount of ACTH released between two successive time points. The data are the means ± SEM of three experiments (A and B) or means of two experiments (C and D). In panel A, 2 mM $CaCl_2$ was added 90 min after the initial incubation in calcium-free medium (dotted lines). *$p<0.05$ compared to control.

effect of CRF is also affected by calcium influx. Thus, the voltage-dependent calcium channel agonist, BK 8644, increases basal ACTH release and the plateau stimulated by CRF, while the calcium channel antagonist, nitrendipine, is inhibitory. Also, CRF and cAMP analogs have been shown to increase cytosolic calcium in AtT-20 cells, an effect which is blocked by calcium channel antagonists (41).

Arachidonic acid has also been implicated in the release of ACTH and the actions of CRF and other ACTH secretagogues in normal pituitary cells. Addition of arachidonic acid or melittin, which activates phospholipase A_2 and promotes endogenous arachidonic acid release, stimulates ACTH release in cultured pituitary cells (42). The effects of arachidonic acid and melittin on ACTH release are not additive to those of CRF, VP, PMA or 8-Br-cAMP. On the other hand, CRF, VP, PMA and 8-Br-cAMP increase [^{3}H]arachidonic acid release from prelabeled anterior pituitary cells, suggesting that production of arachidonic acid is part of the mechanism of ACTH release by these secretagogues. This could be primarily due to activation of phospholipase A_2 or may be secondary to phospholipid breakdown followed by the action of diglyceride lipase. The latter mechanism could be relevant for VP and other cAMP-independent stimuli which are likely to be mediated by activation of phospholipase C, with formation of diacylglycerol and inositol trisphosphate. ACTH regulators could promote arachidonic acid release directly by receptor-coupled activation of phospholipase A_2 or diglyceride lipase, or more likely through the actions of a second messenger, as suggested by the ability of 8-bromo-cAMP and PMA to release radioactivity from cells prelabeled with [^{3}H]arachidonic acid.

The lack of parallelism between the ability of the stimuli to promote arachidonic acid release and hormone secretion suggests that the physiological regulators may also influence the activity of the enzymes involved in the metabolism of arachidonic acid. In this regard, lipoxygenase inhibitors such as NDGA (nordihydroguaiaretic acid) and ETYA (5,8,11,14-eicosatetraynoic acid) inhibit CRF- arachidonic acid- and melittin-stimulated ACTH release, whereas cyclooxygenase inhibition by indomethacin has the opposite effect (42). The effects of inhibitors of the different metabolic pathways of arachidonic acid suggest a dual modulatory effect of arachidonic acid metabolism on ACTH release, with inhibition by cyclooxygenase products and stimulation by lipoxygenase products (Figure 9). Thus, inhibition of the cyclooxygenase pathway by indomethacin enhanced not only the ACTH-releasing effect of arachidonic acid and melittin on ACTH, but also the effect of CRF, suggesting that a cyclooxygenase product exerts a local inhibitory action on ACTH release. This is consistent with the observation that prostaglandin E_2 inhibits VP-stimulated ACTH release from quartered rat pituitaries and CRF-induced ACTH release from rat pituitary cell cultures (Abou Samra, A.-B. and Aguilera, G., unpublished observations). On the other hand, the lipoxygenase inhibitors, NDGA and ETYA, decrease the stimulatory effect of arachidonic acid and melittin on ACTH release, suggesting that a lipoxygenase metabolite mediates the stimulatory effect of arachidonic acid on ACTH release. In pituitary quarters and isolated cells, NDGA and ETYA also decrease hormone release stimulated by CRF and VP. This suggests that formation of lipoxygenase products of arachidonic acid is necessary for full expression of the stimulatory activity of the physiological regulators of ACTH secretion. If so, the potentiation of ACTH release by indomethacin, in addition to the blockade of formation of an inhibitory cyclooxygenase metabolite, could reflect the increased availability of substrate for metabolism via the lipoxygenase pathway. Such an action has been suggested for the potentiating effects of indomethacin on PRL and LH release. The possibility that arachidonic acid metabolism is a component of the mechanism of action of ACTH secretagogues is also supported by the lack of additivity of the effects of arachidonic acid agonists and ACTH regulators on ACTH release.

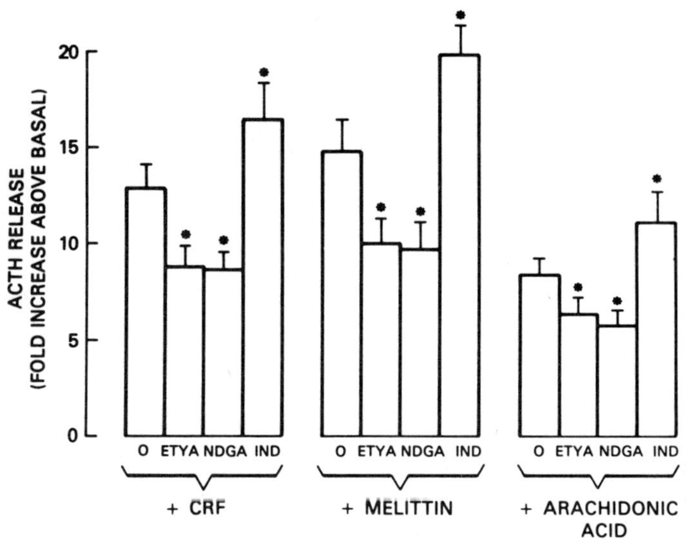

Figure 9. Effects of cyclooxygenase and lipoxygenase inhibitors on CRF, arachidonic acid, and mellitin-induced ACTH release. The cells were incubated with 10 μM arachidonic acid for 3 h. The data are means ± SEM of three experiments. *p<0.05.

INTERACTIONS BETWEEN CRF AND OTHER ACTH REGULATORS

The interaction of CRF with other ACTH secretagogues plays an important role in the physiological control of ACTH secretion. VP, NE and AII alone are minor stimuli of ACTH secretion, but they markedly potentiate the stimulatory action of CRF, an effect that is probably mediated by protein kinase C (27). VP and phorbol esters have also been shown to enhance the effect of CRF on cAMP production, which probably contributes to the potentiation of ACTH release by these factors (43,44) (Figure 10). The potentiating action of VP on CRF stimulation of cAMP accumulation has been shown to involve stimulation of adenylate cyclase, and a decrease in cAMP degradation by inhibition of phosphodiesterase. There is evidence that both of these effects may be mediated by activation of protein kinase C. First, the effects of VP are mimicked by phorbol esters and the synthetic diacyglycerol, dioctanoylglycerol, and second, inhibition of protein kinase C activation by retinal attenuates the potentiating effects of VP and phorbol esters (44).

The possibility that the potentiating effects of VP (and PMA) on CRF-induce cAMP accumulation could be due to inhibition of cAMP degradation was first suggested by the ability of phosphodiesterase inhibitors to increase cAMP responses in the presence of these agents (44). Direct measurement of phosphodiesterase activity has revealed significant inhibition of the low affinity, but not the high affinity, enzyme during potentiation of cAMP production by VP and PMA (Table 2). The lack of effect on the high affinity enzyme may explain the ability of VP and PMA to enhance cAMP production only in the presence of increased intracellular cAMP levels during stimulation by CRF. This is in contrast to the effect of MIX, which inhibits both low and high affinity phosphodiesterases and increases basal cAMP production (45). Inhibition of phosphodiesterase partially accounts for the effects of VP and phorbol esters, but their ability to potentiate CRF action in the presence of phosphodiesterase inhibitors suggests that stimulation of adenylate cyclase, in addition to inhibition of phosphodiesterase, is responsible for their effects on cAMP production.

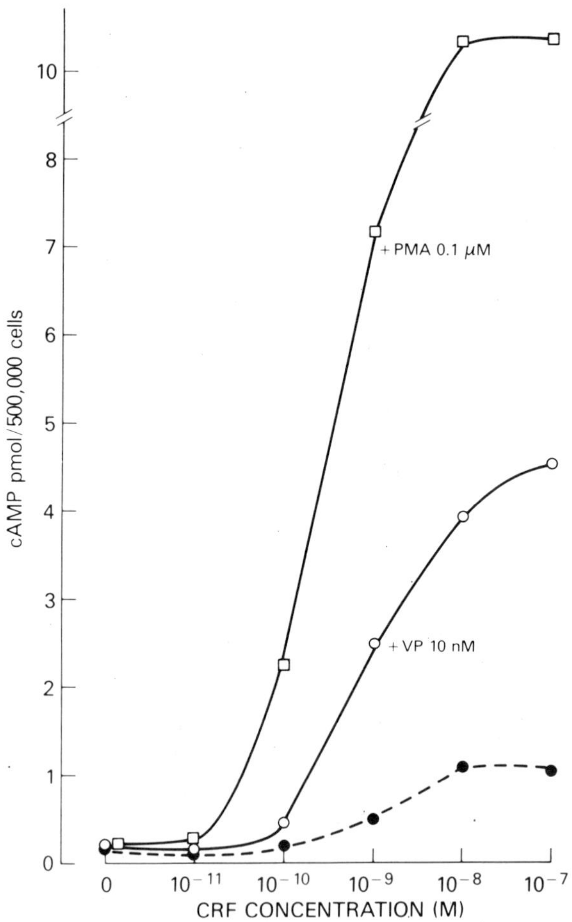

Figure 10. Effects of VP and phorbol 12-myristate 13-acetate (PMA) on the dose dependent stimulation of cAMP production by CRF. 48 hr cultured rat pituitary cells were incubated for 15 min with increasing concentrations of CRF in the presence and in the absence of 10 μM VP or 100 μM PMA. Acid extracted cAMP was measured by radioimmunoassay.

Table 2. Effect of CRF, VP, and PMA, on cAMP phosphodiesterase activity. The pituitary cells were incubated with 10 nM CRF, 10 nM VP, 100 nM PMA, or combinations thereof for 15 min and then homogenized. Phosphodiesterase activity was measured at two substrate concentrations, 0.5 and 50 μM cAMP for the high and low affinity enzymes, respectively. The data are means ± S.E. of three experiments. [a]p<0.01 versus control, [b]p<0.05 versus control.

Treatment	Low affinity cAMP phosphodiesterase		High affinity cAMP phosphodiesterase	
	Cytosol	Membranes	Cytosol	Membranes
	pmol/min/mg protein			
Control	136 ± 26	182 ± 33	45 ± 7	56 ± 8
CRF	129 ± 31	168 ± 29	36 ± 8	44 ± 6
VP	96 ± 22[a]	152 ± 35[b]	41 ± 6	53 ± 5
PMA	92 ± 19[a]	148 ± 30[b]	40 ± 5	50 ± 5
VP + CRF	95 ± 28[a]	150 ± 25[b]	38 ± 7	43 ± 5
PMA + CRF	88 ± 18[a]	145 ± 28[b]	36 ± 6	42 ± 6

Furthermore, direct measurements of adenylate cyclase activity in pituitary membranes suggest complex interactions of VP and PMA with the enzyme, since both stimulatory and inhibitory actions are demonstrable according to the experimental conditions. In intact cells, VP and phorbol esters enhance agonist-stimulated adenylate cyclase activity (Figure 11). In contrast, VP, but not PMA, has a paradoxical inhibitory effect in broken cell preparations. This inhibitory effect is evident only in the presence of GTP, suggesting the participation of a guanyl nucleotide regulatory protein (Table 3). The divergent effects of VP on potentiation of CRF-stimulated cAMP accumulation in intact cells, and inhibition of adenylate cyclase activity in broken cell preparations, can be interpreted in terms of the current concept of adenylate cyclase regulation. The activity of the catalytic subunit is regulated by stimulatory (Ns) and inhibitory (Ni) guanyl nucleotide binding proteins. In addition to its role in inhibiting adenylate cyclase, Ni may also couple receptors of certain cAMP-independent hormones to phospholipid turnover (46-48) or ion channels (49,50). Such a role has also been postulated for other pertussis toxin-sensitive regulatory proteins (No) (47,48), and as yet uncharacterized pertussis toxin-insensitive guanyl nucleotide binding proteins (51,52) which, although not directly linked with adenylate cyclase have a common β subunit with Ni and have the potential to inhibit adenylate cyclase. In rat hepatocytes, a pertussis toxin-insensitive guanyl nucleotide binding protein is involved in VP-receptor mediated hydrolysis of PIP_2, production of inositol trisphosphate (IP_3), mobilization of Ca^{++}, activation of phosphorylase, and glycogenolysis (53). In the pituitary actions of VP, coupling of the hormone-activated receptor to Ni, or another guanyl nucleotide binding proteins primarily associated with phospholipid turnover, could subsequently inhibit adenylate cyclase, producing the effect seen in broken cell preparation.

On the other hand, activation of protein kinase C in the intact cell could cause phosphorylation of a component of the adenylate cyclase complex with consequent activation of the catalytic moiety resulting in increased cAMP formation. In this regard, purified protein kinase C has been reported to phosphorylate the purified α-subunit of the Ni protein (54). Such phosphorylation might block the activity of the inhibitory subunit of the adenylate cyclase complex, and augment the action of a stimulatory ligand such as CRF. Also, PMA has been shown to enhance adenylate cyclase stimulation by factors that directly interact with the stimulatory Ns protein such as cholera toxin, or the catalytic subunit such as Mn^{++}, suggesting an action distal to the N protein (55). Pertussis toxin also potentiates CRF-stimulated cAMP accumulation, an effect which is not additive to those of VP and PMA. This finding

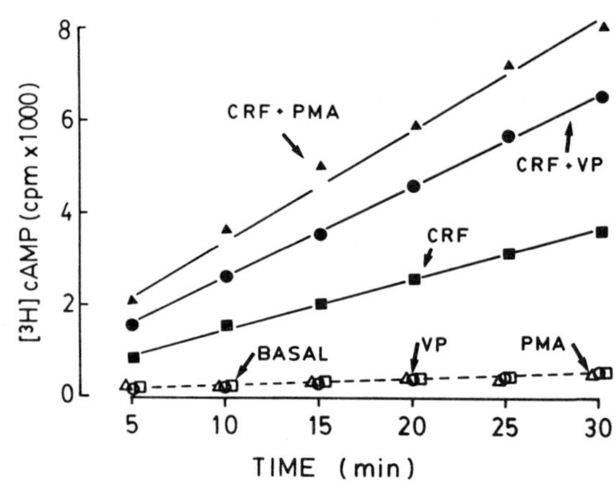

Figure 11. Effect of VP and PMA on CRF-stimulated adenylate cyclase activity in intact cultured anterior pituitary cells. Cells were incubated with [^{3}H]adenine for 18 h, washed with serum-free medium, and incubated with 10 nM CRF, 10 nM VP, 100 nM PMA or their combinations in the presence of 5 mM MIX. [^{3}H]cAMP formed was separated using Dowex and alumina columns after addition of tracer amounts of [^{14}C]cAMP to calculate recovery (44).

Table 3. Effect of VP on CRF stimulation of adenylate cyclase activity measured by conversion of α[^{32}P]ATP to [^{32}P]cAMP in rat pituitary homogenates.

Incubation Conditions	Adenylate cyclase activity cAMP pmol/mg/min	% Basal
Basal	62 ± 5	100
CRF 10^{-7}M	98 ± 8*	158
VP 10^{-8}M	59 ± 4	95
CRF + VP	58 ± 5	93

* $p<0.01$ compared to basal.

indicates that blockade of the inhibitory component of adenylate cyclase by ADP ribosylation can, in fact, potentiate the agonist-activated enzyme, and suggests that Ni is the site regulated by protein kinase C (Figure 12). However, it is not possible to exclude the participation of other guanyl nucleotide binding proteins, or that multiple components of the adenylate cyclase system could be subject to protein kinase C-dependent phosphorylation. Recent studies using purified protein kinase C have demonstrated that the catalytic unit of adenylate cyclase is a substrate for the kinase, but it has yet to be shown that phosphorylation modulates the activity of the enzyme (56).

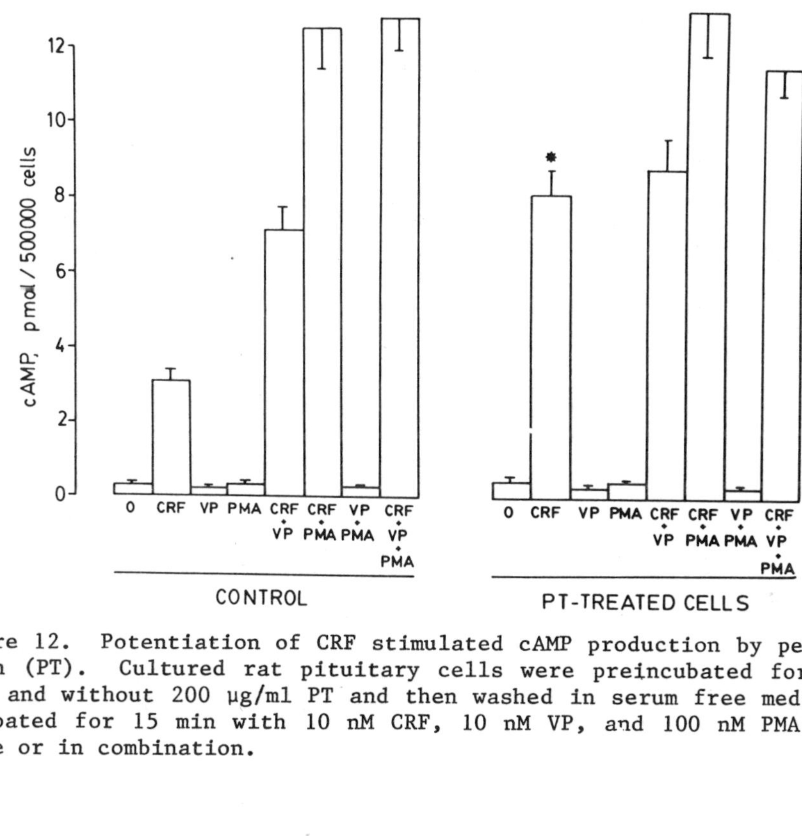

Figure 12. Potentiation of CRF stimulated cAMP production by pertussis toxin (PT). Cultured rat pituitary cells were preincubated for 4 hrs with and without 200 μg/ml PT and then washed in serum free medium and incubated for 15 min with 10 nM CRF, 10 nM VP, and 100 nM PMA either alone or in combination.

The CRF receptor is another potential site of modulation for the effect of VP on CRF-stimulated cAMP accumulation. However, it has been shown that β-adrenergic receptor function is negatively modulated by cAMP-dependent and -independent phosphorylation (57). Thus, it seems unlikely that up-regulation of receptor function would be a significant factor in the VP/CRF system, particularly since VP treatment has been shown to cause a decrease rather than an increase in CRF receptors in intact rats and preincubation of pituitary quarters (25).

In contrast to VP, AII does not potentiate CRF-stimulated cAMP production, indicating that the octapeptide enhances the ACTH-releasing activity of CRF by different mechanisms. The mode of action of AII in other target cells involves changes in calcium fluxes and phospholipid turnover (40,52). Since the effect of AII in the corticotroph is not additive to that of phorbol esters, it may also causes activation of protein kinase C as in other systems. However, the lack of potentiating effect on CRF-stimulated cAMP production indicates that the AII receptor is coupled to regulatory units other than those associated with VP action. In the case of VP, it is important to note that potentiation of CRF-stimulated cAMP production can only partially account for increased ACTH release. The major component of the potentiating action of VP is an increase in the maximum response to CRF, and this cannot be explained by enhancement of CRF-induced cAMP formation by VP, as indicated by the additivity data. Thus, addition of 8-Br-cAMP to CRF-treated cells does not further increase CRF-induced ACTH release, while 8-Br-cAMP added to VP-treated cells increases ACTH release to values similar to those produced by VP/CRF combinations.

CONCLUSIONS

The development of labeled analogs of CRF has made possible the identification of CRF binding sites and their characterization as true receptors which mediate the biological actions of the peptide. Consistent with the recognized actions of CRF, receptors for the peptide have been found in the anterior and intermediate lobes of the pituitary, and in the nervous system at specific sites related to the stress response. In keeping with the characteristics of most peptide hormone receptors, CRF binding sites are saturable, have a high affinity for the ligand and are highly specific for CRF-related peptides. In the anterior pituitary, the affinities of a series of CRF analogs parallel the ACTH secretory activity of the peptides. In pituitary and brain membrane fractions, CRF binding is enhanced by divalent cations and is inhibited by guanyl nucleotides, consistent with coupling of the receptors to adenylate cyclase.

CRF receptors in the anterior pituitary but not in the intermediate lobe, undergo pronounced regulatory changes during alterations in corticotroph activity. Following adrenalectomy and chronic stress, increases in ACTH secretion are accompanied by decreases in pituitary CRF receptors as a result of the coordinated actions of CRF and vasopressin on the corticotroph. During glucocorticoid administration, CRF receptors decrease in parallel with inhibition of ACTH secretion suggesting that CRF receptor down-regulation may contribute to the physiological effects of the steroid.

The interaction of CRF and other regulators with membrane receptors trigger the formation of intracellular messengers that lead to release and/or synthesis of ACTH. Several lines of evidence indicate that cAMP is the main mediator of CRF action upon ACTH release. CRF increases cAMP accumulation in pituitary cells, stimulates adenylate cyclase activity, and activates cAMP-dependent protein kinase at concentrations which are in the range of its ACTH stimulating activity. In addition to cAMP, the mechanisms of action of CRF may involve other messenger systems such as arachidonic acid and its metabolites.

Other stimulators of ACTH release do not stimulate cAMP, but probably exert their actions through calcium-phospholipid dependent mechanisms. Although there is no direct evidence for these mechanisms in the corticotroph, the fact that the actions of VP, AII and NE are mimicked by phorbol esters, with no additivity between these effects, suggests the involvement of protein kinase C in their activation mechanisms. As with CRF, the effects of cAMP-independent regulators may include arachidonic acid release. Arachidonic acid metabolism by the lipoxygenase and cyclooxygenase pathways results in the formation of stimulatory and inhibitory metabolites, respectively. Indirect evidence indicates that the several ACTH stimulators cause an increase in production of lipoxygenase products by pituitary cells.

Although the actions of both CRF and cAMP-independent regulators require calcium, the latter are more dependent on the extracellular calcium concentration. Intracellular calcium is required for the optimal functioning of signal generating systems such as adenylate cyclase and phospholipase A_2, whereas influx calcium, including entry through voltage-sensitive channels, is required for maintenance of the sustained phase of ACTH secretion.

An important aspect of the control of ACTH secretion is the synergistic interaction between CRF and cAMP-independent regulators. Although CRF is the major regulator, it is clear that interactions with other factors are necessary for the sustained increases of ACTH secretion in conditions such as stress. The mechanism of the potentiating effect of cAMP-independent regulators appears to involve protein kinase C, and at least for VP and NE, enhancement of CRF-stimulated cAMP production is partially responsible for such effects. In the intact corticotroph, VP interacts with CRF-stimulated adenylate cyclase indirectly, probably via protein kinase C dependent phosphorylation of a pertussis toxin sensitive regulatory protein.

In addition to the normal circadian fluctuations, ACTH secretion undergoes rapid regulatory changes under physiological conditions to adapt to alterations in the external and internal environments. The multifactorial regulation of ACTH secretion by various hormones, acting through different intracellular signals, provides a system by which the corticotroph can respond appropriately to the complex neuroendocrine changes that occur during the response to stressful stimuli.

REFERENCES

1. Vale, W., C. Rivier, M. R. Brown, J. Spiess, G. Koob, L. Swanson, L. Bilizikjian, F. Bloom, and J. Rivier, 1983, Chemical and biological characteristization of corticotropin releasing factor, Rec. Progr. Horm. Res., 39:245-270.

2. Rivier, C. L., and P. M. Plotsky, 1986, Mediation by corticotropin releasing factor (CRF) of adenohypophysial hormone secretion, Ann. Rev. Physiol., 48:475-494.

3. Wynn, P. C., Aguilera, G., Morell, J. and Catt, K. J., 1983, Properties and regulation of high-affinity pituitary receptors for corticotropin-releasing factor, Biochem. Biophys. Res. Commun., 110-602-609.

4. Holmes, M. C., Antoni, F. A., Szentendrei, T., 1984, Pituitary receptors for corticotropin-releasing factor: no effect of vasopressin on binding or activation of adenylate cyclase, Neuroendocrinology, 39:162-169.

5. De Souza, E. B., and M. J. Kuhar, 1986, Corticotropin-releasing factor receptors in the pituitary gland and central nervous system: Methods and overview, Methods in Enzymology, 124:560-590.

6. Wynn, P. C., R. L. Hauger, M. C. Holmes, M. A. Millan, K. J. Catt, and G. Aguilera, Brain and pituitary receptors for corticotropin

releasing factor: Localization and differential regulation after adrenalectomy, 1984, Peptides, 5:1077-1084.

7. Millan, M. A., Abou Samra, A-B, Wynn, P. C., Catt, K. J., and Aguilera, G., 1987, Receptors and actions of corticotropin releasing hormone in the primate pituitary gland, J. Clin. Endocrinol. Metab., 64:1036-1041.

8. Leroux, P., Pelletier, G., 1984, Radioautographic study of binding and internalization of corticotropin-releasing factor by rat anterior pituitary corticotrophs, Endocrinology, 114:14-21.

9. Childs, G. V., Morell, J. L., Niendorf, A., Aguilera, G., 1986, Cytochemical studies of CRH receptors in anterior lobe corticotropes: Binding, glucocorticoid regulation and endocytosis of [Biotinyl-Ser^{1}]CRH, Endocrinology, 119:2129-2142.

10. Schwartz, J., Billestrup, N., Perrin, M., Rivier, J., Vale, W. W., 1986, Identification of corticotropin-releasing factor (CRF) target cells and effects of dexamethasone on binding in anterior pituitary using a fluorescent analog of CRF, Endocrinology, 119:2376-2382.

11. Gillies, G. E., Linton, E. A., Lowry, P. J., 1982, Corticotropin-releasing activity of the new CRF is potentiated several times by vasopressin, Nature, 299:355-357.

12. Aguilera, G., Harwood, J. P., Wilson, J. X., Morell, J., Brown, J. H., and Catt, K. J., 1983, Mechanisms of action of corticotropin release in rat pituitary cells, J. Biol. Chem. 258:8039-8044.

13. Aguilera, A., Wynn, P. C., Harwood, J. P., Hauger, R. L., Millan, M. A., Grewe, C., and Catt, K. J., 1986, Receptor-mediated actions of corticotropin-releasing factor in pituitary gland and nervous system, Neuroendocrinology, 43:79-88.

14. Perrin, M. H., Haas, Y., Rivier, J., and Vale, W. W., 1986, Corticotropin-releasing factor binding to the anterior pituitary receptor is modulated by divalent cations and guanyl nucleotides, Endocrinology, 118:1171-1179.

15. Wynn, P. C., Harwood, J. P., Catt, K. J., and Aguilera, G., 1985, Regulation of corticotropin-releasing factor (CRF) receptors with the rat pituitary gland: Effect of adrenalectomy on CRF receptors and corticotroph responses, Endocrinology, 116:1653-1659.

16. Catt, K. J., Harwood, J. P., Aguilera, G., and Dufau, M. L., 1979, Hormonal regulation of peptide receptors and target cell responses, Nature, 280-109-116.

17. Holmes, M. C., Antoni, F. A., Catt, K. J., and Aguilera, G., 1985, Predominant release of vasopressin vs. corticotropin-releasing factor from the isolated median eminence after adrenalectomy, Neuroendocrinology, 43:245-251.

18. Plotsky, P. M., Sawchenko, P. E., 1987, Hypophysial-portal plasma levels, median eminence content, and immunohistochemical staining of corticotropin-releasing factor, arginine vasopressin, and oxytocin after pharmacological adrenalectomy, Endocrinology, 120:1361-1369.

19. Oliver, C., Conte-Devolx, B., Rey, M., Boudouresque, F., Giraud, P., Castanas, E., and Porter, J. C., 1983, Immunoreactive 41-CRF in

hypophysial portal blood of intact and adrenalectomized rats. Acta Endocrinol. 103:98 abstract.

20. Leroux, P., Pelletier, G., 1984, Radioautographic study of binding and internalization of corticotropin-releasing factor by rat anterior pituitary corticotrophs. Endocrinology, 114:14-21.

21. Wynn, P.C., Harwood, J.P., Catt, K.J., Aguilera, G., 1988, Corticotropin releasing factor (CRF) induces desensitization of the rat pituitary CRF receptor-adenylate cyclase complex, Endocrinology, Vol. 122, in press.

22. Hoffman, A.R., Ceda, G., Reisine, T.D., 1985, Corticotropin-releasing factor desensitization of adrenocorticotropic hormone release is augmented by arginine vasopressin, J. Neuroscience, 5:234-242.

23. Kiss, J. Z., Mezey, E., Skirboll, L., 1984, Corticotropin-releasing factor-immunoreactive neurons of the paraventricular nucleus become vasopressin positive after adrenalectomy. Proc. Natl. Acad. Sci. USA, 81:1854-1858.

24. Koenig, J. I., Meltzer, H. Y., Devane, G. D., and Gudelsky, G. A., 1986, The concentration of arginine vasopressin in pituitary stalk plasma of the rat after adrenalectomy or morphine, Endocrinology, 118:2534-2539.

25. Holmes, M. C., Catt, K. J., Aguilera, G., 1987, Involvement of vasopressin in the down-regulation of pituitary corticotropin-releasing factor receptors following adrenalectomy, Endocrinology, Vol. 121, in press.

26. Hauger, R. L., Millan, M. A., Catt, K. J., and Aguilera, G., 1987, Differential regulation of brain and pituitary corticotropin-releasing factor receptors by corticosterone, Endocrinology, 120:1527-1533.

27. Abou-Samra, A-B, Catt, K. J., and Aguilera, G., 1986, Involvement of protein kinase C in the regulation of adrenocorticotropin release from rat anterior pituitary cells, Endocrinology, 118:212-217.

28. Eberwine, J. H., Roberts, J. L., 1984, Glucocorticoid regulation of proopiomelanocortin gene transcription in the rat pituitary, J. Biol. Chem., 259:2166-2170.

29. Knepel, W., Nutto, D., Meyer, D. K., Vlaskovska, M. 1984, Vasopressin release from rat medial basal hypothalamus after adrenalectomy or lesions of the paraventricular nuclei. Neurosci. Lett., 48:321-326.

30. Suda, T., Tomori, N., Tozawa, F., Mori, T., Demura, H., and Shizuma, K., 1983, Effects of bilateral adrenalectomy on immunoreactive corticotropin-releasing factor in the rat median eminence and intermediate-posterior pituitary, Endocrinology, 113:1182-1184.

31. Giguere, V., Labrie, F., Cote, J., Coy, D. H., Sueiras-Diaz, J., Schally, A. V., 1982, Stimulation of cyclic AMP accumulation and corticotropin release by synthetic ovine corticotropin-releasing factor in rat anterior pituitary cells: Site of glucocorticoid action, Proc. Natn. Acad. Sci. USA, 79:3466-3469.

32. Hook, V. Y. H., Heisler, S., and Axelrod, J., 1982, Corticotropin-releasing factor stimulates phospholipid methylation and corticotropin

secretion in mouse pituitary tumor cells, Proc. Natl. Acad. Sci. USA, 79:6220-6223.

33. Heisler, S., Hook, V. Y. H., and Axelrod, J., 1983, Corticotropin releasing factor stimulation of protein carboxylmethylation in mouse pituitary tumor cells, Biochem. Pharmacol., 32:1295-1299.

34. Hirata, F., and Axelrod, J., 1980, Phospholipid methylation and biological signal transmission, Science, 209:1082-1084.

35. Giguere, V., Lefevre, G., and Labrie, F., 1982, Site of calcium requirement for stimulation of ACTH release in rat anterior pituitary cells in culture by synthetic ovine corticotropin-releasing factor, 1982, Life Sci., 31:3057-3062.

36. Murakami, K., Hashimoto, K., and Ota, Z., 1985, The effects of nifedipine on CRF-41 and AUP-induced ACTH released _in vitro_, 1985, Acta Endocrinol. (Copenh), 109:32-37.

37. Abou-Samra, A.-B., Catt, K. J., and Aguilera, G., 1987, Calcium-dependent control of corticotropin release in rat anterior pituitary cell cultures, Endocrinology, 121:965-971.

38. Piascik, M. T., Bibich, M., Rush, M. E., 1983, Calmodulin stimulation and calcium regulation of smooth muscle adenylate cyclase activity, J. Biol. Chem., 258:10913-10918.

39. Resink, T. J., Stucki, S., Grigorian, G. Y., Zschauer, A., Buhler, F. R., 1986, Biphasic Ca^{2+} response of adenylate cyclase. The role of calmodulin in its activation by Ca^{2+} ions, Eur. J. Biochem., 154:451-459.

40. Rasmussen, H., and Barrett, P. Q., Calcium messenger system: An integrated view, Physiol. Rev. 64:938-940.

41. Luini, G., Lewis, D., Guild, S., Corda, D., and Axelrod, J., 1985, Hormone secretagogues increase cytosolic calcium by increasing cAMP in corticotropin-secreting cells, Proc. Natl. Acad. Sci. USA, 82:8034-8038.

42. Abou-Samra, A.-B., Catt, K. J., Aguilera, G., 1986, Role of arachidonic acid in the regulation of adrenocorticotropin release from rat anterior pituitary cell cultures, Endocrinology, 119:1427-1431.

43. Giguere, V., and Labrie, F., Vasopressin potentiates cyclic AMP accumulation and ACTH release induced by corticotropin-releasing factor (CRF) in rat anterior pituitary cells in culture, Endocrinology, 111:1752-1754.

44. Abou-Samra, A.-B., Harwood, J., Manganiello, V., Catt, K. J., and Aguilera, G., 1986, Phorbol 12-myristate 13-acetate and vasopressin potentiate the effect of corticotropin-releasing factor on cyclic AMP production in rat anterior pituitary cells, J. Biol. Chem., 262:1129-1136.

45. Weishaar, R. E., Cain, M. H., and Bristol, J. A., 1985, A new generation of phosphodiesterase inhibitors: Multiple molecular forms of phosphodiesterase and the potential for drug selectivity. J. Med. Chem., 28:537-545.

46. Nakamura, T., and Ui, M., 1985, Simultaneous inhibition of inositol phospholipid breakdown, arachidonic acid release, and histamine

secretion in mast cells by islet-activating protein, pertussis toxin, J. Biol. Chem., 260:3584-3593.

47. Higashida, H., Streaty, R. A., Klee, W., Nirenberg, M., 1986, Brandykinin-activated transmembrane signals are coupled via No or Ni to production of inositol 1,4,5-trisphosphate, a second messenger in NG108-15 neuroblastomg-glioma hybrid cells, Proc. Natl. Acad. Sci. USA, 83:942-946.

48. Lad, P. M., Olson, C. V., Grewal, I. S., and Scott, S. J., 1985, A pertussis toxin-sensitive GTP-binding protein in human neutrophil regulates multiple receptors, calcium mobilization, and lectin-induced capping, Proc. Natl. Acad. Sci. USA, 82:8643-8647.

49. Pfaffinger, P. J., Martin, J. M., Hunter, D. D., Mathanson, N. M., and Hillye, B., 1985, GTP-binding proteins couple cardiac muscarinic receptors to a K channel, Nature, 317:536-538.

50. Holziv, G. G., Rane, S. G., Dunlap, K., 1986, GTP-binding proteins mediate transmitter inhibition of voltage-dependent calcium channels, Nature, 319:670-672.

51. Litosch, I., Wallis, C., and Fain, J. N., 5-hydroxytryptamine stimulates inositol phosphate production in a cell-free system from blowfly salivary gland, J. Biol. Chem., 260 (9) 5464-5471.

52. Lynch, C. J., Prpic, V., Blackmore, P. F., Exton, J. H., Effect of islet-activating pertussis toxin on the binding characteristics of Ca^{2+}-mobilizing hormones and on agonist activation of phosphorylase in hepatocytes, Mol. Pharmacol., 29:196-203.

53. Uhing, R., Prpec, V., Juang, H., Exton, J., 1986, Hormone stimulated polyphosphoinositide breakdown in rat liver plasma membranes, J. Biol. Chem. 261:2140-2146.

54. Katada, T., Gilman, A. G., Watanabe, Y., Banor, S., Jakobs, K. H., 1985, Protein kinase C phosphorylates the inhibitory guanine-nucleotide-binding regulatory component and apparently suppresses its function in hormonal inhibition of adenylate cyclase, Eur. J. Biochem., 151:431-436.

55. Bell, J. D., Buxton, I. L. O, Brunton, L. L., 1985, Enhancement of adenylate cyclase activity in S49 lymphoma cells by phorbol esters, J. Biol. Chem., 260:2625-2628.

56. Yoshimasa, T., Bouvier, M., Benovic, J. L., Amlaiky, N., Lefkowitz, R. J., Caron, M. G., 1986, Catalytic unit of adenylate cyclase: Purification and phosphorylation by cyclic AMP-dependent protein kinase and protein kinase C., Clin. Res., 34:689A.

57. Benovic, J. L., Strasser, R. H., Mayor, F. Jr., Caron, M. G., Lefkowitz, R. J., 1986, β-adrenergic receptor kinase: A novel cAMP-independent protein kinase which phosphorylates and desensitizes and agonist-occupied form of the receptor, Clin. Res., 34:642A.

REGULATION OF PITUITARY CYCLIC AMP, PLASMA PROLACTIN AND POMC-DERIVED PEPTIDE RESPONSES TO STRESSFUL CONDITIONS [a]

J.L. Meyerhoff, G.J. Kant, B.N. Bunnell, and E.H. Mougey

Neurochemistry & Neuroendocrinology Branch
Department of Medical Neurosciences
Division of Neuropsychiatry
Walter Reed Army Institute of Research
Washington, DC 20307-5100

INTRODUCTION

We have been investigating the role of pituitary cyclic AMP in regulating the stress-induced release of pituitary hormones <u>in vivo</u>. This paper will review a series of experiments describing the effects of both physical and psychological stressors on these responses and exploring possible mechanisms of regulation. Exposure to stressful conditions elevates plasma levels of corticosterone (CS) (1-3), and increases plasma levels of several hormones secreted from the anterior pituitary, including adrenocorticotrophic hormone (ACTH), beta-endorphin (β-EP), beta-lipotropin (β-LPH), and prolactin (PRL) (3-8). ACTH, β-LPH and β-EP are derived from a common precursor, a 31,000 dalton glycoprotein currently referred to as proopiomelanocortin (POMC) (9,10). β-LPH is a 91 amino acid peptide (11) which contains β-EP as the 61-91 segment (12). In the rat, the three POMC-derived peptides are co-localized in the same secretory granules in anterior pituitary corticotrophs (13,14), while the intermediate lobe contains and releases predominantly β-EP (15-18). Plasma levels of β-EP are high relative to β-LPH (8,19), but exposure to stressors markedly increases plasma levels of all three POMC-derived peptides (5,8,19).

The release of the POMC-derived peptides from the pituitary gland is believed to be chiefly regulated by corticotropin-releasing factor (CRF), a 41 amino-acid peptide (20) secreted from the hypothalamus into the portal circulation. In cultured anterior pituitary cells, CRF stimulates cyclic AMP accumulation, cyclic AMP-dependent protein kinase activity, and release of ACTH into the culture medium (21-25). The incubation of pituitaries with cyclic AMP releases ACTH into the medium (26), and cyclic AMP is believed to function as a second messenger (27) involved in the synthesis of anterior pituitary POMC (28) as well as the release of the peptide hormones derived from POMC (26,29). In the sections which follow, we will present data which

---

[a] This material has been reviewed by the Walter Reed Army Institute of Research, and there is no objection to its presentation or publication. The opinions or assertions contained herein are the private views of the authors and are not to be construed as official or as reflecting the views of the Department of the Army or the Department of Defense.

suggest that cyclic AMP is a mediator *in vivo* of CRF-stimulated neuroendocrine responses to stressful conditions.

PHYSICAL STRESSORS INCREASE PITUITARY CYCLIC AMP AND RELEASE PITUITARY HORMONES *IN VIVO*

We have reported that exposure to physical stressors, such as forced running, immobilization, or footshock produces elevations of pituitary cyclic AMP *in vivo* in rats (8,30-35).[b] These increases were localized to the anterior lobe of the pituitary (36), persisted at least 15 minutes following termination of footshock (37), and were associated with increased plasma levels of ACTH, β-EP and β-LPH.

To determine whether the cyclic AMP response was proportional to the intensity of the stressor, we subjected rats to increasing intensities of footshock for 15 minute periods and measured pituitary cyclic AMP and plasma hormonal responses (31). The data are displayed in Figures 1 and 2. On the basis of these data, 1.6 mA was chosen as an optimal shock intensity to be used in subsequent experiments using footshock.

PSYCHOLOGICAL STRESSORS INCREASE PITUITARY CYCLIC AMP AND RELEASE PITUITARY HORMONES *IN VIVO*

If the pituitary cyclic AMP response to physical stressors represents a critical step in mediating hypothalamic-pituitary-adrenal responses in general, then pituitary cyclic AMP levels should also increase in animals subjected to purely psychological stressors. To examine this question, we

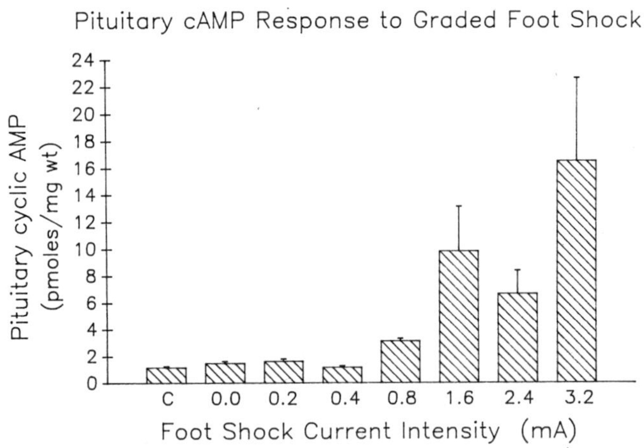

Figure 1. Pituitary cyclic AMP response to increasing intensities of footshock for 15 minutes (31). Values represent mean ± SEM of 6 rats per group. Controls (C) were sacrificed upon removal from home cage.

[b] In conducting the research described in this report, the investigators adhered to the "Guide for the Care and Use of Laboratory Animals" as promulgated by the Committee on Care and Use of Laboratory Animals of the Institute of Laboratory Animal Resources, National Research Council.

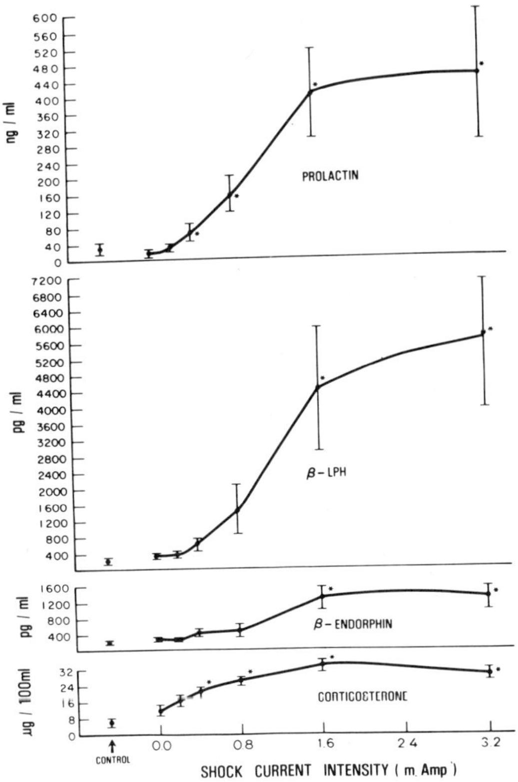

Figure 2. Responses of plasma hormones to increasing intensities of footshock for 15 min (31). Values represent mean ± SEM of 6 rats per group. Controls were sacrificed upon removal from home cage.

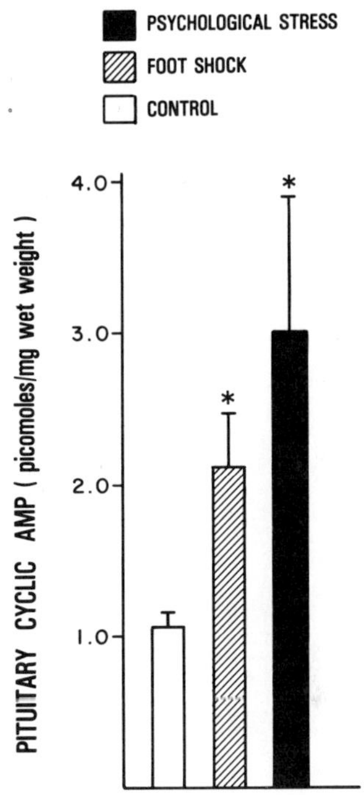

Figure 3. Pituitary cyclic AMP in the psychological stress, footshock and control groups after 10 min in the test chambers (37). Values represent mean ± SEM of 6 rats per group. * = $p < 0.05$, Duncan's multiple range test.

exposed rats to a situation previously associated with an aversive stimulus (footshock). This was accomplished by exposing rats to footshock in a chamber for 20 minutes on four consecutive days (training), and returning them to the chamber for 10 minutes without footshock on the fifth day (test). These rats were designated the "psychological stress" group. For comparison, "footshock" and "control" groups were also studied. They were subjected to either footshock on five consecutive days (20 minutes per day for 4 days, 10 minutes on 5th day), or five days of being placed in the chamber without ever receiving footshock (control group). Rats in all three groups were killed on day five after 10 minutes in the chamber. As shown in Figures 3-5, pituitary cyclic AMP, plasma corticosterone (CS) and PRL were elevated in "psychological stress" as well as "footshock" groups, compared to control animals (37).

Since plasma β-EP is reported to be increased after exposure to such physical stressors as bone fracture (5) or inescapable footshock (38), we examined the effect of a psychological stressor on this peptide as well. We used the same 5-day paradigm as above, except that training (days 1 through 4) and test (day 5) sessions were 15 minutes each. As shown in Figure 6, a psychological stressor, as well as footshock, elicited increases in plasma levels of β-EP after 15 minutes of exposure (39). When rats were left in the chamber for 30 minutes, these elevations persisted in both the psychological stress and footshock groups (data not shown).

CRF ANTISERA ATTENUATES STRESS-INDUCED INCREASES IN PITUITARY CYCLIC AMP AND PLASMA ACTH

As noted, the release of the POMC-derived peptide hormones from the pituitary gland is thought to be regulated primarily by CRF. Administration of exogenous CRF *in vivo* elicited increased plasma levels of ACTH (34,40,41). Moreover, CRF has been specifically linked with the stress-induced activation of the pituitary-adrenal axis *in vivo* by reports that antibodies to CRF attenuate the stress-induced release of ACTH (34,42,43). Since CRF has been shown to stimulate cyclic AMP accumulation in pituitary cells *in vitro*, as well as release of ACTH (21-25), we attempted to block the stress-induced increase in pituitary cyclic AMP *in vivo* via pretreatment with antibodies to CRF. The antisera were administered intravenously 6 minutes prior to exposure to a stress condition lasting 15 minutes. We found that this pretreatment attenuated both the stress-induced increases in

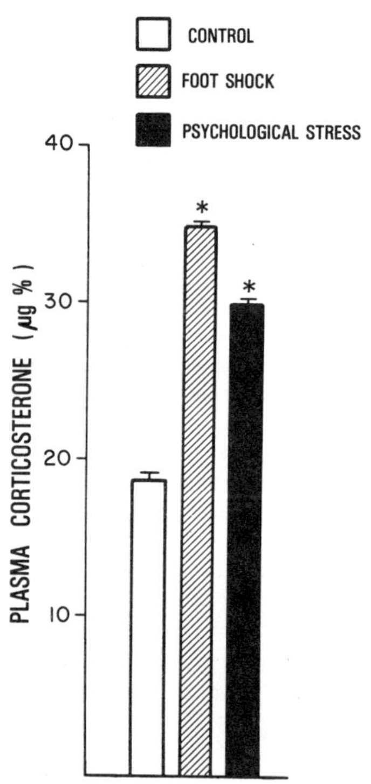

Figure 4. Plasma corticosterone levels in psychological stress, footshock and control groups after 10 min in the test chambers (37). Values represent mean ± SEM of 6 rats per group. * = $p < 0.01$, Duncan's multiple range test.

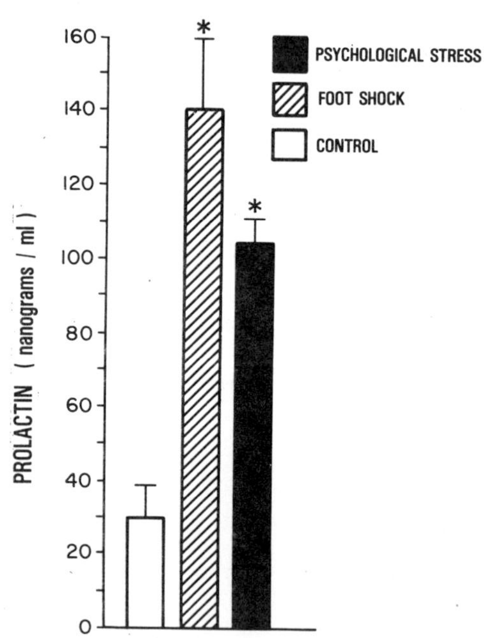

Figure 5. Plasma prolactin levels in the psychological stress, footshock and control groups after 10 min in the test chambers (37). Values represent mean ± SEM of 6 rats per group. * = $p < 0.01$, Duncan's multiple range test.

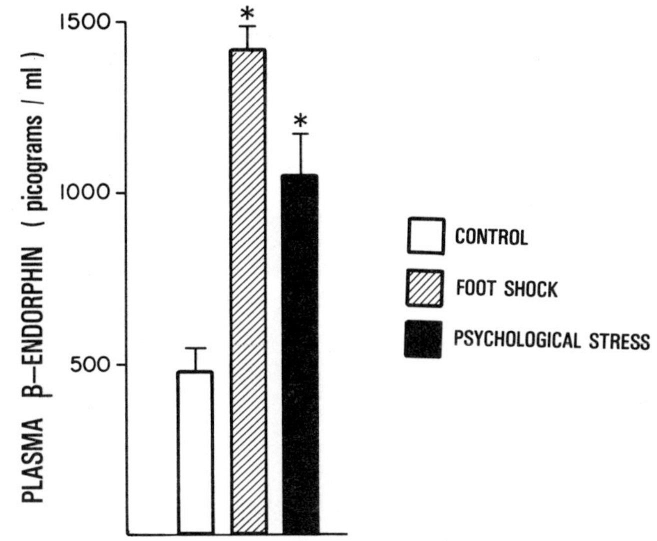

Figure 6. Plasma β-EP levels in psychological stress, footshock and control groups after 15 min in the test chambers (39). Values represent mean ± SEM of 7 rats per group. * = $p < 0.05$, Duncan's multiple range test.

pituitary cyclic AMP (Figure 7) and the stress-induced increases in plasma ACTH (34). The incomplete blockade of stress-induced increases in cyclic AMP is similar to reported incomplete blockade of ACTH responses to stressful conditions following pretreatment with antisera to CRF (34,43).

ADRENALECTOMY ELIMINATES THE STRESS-INDUCED INCREASE IN PITUITARY CYCLIC AMP

Adrenalectomy (ADX), by eliminating the negative feedback of corticosterone, produces elevated resting plasma levels of ACTH and enhanced responsiveness to stressful conditions (44). This may be due to increased hypothalamic release of CRF (45,46). CRF is increased in the median eminence and the paraventricular nucleus of the hypothalamus following ADX (47-49). Given these findings, one might have expected that ADX would enhance the pituitary cyclic AMP response to stressful conditions. As shown in Figure 8, however, we found that ADX completely prevented this response (33), but had no effect on resting levels. The mechanism for the loss of this response is likely to be the 80% down-regulation of pituitary CRF receptors reported to occur after ADX (50,51), which also markedly decreased the adenylate cyclase response to CRF *in vitro* (51,52). The persistance of ACTH responsiveness after ADX is presumably mediated via one of the the non-cyclic AMP dependent mechanisms of ACTH stimulation (50) discussed below.

PARAVENTRICULAR LESIONS ABOLISH THE STRESS-INDUCED RISE IN PITUITARY CYCLIC AMP AND ATTENUATE THE INCREASES IN PLASMA LEVELS OF POMC-DERIVED PEPTIDES AND PROLACTIN

CRF has been demonstrated immunocytochemically in the parvocellular neurons of the paraventricular nucleus (PVN) of the hypothalamus (48,53,54) which project to the median eminence (ME) (47), and CRF is released into the portal system during stress (55). The PVN has been identified as a major

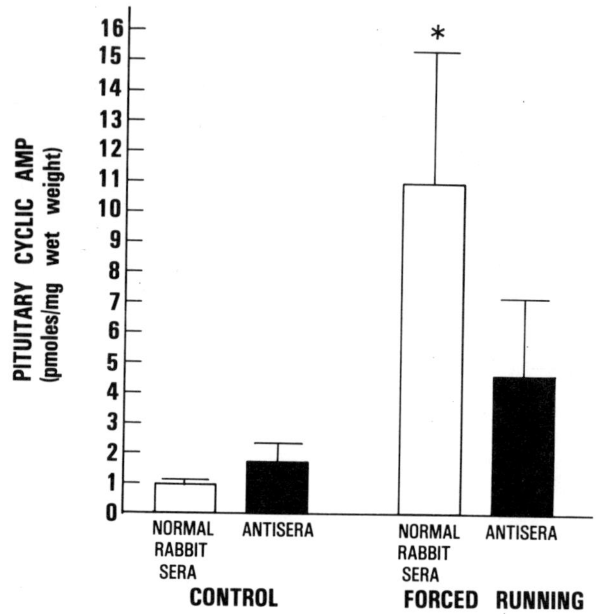

Figure 7. Intravenous pretreatment with antibody to CRF attenuates the stress-induced increase in pituitary cyclic AMP (34). Values represent the mean ± SEM of 6 rats. * Differs significantly from control.

center involved in the regulation of ACTH secretion from the anterior pituitary gland. Stimulation of the PVN in anesthetized animals has been reported to elevate plasma levels of ACTH (56,57) and CS (58), whereas lesioning the PVN depletes CRF-like immunoreactivity in the ME (59) and attenuates the stress-induced rise in plasma ACTH (59-61). We found that lesions of the PVN prevented the stress-induced increases of pituitary levels of cyclic AMP in vivo (8,62). As shown in Figure 9, footshock increased pituitary cyclic AMP more than twofold in the sham-operated rats. Lesioning the PVN completely abolished this stress-induced effect without affecting resting levels in unstressed rats. Pituitary levels of cyclic AMP were practically identical in sham-operated/unshocked rats and in PVN-lesioned rats, both shocked and unshocked. We attributed the effect of the PVN lesions to destruction of CRF-containing neurons, but were not able to rule out a contribution of vasoactive intestinal peptide (VIP) (8).

In contrast to the complete elimination of the pituitary cyclic AMP response, the PVN lesions merely attenuated the responses of prolactin and the three POMC-derived peptides. As seen in Table 1, footshock increased plasma levels of ACTH, β-EP, β-LPH and PRL by sixfold, fourfold, ninefold and ninefold, respectively. PVN lesions, while having no effect on resting plasma levels of any of these hormones, attenuated the stress-induced increases by 64%, 33%, 61% and 62% respectively (8,63). The β-EP response was less affected by these lesions than were the responses of the other two POMC-derived peptides. This might be explained by the observation that while β-EP is present in both the anterior and intermediate lobes of the pituitary (15,16), ACTH and β-LPH are major products only in the anterior, but not the intermediate lobe (16). Presumably the PVN lesions affected secretion from the anterior lobe preferentially, with a resulting greater attenuation of plasma ACTH and β-LPH increases. There are at least three possible mechanisms for the persistence of residual POMC peptide responses to footshock in the lesioned animals: the effects of residual CRF, the effects of residual vasopressin (AVP) and the proposed effects of epinephrine (EPI).

REGULATION OF THE STRESS-INDUCED INCREASE IN PITUITARY CYCLIC AMP

One might expect a neurohumoral mediator of the stress-induced increase in pituitary cyclic AMP to be present in the ME and portal blood, and also in the PVN or the arcuate nucleus of the hypothalamus; it should also stimulate pituitary cyclic AMP accumulation *in vitro*. Likely candidates, in addition to CRF, might include AVP, VIP and EPI. We will also discuss the possibility of mediators present in the general circulation.

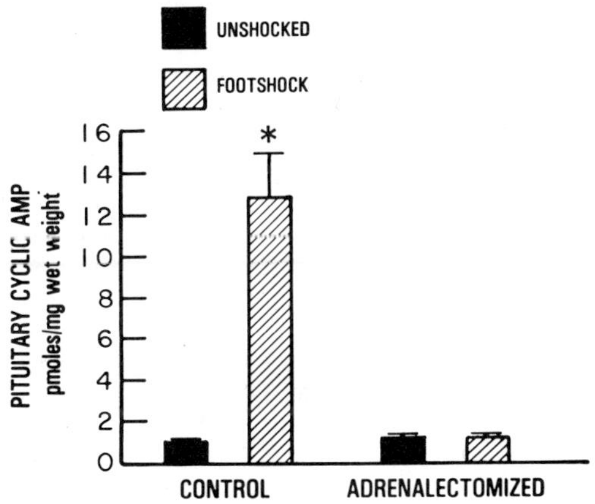

Figure 8. Adrenalectomy completely eliminates the pituitary cyclic AMP response to stress (33). Values represent the mean ± SEM of 6 rats. * Differs significantly from control.

In normal pituitary cells in culture, AVP had no direct effect on cyclic AMP formation, although it significantly enhanced the stimulatory effect of CRF (64). Since AVP did not stimulate cyclic AMP accumulation *in vitro*, it is unlikely to be a direct mediator of the stress-induced rise in pituitary cyclic AMP *in vivo*, but could enhance the effect of CRF by a cyclic AMP-independent mechanism.

VIP was reported to stimulate cyclic AMP accumulation in cultured pituitary tumor cells (65,66) as well as adenylate cyclase activity in pituitary homogenates (67). Also VIP was found in high concentrations in portal blood in both Long-Evans (68) and Wistar rats (69). On the other hand, in the PVN of colchicine-treated Sprague-Dawley rats, neurons staining for VIP were reported as either not visualized (70) or as faintly visualized (71).

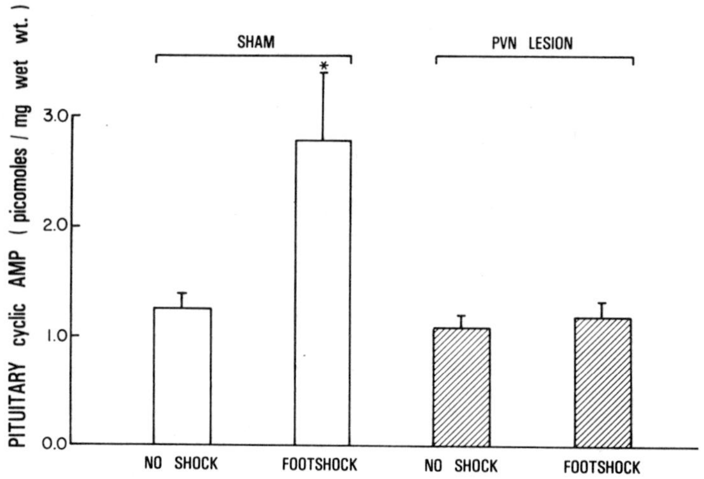

Figure 9. Paraventricular lesions abolish pituitary cyclic AMP response to stress (8). Values represent the mean ± SEM of 10 rats. * Differs significantly from control.

It is well established that exposure to various stressors increases circulating epinephrine (EPI) and norepinephrine (NE) (72,73) and stressful conditions have also been reported to increase turnover of NE in the PVN (74) and to elevate EPI levels in portal blood (55,75). EPI, however, had no direct effect on cyclic AMP levels in normal pituitary cells, but like AVP, it markedly enhanced the stimulatory effect of CRF on both cyclic AMP formation and ACTH secretion (22). We found that splanchnic denervation, which reduces EPI release from the adrenal medulla, had no effect on the stress-induced increase in pituitary cyclic AMP (76). Moreover, the cyclic AMP response *in vivo* was not blocked by pretreatment with either the beta-adrenergic blocker practolol or the alpha-adrenergic blocker prazocin (Meyerhoff and Kant, unpublished data). We found that EPI injections increase pituitary cyclic AMP *in vivo*, but these increases as well as those elicited by footshock were completely prevented by transections of the pituitary stalk. Intravenous injection of CRF, however, did elicit increases in pituitary cyclic AMP in the stalk-transected rats (34). Thus, the effects of injected EPI were probably mediated centrally via CRF.

The abundant evidence linking CRF with pituitary cyclic AMP and POMC peptide responses to stressors has been reviewed above and will be summarized here. CRF is a potent stimulator of pituitary cyclic AMP accumulation *in vitro* and *in vivo*. Also, both CRF and cyclic AMP stimulate the release of ACTH from pituitary cells, as well as synthesis of pituitary POMC mRNA. Moreover, CRF has been localized in neurons in the PVN which project to the ME. Lesions of the PVN reduce CRF levels in the ME and attenuate the ACTH and CS responses to stressful conditions, and stimulation of the PVN elicits release of ACTH. Furthermore, a stressful challenge was reported to increase CRF in portal blood. Finally, we have recently shown that antibodies to CRF attenuate, and PVN lesions completely abolish the stress-induced increases in pituitary cyclic AMP *in vivo*. Taken together, these data strongly suggest that the stress-induced increases in pituitary cyclic AMP levels are probably mediated via CRF-containing neurons projecting from the PVN to the ME, but a contribution of VIP has not been ruled out. Although neither AVP nor circulating EPI seem likely to be direct mediators of the cyclic AMP response, they might augment responses to CRF.

Table 1. PVN Lesions Block the Stress-Induced Increase in Pituitary cyclic AMP Levels and Attenuate Increases in Plasma ACTH, β-EP and β-LPH.

	NO SHOCK	FOOTSHOCK	% INCREASE
cAMP (pm/mg)			
SHAM	1.24 ± 0.11	2.78 ± 0.63 **	124%
PVN LESION	1.07 ± 0.09	1.17 ± 0.12 *	NS
ACTH (pg/ml)			
SHAM	91.1 ± 8.7	588.7 ± 64.2 **	547%
PVN LESION	76.9 ± 5.0	216.3 ± 38.3 *	181%
β-EP (pg/ml)			
SHAM	122.0 ± 9.6	488.5 ± 36.1 **	301%
PVN LESION	119.1 ± 14.7	329.7 ± 61.9 *	177%
β-LPH (pg/ml)			
SHAM	77.3 ± 6.9	712.2 ± 128.3 **	825%
PVN LESION	88.0 ± 7.7	274.8 ± 54.4 *	213%
PRL (ng/ml)			
SHAM	8.3 ± 1.9	74.1 ± 14.3 **	893%
PVN LESION	10.9 ± 2.0	27.9 ± 4.8 *	256%

** Significantly greater than "sham/no shock" group, $p < 0.001$.
* Significantly less than "sham/footshock" group, $p < 0.01$.
N = 10 per group; values are mean ± SEM. Data analyzed by 2-way ANOVA.

REGULATION OF THE STRESS-INDUCED SECRETION OF POMC-DERIVED PEPTIDES

As we have noted above, there is now a large body of biochemical, physiological and anatomical evidence that CRF is a critical regulator of ACTH secretion; CRF is a potent secretogogue both *in vitro* (24,25) and *in vivo* (34,40), and the stress-induced increase in plasma ACTH is attenuated by administration of antibodies to CRF (34,42,43). PVN lesions were reported to deplete CRF in the ME by 90% (59), which implies persistance of 10% residual CRF. It has been reported that ACTH was released *in vitro* by concentrations of CRF which were insufficient to stimulate cyclic AMP accumulation (77). Thus, a CRF-stimulated component of the POMC peptide response to stressors might persist in the PVN-lesioned animals (perhaps combined with enhancement via AVP) despite the apparent elimination of the cyclic AMP response.

While CRF plays a major role in mediating the POMC peptide responses to stressful conditions, its effects are enhanced by AVP, which itself is an ACTH secretogogue (64). AVP and oxytocin (OT) have been identified immunocytochemically in magnocellular neurons of the PVN (48,49,78,79). These neurons are believed to be the source of vasopressinergic and oxytocinergic fibers in the ME, which disappear after lesions of the PVN (78). Stimulation of the PVN in anesthetized animals released AVP (56), and hemorrhagic stress elicited increased AVP levels in portal blood (55). The ACTH response to stressors is attenuated by pretreatment with antibodies to AVP (42), as well as antibodies to CRF (34,43). In rats given both antibodies, the ACTH response to formalin injection was reduced further than after either antibody given singly (42). Moreover, administration of a pharmacological antagonist to AVP was reported to attenuate the ACTH response to stressful conditions (59). While PVN lesions alone merely attenuated the ACTH response to stress, the combination of lesions and pretreatment with the antagonist virtually eliminated the ACTH response (59). PVN lesions may spare some laterally placed magnocellular neurons (61), permitting AVP to contribute to the residual ACTH response in the lesioned rats. In addition to AVP, angiotensin II may release ACTH by cyclic AMP-independent mechanisms (50).

It is unclear whether catecholamines act directly on the normal pituitary gland to release ACTH in vivo. Chlorisondamine, which blocks the release of EPI from the adrenal medulla, partially inhibited the stress-induced increase in plasma ACTH (59), and the combination of PVN lesions with chlorisondamine virtually eliminated the ACTH response. We reported that EPI injection or footshock increased plasma ACTH in normal rats and that section of the pituitary stalk attenuated but did not completely eliminate these increases (34). This is consistent with the report that isoproterenol or salmefalol (a β-2 adrenergic agonist) increased plasma ACTH in stalk-sectioned rats (80). On the other hand, others have reported that plasma catecholamines do not contribute to the ACTH response to forced immobilization (81).

REGULATION OF THE STRESS-INDUCED SECRETION OF PROLACTIN

Exposure to stressors increases plasma levels of PRL, an anterior pituitary hormone not derived from POMC. There are at least 4 possible regulators of the stress-induced increase in plasma PRL: VIP, thyrotropin releasing hormone (TRH), oxytocin (OT) and β-EP. There is now considerable evidence that VIP may serve as a physiological prolactin-releasing factor. Although VIP was reported to stimulate release of ACTH from pituitary tumor cells (65,66), in cells from normal pituitaries it stimulated the release of PRL, not ACTH (82). VIP was found in high concentrations in portal blood (68,69). Lactation enhanced VIP staining in PVN neurons and ME fibers (71), PVN lesions attenuated the suckling-induced rise in plasma PRL levels (83), and antiserum to VIP completely blocked the increase in plasma PRL elicited by ether exposure (84). VIP can stimulate the release of PRL at doses less than one tenth of those required to stimulate cyclic AMP formation, and may stimulate PRL release via cyclic AMP-independent as well cyclic AMP-dependent mechanisms (85). Thus VIP might contribute to the stress-induced rise in plasma PRL while having only minimal effect on the stress-induced increase in pituitary cyclic AMP.

TRH has been demonstrated immunocytochemically in the PVN (86); lesions of the PVN decrease TRH levels in the ME (87). TRH has been measured in portal blood (88) and is thought to stimulate PRL secretion (89). It has been reported that β-EP can overcome dopaminergic inhibition of PRL release from anterior pituitary cells, and it was suggested that this endogenous opioid may partially mediate the PRL response to stressors (90). Two other peptides, OT and PHI, are found in the PVN (78,91) and can release PRL from dispersed anterior pituitary cells (91,92). OT has been detected in portal blood (55,68), and its concentration was increased by hemorrhagic stress (55).

As shown in Table 1, PVN lesions attenuated the PRL response by 62% (8). This decrement might be due to the elimination of neurons containing VIP, TRH or OT. Another possible mechanism for the PVN lesion-induced attenuation in the plasma PRL response to stressful conditions could be the attenuation we found in the stress-induced release of β-EP.

CONCLUSIONS

In conclusion, we have demonstrated that a purely psychological stressor (CER), as well as physical stressors, increase pituitary cyclic AMP and plasma levels of PRL and POMC-derived peptides in-vivo. We have shown that an intact PVN is required for the stress-induced pituitary cyclic AMP response, that the PVN contributes to the hormonal responses to stressful conditions, and that the hormonal as well as the pituitary cyclic AMP responses are attenuated by administration of antibodies to CRF. We have reviewed several lines of research linking CRF, pituitary cyclic AMP and secretion of ACTH. These studies provide strong evidence for the role of pituitary cyclic AMP as a mediator of the CRF-stimulated neuroendocrine response to stressors *in vivo*.

ACKNOWLEDGEMENTS

The authors wish to thank Lori Wittig, David Collins, Clint Wormley, Lee Pennington, Clyde Kenion, Golden Driver, Willie Gamble, Terry Eggleston, Leigh Roberts, Tad Yates, Walter Franz and Charles Gernand for excellent technical assistance, as well as Cyril Wingfield and Michie Vane for outstanding histological support. Expert statistical consultation was generously provided by John R. Burge.

REFERENCES

1. R. Ader and S. B. Freidman, Plasma corticosterone response to environmental stimulation: Effects of duration of stimulation and the 24-hour adrenocortical rhythm, Neuroendo. 3:378-386 (1968).
2. G. M. Brown and J. B. Martin, Corticosterone, prolactin, and growth hormone response to handling and new environment in the rat, Psychosom. Med. 36:241-247 (1974).
3. R. H. Lenox, G. J. Kant, G. R. Sessions, L. L. Pennington, E. H. Mougey, and J. L. Meyerhoff, Specific hormonal and neurochemical responses to different stressors, Neuroendo. 30:300-308 (1980).
4. B. N. Bunnell, W. E. Hills, D. Terrell, J. H. Tigges, E. H. Mougey, L. L. Pennington, and J. L. Meyerhoff, Plasma beta-endorphin and beta-lipotropin response to conditioned stimuli, Neurosci. Abstr. 9:1124 (1983).
5. R. Guillemin, T. Vargo, J. Rossier, S. Minick, N. Ling, C. Rivier, W. Vale, and F. Bloom, β-Endorphin and adrenocorticotropin are secreted concomitantly by the pituitary gland, Science 197:1367-1369 (1977).
6. L. Krulich, E. Hefco, P. Illner, and C. B. Read, The effects of acute stress on the secretion of LH, FSH, prolactin and GH in the normal male rat, with comments on their statistical evaluation, Neuroendo. 16:293-311 (1974).
7. G. J. Kant, E. H. Mougey, and J. L. Meyerhoff, Diurnal variation in neuroendocrine response to stress in rats: Plasma ACTH, β-endorphin, β-LPH, corticosterone, prolactin, and pituitary cyclic AMP responses, Neuroendo. 43:383-390 (1986).
8. J. L. Meyerhoff, E. H. Mougey, and G. J. Kant, Paraventricular lesions abolish the stress-induced rise in pituitary cyclic AMP and attenuate the increases in plasma levels of POMC peptides and prolactin, Neuroendo. 43(3):222-230 (1987).
9. R. E. Mains, B. A. Eipper, and M. Ling, Common precursors to corticotropins and endorphins, Proc. natn. Acad. Sci. USA 74:3014-3018 (1977).
10. J. L. Roberts and G. Herbert, Characterization of a common precursor to corticotropin and β-lipotropin: Cell-free synthesis of the precursor and identification of corticotropin peptides in the molecule, Proc. natn. Acad. Sci. USA 74:4826-4830 (1977).
11. C. H. Li, L. Barnafi, M. Chretien, and D. Chung, Identification and amino acid sequence of beta-LPH from sheep pituitary glands, Nature 208:1093-1094 (1965).
12. L. H. Lazarus, H. Ling, and R. Guillemin, Beta-lipotropin as a prohormone for the morphinomimetic peptides, endorphins and

enkephalins, Proc. natn. Acad. Sci. USA 73:2156-2159 (1976).

13. G. Pelletier, R. Leclerc, F. Labrie, J. Cote, M. Chretien, and M. Lis, Immunohistochemical localization of β-lipotropic hormone in the pituitary gland, Endocrinology 100:770-776 (1977).
14. E. Weber, K. H. Voigt, and R. Martin, Concomitant storage of ACTH- and endorphin-like immunoreactivity in secretory granules of anterior pituitary corticotrophs, Brain Res. 157:385-390 (1978).
15. F. E. Bloom, E. Battenberg, J. Rossier, N. Ling, J. Leppaluoto, T. M. Vargo, and R. Guillemin, Endorphins are located in the intermediate and anterior lobes of the pituitary gland, not in the neurohypophysis, Life Sciences 20:43-48 (1977).
16. B. A. Eipper and R. E. Mains, Structure and biosynthesis of pro-adrenocorticotropin/endorphin and related peptides, Endocrine Reviews 1(1):1-27 (1980).
17. R. Przewlocki, V. Hollt, K. H. Voigt, and A. Herz, Modulation of in vitro release of β-endorphin from the separate lobes of the rat pituitary, Life Sciences 24:1601-1608 (1979).
18. I. Vermes, G. H. Mulder, P. G. Smelik, and F. J. H. Tilders, Differential control of β-endorphin/β-lipotropin secretion from anterior and intermediate lobes of the rat pituitary gland in vitro, Life Sciences 27:1761-1768 (1980).
19. E. B. De Souza and G. R. Van Loon, Differential plasma β-endorphin, β-lipotropin, and adrenocorticotropin responses to stress in rats, Endocrinology 116(4):1577-1586 (1985).
20. W. Vale, J. Spiess, C. Rivier, and J. Rivier, Characterization of a 41-residue ovine hypothalamic peptide that stimulates secretion of corticotropin and β-endorphin, Science 213:1394-1397 (1981).
21. J. Axelrod and T. D. Reisine, Stress hormones: Their interaction and regulation, Science 224:452-459 (1984).
22. V. Giguere, and F. Labrie, Additive effects of epinephrine and corticotropin-releasing factor (CRF) on adrenocorticotropin release in rat anterior pituitary cells, Biochem. & Biophys. Res. Commun. 110(2):456-462 (1983).
23. V. Giguere, F. Labrie, J. Cote, D. H. Coy, J. Sueiras-Diaz, and A. V. Schally, Stimulation of cyclic AMP accumulation and corticotropin release by synthetic ovine corticotropin-releasing factor in rat anterior pituitary cells: Site of glucocorticoid action, Proc. natl. Acad. Sci. USA 79:3466-3469 (1982).
24. F. Labrie, R. Veilleux, G. Lefevre, D. H. Coy, J. Sueiras-Diaz, and A. Schally, Corticotropin-releasing factor stimulates accumulation of adenosine 3',5'-monophosphate in rat pituitary corticotrophs, Science 216:1007-1008 (1982).
25. G. Aguilera, J. P. Harwood, J. X. Wilson, J. Morell, J. H. Brown, and K. J. Catt, Mechanisms of action of corticotropin-releasing factor and other regulators of corticotropin release in rat pituitary cells, J. Biol. Chem. 258(13):8039-8045 (1983).
26. W. Vale, C. Rivier, L. Yang, S. Minick, and R. Guillemin, Effects of purified hypothalamic corticotropin-releasing factor and other substances on the secretion of adrenocorticotropin and β-endorphin-like immunoactivities in vitro, Endocrinology 103(5):1910-1915 (1978).
27. P. Greengard and J. W. Kebabian, Role of cyclic AMP in synaptic transmission in the mammalian peripheral nervous system, Fed. Proc. 33:1059-1067 (1974).
28. H. A. Affolter and T. Reisine, Corticotropin releasing factor increases proopiomelanocortin messenger RNA in mouse anterior pituitary tumor cells, J. Biol. Chem. 260(29):15477-15481 (1985).
29. U. Zor, T. Kaneko, H. P. G. Schneider, S. M. McCann, I. P. Lowe, G. Bloom, B. Borland, and J. B. Field, Stimulation of anterior pituitary adenyl cyclase activity and adenosine 3':5'-cyclic phosphate by hypothalamic extract and prostaglandin E_1 Proc. natl. Acad. Sci. USA 63:918-925 (1969).

30. G. J. Kant, J. L. Meyerhoff, B. N. Bunnell, and R. H. Lenox, Cyclic AMP and cyclic GMP response to stress in brain and pituitary: Stress elevates pituitary cyclic AMP, Pharmacol., Biochem. & Behavior 17:1067-1072 (1982).
31. G. J. Kant, E. H. Mougey, L. L. Pennington, and J. L. Meyerhoff, Graded footshock stress elevates pituitary cyclic AMP and plasma β-endorphin, β-LPH, corticosterone and prolactin, Life Sciences 33:2657-2663 (1983).
32. B. N. Bunnell, G. J. Kant, R. H. Lenox, L. L. Pennington, D. R. Collins, E. H. Mougey, and J. L. Meyerhoff, Pituitary cyclic AMP in rats is increased by psychological stress, Neurosci. Abstr. 7:869 (1981).
33. J. L. Meyerhoff, G. J. Kant, C. J. Nielsen, and E. H. Mougey, Adrenalectomy abolishes the stress-induced increase in pituitary cyclic AMP, Life Sciences 34:1959-1965 (1984).
34. G. J. Kant, M. A. Oleshansky, D. D. Walczak, E. H. Mougey, and J. L. Meyerhoff, Comparison of the effects of CRF and stress on levels of pituitary cyclic AMP and plasma ACTH in vivo, Peptides 7:1153-1158 (1986).
35. J. L. Meyerhoff, G. J. Kant, G. R. Sessions, E. H. Mougey, L. L. Pennington, and R. H. Lenox, Brain and pituitary cyclic nucleotide response to stress, in: "Perspectives in Behavioral Medicine," Vol. II, Redford B. Williams, Jr., ed., Academic Press, New York (1985).
36. E. H. Mougey, J. L. Meyerhoff, L. L. Pennington, W. L. Gamble, G. C. Driver, and G. J. Kant, Stress-induced increases in pituitary cyclic AMP occur in the anterior but not the posterior lobe, Neurosci. Abstr. 12:1059 (1986).
37. B. N. Bunnell, J. L. Meyerhoff, and G. J. Kant, Psychological stress increases pituitary cyclic AMP, Pharmacol. Biochem. & Behavior (in press).
38. J. Rossier, E. French, C. Rivier, N. Ling, R. Guillemin, and F. Bloom, Foot-shock induced stress increases β-endorphin levels in blood but not brain, Nature 270:618-620 (1977).
39. J. L. Meyerhoff, B. N. Bunnell, and E. H. Mougey, Plasma beta-endorphin in rats is increased by psychological stress, Neurosci. Abstr. 7:166 (1981).
40. C. Rivier, M. Brownstein, J. Spiess, J. Rivier, and W. Vale, In vivo corticotropin-releasing factor-induced secretion of adrenocorticotropin, β-endorphin, and corticosterone, Endocrinology 110(1):272-278 (1982).
41. G. P. Chrousos, H. M. Schulte, E. H. Oldfield, P. W. Gold, G. B. Cutler, and D. L. Loriaux, The corticotropin-releasing factor stimulation test, New England J. Med. 310:622-626 (1984).
42. E. A. Linton, F. J. H. Tilders, S. Hodgkinson, F. Berkenbosch, I. Vermes, and P. J. Lowry, Stress-induced secretion of adrenocorticotropin in rats is inhibited by administration of antisera to ovine corticotropin-releasing factor and vasopressin, Endocrinology 116(3):966-970 (1985).
43. C. Rivier, J. Rivier, and W. Vale, Inhibition of adrenocorticotropic hormone secretion in the rat by immunoneutralization of corticotropin-releasing factor, Science 218:377-379 (1982).
44. A. Ruhmann-Wennhold and D. H. Nelson, Plasma ACTH levels in stressed and nonstressed adrenalectomized rats, Annals New York Acad. Sci. 297:498-508 (1977)
45. T. Suda, N. Tomori, F. Tozawa, T. Mouri, H. Demura, and K. Shizume, Effects of bilateral adrenalectomy on immunoreactive corticotropin-releasing factor in the rat median eminence and intermediate-posterior pituitary, Endocrinology 113(3):1182-1184 (1983).
46. E. W. Hillhouse and M. T. Jones, Effect of bilateral adrenalectomy and corticosteroid therapy on the secretion of corticotrophin-releasing factor activity from the hypothalamus of the rat in vitro, Endocrinology 71:21-30 (1976)

47. I. Merchenthaler, S. Vigh, P. Petrusz, and A. V. Schally, The paraventriculo-infundibular corticotropin releasing factor (CRF) pathway as revealed by immunocytochemistry in long-term hypophysectomized or adrenalectomized rats, Regulatory Peptides 5:295-305 (1983).
48. P. E. Sawchenko, L. W. Swanson, and W. W. Vale, Co-expression of corticotropin-releasing factor and vasopressin immunoreactivity in parvocellular neurosecretory neurons of the adrenalectomized rat, Proc. natl. Acad. Sci. USA 81:1883-1887 (1984).
49. P. E. Sawchenko and L. W. Swanson, Localization, colocalization, and plasticity of corticotropin-releasing factor immunoreactivity in rat brain, Fed. Proc. 44(1):221-227 (1985).
50. G. Aguilera, P. C. Green, J. P. Harwood, R. L. Hauger, M. A. Millan, C. Grewe, and K. J. Catt, Receptor-mediated actions of corticotropin-releasing factor in pituitary gland and nervous system, Neuroendo. 43:79-88 (1986).
51. P. C. Wynn, J. P. Harwood, K. J. Catt, and G. Aguilera, Regulation of corticotropin-releasing factor (CRF) receptors in the rat pituitary gland: Effects of adrenalectomy on CRF receptors and corticotroph responses, Endocrinology 116(4):1653-1659 (1985).
52. M. A. Oleshansky, G. J. Kant, and J. L. Meyerhoff, Effects of adrenalectomy on CRF stimulation of rat anterior pituitary adenylate cyclase, Neurosci. Abstr. 10:84 (1984)
53. F. E. Bloom, E. L. F. Battenberg, J. Rivier, and W. Vale, Corticotropin-releasing factor (CRF): Immunoreactive neurones and fibers in rat hypothalamus, Regulatory Peptides 4:43-48 (1982).
54. S. Cummings, R. Elde, J. Ells, and A. Lindall, Corticotropin-releasing factor immunoreactivity is widely distributed within the central nervous system of the rat: An immunohistochemical study, J. Neuroscience 3(7):1355-1368 (1983).
55. P. M. Plotsky, T. O. Bruhn, and W. Vale, Evidence for multifactor regulation of the adrenocorticotropin secretory response to hemodynamic stimuli, Endocrinology 116(2):633-639 (1985).
56. A. Dornhorst, D. E. Carlson, S. M. Seif, A. G. Robinson, E. A. Zimmerman, and D. S. Gann, Control of release of adrenocorticotropin and vasopressin by the supraoptic and paraventricular nuclei, Endocrinology 108(4):1420-1424 (1981).
57. M. Rey, B. Conte-Devoix, J. M. Grillo, E. Castanas, G. Mercier, F. Boudouresque, P. Giraud, Y. Millet, and C. Oliver, Effect of electrical stimulation of the paraventricular nucleus (PVN) on ACTH and corticosterone secretion in the Long-Evans and Brattleboro male rat, Comptes rendus de l'Academie des sciences, Paris 295(12):697-699 (1982).
58. J. Dunn and V. Critchlow, Electrically stimulated ACTH release in pharmacologically blocked rats, Endocrinology 93(4):835-842 (1973).
59. T. O. Bruhn, P. M. Plotsky, and W. W. Vale, Effect of paraventricular lesions on corticotropin-releasing factor (CRF)-like immunoreactivity in the stalk-median eminence: Studies on the adrenocorticotropin response to ether stress and exogenous CRF, Endocrinology 114(1):57-62 (1984).
60. G. Ixart, G. Alonso, A. Szafarczyk, F. Malaval, J. Nouguier-Soule, and I. Assenmacher, Adrenocorticotropic regulations after bilateral lesions of the paraventricular or supraoptic nuclei and in Brattleboro rats, Neuroendo. 35:270-276 (1982).
61. G. B. Makara, E. Stark, M. Karteszi, M. Palkovits, and G. Rappay, Effects of paraventricular lesions on stimulated ACTH release and CRF in stalk-median eminance of the rat, Amer. J. Physiol. 240(4):E441-E446 (1981).
62. J. L. Meyerhoff, E. H. Mougey, C. B. Wormley, L. L. Pennington, and G. J. Kant, Lesions of the paraventricular nucleus of the hypothalamus prevent stress-induced increases in pituitary cyclic AMP, Neurosci.

Abstr. 11:361 (1985).
63. J. L. Meyerhoff, G. J. Kant, and E. H. Mougey, PVN lesions completely abolish stress-induced increases in pituitary cyclic AMP in vivo, Psychiatry Research 16(4), Special Suppl.,:56 (1986)
64. V. Giguere and F. Labrie, Vasopressin potentiates cyclic AMP accumulation and ACTH release induced by corticotropin-releasing factor (CRF) in rat anterior pituitary cells in culture, Endocrinology 111(5):1752-1754 (1982).
65. S. Heisler, T. D. Reisine, V. Y. H. Hook, and J. Axelrod, Somatostatin inhibits multireceptor stimulation of cyclic AMP formation and corticotropin secretion in mouse pituitary tumor cells, Proc. natl. Acad. Sci. USA 79:6502-6506 (1982).
66. T. D. Reisine, S. Heisler, V. Y. H. Hook, and J. Axelrod, Multireceptor-induced release of adrenocorticotropin from anterior pituitary tumor cells, Biochem. & Biophys. Res. Commun. 108(3):1251-1257 (1982).
67. C. Borghi, S. Nicosia, A. Giachetti, and S. I. Said, Adenylate cyclase of rat pituitary gland. Stimulation by vasoactive intestinal polypeptide (VIP), FEBS Letters 108(2):403-406 (1979).
68. S. I. Said and J. C. Porter, Vasoactive intestinal polypeptide. Release into hypophyseal portal blood, Life Sciences 24:227-230 (1979).
69. A. Shimatsu, Y. Kato, N. Matsushita, H. Katakami, N. Yanaihara, and H. Imura, Immunoreactive vasoactive intestinal polypeptide in rat hypophysial portal blood, Endocrinology 108(2):395-398 (1981).
70. E. Mezey, Vasoactive intestinal polypeptide immunopositive neurons in the paraventricular nucleus of homozygous Brattleboro rats, Neuroendo. 42:88-90 (1986).
71. E. Mezey and J. Z. Kiss, Vasoactive intestinal peptide-containing neurons in the paraventricular nucleus may participate in regulating prolactin secretion, Proc. natl. Acad. Sci. USA 82:245-247 (1985).
72. B. H. Natelson, W. N. Tapp, J. E. Adamus, J. C. Mittler, and B. E. Levin, Humoral indices of stress in rats, Physiology & Behavior 26:1049-1054 (1981).
73. R. Kvetnansky, C. L. Sun, C. R. Lake, N. Thoa, T. Torda, and I. J. Kopin, Effect of handling and forced immobilization on rat plasma levels of epinephrine, norepinephrine and dopamine beta-hydroxylase, Endocrinology 103:1868-1874 (1978)
74. C. A. Johnston, E. J. Spinedi, and A. Negro-Vilar, Effect of acute ether stress on monoamine metabolism in median eminence and discrete hypothalamic nuclei of the rat brain and on anterior pituitary hormone secretion, Neuroendo. 41:83-88 (1985).
75. D. M. Gibbs, Hypothalamic epinephrine is released into hypophysial portal blood during stress, Brain Research 335:360-364 (1985).
76. G. J. Kant, C. J. Nielsen, M. A. Oleshansky, E. H. Mougey, L. L. Pennington, and J. L. Meyerhoff, Subsensitive pituitary cyclic AMP response to stress following adrenalectomy is not caused by loss of adrenal epinephrine, Life Sciences 36:2421-2428 (1985).
77. L. M. Bilezikjian and W. W. Vale, Glucocorticoids inhibit corticotropin-releasing factor-induced production of adenosine 3',5'-monophosphate in cultured anterior pituitary cells, Endocrinology 113(2):657-662 (1983).
78. F. Vandessande, K. Diedrickx, and J. De Mey, The origin of the vasopressinergic and oxytocinergic fibres of the external region of the median eminence of the rat hypophysis, Cell Tissue Research 180:443-452 (1977).
79. E. A. Zimmerman, M. A. Stillman, L. D. Recht, J. L. Antunes, and P. W. Carmel, Vasopressin and corticotropin-releasing factor: An axonal pathway to portal capillaries in the zona externa of the median eminence containing vasopressin and its interaction with adrenal corticoids, Annals New York Acad. Sci. 297:405-419 (1977).
80. E. Mezey, T. D. Reisine, M. Palkovits, M. J. Brownstein, and J.

Axelrod, Direct stimulation of β_2-adrenergic receptors in rat anterior pituitary induces the release of adrenocorticotropin in vivo, Proc. Natl. Acad. Sci. USA 80:6728-6731 (1983).

81. G. B. Makara, R. Kvetnansky, D. Jezova, A. Jindra, I. Kakucska, and Z. Oprsalova, Plasma catecholamines do not participate in pituitary-adrenal activation by immobilization stress in rats with transection of nerve fibers to the median eminence, Endocrinology 119(4):1757-1762 (1986).
82. W. H. Rotsztejn, L. Benoist, J. Besson, G. Beraud, M. T. Bluet-Pajot, C. Kordon, G. Rosselin, and J. Duval, Effect of vasoactive intestinal peptide (VIP) on the release of adrenohypophyseal hormones from purified cells obtained by unit gravity sedimentation, Neuroendo. 31:282-286 (1980).
83. J. Z. Kiss, B. Kanyicska, and G. Y. Nagy, The hypothalamic paraventricular nucleus has a pivotal role in regulation of prolactin release in lactating rats, Endocrinology 119(2):870-873 (1986).
84. H. Abe, D. Engler, M. E. Molitch, J. Bollinger-Gruber, and S. Reichlin, Vasoactive intestinal peptide is a physiological mediator of prolactin release in the rat, Endocrinology 116(4):1383-1390 (1985).
85. P. Onali, C. Eva, M. C. Olianas, J. P. Schwartz, and E. Costa, In GH^3 pituitary cells, acetylcholine and vasoactive intestinal peptide antagonistically modulate adenylate cyclase, cyclic AMP content, and prolactin secretion, Molecular Pharmacology 24:189-194 (1983).
86. R. M. Lechan and I. M. D. Jackson, Immunohistochemical localization of thyrotropin-releasing hormone in the rat hypothalamus and pituitary, Endocrinology 111(1):55-65 (1982).
87. M. J. Brownstein, R. L. Eskay, and M. Palkovits, Thyrotropin releasing hormone in the median eminence is in the processes of paraventricular nucleus neurons, Neuropeptides 2:197-201 (1982).
88. G. Fink, Y. Koch, and N. B. Aroya, Release of thyrotropin releasing hormone into hypophysial portal blood is high relative to other neuropeptides and may be related to prolactin secretion, Brain Research 243:186-189 (1982).
89. A. Enjalbert, J. Epelbaum, S. Arancibia, L. Tapia-Arancibia, M. T. Bluet-Pajot, and C. Kordon, Reciprocal interactions of somatostatin with thyrotropin-releasing hormone and vasoactive intestinal peptide on prolactin and growth hormone secretion in vitro, Endocrinology 111(1):42-47 (1982).
90. K. H. Voight, D. Frank, E. Duker, R. Martin, and W. Wuttke, Dopamine-inhibited release of prolactin and intermediate lobe-POMC-peptides: Different modulation by opioids, Life Sciences 33(1):507-510 (1983).
91. T. Hokfelt, J. Fahrenkrug, K. Tatemoto, V. Mutt, S. Werner, A. L. Hulting, L. Terenius, and K. J. Chang, The PHI (PHI-27)/corticotropin-releasing factor/enkephalin immunoreactive hypothalamic neuron: Possible morphological basis for integrated control of prolactin, corticotropin, and growth hormone secretion, Proc. natl. Acad. Sci. USA 80:895-898 (1983).
92. W. K. Samson, M. D. Lumpkin, and S. M. McCann, Evidence for a physiological role for oxytocin in the control of prolactin secretion, Endocrinology 119(2):554-560 (1986).

CORTICOTROPIN-RELEASING HORMONE (CRH) RECEPTORS IN BRAIN

Errol B. De Souza and George Battaglia

Neuropeptide Unit, Neuroscience Branch
Addiction Research Center
National Institute on Drug Abuse
Baltimore, Maryland 21224

INTRODUCTION

Corticotropin-releasing hormone (CRH), a 41-amino acid peptide originally isolated from ovine hypothalami[1], stimulates the release of proopiomelanocortin (POMC)-derived peptides from the pituitary gland (reviewed in [2,3]). More recently, the amino acid sequences of both rat[4] and human[5] CRH have been elucidated and were found to be identical to one another but to differ from ovine CRH by seven amino acid residues. In addition to its endocrine role at the pituitary, CRH has been demonstrated to have a broad extra-hypothalamic distribution in the central nervous system (CNS)[6-8], and to produce a wide spectrum of autonomic[9-11] and behavioral [12-15] effects. The results of these studies have led to the suggestion that CRH may act as a neurotransmitter in the CNS, where it appears to play a role in integration of the organism's response to stress. A neurotransmitter role for CRH in the rat CNS is further supported by several observations at the cellular level including its calcium-dependent release from brain slices following potassium stimulation[16], the ability of CRH to alter neuronal firing rates following iontophoretic application [17-19], and its ability to promote the formation of cyclic AMP (cAMP)[20-22]. In addition, recent clinical data suggest that this novel neuropeptide may be of relevance to endocrine, psychiatric and neurologic illnesses[23]. The actions of CRH in the CNS are presumed to be initiated by binding to high-affinity receptors for the neuropeptide.

In this chapter, we summarize some of the data from our recent studies in which we have identified, characterized and localized CRH receptors in rat and human brain. The characteristics of the second messenger systems mediating the effects of CRH in brain will be described. The data will be discussed in the context of the role of CRH as a neurotransmitter regulating stress responses in brain. Finally, the role of CRH and CRH receptors in neurologic diseases will be discussed.

METHODS

The details of the procedure for iodinating CRH[24,25], the optimal conditions used for labeling CRH receptors in homogenates[25-27] and slide-mounted sections[24,25,28-30], and the optimal conditions for characterizing CRH-mediated cAMP production[22] have all been described elsewhere.

CRH RECEPTORS IN THE CNS

Kinetics and Pharmacology

A stable, iodine-125-labeled analog of rat/human CRH or ^{125}I-Tyr^{o}-rat/human CRH (^{125}I-rCRH) was used to define the characteristics of CRH receptors in rat[26] and human[27] brain. ^{125}I-rCRH binding was dependent on time and temperature, was sensitive to the pH, ionic strength and cationic composition of the incubation buffer, and was linear with membrane protein concentration. Receptor binding was optimal in 50 mM Tris HCl buffer, pH 7.2 and was maximal at room temperature for 90-180 minutes with minimal degradation of the radioiodinated peptide occurring under these conditions. ^{125}I-rCRH binding to brain membranes was saturable, reversible, and Scatchard analysis of the saturation data revealed a high-affinity component with an apparent K_D of 0.2 nM and a low-affinity binding site with a K_D of approximately 20 nM. ^{125}I-rCRH binding sites were primarily localized to tissue preparations enriched in synaptic membranes (i.e. the total number and density of specific ^{125}I-rCRH binding sites was greater in the crude mitochondrial/synaptosomal P_2 pellet than in the P_1 fraction representing nuclei and cellular debris; no significant binding was found in the microsomal P_3 pellet). The pharmacological characteristics of the binding site were examined by determining the relative potencies of a variety of CRH-related and -unrelated peptides in inhibiting specifically bound ^{125}I-rCRH. The results of these studies are summarized in Figure 1. Overall, the binding sites for ^{125}I-rCRH in brain exhibit a pharmacological specificity for CRH analogs and fragments which corresponds well with their relative intrinsic potency in stimulating or inhibiting anterior pituitary secretion of adrenocorticotropic hormone (ACTH) in vitro. The specificity of the CRH binding site is further strengthened by the lack of inhibition of ^{125}I-rCRH binding by peptides such as angiotensin II, arginine vasopressin and vasoactive intestinal peptide, which have been shown to also stimulate ACTH secretion through actions on their own respective receptors. The kinetic and pharmacological characteristics of ^{125}I-rCRH binding sites in rat brain are comparable to the characteristics previously observed in rat[31-33], bovine[25,29,31] and human[34] pituitary and in monkey[35] and human[27] brain using a variety of ligands. These data substantiate earlier suggestions[19,36,37] that the structural requirements for CRH activity are shared by brain and pituitary receptors.

CHARACTERIZATION OF CRH RECEPTOR-MEDIATED ADENYLATE CYCLASE ACTIVITY IN RAT BRAIN

Evidence from Radioligand Binding Studies

Agonist binding to receptors mediating their effects through adenylate cyclase activity have been shown to be influenced by divalent cations and guanine nucleotides. Magnesium ions enhance agonist binding to adenylate cyclase-coupled receptors by stabilizing the agonist high-affinity state of the receptor. In contrast, guanine nucleotides selectively decrease the affinity of agonists for their receptors by promoting the dissociation of the agonist high-affinity form of the receptor. In order to determine whether CRH receptors in brain might represent two-state receptors coupled to adenylate cyclase, we carried out a series of studies to examine the effects of magnesium ions and nucleotides on ^{125}I-rCRH binding in homogenates of rat brain. The

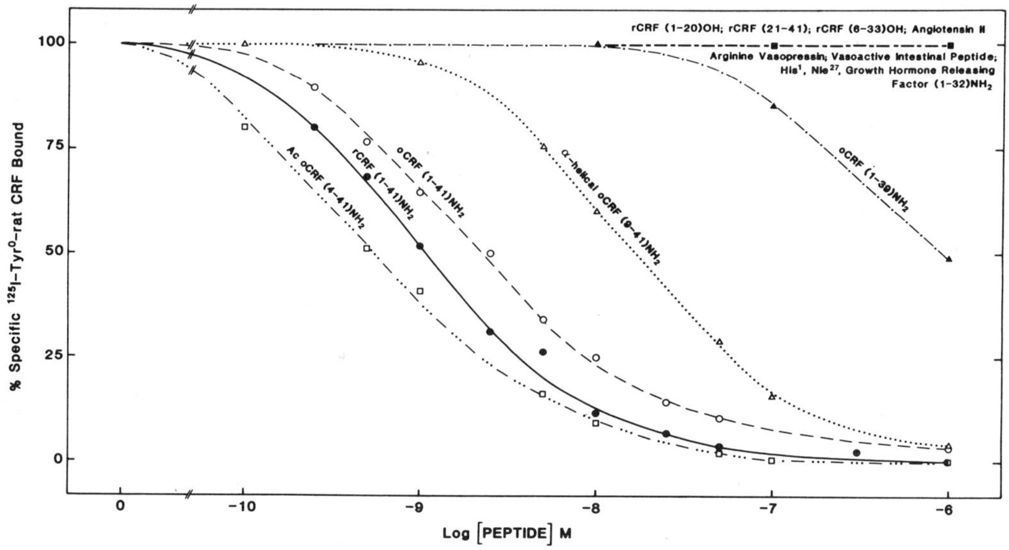

Fig. 1. Characterization of the pharmacological specificity of ^{125}I-Tyr^{0}-rat/human CRH binding in rat olfactory bulb membranes. Crude membranes were incubated for 120 min at room temperature in the presence of 0.1 nM ^{125}I-Tyr^{0}-rat/human CRH and varying concentrations of CRH-related peptides. Nonspecific binding was determined in the presence of 1 uM rat CRH (from De Souza, Ref. 26, with permission).

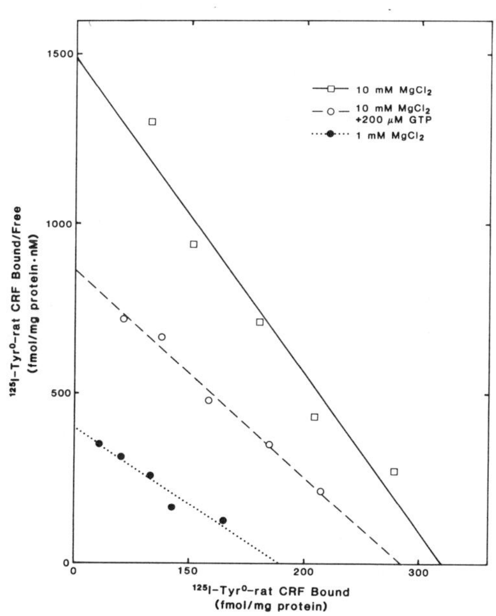

Fig. 2. Effects of magnesium ions and guanosine-5'-triphosphate (GTP) on ^{125}I-Tyr^{0}-rat/human CRH binding in rat olfactory bulb membranes. Scatchard plots of the specific binding of increasing concentrations (0.05 - 1nM) of ^{125}I-Tyr^{0}-rat/human CRH to crude membranes in the presence of 1 mM $MgCl_2$ (····•····), 10 mM $MgCl_2$ (—□—), or 10 mM $MgCl_2$ plus 200 uM GTP (--o--) (from De Souza, Ref. 26, with permission).

binding of ^{125}I-rCRH was dependent on the presence of magnesium ions; linear increases in specific binding were observed up to a concentration of 10 mM $MgCl_2$. This effect was specific to magnesium, as slight increases were observed in the presence of a comparable concentration of $CaCl_2$ while comparable concentrations of NaCl had no effect on ^{125}I-rCRH binding. Increasing concentrations of guanosine-5'-triphosphate (GTP) added to the incubation medium resulted in an increasing amount of inhibition of ^{125}I-rCRH binding; the non-hydrolyzable GTP analog, 5'-guanylimidodiphosphate (GppNHp), was even more potent. This effect appeared to be guanine nucleotide-specific, as equimolar concentrations of ATP had no effect on ^{125}I-rCRH binding. To determine whether the effects of guanine nucleotides and magnesium ions reflected changes in receptor density and/or affinity, we carried out a series of parallel ^{125}I-rCRH saturation assays in the presence of 1 mM $MgCl_2$, 10 mM $MgCl_2$ and 10 mM $MgCl_2$ plus 200 uM NaGTP. Scatchard analyses of these saturation data are presented in Figure 2. GTP increased the apparent K_D of ^{125}I-rCRH by 35%; this decrease in affinity was accomplished by a small reduction in the B_{max} value. Conversely, the effect of magnesium to increase ^{125}I-rCRH binding was the result of both an increase in the affinity (50% decrease in K_D from 1 mM to 10 mM $MgCl_2$) and in the number of binding sites (Figure 2). These results are consistent with the characteristic effects of adenylate cyclase modulators on agonist binding to receptors mediating their effects through guanine nucleotide regulatory proteins which stimulate cAMP production.

Characterization of CRH Receptor-Mediated Adenylate Cyclase Activity in Rat Brain

We examined the influence of various incubation parameters on CRH-stimulated adenylate cyclase activity in homogenates of cerebral cortex. CRH-stimulated adenylate cyclase activity increased linearly with respect to time and tissue protein concentration. In addition, the CRH stimulation of adenylate cyclase activity was temperature-dependent with greater activity observed at 37^oC compared with that at 30° or 22^oC. The rank order of potency (EC_{50}; nM; mean $\pm$ S.E.M.) for CRH analogs and fragments in stimulating adenylate cyclase activity was ($Nle^{21,38}$) rat CRH (16 $\pm$ 5) > rat CRH (36 $\pm$ 4) ≃ acetyl ovine CRH (4-41) (38 $\pm$ 2) ≃ alpha helical ovine CRH (43 $\pm$ 12) > ovine CRH (100 $\pm$ 21) >> ovine CRH (1-39) (>1000) ≃ ovine CRH (7-41) (> 1000). The rank order of potency of various CRH-related peptides in stimulating adenylate cyclase activity was consistent with their affinities for CRH receptors in brain (see above) and their relative potencies in stimulating pituitary ACTH secretion in vitro[1,38]. While most of the biologically active CRH compounds elicited maximal stimulation of adenylate cyclase activity at 1 uM concentrations and had comparable V_{max} values (100- 120 pmol cAMP/min/mg protein), the maximal stimulation elicited by alpha helical ovine CRH (36$\pm$7 pmol cAMP/min/mg protein) was significantly lower. These data suggest that not all CRH-related peptides exhibit comparable intrinsic activities with respect to the stimulation of adenylate cyclase activity in brain. In order to examine the receptor specificity of these peptides, we investigated the effects of the weak receptor antagonist, alpha helical ovine CRH (9-41), on stimulation of enzyme activity by rat CRH. The putative CRH receptor antagonist, did not stimulate cAMP production but was able to attenuate the stimulation by various concentrations of rat CRH resulting in an apparent 5-fold shift in the EC_{50} for rat CRF stimulation, from 36 to 190 nM in the absence versus the presence of 100 nM alpha helical ovine CRH (9-41), respectively. These data indicate that both compounds are interacting in a competitive manner with the same receptor population.

Table 1. Distribution of CRH receptors and CRH-stimulated adenylate cyclase activity in discrete regions of rat CNS

Region	CRH Receptors (fmol/mg protein)	CRH-stimulated Adenylate Cyclase Activity (pmol cAMP/min/mg protein)	"CRH Activity Index" (fmol cAMP/min/ fmol receptor)
Cerebral Cortex	50±4	95±10	1900±285
Olfactory Bulb	112±6	54±6	482±95
Cerebellum	48±7	47±5	979±95
Midbrain	16±1	25±1	1562±38
Hippocampus	10±1	20±1	1905±97
Striatum	18±2	17±3	971±228
Hypothalamus	22±2	13±3	591±76
Spinal Cord	8±2	8±4	941±57

Distribution of CRH receptors labeled by ^{125}I-Tyr^{o}-ovine CRH and the stimulation of adenylate cyclase activity by 1 uM rat CRH in discrete regions of rat CNS was determined as previously described[22]. "CRH Activity Index", defined as the stimulation of adenylate cyclase activity per unit receptor, was calculated from values obtained in the same homogenate preparation. Data represent the mean and SEM of three independent experiments.

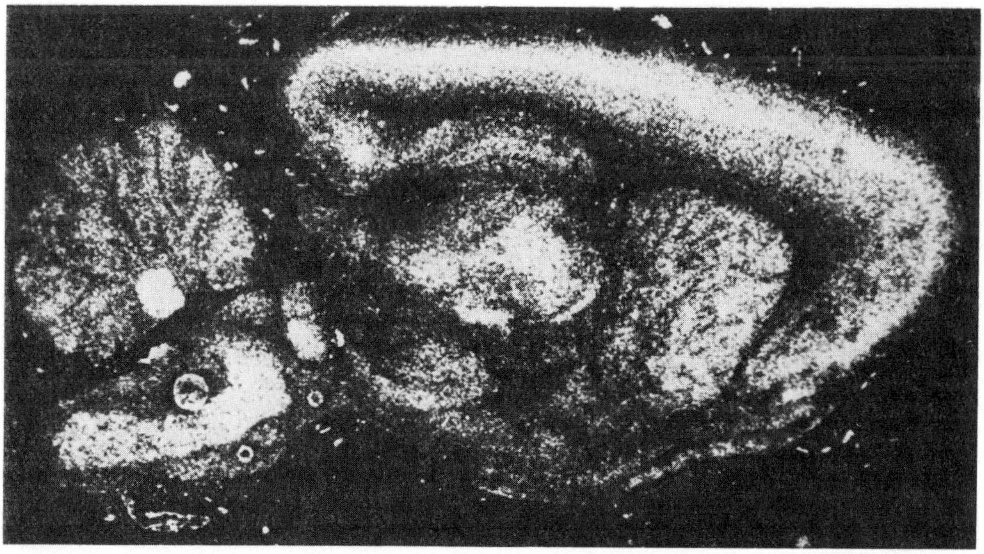

Fig. 3. Autoradiographic distribution of CRH receptor-binding sites in rat brain, labeled with Nle^{21}, ^{125}I-Tyr^{32}-ovine CRH. Darkfield photomicrograph (^{3}H-Ultrofilm), showing the distribution of autoradiographic grains in a saggital section of rat brain. In darkfield illumination, the autoradiographic grains, i.e. binding sites, appear as white spots and the tissue is not visible. Thus, the brightest areas have the highest concentration of binding sites (from De Souza et al., Ref. 24, with permission).

REGIONAL DISTRIBUTION AND RELEVANCE OF CRH RECEPTORS IN THE CNS

The regional distribution of CRH receptors and CRH-stimulated adenylate cyclase activity in homogenates of discrete areas of the rat CNS will be described . In addition, the autoradiographic distribution of ^{125}I-CRH binding sites in rat CNS will be described . This section of the chapter will involve a discussion of the distribution of CRH receptors in the context of their correspondence with the immunohistochemical distribution of CRH-like immunoreactivity (CRH-IR) and the anatomical loci mediating the pharmacological actions of CRH in rat brain. Also, the implications of endogenous CRH as a "stress" neurotransmitter in the brain will be discussed.

In order to assess the correspondence between CRH receptor number and the degree of CRH-stimulated cAMP production, CRH-stimulated adenylate cyclase activity and ^{125}I-Tyr^{o}-ovine CRH (^{125}I-oCRH) binding were measured in parallel in various regions of the rat CNS. As shown in Table 1, the highest concentration of CRH receptors was localized to the olfactory bulb followed by cerebral cortex and cerebellum, and progressively lower binding was present in hypothalamus, striatum, midbrain, hippocampus and spinal cord, respectively. The highest level of CRH-stimulated adenylate cyclase activity was observed in the cerebral cortex. The olfactory bulb and cerebellum exhibited the next greatest degree of CRH-stimulated adenylate cyclase activity followed by progressively lower stimulation in midbrain, hippocampus, striatum, hypothalamus and spinal cord, respectively (Table 1). Clearly, the regional distribution of CRH receptors does not correspond with the regional degree of CRH receptor-mediated stimulation of adenylate cyclase. The greatest disparity is observed in the olfactory bulb which contains the highest concentration of CRH receptors, nearly twice the amount found in the cerebral cortex, yet the level of CRH-stimulated adenylate cyclase activity is only 60% of that found in the cerebral cortex. Discrepancies between CRH receptor number and the amount of CRH stimulation of enzyme activity in olfactory bulb and possibly other brain regions may be explained by (1) a large reserve of spare receptors, (2) a subpopulation of CRH receptors which may be coupled to the catalytic unit in an inhibitory fashion and/or (3) a subpopulation of CRH receptors which may be coupled to an entirely distinct signal transduction mechanism. To further discern the extent of "correspondence" or "match" between the regional concentration of CRH receptors and CRH receptor-stimulated enzyme activity, we have determined a "functional" index of CRH activity (i.e. CRH Activity Index) by calculating the amount of cAMP producted per unit CRH receptor in the various brain regions. As shown in Table 1, in olfactory bulb, the mismatch between CRH receptor number and cAMP production is even more accentuated using this measure. Given the suggestion of multiple binding sites for CRH in brain (see above), it would be of interest to examine in greater detail the roles of other signal transduction mechanisms in addition to CRH-stimulated adenylate cyclase activity in mediating the actions of CRH.

The autoradiographic distribution of ^{125}I-CRH binding sites in rat CNS is shown in Figure 3. Details of the quantities and patterns of distribution of CRH receptors in the rat CNS have been described elsewhere[24,25,29,30].

There are areas in the rat CNS in which the densities and sites of localization of CRH receptors correspond with the relative distribution of CRH-IR terminals. The best example is provided in the neocortex in which CRH cell bodies are concentrated in laminae II and III with projections to laminae I and IV[6-8], areas which are rich in CRH receptors. Additional areas of good correspondence between the distribution of CRH-IR nerve terminals and CRH binding sites include the olfactory tubercle, caudate-putamen, nucleus of the diagonal band of Broca, medial and lateral septal nuclei, the cranial nerve nuclei in the brain stem, and in other brain stem regions including the central gray, inferior olive, reticular formation and the cuneate and parabrachial nuclei. Low concentrations of CRH receptors are present in most of the hypothalamus with binding sites present in the external layer of the median emminence, where the greatest concentration of CRH fibers is found[6-8]. In contrast, there are areas in the rat CNS in which the distributions of CRH-IR nerve fibers and CRH receptors do not correlate as well. Some of these areas include the olfactory bulb, amygdala, cerebellum, and spinal cord. In the amygdala, CRH cell bodies and nerve terminals appear confined primarily to the central nucleus[6-8], whereas the highest concentration of CRH receptors is present in the basolateral nucleus. Additional examples of this disparity involve the olfactory bulb and cerebellum. In the rat cerebellum, moderate[7], low[6] and undetectable[8] densities of CRH-IR fibers have been reported, while very high concentrations of CRH binding sites and CRH-stimulated adenylate cyclase activity are detected in this brain region. Similarly, the olfactory bulb has the highest concentrations of CRH receptors in the CNS but only low concentrations of CRH-IR in olfactory bulb have been reported[6-8]. This phenomenon does not appear to be unique to CRH, as similar mismatches between the levels of neurotransmitter and receptor have been noted in other neuronal systems. The "mismatch problem" has been discussed extensively in a recent report[39].

The hypothalamus represents a primary site of action for integrating the endocrine, autonomic and behavioral effects of CRH. Although the concentration of CRH receptors is low in this brain region, the receptors appear to be primarily coupled in a stimulatory fashion to adenylate cyclase activity as evidenced by the relatively high "CRH Activity Index" (see Table 1). Autoradiographic data have localized CRH receptors in moderate concentrations in the paraventricular nucleus[24] which may represent presynaptic autoreceptors and/or receptors for regulating peptide release. Microinfusions of CRH into the hypothalamus suppress sexual behavior[40] and gastric acid secretion[11]. In addition intracerebroventricular injections of CRH have been shown to inhibit the release of growth hormone[41,42], luteinizing hormone[42,43], vasopressin and oxytocin[44] and to have deleterious effects on reproductive functions[43]. CRH can also stimulate the secretion of somatostatin in cultured rat hypothalamic cells[37]. A role has been proposed for CRH in the control of food intake[13]; the medial lateral hypothalamus represents an important brain area that is involved in regulating food intake[45].

CRH injected intracerebroventricularly produces dose-dependent locomotor activation[14], and generalized arousal[12,15]. Brain structures including the caudate-putamen, cerebellum, and the facial and hypoglossal motor crainal nerve nuclei contain high concentrations of CRH receptors that may be involved in subserving the motor effects of this neuropeptide. The localization of CRH binding sites in the ventral horn of the spinal cord may provide another potential site for CRH to centrally affect motor activity. The autonomic effects of CRH, similar to those changes produced by many forms of stress, include increased plasma concentrations of epinephrine, norepinephrine, glucose, glucagon, cardiovascular parameters such as heart rate and mean arterial pressure, and decreased gastric acid secretion[9-11]. CRH may produce some of its effects on autonomic activity via receptors in the parabrachial nucleus, medullary reticular formation, nucleus of the solitary tract, or the brain stem areas known to participate in autonomic regulation. Forebrain regions including the hypothalamus, especially the paraventricular nucleus, the bed nucleus of the stria terminalis, amygdala, septum, hippocampus, and cingulate cortex exert powerful influences on autonomic outflow and contain moderate to high concentrations of CRH receptors. The autonomic effects of CRH administration may also be mediated by receptors in the spinal cord as suggested by the effects of intrathecal injections of CRH at the lumbar level to inhibit gastric acid secretion[46].

Brain areas including the cerebral cortex, hypothalamus, thalamus and lateral septal area have been demonstrated to be electrophysiologically responsive to iontophoretic application of CRH[18], as have discrete brain stem areas such as the locus coeruleus and parabrachial nucleus[19]; these brain areas contain moderate to high concentrations of CRH receptors. Hippocampal slice preparations have been employed to demonstrate the electrophysiological effects of CRH *in vitro*[17], and intracerebroventricular administration of high doses of CRH have been shown to elicit epileptiform seizures that appear on the basis of associated EEG patterns to originate an amygdala[47].

The immunocytochemical, electrophysiologic, behavioral, pharmacologic, and receptor binding studies provide strong evidence that endogenous CRH may function as a neurotransmitter in the CNS. In addition, the data discussed above suggest a key role for this neuropeptide in integrating the endocrine, autonomic, and behavioral responses to stress. Preliminary evidence demonstrating that footshock stress can increase the cerebrospinal fluid (CSF) levels of CRH-IR in rats[48], and that central administration of a CRH receptor antagonist produces dose-dependent inhibition of the plasma epinephrine responses to ether stress or to insulin-induced hyperglycemia[49], support a role for endogenous brain CRH in mediating autonomic responses to stress. Future studies using more potent receptor antagonists and techniques to immunoneutralize CRH should be useful in elucidating the role of CRH as a "stress" neurotransmitter in the CNS.

MODULATION OF CRH RECEPTORS IN BRAIN

Modulation of CRH Receptors in the Cerebral Cortex in Alzheimer's Disease

There is increasing evidence that neuropeptides in brain may be important in neurological diseases and psychiatric disorders. The role of several neuropeptides has been examined in Alzheimer's disease. Although cortical levels of somatostatin and somatostatin receptors are reduced in Alzheimer's disease, no consistent changes have been reported in other neuropeptide systems (see Ref. 27). Recently, we have examined pre- and post-synaptic markers of CRH in postmortem human brain tissues from patients with Alzheimer's disease and age-matched controls[27]. In individuals with Alzheimer's disease, the concentrations of CRH-IR (Figure 4a) were decreased and CRH receptors (Figure 4b) were reciprocally increased in the temporal, frontal and occipital cortex. Neither CRH-IR nor CRH receptors were changed significantly in the cingulate cortex. A highly significant negative correlation exists between the increase in CRH receptors and the decrease in CRH-IR in each cortical area ($r = -0.65$; $p < 0.005$). These data strongly support a neurotransmitter role for CRH in brain and demonstrate, for the first time, the modulation of CRH receptors associated with altered CRH content. Additional evidence for a role for CRH in Alzheimer's disease is provided by observations demonstrating decreases in CRH-IR in other brain areas including the caudate[50], decreased concentrations of CRH-IR in CSF obtained from patients with Alzheimer's disease[51], and

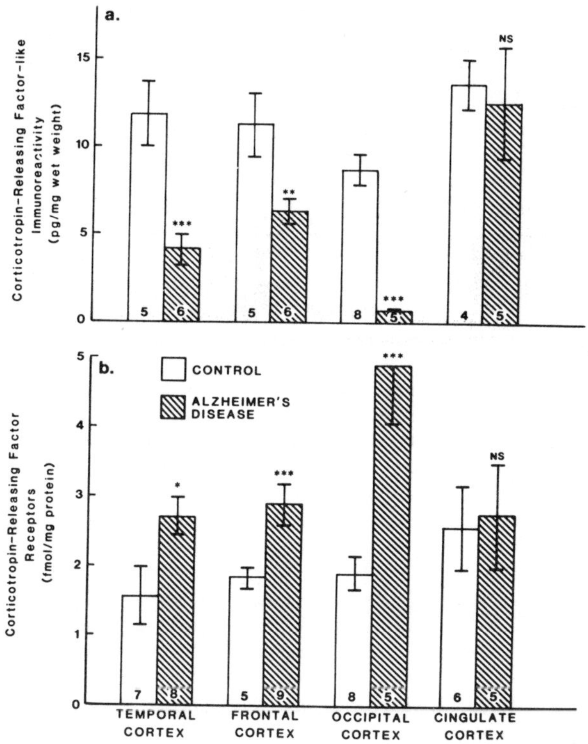

Fig. 4. CRH-like immunoreactivity (a) and CRH receptor binding (b) in discrete regions of the cerebral cortex of Alzheimer's patients and controls. All values are means ± S.E.M. The number of subjects in each group is given at the bottom of each histogram. Significant differences from control group at $p < 0.05$, $p < 0.025$ and $p < 0.005$ are denoted by *, ** and ***, respectively. NS, not significant (from De Souza et al., Ref. 27, with permission).

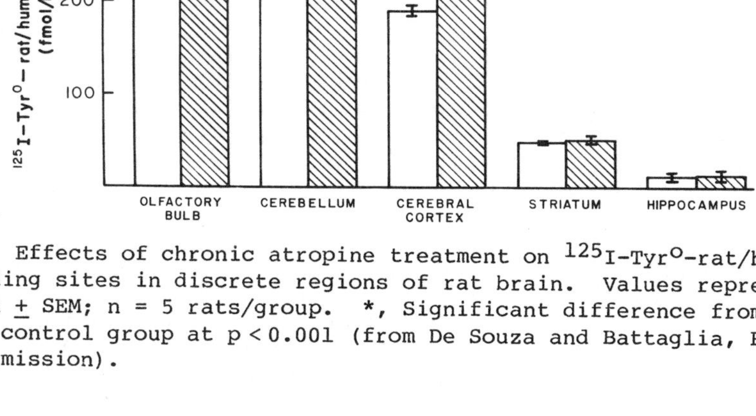

Fig. 5. Effects of chronic atropine treatment on ^{125}I-Tyr^{0}-rat/human CRH binding sites in discrete regions of rat brain. Values represent the mean ± SEM; n = 5 rats/group. *, Significant difference from saline-treated control group at $p < 0.001$ (from De Souza and Battaglia, Ref. 54, with permission).

the documentation of the presence of CRH-IR fibers abnormalities in senile plaques in amygdalae from individuals with Alzheimer's disease[52]. These observations suggest a possible role for CRH in the pathophysiology of Alzheimer's disease. These initial findings should stimulate further studies to investigate additional roles for brain CRH in other endocrine, psychiatric and neurologic disorders.

Interactions of CRH and Acetylcholine in Brain

Recent anatomical and behavioral studies have indicated interactions between CRH and acetylcholine in brain[53]. Specifically, colocalization of CRH and acetylcholinesterase was detected by retrograde tracing and immunocytochemical staining in some brain stem nuclei projecting to the frontal cortex[53]. In behavioral studies, CRH injected into the rat frontal cortex did not initiate, but significantly inhibited the effects of carbachol, a muscarinic cholinergic receptor agonist, on a sterotyped motor "boxing" behavior[53]. At present, the cerebral cortical cholinergic deficiency appears to be one of the most severe and consistent neurochemical deficits associated with Alzheimer's disease[54]. In order to examine possible interactions between cholinergic deficits in Alzheimer's disease and the alterations in CRH markers described above, choline acetyltransferase (ChAT) activity was correlated with CRH-IR and with CRH receptors in each cerebral cortical area examined. Significant positive relationships existed between ChAT activity and CRH-IR levels in the frontal ($r = +0.68$; $p < 0.05$), temporal ($r = +0.79$; $p < 0.05$) and occipital ($r = +0.74$; $p < 0.01$) cortex, and significant negative correlations were detected between ChAT activity and CRH receptors in the temporal ($r = -0.61$; $p < 0.05$) and occipital ($r = -0.77$; $p < 0.01$) cortex. The demonstration of significant correlations between changes in CRH and ChAT activity in human cortex, complimented by similar preliminary studies in rodents described above[53], suggests that CRH and cholinergic systems may interact.

To investigate the functional consequence of cholinergic deficits on the regulation of brain CRH neuronal activity, we designed a study in which rats were treated chronically with atropine (14 days, 20 mg/kg/day, s.c.), a muscarinic cholinergic receptor antagonist, and CRH receptors were measured in various regions of brain[55]. Chronic atropine treatment produced significant increases in muscarinic cholinergic receptors in the frontoparietal cortex (30% increase) and hippocampus (20% increase). No significant changes in the concentration of CRH binding sites were observed in olfactory bulb, cerebellum, striatum and hippocampus (Figure 5). In contrast, there was a significant and selective increase in CRH receptors in the frontoparietal cortex of atropine-treated rats (Figure 5). These results demonstrating a selective upregulation of CRH receptors in the cerebral cortex of atropine-treated rats suggests that the reported reciprocal changes in pre- and post-synaptic markers of CRH in the cerebral cortex of patients with Alzheimer's disease (see above) may be, in part, a consequence of deficits in the cholinergic projections to the cerebral cortex in Alzheimer's disease. Additional studies are necessary to determine the precise population(s) of cerebral cortical CRH receptors that may be modulated by acetylcholine, and to determine the functional significance of the interaction between CRH and cholinergic systems.

SUMMARY AND CONCLUSIONS

Studies with iodine-125-labeled analogues of CRH have identified, characterized, and localized binding sites for CRH in rat and human brain. In addition, we have demonstrated that CRH stimulates cAMP production in various regions of the rat CNS. The significant regional differences in the stoichiometric relationship between receptor number and receptor-mediated cAMP production suggests that some populations of CRH receptors in brain may be functionally coupled to alternative signal transduction mechanisms. The autoradiographic data demonstrate that the distribution of CRH binding sites in the rat CNS corresponds well with the immunohistochemical distribution of CRH pathways and pharmacological sites of action of CRH in some areas of brain but not in others. The demonstration of an upregulation of CRH receptors following a decrease in CRH-IR in the cerebral cortex in Alzheimer's disease indicates a physiological relevance of the receptor site and is consistent with the concept that CRH acts as a neurotransmitter in regulating normal cortical functions. In addition, the data suggest that disease of this peptidergic system may be important in certain clinical manifestations of Alzheimer's disease. The effects of chronic atropine treatment to selectively upregulate CRH receptors in the cerebral cortex suggests an interaction between CRH and acetylcholine. These data provide further support for the proposed role of CRH as a neurotransmitter in the CNS. Furthermore, these studies demonstrating the charactertistics of CRH receptors and CRH receptor-mediated signal transduction mechanisms in brain provide a means for better understanding the various functions of this neuropeptide under physiological and pathological conditions.

ACKNOWLEDGEMENTS

The work described in this chapter was conducted in collaboration with Michael J. Kuhar, Elizabeth L. Webster, Thomas R. Insel, Peter J. Whitehouse, Donald L. Price, Marilyn Perrin, Jean Rivier and Wylie Vale. We thank Terrie Pierce and Mary Flutka for manuscript preparation.

REFERENCES

1. W. Vale, J. Spiess, C. Rivier, and J. Rivier, Characterization of a 41-residue ovine hypothalamic peptide that stimulates secretion of corticotropin and beta-endorphin, Science 213:1394-1397 (1981).
2. W. Vale, C. Rivier, M. R. Brown, J. Spiess, G. Koob, L. Swanson, L. Bilezikjian, F. Bloom, and J. Rivier, Chemical and biolgical characterization of corticotropin-releasing factor, Rec. Prog. Horm. Res. 39:245-270 (1983).
3. E. B. De Souza, Corticotropin-releasing factor, Comp. Ther. 11:3-5 (1985).
4. J. Rivier, J. Spiess, and W. Vale, Characterization of rat hypothalamic corticotropin-releasing factor, Proc. Natl. Acad. Sci. USA 80:4851-4855 (1983).
5. S. Shibihara, Y. Morimoto, Y. Furutani, M. Notake, H. Takahashi, S. Shimizu, S. Horikawa, and S. Numa, Isolation and sequence analysis of the human corticotropin-releasing factor precursor gene, EMBO J. 2:775-779 (1983).
6. J. A. Olschowka, T. L. O'Donohue, G. P. Mueller, and D. M. Jacobowitz, The distribution of corticotropin-releasing factor-like immunoreactive neurons in rat brain, Peptides 3:995-1015.
7. S. Cummings, R. Elde, J. Ells, and A. Lindall, Corticotropin-releasing factor immunoreactivity is widely distributed within the central nervous system of the rat: An immunohistochemical study, J. Neurosci. 3:1355-1368 (1983).
8. L. W. Swanson, P.E. Sawchenko, J. Rivier, and W. W. Vale, Organization of ovine corticotropin-releasing factor immunoreactive cells and fibers in the rat brain: An immunohistochemical study, Neuroendocrinology 36:165-186 (1983).

9. M. R. Brown, L. A. Fisher, J. Spiess, C. Rivier, J. Rivier, and W. Vale, Corticotropin-releasing factor: Actions on the sympathetic nervous system and oxygen consumption, Life Sci. 30:207-210 (1982).
10. L. A. Fisher, G. Jessen, and M. R. Brown, Corticotropin-releasing factor (CRF): Mechanism to elevate mean arterial pressure and heart rats, Regul. Pept. 5:153-161 (1983).
11. Y. Tache, Y. Goto, M. W. Gunion, W. W. Vale, J. Rivier, and M. Brown, Inhibition of gastric acid secretion in rats by intracerebral injection of corticotropin-releasing factor, Science 222:935-937 (1983).
12. D. R. Britton, G. F. Koob, J. Rivier, and W. Vale, Intraventricular corticotropin-releasing factor enhances behavioral effects of novelty, Life Sci. 31:363-367 (1982).
13. J. E. Morley, and A. S. Levine, Corticotropin-releasing factor, grooming and ingestive behavior, Life Sci. 31:1459-1464 (1982).
14. R. E. Sutton, G. F. Koob, M. LeMoal, J. Rivier, and W. W. Vale, Corticotropin-releasing factor produces behavioral activation in rats, Nature 297:331-333 (1982).
15. H. D. Veldhuis, and D. DeWied, Differential behavioral actions of corticotropin-releasing factor (CRF), Pharmacol. Biochem. Behav. 21:707-713 (1984).
16. M. Smith, G. Bisette, T. A. Slotkin, D. L. Knight and C. B. Nemeroff, Release of corticotropin-releasing factor from rat brain regions in vitro, Endocrinology 118:1997-2001 (1986).
17. J. B. Aldenhoff, D. L. Gruol, J. Rivier, W. Vale, and G. R. Siggins, Corticotropin-releasing factor decreases post-burst hyperpolarizations and excites hippocampal pyramidal neurons in vitro, Science 221:875-876 (1983).
18. L. B. Eberly, C. A. Dudley, and R. L. Moss, Iontophoretic mapping of corticotropin-releasing factor (CRF) sensitive neurons in the rat forebrain, Peptides 4:837-841 (1983).
19. R. J. Valentino, S. L. Foote, and G. Aston-Jones, Corticotropin-releasing factor activates noradrenergic neurons of the locus ceruleus, Brain Res. 270:363-367 (1983).
20. P. C. Wynn, R. L. Hauger, M. C. Holmes, M. A. Millan, K. J. Catt, and G. Aguilera, Brain and pituitary receptors for corticotropin- releasing factor: Localization and differential regulation after adrenalectomy, Peptides 5:1077-1084 (1984).
21. F. M. Chen, L. M. Bilezikjian, M. H. Perrin, J. Rivier, and W. Vale, CRF receptor mediated stimulation of adenylate cyclase activity in rat brain, Brain Res. 381:49-57 (1986).
22. G. Battaglia, E. L. Webster, and E. B. De Souza, Characterization of corticotropin-releasing factor (CRF) receptor-mediated adenylate cyclase in rat central nervous system, Molecular Pharmacology (submitted).
23. P. W. Gold, G. Chrousos, C. Kellner, R. Post, A. Roy, P. Augerino, H. Schulte, E. Oldfield, and D. L. Loriaux, Psychiatric implications of basic and clinical studies with corticotropin-releasing factor, Am. J. Psychiatry 141:619-627 (1984).
24. E. B. De Souza, T. R. Insel, M. H. Perrin, J. Rivier, W. W. Vale, and M. J. Kuhar, Corticotropin-releasing factor receptors are widely distributed within the rat central nervous system: An autoradiographic study, J. Neurosci. 5:3189-3203 (1985).
25. E. B. De Souza, and M. J. Kuhar, Corticotropin-releasing factor receptors in the pituitary gland and central nervous system: methods and overview, Methods in Enzymol. 124:560-590 (1986).
26. E. B. De Souza, Corticotropin-releasing factor receptors in the rat central nervous system: Characterization and regional

distribution, J. Neurosci. (in press) (1987).

27. E. B. De Souza, P. J. Whitehouse, M. J. Kuhar, D. L. Price and W. W. Vale, Reciprocal changes in corticotropin-releasing factor (CRF)-like immunoreactivity and CRF receptors in cerebral cortex of Alzheimer's disease, Nature 319:593-595, 1986.
28. E. B. De Souza, T. R. Insel, M. H. Perrin, J. Rivier, W. W. Vale and M. J. Kuhar, Differential regulation of corticotropin-releasing factor receptors in anterior and intermediate lobes of pituitary and in brain following adrenalectomy in rats, Neurosci. Lett. 56:121-128 (1985).
29. E. B. De Souza, and M. J. Kuhar, Corticotropin-releasing factor receptors: Autoradiographic identification, in: "Neuropeptides in Neurologic and Psychiatric Disease", J. B. Martin and J. Barchas, Eds., Raven Press, New York, pp. 179-198 (1986).
30. E. B. De Souza, M. H. Perrin, T. R. Insel, J. Rivier, W. W. Vale, and M. J. Kuhar, Corticotropin-releasing factor receptors in rat forebrain: Autoradiographic Idendification, Science 224:1449-1451 (1984).
31. E. B. De Souza, M. H. Perrin, J. E. Rivier, W. W. Vale, and M. J. Kuhar, Corticotropin-releasing factor receptors in rat pituitary gland: Autoradiographic localization, Brain Res. 296:202-207 (1984).
32. M. C. Holmes, F. A. Antoni, and T. Szeintendnei, Pituitary receptors for corticotropin-releasing factor: No effect of vasopressin on binding or activation of adenylate cyclase, Neuroendocrinology 39:1001-1007 (1984).
33. P. C. Wynn, G. Aguilera, J. Morell, and K. J. Catt, Properties and regulation of high-affinity pituitary receptors for corticotropin- releasing factor, Biochem. Biophys. Res. Commun. 110:602-608 (1983).
34. E. B. De Souza, M. H. Perrin, P. J. Whitehouse, J. E. Rivier, W. W. Vale, and M. J. Kuhar, Corticotropin-releasing factor receptors in human pituitary gland: Autoradiographic localization, Neuroendocrinology 40:419-422 (1985).
35. M. A. Millan, D. M. Jacobowitz, R. L. Hauger, K. J. Catt and G. Aguilera, Distribution of corticotropin-releasing factor receptors in primate brain, Proc. Natl. Acad. Sci.(USA) 83:1921-1925 (1986).
36. D. R. Britton, D. K. Hoffman, K. Lederis, and J. Rivier, A comparison of the behavioral effects of CRF, sauvagine and urotensin I, Brain Res. 304:201-205 (1984).
37. R. A. Peterfreund, and W. W. Vale, Ovine corticotropin-releasing factor stimulates somatostatin secretion from cultured brain cells, Endocrinology 112:1275-1278 (1983).
38. J. Rivier, J. Spiess, C. Rivier, R. Galyean, and W. W. Vale, Solid phase synthesis of amunine (CRF), sauvagine and urotensin I, in: "Peptides", K. Blaha and P. Malon, eds., Wide Gruyter, Berlin, pp. 597-602 (1982).
39. M. J. Kuhar, The mismatch problem in receptor mapping studies, Trends Neurosci. 8:190-191 (1985).
40. D. J. S. Sirinathsinghji, L. H. Rees, J. Rivier, and W. W. Vale, Corticotropin-releasing factor is a potent inhibitor of sexual receptivity in the female rat, Nature 305:232-235 (1983).
41. C. R. Rivier, and W. Vale, Corticotropin-releasing factor (CRF) acts centrally to inhibit growth hormone secretion in the rat, Endocrinology 114:2409-2411 (1984).
42. N. Ono, M. D. Lumpkin, W. K. Samson, J. K. McDonald and S. M. McCann, Intrahypothalamic action of corticotropin-releasing factor (CRF) to inhibit growth hormone and LH release in the rat, Life Sci. 35:1117-1123 (1984).

43. C. R. Rivier, and W. Vale, Influence of corticotropin-releasing factor on reproductive functions in the rat, Endocrinology 114:914-921 (1984).

44. P. M. Plotsky, T. O. Bruhn, and S. Otto, Central modulation of immunoreactive arginine vasopressin and oxytocin secretion into the hypophysial-portal circulation by corticotropin-releasing factor, Endocrinology 116:1669-1671 (1985).

45. S. P. Grossman, Role of the hypothalamus in the regulation of food and water intake, Psychol. Rev. 82:200-224 (1975).

46. D. Hamel, and Y. Tache, Intrathecal (I.T.) injection of bombesin and rat CRF inhibits gastric acid secretion in rats, Soc. Neurosci. Abstr. 10:812 (1984).

47. C. L. Ehlers, S. J. Henriksen, M. Wang, J. Rivier, W. W. Vale, and F. E. Bloom, Corticotropin-releasing factor produces increases in brain excitability and convulsive seizures in rats, Brain Res. 278:332-336 (1983).

48. K. T. Britton, M. Lyon, W. Vale, and G. F. Koob, Stress-induced secretion of corticotropin-releasing factor immunoreactivity in rat cerebrospinal fluid, Soc. Neurosci. Abstr. 10:94 (1984).

49. M. R. Brown, L. A. Fisher, V. Webb, W. W. Vale, and J. E. Rivier, Corticotropin-releasing factor: A physiological regulator of adrenal epinphrine secretion, Brain Res. 328:355-357 (1985).

50. G. Bissette, G. P. Reynolds, C. D. Kilts, E. Widerlov, and C. B. Nemeroff, Corticotropin-releasing factor-like immunoreactivity in senile dementia of the Alzheimer type, JAMA 254:3067-3069 (1985).

51. M. M. Mouradian, J. M. Farah, T. L. O'Donhue, and T. N. Chase, Decreased immunoreactive corticotropin-releasing factor (CRF) in CSF of Alzheimer's disease, Soc. Neurosci. Abstr. 12:98 (1986).

52. D. L. Price, R. E. Powers, L. C. Walker, R. G. Struble, P. J. Whitehouse, W. W. Vale, and E. B. De Souza, Corticotropin-releasing factor immunoreactivity in senile plaques, Soc. Neurosci. Abstr. 12:98 (1986).

53. J. N. Crawley, J. A. Olschowka, D. I. Diz, and D. M. Jacobowitz, Behavioral investigation of the coexistence of substance P, corticotropin-releasing factor, and acetylcholinesterase in lateral dorsal tegmental neurons projecting to the medial frontal cortex of the rat, Peptides 6:891-901 (1985).

54. J. T. Coyle, D. L. Price, and M. R. De Long, Alzheimer's disease: A disorder of cortical cholinergic innervation, Science 219:1184-1190 (1983).

55. E. B. De Souza and G. Battaglia, Increased corticotropin-releasing factor receptors in rat cerebral cortex following chronic atropine treatment, Brain Res. 397:401-404 (1986).

STUDIES ON THE SECRETION OF CORTICOTROPIN-RELEASING FACTOR INTO HYPOPHYSIAL PORTAL BLOOD

V. Guillaume, M. Grino, G. Peyre, B. Conte-Devolx, and C. Oliver

Laboratoire de Neuroendocrinologie Expérimentale
Faculté de Médecine Nord, INSERM U 297, Bd P. Dramard
13326 Marseille, Cedex 15, France

ACTH secretion by the anterior pituitary gland is controlled by several regulatory factors of hypothalamic, neurohypophysial and peripheral origins. The ACTH-regulatory factors include corticotropin-releasing factor (CRF), arginine-vasopressin (AVP), oxytocin, epinephrine and angiotensin II[1]. There is now evidence that CRF is the major physiologic regulator of ACTH secretion[2]. Therefore, the measurement of CRF release into hypophysial portal blood (HPB) may provide important informations on the mechanisms of regulation of ACTH secretion. However, HPB is collected under special experimental conditions that may alter the physiologic release of CRF. In this report, we will first discuss technical problems in CRF measurements in HPB and secondly review data available on the secretion of CRF into HPB including data from our laboratory.

METHODOLOGICAL ANALYSIS OF CRF MEASUREMENT INTO HYPOPHYSIAL PORTAL BLOOD

HPB has been first collected in dogs in 1956, blood from the pituitary stalk being aspirated in order to measure CRF bioactivity[3]. Thereafter, HPB has been collected mostly in rats either from the entire pituitary stalk or from a single portal vessel, using a glass microcannula. The latter method has not been used for studying CRF secretion into HPB since it does not allow the collection of a sufficient amount of blood. HPB has also been collected in other species such as rabbits, monkeys and sheep.

The procedure for collecting HPB in the rat is briefly described. Rats (250-350 grams, b.w.) are firts anesthetized; then, a cannula is inserted into the trachea and the rat's head is place into a special holder with adjustable ear bars. The femoral artery and vein are catheterized. All subsequent manipulations are performed with the aid of a binocular microscope. After dissection of all cervical muscles, the basisphenoid bone which covers the ventral surface of the hypothalamo-pituitary complex is exposed with the help of several retractors. Thereafter, the basisphenoid bone is progressively thinned using a drill and removed, allowing the exposure of the median eminence, pituitary stalk and anterior pituitary gland. The dura covering the pituitary gland is cut and flop over. After breaking the venous sinuses, the entire pituitary gland is carefully removed from the turcica fossa. After an acute injection of heparin (100 units into the femoral vein), the pituitary stalk is cut and the pituitary gland taken out[4]. At this stage, there are two ways for collecting HPB from the pituitary stalk:

- the cannulation method : a cannula made of polystyrene tubing is placed over the pituitary stalk and blood is collected at the rate of 7µl/min, using an aspiration pump. There is no contamination of HPB by blood from other sources. However, blood is not immediately chilled and remains about 4 min at room temperature, a period that is sufficient for degradation of neuropeptides[5].

- the aspiration method : every 10 seconds, blood is mannually collected from the pituitary stalk with a polystyrene cannula and placed into a chilled tube. This method lowers the risks of inactivation of neuropeptides. However, it may offer the risk of contamination of HPB by blood from other sources (about 32%). The rate of collection ranges between 10 and 14ul per minute. As a consequence, the collection of one ml plasma needs a 5-hour period collection[6].

The surgical procedure needed before starting of HPB collection requires 30 to 45 minutes and is highly stressful. Indeed, at the end of the surgical period, plasma ACTH levels are very high (2636 ± 973 pg/ml), then decrease (654 ± 328 pg/ml) and remain stable at levels that are much higher than basal values in intact, unstressed rats (129 ± 10 pg/ml). Therefore, it is most appropriate to begin the collection of HPB 60 minutes after the end of the preparative surgical procedure. Besides, one must keep in mind that this collection is performed after 30 to 45 minutes period of high and sustained stress.

The anesthesia and the stressful surgical procedure may also influence the metabolic conditions of the animals and especially their blood glucose levels. When the rats have access to food during the night preceeding the collection of portal blood, blood glucose levels increase from 1 g/l to 3-4 g/l during surgery, under any kind of anesthesia. However, glycaemia remains stable and within the normal range (0.8-1.2 g/l), when food is withdrawn from the rats overnight, before the operation (G. Peyre, personnal observations).

During the period of collection of HPB, the concentrations of neuropeptides may be artefactually lowered through inactivation by blood proteolytic enzymes or adsorption on tubes or cannulas. Unlike other neuropeptides, immunoreactive CRF is stable for at least 2 hours, when incubated in blood or plasma at 0°C or 37°C. However, CRF is rapidly adsorbed on glass, polystyrene or polypropylene tubes. In all experiments, HPB must be appropriately treated, right after the end of collection :

- centrifugation at 0°C
- collection and extraction of plasma in 2 vol cold acetone; then, the acetone extract is washed with 2 vol petroleum ether
- evaporation in a Speed-Vac concentrator. The dried extracts are stored at -20°C until the assay which is always performed within 10-days.

This extraction method allows a high and reproducible recovery of synthetic 41-rat CRF (76,5% ± 2,3%)[7].

CRF SECRETION INTO HYPOPHYSIAL PORTAL BLOOD

In 1982, Gibbs and Vale have first reported the presence of CRF in HPB at concentrations much higher (104,9 ± 0,7 pm) than in systemic blood (<3.8 pm). The levels of CRF in portal blood are sufficient for stimulating the release of ACTH. When analyzed on H.P.L.C. immunoreactive CRF in portal plasma extracts shares the same retention time than synthetic and hypothalamic CRF[8]. Four percent of the hypothalamic CRF content are released in HPB within one-hour. The percentage of release of the other hypothalamic peptides into HPB is 5% percent for TRH, 1%-4% percent for GRF, 0,1-0,3 percent for LH-RH and 0,3 percent for Somatostatin[7].

Several studies have been performed on the regulation of CRF secretion in HPB.

1. Effect of monoamines and peptides

The administration of serotonin stimulates ACTH secretion in the rat. Fluoxetine is a serotonin agonist that inhibits serotonin neuronal uptake, then increasing serotoninergic activity. The acute injection of Fluoxetine stimulates the release of ACTH and Gibbs and Vale have shown that this effect is mediated through a CRF increase[9].

The intraventricular administration of AVP (from 0.5 to 10 pmol) lowers the release of CRF. This effect may not only be pharmacological, but also physiological since the injection of an antiserum anti-AVP into the lateral ventricle is followed by an increased CRF release into HPB[10].

The injection of synthetic ovine CRF into the lateral ventricle does not influence the release of endogenous CRF (measured by a specific rat CRF radioimmunoassay which does not cross-react with ovine CRF). This result does not support the existence of an ultra-short feed-back of CRF on CRF-secreting neurons. It must be noticed that the injection of CRF decreases AVP and oxytocin concentrations into HPB of the same rats[11].

2. Effect of various stress

Three different stress have been investigated and surimposed to the surgical procedure.

The acute hemorrhagic stress (loss of 15 percent of blood volume) induces a 4.5-fold increase in plasma ACTH levels and a two-fold increase in CRF concentrations into HPB. Both ACTH- and CRF-increased secretion are blocked by the previous administration of dexamethasone (100 μg/kg i.m. 6 hours before)[12].

In opposition, the changes in plasma ACTH levels after hypothermia and hypoglycemia may not be related to variations in CRF secretion. For example, if the rats are anesthetized there is a diminution of circulating ACTH after hypothermia. Under the same conditions, CRF release does not change and the effect of hypothermia on the corticotrophs has been attributed to the decreased release of AVP and oxytocin[13]. The insulin-hypoglycemia is followed by an increased secretion of ACTH. No changes in CRF and oxytocin concentrations in portal blood have been observed; an 1.9- fold increase in AVP release has been demonstrated by Plotsky et coll. under the same conditions[14].

3. Effect of glucocorticoids

Since changes in plasma glucocorticoid levels are followed by modifications in plasma ACTH, it was of interest to determine the effect of glucocorticoids on the hypothalamic secretion of CRF.

- Adrenalectomy is followed by a rapid increase in plasma ACTH levels and a decrease in the hypothalamic CRF content. Thereafter, hypothalamic CRF levels return to initial values within one week. However, we did not observe any significant changes in CRF levels in HPB at all stages (4 hours, 1 day, 8 days) investigated after adrenalectomy[7]. These data may be explained by one of the following :

- an increased CRF release after adrenalectomy may have been overlooked because of the stimulation of afferent impulses to the CNS and the "artefactual" high CRF secretion due to the stress.

- There is, indeed, no CRF increased secretion after adrenalectomy. It must be noticed that TRH and LH-RH release in HPB do not change after thyroidectomy and gonadectomy, respectively[7]. The effect of the lack of glucocorticoids on ACTH secretion may be exerted mainly at the level of the pituitary gland.

CRF levels in HPB under the stressful conditions of portal blood collection does not decrease when corticosterone (10-100 µg/ml) or betamethasone (1-10 µg/ml) are given in drinking water for several days before the experiment. When a very high dose of dexamethasone (400 µg, 100 grams b.w.) is injected i.p. 6 hours before, a 60% decrease in CRF levels in HPB is observed[7]. Plotsky et al. have recently determined CRF concentrations in HPB of rats receiving exogenous corticosterone I.V., after suppression of their own endogenous adrenal secretion. When the infusion of corticosterone gives plasma corticosterone levels ranging between 8 and 12 µg/dl, the stimulated CRF release induced by nitroprusside is suppressed. Plasma corticosterone levels must reach 40 µg/dl for blocking the basal CRF release[15]. These data indicate that glucocorticoid feedback on ACTH secretion may involve the hypothalamus. High glucocorticoid doses are necessary for attenuating the basal secretion of CRF. The doses of glucocorticoids needed for attenuating CRF release after hemodynamic stress are significantly lower.

4. Effect of the endogenous catecholaminergic system

It has been shown that catecholamines stimulate ACTH release through a direct action at the pituitary level. However, the role of central catecholamines on ACTH secretion remains highly controversial. Several pharmacological studies using catecholamine agonists and antagonists have led to conflicting results, the central effect of catecholamines on ACTH release being either stimulating or inhibiting. Recenlty, Szafarczyk et al. have reported that the depletion of endogenous hypothalamic epinephrine and norepinephrine (after bilateral stereotaxical injection of 6-hydroxydopamine into the ventral noradrenergic ascending bundle) is accompanied by a reduction in both basal and ether-induced ACTH release and by a decrease CRF immunostaining in the paraventricular nucleus[16]. In collaboration with the group of Szafarczyk and Assenmacher, we have found that a similar injection of 6-hydroxydopamine into the ventral noradrenergic ascending bundle induces a reduction in surgical stress- induced ACTH secretion. In the same animals, there is a significant decrease in CRF levels in HPB with a direct correlation between the magnitude of CRF decrease in HPB and the diminution of epinephrine and norepinephrine hypothalamic content[17]. No changes in AVP levels in HPB was observed in these rats (unpublished personnal observations). These results suggest that the endogenous central catecholamines have a stimulating effect on the secretion of CRF by the hypothalamus. The method of collection of HPB should be of help determine the respective role of epinephrine and norepinephrine in the control of catecholaminergic receptors involved in this regulation.

CONCLUSION

Although it has some methodological limitations due to the anesthesia and surgical stress, the measurement of CRF release in hypophysial portal blood has already provided useful informations about the interactions neurotansmitters - CRF neurones, the role of various stress on CRF release and the site(s) of action of glucocorticoid feedback on ACTH secretion. Therefore, the method of HPB collection, which is now the only way to quantify in vivo CRF secretion, remains a valuable tool for investigating the regulation of ACTH secretion.

ACKNOWLEDGMENTS

Our studies on CRF secretion have been made possible with the support of INSERM, Fondation pour la Recherche Médicale Française and University of Medicine of Marseilles. The authors thank Ms R. Quérat for her excellent secretarial assistance.

REFERENCES

1. J. Axelrod and T.D. Reisine, Stress hormones : their interaction and regulation, Science 224 : 452 (1984)
2. C.L. Rivier and P.M. Plotsky, Mediation by corticotropin-releasing factor (CRF) of adenohypophysial hormone secretion, Ann. Rev. Physiol. 48 : 475 (1986)
3. J.C. Porter and J.C. Jones, Effect of plasma from hypophyseal-portal vessel blood or adrenal ascorbic acid, Endocrinology, 58:62 (1956)
4. J.C. Porter, Methods for studying pituitary-hypothalamic axis in situ, in: "Methods in Enzymology", J.G. Hardman and B.W. O'Malley, eds, Academic Press, New York (1975)
5. J.C. Porter and K.R. Smith, Collection of hypophysial stalk blood in rats, Endocrinology, 81 : 1182 (1967)
6. D.M. Gibbs, Collection of pituitary portal blood : a methodologic analysis, Neuroendocrinology, 38 : 97 (1984)
7. B. Conte-Devolx, Rôle de la corticolibérine (CRF) dans la régulation de la fonction corticotrope chez le rat in vivo, Thèse de Doctorat (Sciences) Université de Montpellier II, France (1985)
8. D.M. Gibbs and W. Vale, Presence of corticotropin-releasing factor-like immunoreactivity in hypophysial portal blood, Endocrinology, 111 : 1418 (1982)
9. D.M. Gibbs and W. Vale, Effects of the serotonin receptake inhibitor fluoxetine on corticotropin-releasing factor and vasopressin secretion into hypophyseal portal blood, Brain Res. 280 : 176 (1983)
10. P.M. Plotsky, T.O. Bruhn and W. Vale, Central modulation of immunoreactive corticotropin-releasing factor secretion by arginine vasopressin, Endocrinology, 115 : 1639 (1984)
11. P.M. Plotsky, T.O. Bruhn and S. Otto, Central modulation of immunoreactive arginine vasopressin and oxytocin secretion into the hypophysial-portal circulation by corticotropin-releasing factor, Endocrinology, 116 : 1669 (1985)
12. P.M. Plotsky and W. Vale, Hemorrhage-induced secretion of corticotropin-releasing factor-like immunoreactivity into the rat hypophysial portal circulation and its inhibition by glucocorticoids, Endocrinology, 114 :164 (1984)
13. D.M. Gibbs, Inhibition of corticotropin-release during hypothermia : the role of corticotropin-releasing factor, vasopressin and oxytocin, Endocrinology, 116 : 723 (1985)
14. P.M. Plotsky, T.O. Bruhn and W. Vale, Hypophysiotropic regulation of adrenocorticotropin secretion in response to insulin-induced hypoglycemia, Endocrinology, 117 : 323 (1985)
15. P.M. Plotsky, S. Otto and R.M. Sapolsky, Inhibition of immunoreactive corticotropin-releasing factor secretion into the hypophysial portal circulation by delayed glucocorticoid feedback, Endocrinology, 119 : 1126 (1986)
16. A. Szafarczyk, G. Alonso, G. Ixart, F. Malaval and I. Assenmacher, Diurnal-stimulated and stress-induced ACTH release in rats is mediated by ventral noradrenergic bundle, Am. J. Physiol. 249 : E219 (1985)
17. V. Guillaume, B. Conte-Devolx, A. Szafarczyk, F. Malaval, N. Pares-Herbute, M. Grino, G. Alonso, I. Assenmacher and C. Oliver, CRF release in rat hypophysial portal blood is mediated by brain catecholamines, Neuroendocrinology, (1987) (Accepted for publication)

IS THERE A "FINAL COMMON PATHWAY" IN THE REGULATION OF ACTH RELEASE?

Mark H.Whitnall, with the technical assistance of Sharon Key

Laboratory of Neurochemistry and Neuroimmunology, NICHD
Bethesda, MD[1]

INTRODUCTION

Roles of CRF and AVP in Regulating ACTH Release

Two hypothalamic peptides which appear to play significant roles in the control of ACTH secretion during stress are the 41 amino acid corticotropin-releasing factor (CRF)[1-4] and vasopressin (AVP)[2-4]. The two substances have at least additive, if not synergistic effects in stimulating ACTH release from the anterior pituitary in vivo[5,6] and in vitro[7-11]. Stimulated ACTH release in vivo can be inhibited by the administration of antisera or antagonists to either CRF[12-15] or AVP[6,14]. Distinct sets of high affinity receptors have been characterized in the anterior pituitary for CRF[16-18] and for AVP[19-23]. Although the population of neurons secreting CRF into portal plasma has often been thought of as a "final common pathway" for stimuli influencing ACTH levels[24], the existence of multiple releasing factors (e.g., CRF and AVP) has led to the speculation that these releasing factors may be involved in the independently regulated components of the system regulating ACTH release during stress[24]. This speculation is supported by studies of changes in levels of CRF and AVP in portal plasma during stress.

Levels of CRF and AVP have been measured in portal blood at levels of 0.1 nM and 1 nM, respectively[25-27]. Changes in peripheral ACTH levels caused by stress have been correlated with parallel changes in portal levels of CRF and AVP[26,27]. In some circumstances, however, the portal level of one of these peptides remains constant while the other changes, indicating that there are multiple independently regulated pools of CRF and AVP contributing to portal plasma. For example, in anesthetized rats rendered hypothermic (core temperature brought to 30°C), ACTH levels decreased by 41% within one hour, while portal AVP decreased by 35% and CRF remained constant[28]. Another example is insulin-induced

---

[1] Present address: Laboratory of Neurochemistry, NINCDS, Bldg. 36, Rm. 4D20, Bethesda, MD 20892

hypoglycemia in rats: 30 min after administration of insulin in these experiments, peripheral glucose had decreased by 70%, peripheral ACTH had risen 2.5 times, and portal AVP had increased 1.8 times, while portal CRF did not change[27]. On the other hand, there are situations in which portal CRF levels change while portal AVP levels remain constant: during nitroprusside-induced hypotension in rats, when blood pressure decreased by 50%, peripheral ACTH rose to twice control, and portal CRF increased 1.4 times, while portal AVP levels did not change[29].

Measurements of peptide levels in portal blood must be interpreted with caution, due to the potential artifacts caused by unavoidable damage to the median eminence (ME) and portal vessels during the collection of portal blood[30]. With these caveats in mind, the above studies can be taken as providing preliminary support for the hypothesis of independently regulated pools of CRF and AVP contributing to portal plasma during stress.

Possible Sources of Independently Regulated CRF and AVP

Some theoretical possibilities for the organization of pools of CRF and AVP contributing to portal blood are depicted in figure 1. If both CRF and AVP were contained within the same neurons, but in separate populations of secretory vesicles, their release could be modulated independently if specific action potential frequencies affected exocytosis of the two vesicle populations differently (fig. 1A). Such a mechanism has been proposed for coexisting norepinephrine and neuropeptide Y in the vas deferens[31]. If, on the other hand, every vesicle in a cell contained both CRF and AVP, then the release of CRF and AVP from that

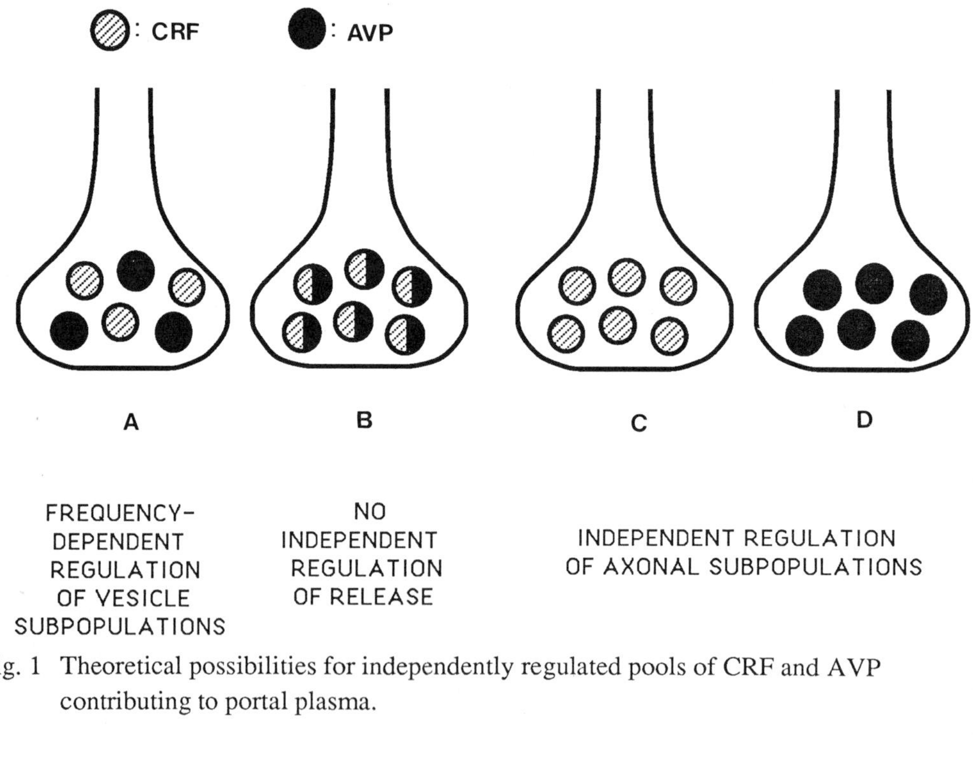

Fig. 1 Theoretical possibilities for independently regulated pools of CRF and AVP contributing to portal plasma.

cell could not be regulated independently: they would always be co-secreted (fig. 1B). However, if subpopulations of cells existed with either CRF or AVP in all of their secretory vesicles (i.e., possibilities C and D in fig. 1), then the release of the two peptides could be controlled independently via separate regulation of the activity of the two axonal subpopulations.

The latter model (C and D) has been proposed by Holmes et al[32]. In this model it is conceived that the two independently regulated pools of CRF and AVP consist of the parvocellular neurosecretory axons in the external zone (EZ) of the ME, and the magnocellular neurosecretory axons which pass through the internal zone of the ME and terminate in the posterior pituitary. According to this proposal, CRF would be released from the parvocellular axons and AVP would be released from the magnocellular axons (i.e., fig. 1C would represent the parvocellular axons and fig. 1D would represent the magnocellular axons). The AVP from the magnocellular axons would reach portal plasma either by backflow through capillaries from the posterior pituitary[33] or by release from preterminal axons in the internal zone of the ME[32]. The appeal of the magnocellular neurons as the source of the independently regulated AVP in portal plasma is that they are spatially and functionally separate from the CRF parvocellular neurons[34,35]. However, some problems with this hypothesis are that backflow of blood from the posterior pituitary to the portal system is difficult to demonstrate convincingly (e.g., removal of the posterior pituitary does not change portal levels of AVP[36]), and the in vitro studies of release from the internal zone of the ME[32] involve damaged MEs under highly unnatural conditions.

An alternative hypothesis is that the source of independently regulated AVP in portal plasma is from the parvocellular CRF neurons themselves. AVP is present in a population of CRF neurons in the paraventricular nucleus of the hypothalamus which project to the portal capillaries in the external zone (EZ) of the ME (see citations in refs. 35 and 37). Although AVP and AVP mRNA have only been observed in the CRF-containing parvocellular perikarya in adrenalectomized rats, i.e., after removal of glucocorticoid feedback inhibition[39-43], the EZ of the ME of normal rats contains immunocytochemically detectable AVP[38,44] and neurophysin (NP)[38,45]. Hence, the AVP in the CRF neuron endings in the EZ of the ME is another potential source of AVP in portal plasma. However, independently regulated release of CRF and AVP from these cells would only be possible if subpopulations of parvocellular neurons could be demonstrated that contained different ratios of CRF and AVP. Preliminary evidence suggested the presence of such subpopulations in normal rats, where approximately half of the CRF axons in the EZ of the ME in normal rats contained AVP and AVP-associated NP (NP-AVP), while the other half of the CRF axons contained little or no detectable AVP or NP-AVP[38]. In the CRF+/AVP+ axons, CRF, AVP and NP-AVP were found to be co-packaged into the same 100 nm neurosecretory vesicles[38]. The CRF+/AVP+ and CRF+/AVP- subpopulations therefore correspond to possibilities B and C, respectively, in fig. 1. In view of the potential roles of these subpopulations in modulating the ratio of AVP to CRF in portal blood, we felt it important to firmly establish their existence in normal rats, and to examine the responses of these subpopulations to manipulations of glucocorticoid levels. We have now quantified the CRF+/AVP+ and CRF+/AVP- axonal subpopulations in normal rats, and after adrenalectomy (ADX), with and without glucocorticoid replacement therapy.

AVP-CONTAINING AND AVP-DEFICIENT COMPONENTS OF THE CRF PROJECTION TO THE PORTAL CAPILLARIES

Quantification of Axonal Subpopulations in Normal Rats

We had previously shown that CRF and AVP (and NP-AVP) are co-packaged into the same neurosecretory vesicles in some of the CRF axons in the EZ of the ME in normal rats[38]. In the present study[46], we used post-embedding electron microscopic (EM) immunocytochemistry (ICC) on serial sections through the normal rat ME, and compared the distribution of CRF immunoreactivity to immunoreactivity for all three portions of the AVP precursor: AVP, NP-AVP, and the carboxy-terminal glycopepide (GP). The use of antibodies to three different portions of the AVP precursor minimized the chance of error due to antibody cross-reactivity, and allowed us to obtain multiple independent measures of the numbers of AVP+ and AVP- axonal profiles. This was important in establishing the validity of the concept of axonal subpopulations (see Discussion).

Adult male rats (Sprague-Dawley), 200-250 g, obtained from Zivic-Miller Labs, Allison Park, PA were used in these studies. Median eminences were fixed in glutaraldehyde[48], embedded in LR White[47], then sectioned and immunostained for EM using the avidin-biotinylated peroxidase method as previously described[48,49].

The CRF antiserum ("rC70"), obtained from W. Vale, has been characterized elsewhere[38,50]. The AVP antisera were donated by S. Glick, and by G. Valiquette and E. Zimmerman, and were described previously[38,49]. The monoclonal antibodies for NP (PS 36 for oxytocin-associated NP, NP-OT, and PS 45 for NP-AVP) were described previously[38,48]. Whereas PS 36 recognized only NP-OT in EM ICC, PS 45 stained both NP-AVP and NP-OT. However, the parvocellular CRF axons described in the present study do not contain OT or NP-OT (see below). The antiserum to OT ("VA 10") was generated in our laboratory, and has been found to be highly specific for OT by RIA and ICC (unpublished data). The GP antiserum ("GlyR1") was donated by D. Smyth, and has been found to be specific by RIA and by LM ICC[51].

Parvocellular axons in the EZ of the ME, and magnocellular (hypothalamo-neurohypophysial) axons in the internal zone of the ME, are readily distinguished at the ultrastructural level because of the fact that the parvocellular axons contain 100 nm dense core vesicles and the magnocellular axons contain 160 nm dense core vesicles[38,52-56]. The axons in the EZ of the ME containing 100 nm vesicles never stained for OT or NP-OT (unpublished results). AVP-containing axons could be unequivocally identified by comparing results obtained with the two AVP antisera, the monoclonal antibody which stained NP-AVP in these axons, and the antiserum to GP. Serial section studies showed that all of these antibodies were labeling all 100 nm vesicles in an identical set of parvocellular axons in the EZ of the ME[46]. The CRF antiserum never stained magnocellular axons in the internal zone of the ME, demonstrating that the CRF antiserum did not crossreact with any of the peptides present within magnocellular axons, including AVP, NP-AVP, GP, OT, and NP-OT[38].

When the distribution of axons labeling for CRF was compared to the distribution of axons labeling for AVP, NP-AVP or GP, it was found that many of the CRF axons contained AVP precursor-derived peptides, with CRF and the AVP precursor-derived peptides copackaged into the same secretory vesicles (fig. 2). In this group of CRF axons, staining for the

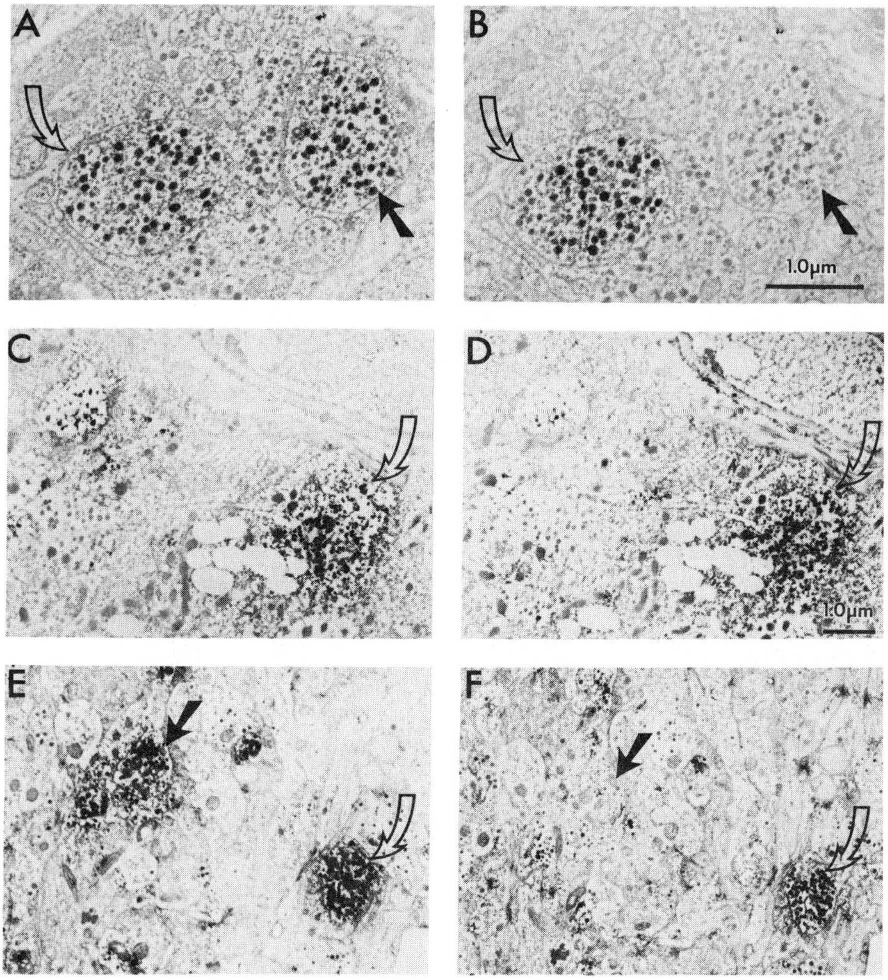

Fig. 2. Serial pairs of thin sections comparing staining for CRF (A,C,E) to that for NP-AVP (B), AVP (D), and GP (F). Although virtually all axons staining for AVP precursor-derived peptides also stain strongly for CRF (open arrows), some CRF-positive axons do not stain for AVP precursor-derived peptides (solid arrows). E and F are printed at the same magnification as C and D. From ref. 46.

Table 1. Percentage of CRF axons in the EZ of the ME labeled with antibodies to AVP precursor-derived peptides

Animal	α AVP	α NP-AVP	α GP	Pooled data
	% CRF axons labeled (total number of CRF axons analyzed)			
1	45.8 (120)	50.3 (147)	35.9 (156)	43.7 (423)
2	30.8 (120)	35.5 (183)	42.6 (223)	37.5 (526)
3	68.5 (184)	58.0 (417)	52.4 (466)	57.4 (1067)
4	----	46.3 (108)	----	46.3 (108)
5	----	40.5 (79)	----	40.5 (79)
6	83.9 (149)	75.7 (214)	78.7 (174)	79.0 (537)
7	63.1 (103)	37.8 (185)	42.4 (250)	44.8 (538)
Mean ± SE	58.4 ± 9.2	49.2 ± 5.3	50.4 ± 7.5	49.9 ± 5.4
Total axons	676	1333	1269	3278
Total Animals	5	7	5	7

(data taken from ref. 46)

AVP precursor-derived peptides was intense (fig. 2). However, another group of CRF axons, intermingled with the first group in the EZ of the ME, was unlabeled for AVP precursor-derived peptides (fig. 2). In this AVP-deficient subpopulation of CRF axons, none of the vesicles were stained for AVP, NP-AVP or GP, or one or two vesicles in an axon were so faintly labeled as to make it difficult to decide whether there was true staining present (e.g., the axonal swelling identified with a solid arrow in fig. 2B). Less than 0.5% of the parvocellular axons labeled for AVP, NP-AVP or GP were unlabeled for CRF; these axons were not included in the quantitative analysis shown in table 1.

When these results were quantified in terms of numbers of axonal swellings staining for CRF only or CRF plus AVP, NP-AVP or GP (table 1), about half of the CRF axons in the EZ of the ME were identified as containing AVP precursor-derived peptides. There was little variation between antibodies (58% for AVP, 49% for NP-AVP, and 50% for GP), or between animals (see standard errors in table 1).

Thus, the CRF axons projecting to the portal capillaries in the EZ of the ME can clearly be classified into AVP-containing and AVP-deficient subpopulations. In order to account for the different physiological components regulating ACTH during stress, the subpopulations should be under independent regulation. As a first step in determining whether the CRF+/AVP+ and CRF+/AVP- subpopulations respond differently to physiological stimuli, we studied[57] their responses to a severe challenge: surgical stress combined with total removal of glucocorticoid feedback inhibition, i.e., adrenalectomy (ADX).

Quantification of Axonal Subpopulations in Adrenalectomized Rats, with and without Dexamethasone

The methods used were similar to those outlined in the previous section, except that in some cases animals were bilaterally adrenalectomized during chloral hydrate anesthesia.

Table 2. AVP content of CRF axons in the EZ of the ME

Condition	% CRF-positive axonal profiles containing AVP[a,b]	# axonal profiles analyzed	# animals (n)
Normal	46.8 ± 4.4	4062	10
ADX[c]	91.0 ± 2.5*	2340	9
Normal + 6 h DEX	51.6 ± 9.0	616	2
ADX[d] + 6 h DEX	98.0 ± 1.3*	461	3
Sham[d] + 4 d DEX	56.4	443	1
ADX[d] + 4 d DEX	59.8 ± 5.4	1141	3

[a] Values are the mean ± SE

[b] The results obtained using antibodies to AVP, NP-AVP and GP were similar and were pooled.

[c] The values for 3 d, 4 d, and 14 d ADX animals were similar and were pooled.

[d] Fixed 4 d after surgery

* Values significantly different from Normal in same treatment group ($p < 0.005$)

(data taken from ref. 57)

These animals were given 0.9% saline to drink throughout the period between surgery and fixation. Sham operations were performed by separating the back muscles and closing without exposing the kidneys. Some animals were given daily IP injections of a potent glucocorticoid agonist, dexamethasone (DEX, 600 µg) suspended in mineral oil, starting immediately after the operation.

In these experiments[57], the percentage of CRF+ axons falling into the AVP+ category (where AVP+ is defined as staining positively for AVP, NP-AVP or GP) was again about 50% in normal animals (table 2). Three to 14 days after bilateral ADX, however, the percentage of CRF+ axons which contained AVP had risen to greater than 90% (table 2). Treating ADX animals with DEX thoughout the period between surgery and fixation prevented the change in the percentage of CRF+ profiles falling into the CRF+/AVP+ category (table 2). However, treating ADX animals with only a single injection of DEX, 6 hours before fixation, had no effect (table 2). Treating normal animals with DEX for 6 hours, or performing a sham operation and treating with DEX for 4 days, had no effect on the percentage of CRF+ profiles containing AVP (table 2)

In order to determine whether the total numbers of axonal profiles containing detectable CRF or AVP were affected by ADX, stained profiles were counted on each micrograph (each micrograph covered approximately 500 μm^2). The entire EZ of the ME in one coronal section was analyzed for each animal. The total number of CRF+ profiles fell by about 40% after ADX (table 3). Treating with DEX thoughout the postoperative period blocked the change in numbers of detectable CRF swellings, while a 6 hour DEX treatment had no effect on the ADX-induced change (table 3)

Table 3. Number of detectable CRF axons in the EZ of the ME

Condition	total # of CRF-positive axonal profiles per 500 μm2[a,b]	# 500 μm2 samples analyzed	# animals (n)	% change
Normal	38.5 ± 6.5	20	3	
ADX[c]	24.1 ± 3.1	25	3	-37.4[d]
Normal + 6 h DEX	39.1 ± 1.5	16	2	
ADX[e] + 6 h DEX	24.8 ± 2.5	18	3	-36.6*
Sham[e] + 4 d DEX	40.3	11	1	
ADX[e] + 4 d DEX	38.3 ± 8.6	29	3	-3.7

[a] Values are the mean ± SE.

[b] The results obtained using antibodies to AVP, NP-AVP and GP were similar and were pooled.

[c] The values for 3 d, 4 d, and 14 d ADX animals were similar and were pooled.

[d] $P = 0.1$ (2-tailed t-test)

[e] Fixed 4 d after surgery

* $P < 0.05$ (2-tailed t-test)

(data taken from ref. 57)

DISCUSSION

Validity of the Concept of CRF+/AVP+ and CRF+/AVP- Subpopulations

The classification of the parvocellular CRF axons into AVP+ and AVP- subpopulations rests on the finding only half of the CRF axons in normal animals contained CRF+ neurosecretory vesicles that stained immunocytochemically for AVP precursor-derived peptides. It could perhaps be argued that the vesicles in the CRF axons contained AVP in concentrations that varied continuously across the entire axonal population, and that our limit of detection fell in the middle of this distribution. We feel that this is very unlikely for the following two reasons.

First, approximately the same result was obtained using two different antisera to AVP, one monoclonal antibody to NP-AVP, and one antiserum to GP. If our limit of detection fell within a smooth distribution of AVP concentrations within the CRF axon population, small differences in the affinities and concentrations of the specific IgGs in the primary antibodies would have resulted in appreciable differences in the numbers of axons stained. It is unlikely that the working dilutions of all of our antibodies had nearly identical affinities and concentrations of specific IgGs.

Second, all of the parvocellular CRF+/AVP+ axons stained intensely for AVP, NP-AVP or GP. If our limit of detection fell in the middle of a smooth distribution of AVP concentrations in the CRF axons, large numbers of axons would have stained weakly or faintly. Judging from other types of solid phase immunoperoxidase assays, such as dot-immunobinding assays[58] and enzyme-linked immunosorbent assays (ELISAs)[48], the difference in antigen concentration necessary to go from intense immunoperoxidase staining to no staining is greater than an order of magnitude. Therefore, the CRF+/AVP- subpopulation probably contains less than one tenth the concentration of AVP, NP-AVP or GP in the CRF+/AVP+ axons. If the subpopulations are independently regulated, they could play a role in modulating ratios of AVP to CRF in portal plasma in certain types of stress.

Implications for the Regulation of ACTH Release during Stress

The CRF+/AVP+ and CRF+/AVP- subpopulations in the EZ of the ME in normal rats are relevant to the multicompartmental nature of the system regulating ACTH release during stress[24]. For example, the ACTH responses to certain types of stress can be completely blocked by prior administration of glucocorticoids, while other types of stress activate glucocorticoid-resistant responses[24].

It has been proposed that the glucocorticoid-resistant component of the ACTH response to severe hemorrhage in cats involves AVP[59]. In accordance with this hypothesis, it is possible to suppress portal CRF levels in rats by the administration of doses of glucocorticoids which do not affect portal AVP levels[29]. If the glucocorticoid-resistant pathway does involve preferential release of AVP into portal plasma, the results of in vitro studies[11] indicate that the synergistic effects of AVP and CRF would help to overcome the inhibition of ACTH secretion by glucocorticoids at the level of the anterior pituitary.

Does the CRF+/AVP+ subpopulation of CRF axons in the EZ of the ME represent a glucocorticoid-resistant pathway? The results of the ADX experiments reported here, while not definitive, are consistent with such a role. One interpretation of the results presented here (tables 2 and 3) is that ADX caused a selective disappearance of the CRF+/AVP- subpopulation. Such a loss of immunocytochemically detectable axons is also seen in studies on the hypothalamo-neurohypophysial system during chronic activation resulting from osmotic stress. The magnocellular AVP- and OT-containing axons in the neurohypophysis and internal zone of the ME are drastically depleted of secretory vesicles after a few days of dehydration[49], and there is evidence for a similar depletion of vesicles occurring in CRF terminals after ADX[60]. Since the CRF+/AVP- subpopulation is the one that disappeared after ADX, this is the subpopulation that would be most sensitive to changes in levels of circulating glucocorticoids; the CRF+/AVP+ subpopulation would be more resistant to glucocorticoid regulation, since the detectable numbers of swellings in this subpopulation were relatively unaffected after ADX (tables 2 and 3).

On the other hand, the decrease in numbers of CRF+ profiles after ADX may have been due to a loss of detectable swellings in both subpopulations after ADX. In this case, the fact that almost all of the remaining swellings were AVP+ would have to be explained by an increase in AVP synthesis in the previously AVP-deficient subpopulation. Such an increase in AVP synthesis after ADX would be consistent with the numerous reports of increases in AVP levels and AVP synthesis in the system after ADX[35,37,39-43]. In this model, circulating levels of glucocorticoids in normal animals are sufficient to suppress AVP synthesis below the level of detectablity in the CRF+/AVP- subpopulation, but not in the CRF+/AVP+ subpopulation. Thus, this scenario is also consistent with a lower sensitivity of the CRF+/AVP+ subpopulation to glucocorticoid inhibition.

In order to resolve these questions, we are currently studying coexistence of CRF and AVP in the parvocellular perikarya in the paraventricular nucleus. For the immediate future, we intend to determine the topographical distributions of the two subpopulations of perikarya within the nucleus, whether one or both subpopulations contain glucocorticoid receptors, and what changes in peptide phenotype take place in the two subpopulations after ADX.

Keller-Wood and Dallman concluded their 1984 review[24] by presenting two alternative hypotheses (where "CRF" meant ACTH secretagogues in general): "In this model we have presumed that all inputs have the same population of CRF neurons and corticotropes in their final common pathway. The resistance of some stresses to corticosteroid inhibition in vivo suggests, however, that this may not be the case, or that in vivo CRF neurons and pituitary feedback sites have limited importance." The present results are consistent with the hypothesis that different stimuli affect separate populations of CRF neurons with differing AVP contents and glucocorticoid sensitivity, i.e., that there is no "final common pathway" for all of the inputs regulating ACTH release.

ACKNOWLEDGMENTS

We thank W. Vale for donating the CRF antiserum, S. Glick, G. Valiquette and E. Zimmerman for donating AVP antisera, S. Zakarian and D. Smyth for donating the GP antiserum, and Harold Gainer, Susan Wray, Kevin Conway, and Mary Dallman for helpful discussions.

REFERENCES

1. W. Vale, J. Spiess, C. Rivier, and J. Rivier, Characterization of a 41-residue ovine hypothalamic peptide that stimulates secretion of corticotropin and B-endorphin, Science 213: 1394-1397 (1981).
2. W. W. Vale, C. Rivier, J. Spiess, and J. Rivier, Corticotropin-releasing factor, in: "Brain Peptides," D. Krieger, J. B. Martin, and M. J. Brownstein, eds., John Wiley & Sons, New York, pp. 961-974 (1983).
3. C. L. Rivier and P. M. Plotsky, Mediation by corticotropin-releasing factor (CRF) of adenohypophysial hormone secretion, Ann. Rev. Physiol. 48: 475-494 (1986).
4. F. A. Antoni, Hypothalamic control of adrenocorticotropin secretion: Advances since the discovery of 41-residue corticotropin-releasing factor, Endocrine Rev. 7: 351-378 (1986).
5. F. E. Yates, S. M. Russell, M. F. Dallman, G. A. Hedge, S. M. McCann, and A. P. S. Dhariwal, Potentiation by vasopressin of corticotropin release induced by corticotropin releasing factor, Endocrinol. 88: 2-15 (1971).
6. C. Rivier, J. Rivier, P. Mormede, and W. Vale, Studies of the nature of the interaction between vasopressin and corticotropin-releasing factor in adrenocorticotropin release in the rat, Endocrinol. 115: 882-886 (1984).
7. G. E. Gillies, E. A. Linton, and P. J. Lowry, Corticotropin releasing activity of the new CRF is potentiated several times by vasopressin, Nature 299: 355-357 (1982).
8. G. E. Gillies, A. Puri, E. A. Linton, and P. J. Lowry, Comparative chromatography of

hypothalamic corticotropin-releasing factors, Neuroendocrinol. 38: 17-24 (1984).
9. C. M. Turkelson, C. R. Thomas, A. Arimura, D. Chang, J. K. Chang, and M. Shimizu, In vitro potentiation of the ability of synthetic ovine corticotropin-releasing factor by arginine vasopressin, Peptides 3: 111-113 (1982).
10. W. Vale, J. Vaughan, M. Smith, G. Yamamota, J. Rivier, and C. Rivier, Effects of synthetic ovine corticotropin-releasing factor, glucocorticoids, catecholamines, neurohypophysial peptides, and other substances on cultured corticotropic cells, Endocrinol. 113: 1121-1131 (1983)
11. A. B. Abou-Samra, K. J. Catt, and G. Aguilera, Biphasic inhibition of adrenocorticotropin release by corticosterone in cultured anterior pituitary cells, Endocrinol. 119: 972-977 (1986).
12. C. Rivier, J. Rivier, and W. Vale, Inhibition of adrenocorticotropic hormone secretion in the rat by immunoneutralization of corticotropin-releasing factor, Science 218: 377-379 (1982).
13. J. Rivier, C. Rivier, and W. Vale, Synthetic competitive antagonists of corticotropin-releasing factor: Effects on ACTH secretion in the rat, Science 224: 889-891 (1984).
14. E. A. Linton, F. J. H. Tilders, S. Hodgkinson, F. Berkenbosch, I. Vermes, and P. J. Lowry, Stress-induced secretion of adrenocorticotropin in rats is inhibited by administration of antisera to ovine corticotropin-releasing factor and vasopressin, Endocrinol. 116: 966-970 (1985).
15. T. Nakane, T. Audhya, N. Kanie, and C. S. Hollander, Evidence for a role of endogenous corticotropin-releasing factor in cold, ether, immobilization, and traumatic stress, Proc. Natl. Acad. Sci. U.S.A. 82: 1247-1251 (1985).
16. M. C. Holmes, F. A. Antoni, and T. Szentendrei, Pituitary receptors for corticotropin-releasing factor: No effect of vasopressin on binding or activation of adenylate cyclase, Neuroendocrinol. 39:162-169 (1984).
17. P. C. Wynn, G. Aguilera, J. Morell, and K. J. Catt, Properties and regulation of high-affinity pituitary receptors for corticotropin-releasing factor, Biochem. Biophys. Res. Commun. 110: 602-608 (1983).
18. P. C. Wynn, J. P. Harwood, K. J. Catt, and G. Aguilera, Regulation of corticotropin-releasing factors in the rat pituitary gland: Effects of adrenalectomy on CRF receptors and corticotroph responses, Endocrinol. 116: 1653-1659 (1985).
19. F. A. Antoni, Novel ligand specificity of pituitary vasopressin receptors in the rat, Neuroendocrinol. 39: 186-188 (1984).
20. F. A. Antoni, M. C. Holmes, and J. Z. Kiss, Pituitary binding of vasopressin is altered by experimental manipulations of the hypothalamo-pituitary-adrenocortical axis in normal as well as homozygous (di/di) Brattleboro rats, Endocrinol. 117: 1293-1299 (1985).
21. B. Lutz-Bucher and B. Koch, Characterization of specific receptors for vasopressin in the pituitary gland, Biochem. Biophys. Res. Commun. 115: 492-498 (1983).
22. B. Koch and B. Lutz-Bucher, Specific receptors for vasopressin in the pituitary gland: Evidence for down-regulation and desensitization to adrenocorticotropin-releasing factor, Endocrinol. 116: 671-676 (1985).
23. K. Todd and S. L. Lightman, Vasopressin activation of phosphatidyl-inositol metabolism in rat anterior pituitary in vitro and its modification by changes in the hypothalamo-pituitary adrenal axis, Neuroendocrinol., in press.

24. M. E. Keller-Wood and M. F. Dallman, Corticosteroid inhibition of ACTH secretion, Endocrine Rev. 5: 1-24 (1984).
25. E. A. Zimmerman, M. A. Stillman, L. D. Recht, J. L. Antunes, P. W. Carmel, and P. C. Goldsmith, Vasopressin and corticotropin-releasing factor: An axonal pathway to portal capillaries in the zona externa of the median eminence containing vasopressin and its interaction with adrenal corticoids, Ann. N. Y. Acad. Sci. 297: 405-419 (1977).
26. D. M. Gibbs, Measurement of hypothalamic corticotropin-releasing factors in hypophyseal portal blood, Fed. Proc. 44: 203-206 (1985).
27. P. M. Plotsky, Hypophyseotropic regulation of adenohypophyseal adrenocorticotropin secretion, Fed. Proc. 44: 207-213 (1985).
28. D. M. Gibbs, Inhibition of corticotropin-release during hypothermia: The role of corticotropin-releasing factor, vasopressin, and oxytocin, Endocrinol. 116: 723-727 (1985).
29. P. M. Plotsky, S. Otto, and R. M. Sapolsky, Inhibition of immunoreactive corticotropin-releasing factor secretion into the hypophysial-portal circulation by delayed glucocorticoid feedback, Endocrinol. 119: 1126-1130 (1986).
30. D. M. Gibbs, Collection of pituitary portal blood: A methodologic analysis, Neuroendocrinol. 38: 97-101 (1984).
31. T. Hokfelt, O. Johansson, and M. Goldstein, Chemical anatomy of the brain, Science 225: 1326-1334 (1984).
32. M. C. Holmes, F. A. Antoni, G. Aguilera, and K. J. Catt, Magnocellular axons in passage through the median eminence release vasopressin, Nature 319: 326-329 (1986).
33. A. J. Baertschi, Portal vascular route from hypophysial stalk/neural lobe to adenohypophysis, Amer. J. Physiol. 239: R463-R469 (1980).
34. L. W. Swanson and P. E. Sawchenko, Hypothalamic integration: Organization of the paraventricular and supraoptic nuclei, Ann. Rev. Neurosci. 6: 269-324 (1983).
35. A.-J. Silverman and E. Z. Zimmerman, Magnocellular neurosecretory system, Ann. Rev. Neurosci. 6: 357-380 (1983).
36. L. D. Recht, D. L. Hoffman, J. Haldas, A.-J. Silverman, and E. A. Zimmerman, Vasopressin concentrations in portal plasma: Insignificant reduction following removal of the posterior pituitary gland, Neuroendocrinol. 33: 88-90 (1981).
37. M. H. Whitnall and H. Gainer, Ultrastructural studies of peptide coexistence in corticotropin-releasing factor- and arginine-vasopressin-containing neurons, in: "Neural and Endocrine Peptides and Receptors," T. W. Moody, ed., Plenum Publ. Co., New York, pp. 159-175 (1986).
38. M. H. Whitnall, E. Mezey, and H. Gainer, Co-localization of corticotropin-releasing factor and vasopressin in median eminence neurosecretory vesicles, Nature 317: 248-250 (1985).
39. G. Tramu, C. Croix, and A. Pillez, Ability of the CRF immunoreactive neurons of the paraventricular nucleus to produce a vasopressin-like material, Neuroendocrinol. 37: 467-469 (1983).
40. J. Z. Kiss, E. Mezey, and L. Skirboll, Corticotropin-releasing factor-immunoreactive neurons of the paraventricular nucleus become vasopressin-positive after adrenalectomy, Proc. Natl. Acad. Sci. U.S.A. 81: 1854-1858 (1984).
41. P. E. Sawchenko, L. W. Swanson, and W. Vale, Co-expression of corticotropin-

releasing factor and vasopressin immunoreactivity in parvocellular neurosecretory neurons of the adrenalectomized rat, Proc. Natl. Acad. Sci. U.S.A. 81: 1883-1887 (1984).

42. B. Wolfson, R. W. Manning, L. G. Davis, R. Arentzen, and F. Baldino,Jr., Co-localization of corticotropin-releasing factor and vasopressin mRNA in neurones after adrenalectomy, Nature 315: 59-61 (1985).

43. W. S. Young, E. Mezey, and R. E. Siegel, Vasopressin and oxytocin mRNAs in adrenalectomized and Brattleboro rats: Analysis by quantitative in situ hybridization histochemistry, Molec. Brain Res. in press.

44. C. Bugnon, D. Fellmann, and A. Gouget, Changes in corticoliberin and vasopressin-like immunoreactivities in the zona externa of the median eminence in adrenalectomized rats. Immunocytochemical study, Neurosci. Lett. 37: 43-49 (1983).

45. F. A. Antoni, M. Palkovits, G. B. Makara, E. A. Linton, P. J. Lowry, and J. Z. Kiss, Immunoreactive corticotropin-releasing hormone in the hypothalamo-infundibular tract, Neuroendocrinol. 36: 415-423 (1983).

46. M. H. Whitnall, D. Smyth, and H. Gainer, Vasopressin coexists in half of the CRF axons present in the external zone of the median eminence in normal rats Neuroendocrinol., in press.

47. G. R. Newman, B. Jasani, and E. D. Williams, A simple post-embedding system for the rapid demonstration of tissue antigens under the electron microscope, Histochem. J. 15: 543-555 (1983).

48. Y. Ben-Barak, J. T. Russell, M. H. Whitnall, K. Ozato, and H. Gainer, Neurophysin in the hypothalamo-neurohypophysial system. I. Production and characterization of monoclonal antibodies, J. Neurosci. 5: 81-97 (1985).

49. M. H. Whitnall and H. Gainer, Ultrastructural immunolocalization of vasopressin and neurophysin in neurosecretory cells of dehydrated rats, Brain Res. 361: 400-404 (1985).

50. P. E. Sawchenko, L. W. Swanson, and W. Vale, Corticotropin-releasing factor: Co-expression within distinct subsets of oxytocin-, vasopressin-, and neurotensin-immunoreactive neurons in the hypothalamus of the male rat, J. Neurosci. 4: 1118-1129 (1984).

51. S. Dubois-Dauphin and S. Zakarian, Distribution of the C-terminal glycopeptide of the vasopressin prohormone in rat brain: An immunohistochemical study, Neuroscience, in press.

52. A.-J. Silverman and E. A. Zimmerman, Ultrastructural immunocytochemical localization of neurophysin and vasopressin in the median eminence and posterior pituitary of the guinea pig, Cell Tissue Res. 159: 291-301 (1975).

53. D. Dube, R. Leclerc, and G. Pelletier, Electron microscopic localization of vasopressin and neurophysin in the median eminence of normal and adrenalectomized rats, Am. J. Anat. 147: 103-108 (1976).

54. G. Pelletier, L. Desy, J. Cote, G. Lefevre, G. Vaudry, and F. Labrie, Immunoelectron microscopic localization of corticotropin-releasing factor in the rat hypothalamus, Neuroendocrinol. 35: 402-404 (1982).

55. C. Leranth, F. A. Antoni, and M. Palkovits, Ultrastructural demonstration of ovine CRF-like immunoreactivity (oCRF-LI) in the rat, Regul. Peptides 6: 179-188 (1983).

56. Z. Liposits, T. Gorcs, G. Setalo, I. Lenguari, B. Flerko, S. Vigh, and A.V. Schally,

Ultrastructural characteristics of immunolabeled, corticotropin-releasing factor (CRF)-synthesizing neurons in the rat brain, Cell Tissue Res. 229: 191-196 (1983).

57. M. H. Whitnall, S. Key, and H. Gainer, Vasopressin-containing and vasopressin-deficient subpopulations of corticotropin-releasing factor axons are differentially affected by adrenalectomy, Endocrinol, in press.

58. R. Hawkes, E. Niday, and J. Gordon, A dot-immunobinding assay for monoclonal and other antibodies, Anal. Biochem. 119: 142-147 (1982).

59. D. E. Carlson and D. S. Gann, Effects of vasopressin on the response of adrenocorticotropin and cortisol to hemorrhage, Endocrinol. 114: 317-324 (1984).

60. Z. Liposits and W. K. Paull, Ultrastructural alterations of the paraventriculo-infundibular corticotropin releasing factor (CRF)-immunoreactive neuronal system in long term adrenalectomized rats, Peptides 6: 1021-1036 (1985).

REGULATION OF VASOPRESSIN AND OXYTOCIN RELEASE IN VITRO

Celia D. Sladek and John A. Olschowka

Departments of Neurology and Neurobiology and Anatomy
University of Rochester School of Medicine
Rochester, NY 14642

Numerous studies have demonstrated a role for vasopressin (VP) and oxytocin (OT) in the regulation of adrenocorticotropin (ACTH) release. Both VP and OT release potentiate the effect of corticotropin releasing factor (CRF) on ACTH release (1), and they are present in portal plasma at the concentration required for this action (2). VP and CRF are co-localized in parvicellular neurons in the paraventricular nucleus which project to the median eminence (3-6), and both of these hormones are released from the isolated median eminence following depolarization (7). Thus, this appears to be one source of VP for regulation of ACTH release. OT is also present in some CRF neurons (8), but it has not been established whether these neurons project to the median eminence. Thus, it is not clear which neurons release OT into the portal circulation, but there are OT fibers in the external zone of the median eminence (9) suggesting a specific projection for control of the anterior pituitary.

Although relatively little specific information is available about the mechanisms which regulate release of VP, OT, and CRF into the portal plasma, numerous studies have provided information about the regulation of peripheral plasma VP (10), and the mechanisms involved in regulating VP release from the posterior pituitary (11). Recent work is also beginning to provide insights into the mechanisms regulating peripheral OT (12). Since certain stimuli cause parallel increases in peripheral plasma concentrations of either VP and ACTH or OT and ACTH, it is possible that there are some mechanisms which are common to the regulation of VP and OT release from the median eminence and the posterior pituitary. Specifically, both pVP and ACTH are stimulated by hemorrhage and hypoglycemia (10,13). Stressors which release both OT and ACTH are restraint, swimming, and ether (2). Thus, it is possible that common pathways and neurotransmitters relay these signals to the anterior and posterior pituitary.

Logical candidates for the transmitters in the common pathways are those transmitters which have been shown to influence both ACTH and VP or OT release. Of the numerous neurotransmitters which have been identified in the supraoptic nucleus and which have been shown to influence either VP or OT release, several have also been implicated in the central control of ACTH release. These are acetylcholine, norepinephrine, and gamma aminobutyric acid (GABA). The remainder of this review will focus on these three transmitters.

NOREPINEPHRINE

The supraoptic and paraventricular nuclei are densely innervated by noradrenergic afferents (14). In the supraoptic nuclei this arises primarily from the A1 noradrenergic neurons located in the ventrolateral medulla, but the paraventricular nucleus receives afferents from the A2 neurons in the nucleus of the tractus solitarius (NTS) and locus coeruleus (A6) as well as the A1 neurons (15). The noradrenergic input to the supraoptic nucleus is concentrated in the ventral region where the VP neurons are located and in ventral dendritic zone (16). In the paraventricular nucleus, noradrenergic afferents to the magnocellular VP neurons are present. There is also a dense noradrenergic input to the parvicellular region, and dopamine beta hydroxylase-positive synapses can be demonstrated on CRF neurons (Fig. 1). Thus, morphological evidence indicates direct noradrenergic control of the VP neurons projecting to the neurohypophysis and to the median eminence.

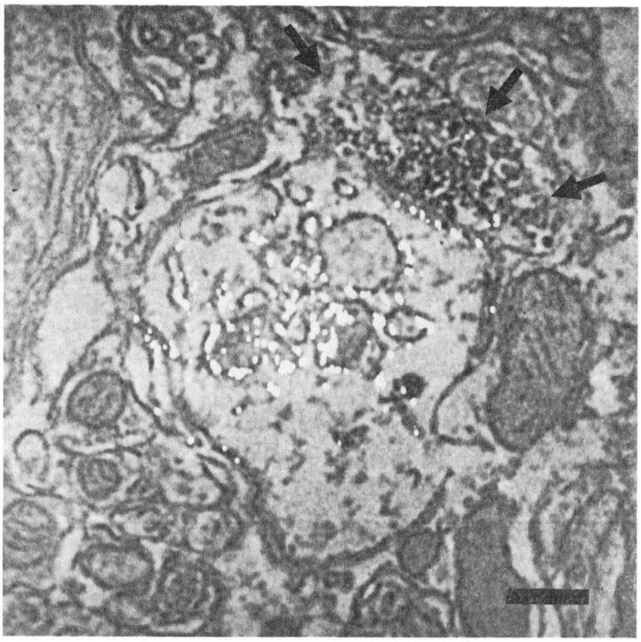

Figure 1. A synaptic contact of a norepinephrine terminal (arrows) on the dendrite of a CRF neuron of the paraventricular nucleus. The tissue was first stained for dopamine beta hydroxylase using a conventional avidin-biotin-peroxidase method and reaction with diaminobenzidine. The tissue was then incubated in an anti-CRF antiserum which had been previously labeled with fluorine. The localization of the anti-CRF was determined by locating fluorine (white dots) in a Zeiss EM 902 analytical electron microscope using the electron spectroscopic imaging mode. First, a reference image just below the energy loss edge of fluorine was stored in a IBAS image analysis computer (spectrometer: eV = 668±10). From this was subtracted the image formed by the electrons specific for the energy loss edge of fluorine (spectrometer: eV = 694±10). The resulting image represents the distribution of fluorine in the section. This image was then superimposed over a high contrast image of the analyzed field (spectrometer: eV = 150±10) in this unstained 30 nm thick section Bar: 0.5 um.

The role of this noradrenergic input to the magnocellular VP system has been investigated in numerous studies. These studies provide evidence for both excitatory and inhibitory effects of norepinephrine on VP neurons. Early studies with organ cultured hypothalamo-neurohypophyseal (HNS) explants provided evidence that VP release was inhibited by norepinephrine (17), but in acutely prepared HNS explants, norepinephrine and phenylephrine increased the firing rate of putative VP neurons (18,19) and increased VP release (18). Similarly, microinjections of norepinephrine and phenylephrine into the supraoptic nucleus resulted in an increase in plasma VP in unanesthetized rats (20,21). Furthermore, injection of tyramine, which releases norepinephrine from neuronal stores, into the supraoptic nucleus also increased plasma VP (20). Thus, endogenous norepinephrine appears capable of stimulating VP release following activation of alpha-adrenergic receptors. This excitatory effect of norepinephrine on VP release is consistent with the evidence which indicates that the A1 noradrenergic projection to the VP neurons transmits baroreceptor initiated signals which elicit VP release in response to hypotension (22).

The possibility that norepinephrine can also have inhibitory effects on VP release deserves further discussion, because both excitatory and inhibitory actions of norepinephrine have been reported in numerous systems including the CRF-ACTH axis (see 23 and 24 for review). In the organ-cultured HNS explant, the effect of norepinephrine on VP release is dependent on other concomitant regulatory signals. As shown in Fig. 2, under isotonic conditions, VP release is stimulated by norepinephrine, but inhibition of VP release is observed when

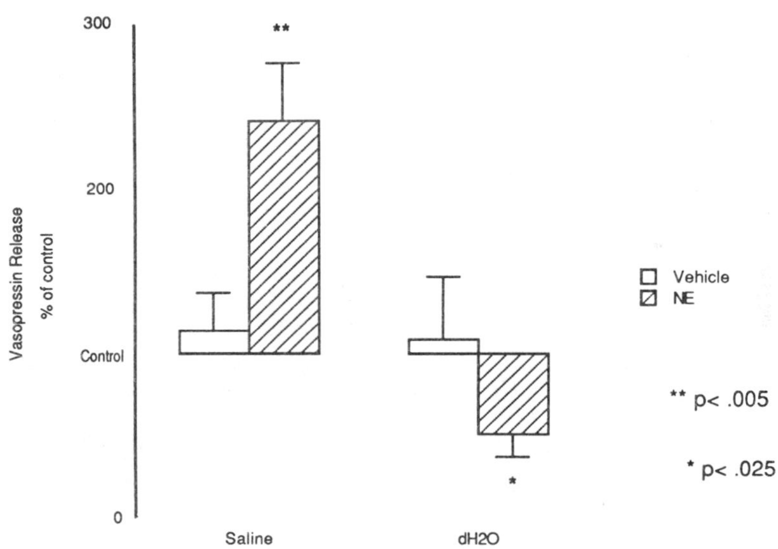

Figure 2. The effects of norepinephrine on VP release from organ-cultured HNS explants. Explants were prepared from adult, male rats and maintained in culture as described previously (17). After 48 or 72 hours of culture, VP degradation was controlled by the addition of bacotracin (1.7×10^{-4}M) and basal VP release was determined during a control hour. At the end of the control hour, norepinephrine (10^{-5}M) was added in either 10 ul of dH_2O or isotonic saline or the vehicle alone was added, and VP release was determined during the test hour. VP release during the test hour is presented as the % of VP release observed during the control hour. $^*p < .025$, $^{**}p < .005$ different from control release by paired-t analysis.

norepinephrine is applied concomitantly with a small decrease in osmolality (3-5 mOsm). This decrease in osmolality is insufficient to alter VP release in the absence of norepinephrine. The receptor, second messenger, and electrophysiological events responsible for these bimodal actions of norepinephrine on VP release remain to be determined, but similar bimodal actions of norepinephrine in other areas of the CNS reflect differential activation of alpha and beta adrenergic receptors (25) and altered postsynaptic responsiveness to other transmitter systems (26). Relative to the physiological implications of these observations, it is possible that bimodal actions of norepinephrine underlie the ability of hypotonicity to eliminate baroreceptor mediated stimulation of VP release (27).

GABA

Morphological and functional evidence exists which implicates GABA as a neurotransmitter regulating both VP and CRF neurons. The synthetic enzyme for GABA, glutamic acid decarboxylase (GAD), is localized in the supraoptic and paraventricular nuclei and nerve terminals containing GAD synapse directly on neurosecretory neurons in the supraoptic nucleus (28) and on CRF positive neurons in the paraventricular nucleus (Fig. 3). GABA inhibits release of acetylcholine and serotonin induced release of bioactive CRF from hypothalamic fragments (29). In organ cultured HNS explants exogenous GABA did not alter either basal or

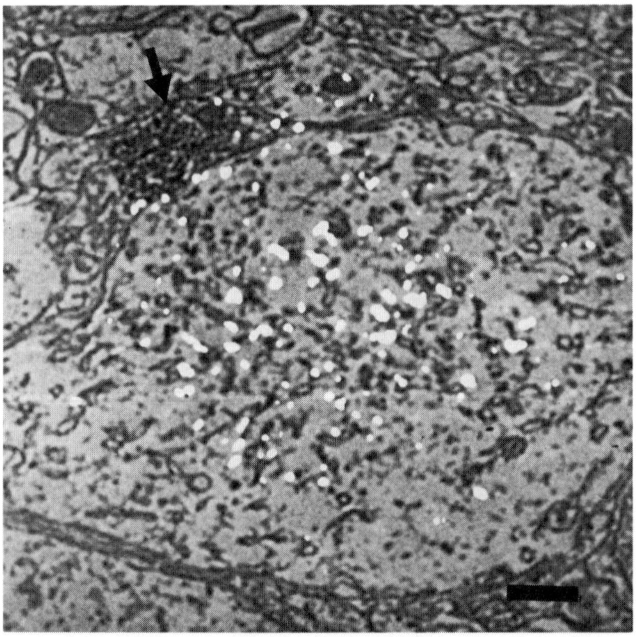

Figure 3. The synaptic contact of a GAD-positive terminal (arrows) with a large CRF-positive dendrite (white dots) in the rat paraventricular nucleus. The staining method was identical to the described for figure 1 except that the tissue was first stained for GAD using the avidin-biotin-peroxidase method, and subsequently incubated in the fluorine labeled anti-CRF. Note that the GAD terminal is contacting a large primary dendrite of the CRF neuron. GAD terminals were frequently observed contacting cell bodies and proximal dendrites within the PVN. This differs from the distribution of the noradrenergic terminals which more commonly innervated the more distal parts of the dendritic tree. The spectrometer settings are identical to those described in figure 1. Bar: 0.5 um.

cholinergically stimulated VP release, but the GABA antagonists, bicuculline and picrotoxin, stimulated VP release in a concentration dependent manner (30). These observations suggest that VP release from HNS explants is tonically inhibited by GABA.

The GABAergic innervation of the supraoptic nucleus arises from two sources. Approximately 60% of the GAD activity in the supraoptic nucleus is associated with neurons that are either intrinsic or adjacent to the nucleus (31). The remainder arises from areas rostral to the nucleus (31). This is consistent with the observations of the HNS explants which suggest that one role of GABA in the regulation of VP release is that of a potent local inhibitory neurotransmitter (30), and with the observations of Renaud et al (32) which suggest that GABA is a likely candidate for the neurotransmitter mediating baroreflex initiated inhibition of VP release. Since the pathways mediating baroreflex regulation of VP and ACTH release are similar (33), GABA may also participate in baroreflex regulation of CRF release.

Another potential role of the GABAergic inhibition of CRF and VP neurons is mediation of inhibitory glucocorticoid feedback. Both ACTH (34) and neurohypophyseal VP secretion (35) are subject to inhibition by glucocorticoids. The GABAergic receptor is a protein complex, which includes the chloride channel as well as other subcomponents which regulate the chloride channel. Barbiturates enhance chloride uptake by the chloride channel and thereby enhance the action of GABA (36). Metabolites of corticosterone have barbiturate-like modulatory effects on the GABA receptor (37). Furthermore, corticosteroids may induce the enzyme GAD and therefore enhance GABA synthesis (38). Therefore the inhibitory effect of glucocorticoids on ACTH release could be the results of enhanced GABAergic inhibition of the CRF and/or VP neurons. One study in Brattleboro rats suggests that the GABA mediated component of glucocorticoid feedback inhibition of ACTH is a consequence of inhibition of VP release (39). Thus, GABAergic regulation of VP release may be central to corticosteroid feedback inhibition.

ACETYLCHOLINE

Acetylcholine has been implicated in the regulation of both VP and CRF release (23,40). VP release from organ-cultured HNS explants is stimulated by acetylcholine and nicotine (40). Muscarinic receptors located in the neural lobe may also participate in cholinergic stimulation of VP release (41). Similarly, acetylcholine has been reported to stimulate release of bioactive CRF from isolated hypothalamic fragments, and both nicotinic and muscarinic receptors appear to participate in this effect (42). Recent studies in which CRF was measured by specific radioimmunoassay confirm that acetylcholine stimulates CRF release from the hypothalamus in vitro (43).

Cholinergic neurons are located immediately dorsal and lateral to the SON (44), and biochemical studies have demonstrated that the majority of cholinergic input to the SON arises from neurons either within or in the immediate vicinity of SON (45). These observations suggest that the cholinergic afferent to the SON may function to relay and potentially integrate signals regulating VP release. One type of information that may be processed by these neurons is the osmotic status of the extracellular fluid. Experiments with HNS explants demonstrated that nicotinic cholinergic antagonists blocked VP release in response to an increase in culture medium osmolality as well as blocking cholinergically stimulated VP release (46). However, it is likely that this is not the sole function of the cholinergic input nor the sole mechanism of relaying osmotic information. Evidence supporting this latter hypothesis is derived from experiments on the effect of acetylcholine in OT release from HNS explants. In contrast to the effect of acetylcholine on VP and CRF release, acetylcholine at 10^{-5}M inhibited OT release (Fig. 4). While the mechanism responsible for cholinergic inhibition of OT release remains to be elucidated, this observation indicates the complexity of both the cholinergic and osmotic control of posterior pituitary hormone release. It also suggests that cholinergic mechanisms may be responsible for mediating signals which stimulate VP and CRF release without OT release.

CONCLUSION

Since there are a number of stimuli which increase both ACTH and either plasma VP or OT release and since there are several neurotransmitter systems which regulate both the VP projection to the neurohypophysis and the VP/CRF projection to the median eminence, we anticipate that the information we have obtained on the neurotransmitter regulation of VP release from the neurohypophysis may be relevant to the control of VP release involved in regulation of the adenohypophyseal-adrenal axis. Final confirmation of this hypothesis will require specific evaluation of the effect of neurotransmitters on VP and CRF release from the median eminence.

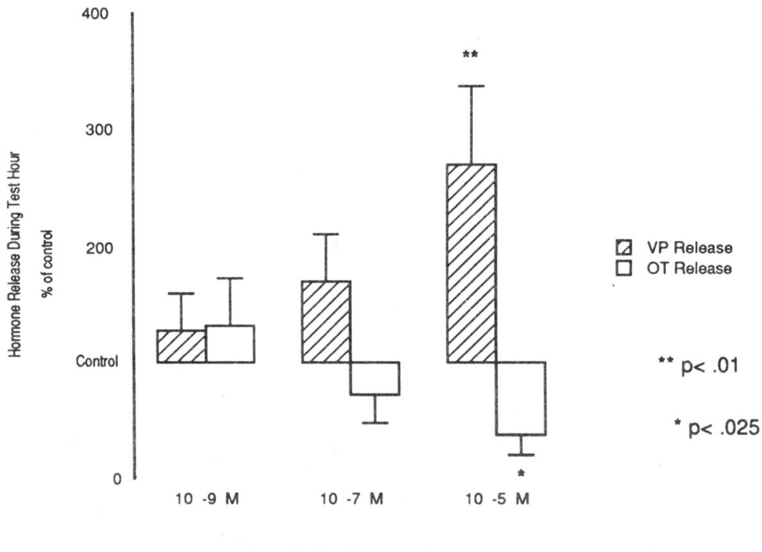

Figure 4. Effects of acetylcholine (10^{-5}M) on VP and OT release from HNS explants. Acetylcholine was added to the medium at the end of the control hour after 48 hours of culture. Mean ± SEM.

REFERENCES

1. Antoni, F.A., Holmes, M.C., and Jones, M.T., 1983, Oxytocin as well as vasopressin potentiate ovine CRF in vitro, Peptides, 4:411-415.
2. Gibbs, D.M., 1985, Measurement of hypothalamic corticotropin-releasing factors in hypophyseal portal blood, Fed. Proc., 44:203-206.
3. Tramu, G., Croix, C., and Pillez, A., 1983, Ability of the CRF immunoreactive neurons of the paraventricular nucleus to produce a vasopressin-like material, Neuroendocin., 37:467-469.
4. Kiss, J.Z., Mezey, E., and Skirboll, L., 1984, Corticotropin-releasing factor-immunoreactive neurons of the paraventricular nucleus become vasopressin positive after adrenalectomy, Proc. Natl. Acad. Sci., 81:1854-1858.
5. Sawchenko, P.E., Swanson, L.W., and Vale, W.W., 1984, Co-expression of corticotropin-releasing factor and vasopressin immunoreactivity in parvocellular neurosecretory neurons of the adrenalectomized rat, Proc. Natl. Acad. Sci., 81:1883-1887.
6. Wolfson, B., Manning, R.W., Davis, L.G., Arentzen, R. and Baldino, F., Jr., 1985, Co-localization of corticotropin releasing factor and vasopressin mRNA in neurons after adrenalectomy, Nature, 315:59-61.
7. Holmes, M.C., Antoni, F.A., Patt, K.J., Aguilera, G., 1986, Predominant release of vasopressin vs. corticotropin-releasing factor from the isolated median eminence after adrenalectomy, Neuroendocrin., 43:245-251.
8. Sawchenko, P.E., Swanson, L.W., and Vale, W.W., 1984, Corticotropin-releasing factor: Co-expression within distinct subsets of oxytocin-, vasopressin-, and neurotensin-immunoreactive neurons in the hypothalamus of the male rat, J. Neurosci., 4:1118-1129.
9. Antunes, J.L. and Zimmerman, E.A., 1978, The hypothalamic magnocellular system of the Rhesis monkey: An immunocytochemical study, J. Comp. Neurol., 181:539-566.
10. Robertson, G.L., 1977, The regulation of vasopressin function in health and disease, Recent Prog. Horm. Res., 33:333-373.
11. Sladek, C.D., 1983, Regulation of vasopressin release by neurotransmitters, neuropeptides, and osmotic stimuli, Prog. Brain Res., 60:71-90.
12. Amico, J.A. and Robinson, A.G., 1985, "Oxytocin," Excerpta Medica, New York.
13. Plotsky, P.M., 1985, Hypophyseotropic regulation of adenohypophyseal adrenocorticotropin secretion, Fed. Proc. 44:207-213.
14. Ungerstedt, J., 1971, Stereotaxic mapping of the monoamine pathways in the rat brain, Acta Physiol. Scand. Suppl., 367:1-48.
15. Swanson, L.W. and Sawchenko, P.E., 1983, Hypothalamic integration: Organization of the paraventricular and supraoptic nuclei, Ann. Rev. Neurosci., 6:269-324.
16. McNeill, T.H. and Sladek, J.R., Jr., 1980, Simultaneous monamine histofluorescence and neuropeptide immunocytochemistry: II. Correlative distribution of catecholamine varicosities and magnocellular neurosecretory neurons in the rat supraoptic and paraventricular nuclei, J. Comp. Neurol., 193:1023-1033.
17. Armstrong, W.E., Sladek, C.D., and Sladek, J.R., 1982, Characterization of noradrenergic control of vasopressin release by the organ cultured rat hypothalamo-neurohypophyseal system, Endocrin., 111:273-279.
18. Armstrong, W.E., Gallagher, M.J., and Sladek, C.D., 1986, Noradrenergic stimulation of supraoptic neuronal activity and vasopressin release in vitro: Mediation by an -receptor, Brain Res., 365:192-197.
19. Randle, J.C.R., Bourque, C.W., and Renaud, L.P., 1984, -Adrenergic activation of rat hypothalamic supraoptic neurons maintained in vitro, Brain Res., 307:374-378.

20. Willoughby, J.O., Jervois, P.M., Menadue, M.F., and Blessing, W.W., 1987, Noradrenaline, by activation of alpha-1-adrenoceptors in the region of the supraoptic nucleus, causes secretion of vasopressin in the unanaesthetized rat, Neuroendocrin., 45:219-226.
21. Spencer, L., Diay, S., Sladek, C., and Leibowity, S.F., 1986, Paraventricular and supraoptic nucleus control of vasopressin release: stimulatory effect of norepinephrine, epinephrine and neuropeptide Y, Soc. for Neurosci. Abs., 12:445.
22. Blessing, W.W. and Willoughby, J.O., 1985, Inhibiting the rabbit caudal ventrolateral medulla prevents baroreceptor-initiated secretion of vasopressin, J. Physiology 367:253-265.
23. Jones, M.T., Gillham, B., DiRenzo, G., Beckford, U., and Holmes, M.C., 1981, Neural control of corticotropin secretion, Front. Horm. Res., 8:12-43.
24. Rivier, C.L. and Plotsky, P.M., 1986, Mediation by corticotropin releasing factor (CRF) of adenohypophysial hormone secretion, Ann. Rev. Physiol., 48:475-494.
25. Waterhouse, B.D., Moises, H.C., Yeh, H.H., and Woodward, D.J., 1982, Norepinephrine enhancement of inhibitory synaptic mechanisms in cerebellum and cerebral cortex: Mediation by beta adrenergic receptors, J. Pharmacol. Exp. Ther., 221:495-506.
26. Woodward, D.J., Moises, H.C., Waterhouse, B.D., Hoffer, B.J., and Freedman, R., 1979, Modulatory actions of norepinephrine in the central nervous system, Fed. Proc., 38:2109-2116.
27. Stricker, E.M. and Verbalis, J.G., 1986, Interaction of osmotic and volume stimuli in the regulation of neurohypophyseal secretion in rats, Am. Phys., 250(2):R267-R275.
28. van den Pol, A.N., 1985, Dual ultrastructural localization of two neurotransmitter-related antigens: Colloidal gold labelled neurophysin immunoreactive supraoptic neurons receive peroxidase labeled glutamate decarboxylase or gold labelled GABA immunoreactive synapses, J. Neurosci., 5:2940-2954.
29. Burden, J., Hillhouse, E.W., and Jones, M.T., 1974, The inhibitory action of GABA and melatonin on the release of corticotropin releasing hormone from the rat hypophysiotropic area in vitro, J. Physiol., 239:116-117.
30. Sladek, C.D. and Armstrong, W.E., 1987, GABA antagonists stimulate vasopressin release from organ-cultured hypothalamo-neurohypophyseal explants, Endocrin., 120(4).
31. Meyer, D.K., Oertel, W.H., and Brownstein, M.J., 1980, Deafferentation studies on the glutamic acid decarboxylase content of the supraoptic nucleus of the rat, Brain Res., 200:165-168.
32. Jhamandes, J.H. and Renaud, L.P., 1986, Pharmacological characterization of an inhibitory baroreceptor-activated projection to supraoptic vasopressinergic neurons, Soc. Neurosci. Abs., 12:538.
33. Gann, D.S., Dallman, M.F., and Engeland, W.C., 1981, Reflex control and modulation of ACTH and corticosteroids, in: "Endocrine Physiology III, International Review of Physiology." S.M. McCann, ed., University Park Press, Baltimore.
34. Keller-Wood, M.E. and Dallman, M.F., 1984, Corticosteriod inhibition of ACTH secretion, Endocrine Rev., 5:1-24.
35. Raff, H., 1987, Glucocorticoid inhibition of neurohypophysial vasopressin secretion, Am. J. Physiol., 252.
36. Olsen, R.W., 1981, GABA-benzodiazepine-barbiturate receptor interactions, J. Neurochemistry, 37:1-13.
37. Majewska, M.D., Harrison, N.L., Schwarty, R.D., Barker, J.L., and Paul, S.M., 1986, Steroid hormone metabolites are barbiturate-like modulators of the GABA receptor, Science, 232:1004-1007.
38. Acs, Z. and Stark, E., 1978, Possible role of gamma-amino butyric acid synthesis in the mechanism of dexamethasone feedback action, J. Endocrine., 77:137-141.

39. Acs, Z., Lutz-Bucher, B., Koch, B., and Stark, E., 1985, Vasopressin secretion as a possible target of the -aminobutyric acid-mediated component of the corticosteroid feedback effect, J. Endocrin., 107:303-307.
40. Sladek, C.D. and Joynt, R.J., 1979, Characterization of cholinergic control of vasopressin release by the organ cultured rat hypothalamo-neurohypophyseal system, Endocrinology, 104:659-663.
41. Gregg, C.M., 1985, The compartmentalized hypothalamo-neurohypophyseal system: Evidence for a neurohypophyseal action of acetylcholine on vasopressin release, Neuroendocrin., 40(5):423-429.
42. Buckingham, J.C. and Hodges, J.R., 1979, Hypothalamic receptors influencing the secretion of corticotrophin-releasing hormone in the rat, J. Physiol., 290:421-431.
43. Calogero, A., et al., this volume.
44. Armstrong, D.M., Saper, C.B., Levey, A.I., Wainer, B.H., and Terry, R.D., 1983, Distribution of cholinergic neurons in rat brain demonstrated by immunocytochemical localization of choline acetyltransferase, J. Comp. Neural., 216:53-68.
45. Meyer, D.K. and Brownstein, M.J., 1980, Effect of surgical deafferentation of the supraoptic nucleus in its choline acetyltransferase content, Brain Res., 193:566-569.
46. Sladek, C.D. and Joynt, R.J., 1979, Cholinergic involvement in osmotic control of vasopressin release by the organ cultured rat hypothalamo-neurohypophyseal system, Endocrinology, 105:367-371.

REGULATION OF RAT HYPOTHALAMIC CORTICOTROPIN-RELEASING HORMONE SECRETION *IN VITRO*: POTENTIAL CLINICAL IMPLICATIONS

Aldo E. Calogero, Renato Bernardini*, Philip W. Gold*, and George P. Chrousos

Developmental Endocrinology Branch, National Institute of Child Health and Human Development and *Biological Psychiatry Branch, National Institute of Mental Health, Bethesda, MD 20892

INTRODUCTION

Corticotropin-releasing hormone (CRH) is located in both the hypothalamus (1) and several extrahypothalamic brain regions (2). Different reports suggest that CRH not only stimulates pituitary adrenocorticotropin (ACTH) secretion (3), but also activates the sympathetic nervous system (4) and causes behavioral changes (5-10). Behavioral studies conducted primarily in rats have indicated that CRH administered intracerebroventricularly (icv) causes anxiety-like behaviors. Such behaviors are, in part, influenced by the animal's familiarity with the environment. In single-caged rats, tested in a familiar environment, CRH increases locomotion, rearing, and self-grooming (10) and decreases food intake (5,8). On the other hand, CRH administered to rats in a novel environment decreases locomotion, rearing, and food intake and increases self-grooming (10). In addition, icv CRH administration increases aggression and mounting behavior of male rats (10). In contrast, microinfusion of CRH into the arcuate hypothalamic nucleus and mesencephalic central gray area of estrogen-primed female rats causes decreased sexual receptivity (9). These effects of CRH are compatible with the concept that this peptide may be an important integrative factor that coordinates endocrine, autonomic, and behavioral responses to stress.

A voluminous literature regarding the regulation of hypothalamic CRH secretion has yielded conflicting and often unclear conclusions. For example, serotonin (5HT) has been reported to stimulate (12), inhibit (13) or have no effect (14) upon CRH secretion. Similarly, norepinephrine (NE) has been reported both to inhibit (15,16) and stimulate (17) the secretion of this peptide. On the other hand, certain other aspects of hypothalamic CRH regulation are poorly understood. These include: (1) the role of other neurotransmitters, such as epinephrine (E) and dopamine (DA); (2) the receptor subtypes mediating the effects of various neurotransmitters; (3) the interactions between the different neurotransmitter systems; (4) the evaluation of possible negative feedback control loops; and (5) the evaluation of interactions between products of the immune system and the hypothalamic CRH neuron. In experiments conducted in animals, factors such as use of relatively nonspecific pharmacologic probes, application of varying routes of drug administration (e.g. peripheral, icv, direct hypothalamic application), the confounding effects of anesthesia, and/or the use of indirect assessments of ACTH secretion (e.g. corticosterone, 11-OH corticosteroids, etc) may have contributed to the conflicting results on the regulation of CRH secretion. Furthermore, it should be noted that the oldest *in vitro* hypothalamic organ culture system employed to study CRH regulation had as end-point corticotropin-releasing bioactivity rather than CRH itself (12,14,16). Several hypothalamic or circulating substances, however, are capable of releasing ACTH, such as vasopressin, oxytocin, E, 5HT, vasoactive intestinal peptide, peptide histidine isoleucine, cholecystokinin, and angiotensin II (18). In addition, the low sensitivity of these bioassays requiring either sample concentration and/or incubation of pooled hypothalami may have contributed to the conflicting results.

To circumvent the shortcomings of previous efforts to elucidate the regulation of hypothalamic CRH secretion, we developed an in vitro rat hypothalamic organ culture system which allowed us to measure immunoreactive CRH (IR-rCRH) secretion from single explanted rat hypothalami.

HYPOTHALAMIC ORGAN CULTURE

A more detailed account of the methods, including validation and quality control has been reported elsewhere (19). Briefly, adult male Sprague-Dawley rats were sacrificed by decapitation and the hypothalami sterilely excised. After an overnight preincubation in medium 199 (M199), single hypothalami were incubated in M199 according to the experimental protocols described below.

The experimental design consisted of serial passages of the hypothalamic explants in 6 different wells every 20 min, for a total period of 120 min. Basal IR-rCRH secretion was 36±2pg/hypothalamus/0.4 ml/20 min of incubation and remained constant for the length of the experiment. We employed three different experimental protocols. Protocol 1 was used when stimulation of IR-rCRH secretion was examined. For this purpose the hypothalami were incubated in plain medium in the first three wells (basal IR-rCRH concentration) and exposed to graded concentrations of the test substance in the next two wells (stimulated IR-rCRH concentration). Controls for protocol 1 were obtained by exposing the hypothalami to medium plus vehicle alone in the fourth and fifth well. Protocol 2 was used when inhibition of stimulation induced by a stimulatory neurotransmitter was examined. For this experimental protocol, the hypothalami were incubated in plain medium in the first two wells (basal IR-rCRH concentration). In the third, fourth, and fifth well the antagonist(s) was present, whereas in the fourth and fifth well the stimulant was added (stimulated IR-rCRH concentration). Finally, protocol 3 was used when inhibition of stimulation was examined. For this purpose, the inhibitor was added during the last hour of preincubation and throughout the experiment conducted as in protocol 1.

The concentration of CRH in the medium was measured directly by RIA, using a specific anti-rCRH rabbit antiserum developed in our laboratory (20).

EXCITATORY NEUROTRANSMITTERS

Acetylcholine

Acetylcholine (ACh) has been proposed by several investigators as an excitatory neurotransmitter of the hypothalamic-pituitary-adrenal (HPA) axis (12,16). The locus of ACh action on the HPA axis has been proposed to be within the central nervous system (CNS). Concordantly, it has been shown that ACh does not affect pituitary ACTH secretion in vitro (21). We examined the hypothesis that ACh stimulates the HPA axis by inducing hypothalamic CRH secretion (22).

ACh induced IR-rCRH secretion in a dose-dependent fashion, at concentrations ranging from 3.3×10^{-10} to 10^{-5} M (Fig 1). The maximal stimulatory effect of ACh on IR-rCRH secretion was antagonized by the simultaneous presence of atropine and hexamethonium, a muscarinic and a nicotinic receptor antagonist, respectively. Further evidence for the cholinergic regulation of the CRH neuron was provided by the findings that both carbachol, a muscarinic receptor agonist, and nicotine, a nicotinic receptor agonist, stimulated IR-rCRH secretion in a dose-dependent fashion, although, their maximal stimulatory effects were lower than that produced by ACh. Carbachol- and nicotine-induced IR-rCRH secretion were antagonized by atropine or hexamethonium, respectively. ACh stimulated IR-rCRH secretion in the presence of ritanserin, a serotonin$_2$ receptor antagonist and phentolamine, an α-adrenergic antagonist suggesting

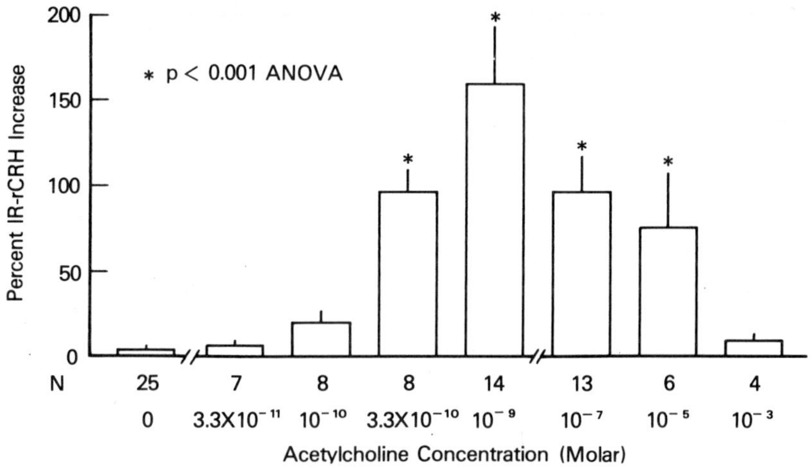

Fig. 1. IR-rCRH secretion by single explanted hypothalami (mean±SE) exposed for 40 min to graded concentrations of acetylcholine, as indicated. Results are expressed as percent increase above baseline. Statistical analysis was done employing one-way analysis of variance (ANOVA) followed by the Ryan-Einot-Gabriel-Welsch multiple range test (REGWQ). Logarithmic transformation of the data was applied to correct for heteroscedasticity, detect by the Bartlett test. Acetylcholine induced a dose-dependent increase of IR-rCRH ($p<0.001$). N=number of hypothalami tested for each dose level (from reference 22).

that the cholinergic stimulation of CRH secretion is not mediated by serotonergic or α-adrenergic interneurons.

Although we cannot rule out a direct effect of systemically administered cholinergic agonists on the adrenal cortex itself (23), our data support the hypothesis that the HPA axis activation mediated by ACh occurs via stimulation of CRH secretion. Although the physiologic role of the stimulatory cholinergic pathway is unknown, there is evidence to suggest that it may play a role in the regulation of the HPA axis during stress. In this regard, it has been shown that central implantation of atropine inhibits stress-induced activation of the HPA axis in rats (24,25) and that the septo-hippocampal cholinergic system, which undergoes rapid activation during acute stress, may play an important role in both endocrine (HPA axis activation) and emotional responsivity to stress (26).

Serotonin

Serotonin has been implicated in the regulation of the HPA axis. In this regard, administration of 5HT precursors, releasers, reuptake inhibitors or receptor agonists increase plasma ACTH, β-endorphin (β-EP) and corticosterone levels in laboratory animals (27,28). Furthermore, it has been reported that various stressors associated with HPA axis activation also enhance brain 5HT turnover (29). A number of experiments have been performed using pituitary cells or hypothalamic explants to determine the site(s) of 5HT action on the HPA axis. Whereas, it seems clear that 5HT is capable of stimulating pituitary ACTH *in vitro* (30), the experiments examining hypothalamic explant release of corticotropin-releasing bioactivity have been less conclusive and 5HT has been reported to stimulate (2), inhibit (13) or to have no effect (14) upon CRH secretion.

In our hands, 5HT stimulated IR-rCRH secretion in a dose-dependent fashion with peak of activity at the 10^{-9} M concentration (19). We attempted also to examine which 5HT receptor subtypes were involved in this response and the possible interactions between 5HT and other neurotransmitter systems. Serotonin-induced IR-rCRH secretion was antagonized by both ketanserin and ritanserin, which are specific $5HT_2$ receptor antagonists, but was not affected by the blockade of cholinergic (atropine plus hexamethonium) or α-adrenergic (phentolamine) receptors (Fig. 2).

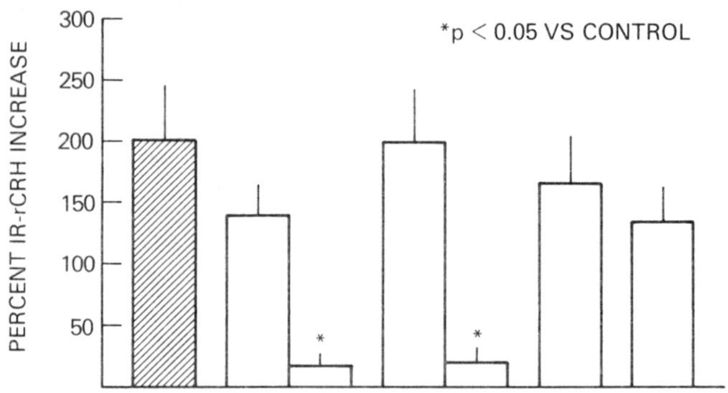

N	17	7	15	7	8	8	15
SEROTONIN (Log M)	−9	−9	−9	−9	−9	−9	−9
KETANSERIN (Log M)	—	−9	−7	—	—	—	—
RITANSERIN (Log M)	—	—	—	−9	−7	—	—
ATROPINE (Log M)	—	—	—	—	—	−7	—
HEXAMETHONIUM (Log M)	—	—	—	—	—	−7	—
PHENTOLAMINE (Log M)	—	—	—	—	—	—	−7

Fig. 2. IR-rCRH secretion by single explanted hypothalami exposed to 10^{-9}M serotonin (shaded bar) or serotonin plus different concentrations of serotonergic, cholinergic and α-adrenergic antagonists. None of these antagonists had any effect on basal IR-rCRH secretion. Results (mean±SE) are expressed as percent increase above baseline. *p<0.05 vs serotonin alone (ANOVA followed by REGWQ test), N=number of hypothalami tested (from reference 19).

Our data suggest that 5HT stimulates hypothalamic CRH secretion primarily via a $5HT_2$ receptor type. This effect does not appear to be mediated by cholinergic or α-adrenergic interneurons, as previously suggested (12). Serotonin appears capable of stimulating the HPA axis not only by releasing hypothalamic CRH and pituitary ACTH, but also by releasing hypothalamic arginine vasopressin (31). In spite of these redundant effects of 5HT on the HPA axis, the physiologic role of this neurotransmitter is unknown. Poorly understood are also its effects on circadian or stress-induced HPA axis activation.

Norepinephrine

The concomitant activation of the CRH and the locus coeruleus-norepinephrine (LC-NE) systems constitutes one of the principal adaptations during either physical or emotional stress (32-34). Hence, an understanding of potential interactions between these two systems may be essential to the study of stress mechanisms. Recent data suggest that CRH significantly increases the locus coeruleus firing rate. However, the effect of NE upon CRH secretion is a matter of considerable debate. Many studies, utilizing indirect assessment of CRH secretion and in the context of widely divergent experimental designs, have failed to convincingly demonstrate whether NE is inhibitory or excitatory to the hypothalamic CRH neuron (15-17). To evaluate the effect of NE on hypothalamic CRH secretion we tested the direct effects of a multiplicity of catecholamine agonists and antagonists upon CRH secretion in vitro (35).

Norepinephrine stimulated IR-rCRH secretion in a dose-dependent fashion with peak of effect in the monomolar range. The effect of NE was antagonized by the mixed α antagonist phentolamine, by the α_1 antagonist prazosin and by the α_2 antagonist yohimbine, but not by the β-blocker L-propanolol (Fig 3).

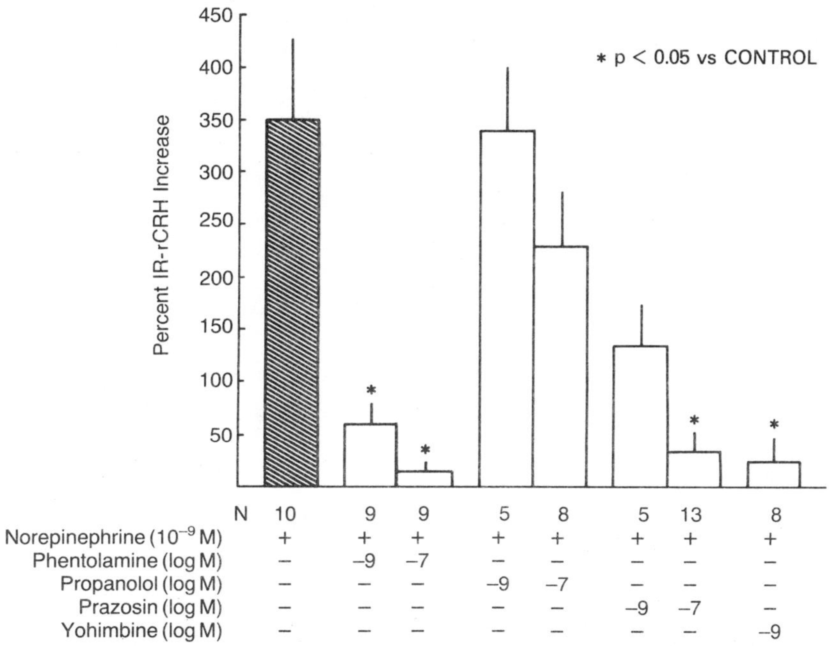

Fig. 3. IR-rCRH secretion by single explanted hypothalami exposed to 10^{-9}M norepinephrine (shaded bar) or norepinephrine plus different concentrations of norepinephrinergic antagonists. None of these antagonists had any effect on basal IR-rCRH secretion. Results (mean±SE) are expressed as percent increase above baseline. *$p<0.05$ vs norepinephrine alone. ANOVA followed by the Duncan multiple range test were employed for the statistical evaluation. N=number of hypothalami tested (from reference 35).

In agreement with these data were the findings that the α_1 agonist phenylephrine and the α_2 agonist clonidine both stimulated IR-rCRH secretion in a dose-dependent fashion. On the other hand, while the β agonist isoproterenol caused a weak, non dose-dependent increase in IR-rCRH secretion, this effect could not be antagonized by L-propanolol. Despite pretreatment with serotonin and ACh receptor antagonists the effect of NE on IR-rCRH secretion was undiminished, suggesting that NE-induced CRH secretion is not mediated by either neurotransmitter. On the other hand, pretreatment with GABA significantly attenuated NE-induced IR-rCRH secretion.

The present data support the hypothesis that NE is excitatory rather than inhibitory upon CRH secretion when it acts directly at a hypothalamic locus. The stimulatory NE effect on hypothalamic CRH secretion may be of interest in light of recent data showing that CRH administration to awake, unrestrained rats markedly increases the LC firing rate (36). These data, taken together, suggest that hypothalamic CRH neurons and NE neurons in regions such as the LC may participate in a mutually reinforcing positive feedback loop. This observation may be pertinent to stressful situations and psychiatric conditions characterized by activation of both the LC-NE and the CRH systems (37).

Epinephrine and Dopamine

Epinephrine has been reported to inhibit the central component of the HPA axis *in vivo* (38), whereas dopamine (DA) has been reported to have no effect upon corticotropin-releasing bioactivity *in vitro* (14). We found that E stimulates IR-rCRH secretion in a dose-dependent fashion. Peak of effect, however, was observed at about 100 times higher concentrations than NE. The stimulatory effect of E could be antagonized by equimolar concentrations of phentolamine, but not by propanolol suggesting that this effect is mediated through α adrenergic receptors.

Dopamine had a weak stimulatory effect on IR-rCRH secretion. This effect was antagonized by the specific D_1 receptor antagonist SCH23390, but not by phentolamine. These data suggest that DA effect on hypothalamic CRH secretion is receptor-mediated and not due to *in vitro* conversion of this neurotransmitter into NE or E (35).

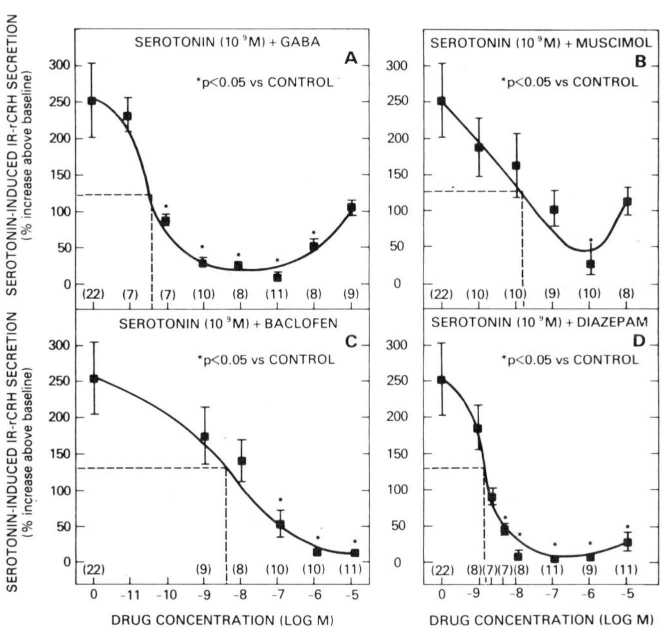

Fig. 4. IR-rCRH secretion by single explanted hypothalami exposed to 10^{-9}M serotonin serotonin plus graded concentrations of GABA (Panel A), serotonin plus graded concentrations of mascimol (Panel B), serotonin plus graded concentrations of baclofen (Panel C), or serotonin plus graded concentrations of diazepam (Panel D). Results (mean±SE) are expressed as percent increase above baseline. *p<0.05 vs serotonin alone (ANOVA followed by Duncan test). Dotted lines intercept curves at ED_{50} level. ED_{50}s were calculated employing the four parameter logistic equation. A computer program, "ALLFIT", developed by DeLean et al., was employed for the computations (43). The number of hypothalami tested for each dose level is indicated in parentheses (from reference 44).

INHIBITORY NEUROTRANSMITTERS

Gamma-Aminobutyric Acid

In contrast to the activating effect of icv administration of CRH on the stress system, GABA and benzodiazepine (BZD) agonists seem capable of not only decreasing excessive responses to stress, such as hypertension (39) and gastric ulceration (40), but also of suppressing the activity of the HPA axis (41). Moreover, activation of the GABA/BZD neurotransmitter system also exerts behavioral effects in opposition to those of CRH by decreasing arousal and inducing effects which can be characterized as anxiolytic (42). In the light of the apparently antithetical effects of CRH and the GABA/BZD system on various physiological and behavioral parameters, we sought to evaluate the effects of the GABA/BZD system on 5HT-induced CRH secretion (43).

GABA inhibited 5HT-induced IR-rCRH secretion from 10^{-10} to 10^{-6} M concentrations (Fig 4, panel A). Muscimol, a $GABA_A$ receptor agonist, inhibited 5HT-induced IR-rCRH secretion only at 10^{-6} M (Fig 4, panel B). Baclofen, a $GABA_B$ receptor agonist was more potent than muscimol, but not than GABA, in suppressing 5HT-induced IR-rCRH secretion (Fig 4, panel C). Diazepam, a classic benzodiazepine which interacts with the BZD site of the $GABA_A$ receptor complex, inhibited 5HT-induced IR-rCRH secretion from 3.3 x 10^{-9} to 10^{-5} M (Fig 4, panel D).

These results suggest that the GABA/BZD system is involved in the regulation of CRH secretion. It appears that both $GABA_A$ and $GABA_B$ receptor types mediate the suppressive effect of GABA upon 5HT-induced CRH secretion. It is not entirely unexpected that activation of the principal inhibitory system in brain (45) would diminish the responsivity to stimulatory inputs of the CRH neuron which is postulated to be a mediator of arousal and of the stress response (5-10). In this regard, it is tempting to postulate that some of the anxiolytic effect of $GABA_A$ and $GABA_B$ receptor agonists may be exerted via inhibition of the CRH neuron.

REGULATORY NEGATIVE "FEEDBACK"INFLUENCES

The activity of the HPA axis is regulated not only by circadian and stress-related excitatory inputs or by inhibitory neural modulation, but also by various negative feedback control loops (46). Negative feedback is exerted by glucocorticoids at the pituitary corticotroph cell, the hypothalamic CRH-secreting neuron and possibly at sites of the limbic system such as the amygdala and the hippocampus (46). ACTH has been proposed to exert negative feedback effects on basal secretion of hypothalamic CRH (47,48) and on its own as well secretion (49). With regard to CRH, it has been suggested that this peptide exerts no ultrashort loop negative feedback under resting conditions but may stimulate its own secretion in stress states (50). Direct evidence for a positive or a negative ultrashort CRH feedback loops would, however, be necessary to clarify this issue. We examined the effects of ovine CRH (oCRH), ACTH, β-EP, α-melanocyte-stimulating hormone (α-MSH), corticotropin-like intermediate lobe peptide (CLIP), ovine β-lipotropin (ovine β-LPH) and dexamethasone (DEX) on basal and neurotransmitter-stimulated CRH secretion (51).

Ovine CRH and DEX inhibited unstimulated IR-rCRH secretion with ED_{50}s at the 10^{-8} M range. ACTH had no detectable suppressive effect at 10^{-8} M. Ovine CRH, ACTH and DEX inhibited 5HT-, ACh- and NE-induced IR-rCRH secretion in a dose-dependent fashion (Fig. 5). β-EP, α-MSH and CLIP also inhibited 5HT-induced IR-rCRH secretion. Of these latter peptides, the strongest inhibitor was β-EP and the weakest CLIP. Ovine β-LPH had only a weak inhibitory effect on 5HT-induced IR-rCRH secretion. Generally, the concentrations required for 50% suppression of neurotransmitter-stimulated IR-rCRH secretion were significantly lower than those required for a similar suppression of unstimulated IR-rCRH secretion.

Our data suggest the presence of multiple regulatory negative, feedback loops operating at the level of the CRH-producing neuron. These include an ultrashort CRH-mediated loop, a short POMC gene-derived peptide loop and a long glucocorticoid-mediated negative feedback loop. The source of ACTH and other fragments of the POMC gene capable to exert negative feedback effects on hypothalamic CRH secretion <u>in vivo</u> has not been definitively elucidated. Several lines of evidence suggest that the most likely source is the arcuate nucleus of the hypothalamus where POMC content is high (52). ACTH-containing cells in the arcuate nucleus receive afferent input from CRH neurons located in the paraventricular nucleus (PVN) (9) and send projections to the PVN (53). In addition, direct application of CRH onto the arcuate nucleus of the rat results in the release of both ACTH and β-EP (54).

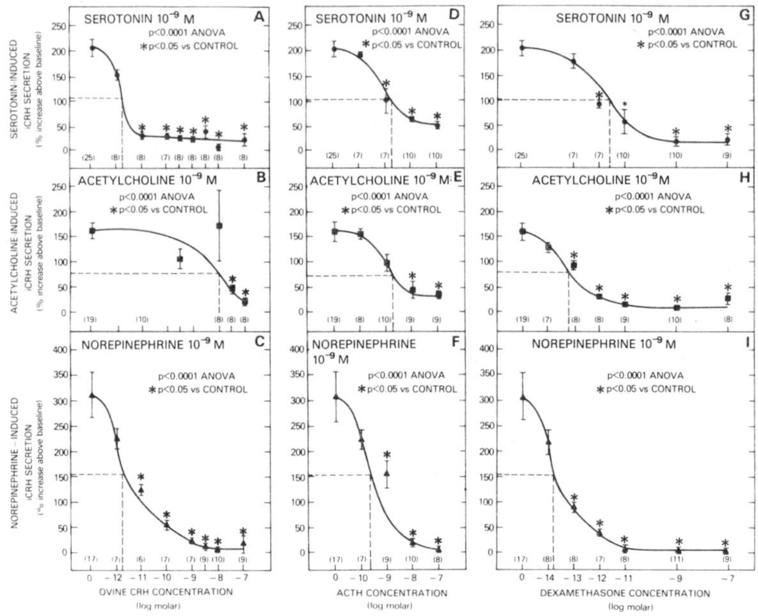

Fig. 5. Effects of graded concentrations of oCRH (left panels), ACTH (central panels) and dexamethasone (right panels) on iCRH secretion by single explanted rat hypothalami stimulated by serotonin (upper panels), acetylcholine (middle panels) or norepinephrine (lower panels). Results (mean±SE) are expressed as percent increase above baseline. *$p<0.05$ vs control (ANOVA followed by Duncan test). Dotted lines intercept curves at ED_{50} level. The number of hypothalami tested at each dose level is indicated in perentheses (from reference 51).

POTENTIAL EXTRAHYPOTHALAMIC INFLUENCES

Activation of the HPA axis is observed during the immune response. Several mediators of immune phenomena, such as inflammation, allergy and anaphylaxis have been shown capable of activating this axis. These include interleukin-1 (IL-1), interleukin-2 (IL-2), thymosins, and lipids (55-60).

A number of studies have been performed to clarify the site of action of IL-1 within the HPA axis. In the rat, IL-1-induced ACTH release was completely abolished by immunoneutralization with CRH antiserum (55,58). Accordingly, IL-1 caused no ACTH secretion by cultured rat anterior pituitary cells (55,58). We have shown that IL-1 stimulated IR-rCRH secretion by cultured rat hypothalami in a dose dependent-fashion down to the concentration of 10^{-15} M (58).

A macrophage-derived cytokine, tumor necrosis factor-α (TNF-α) (Bernardini et al. unpublished information) and IL-2 (58) both stimulated hypothalamic IR-rCRH secretion in vitro. Neither IL-1 nor IL-2 had any effect on pituitary ACTH secretion in vitro. To examine whether the effects of IL-1 on CRH secretion were prostanoid-mediated, we employed the cyclo- and lipooxygenase inhibitors indomethacin (INDO) and eicosatetraynoic acid (ETYA) and the lipooxygenase inhibitor nordihydroguaiaretic acid (NDGA). Both INDO and ETYA inhibited IL-1-induced IR-rCRH secretion, whereas NDGA had no effect. These data suggest that cyclooxygenase products of arachidonic acid, such as prostaglandins (PG) may mediate the IL-1 effects on hypothalamic CRH secretion (61). IL-2 effect on IR-rCRH secretion were not affected by arachidonic acid metabolites. Accordingly, cyclooxygenase metabolites of arachidonic acid, such as $PGF_{2\alpha}$ and thromboxane (TX) B_2, as well as the TXA_2 receptor agonist U-46,619 stimulated hypothalamic IR-rCRH secretion in vitro (62). It is interesting that the mechanism involved in the CRH response to IL-1 is similar to that shown for the fever response to this cytokine.

We have also shown that platelet activating factor (PAF), a glycerolipid with platelet-aggregating properties which is released during the inflammatory response, stimulates ACTH secretion in the rat in vivo and hypothalamic IR-rCRH secretion in vitro (60,63).

We have shown that epidermal growth factor (EGF), a polypeptide mitogen that participates in wound healing, stimulates the HPA axis in primates in a dose-dependent fashion (64). To define whether the locus of stimulation was the hypothalamus and/or the pituitary gland we examined the capacity of mouse EGF to directly stimulate

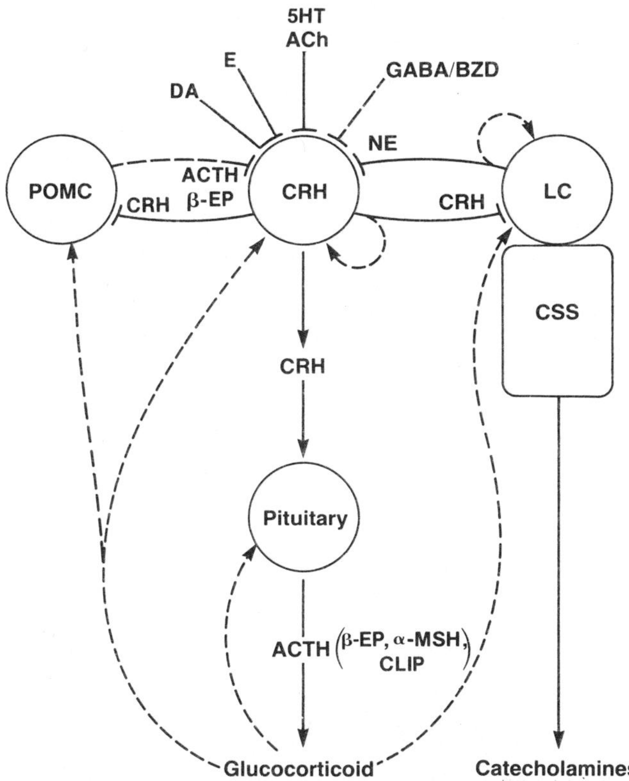

Fig. 6. Schematic representation of the regulation of the hypothalamic-pituitary-adrenal axis and of the functional interrelationships between the hypothalamic CRH neuron and other brain regions, such as the hypothalamic region producing proopiomelanocortin (POMC) gene-derived peptides and the centers of the sympathetic system in the hindbrain. CRH activates the pituitary-adrenocortical axis, the locus coerulus (LC) and the central sympethetic system (CSS) and stimulates secretion of hypothalamic POMC gene-derived peptides. The activity of the CRH neuron appears to be regulated by central stimulatory inputs mediated by serotonin (5HT), acetylcholine (ACh), norepinephrine (NE), epinephrine (E) and dopamine (DA), by central inhibitory inputs mediated by the GABA/benzodiazepine (GABA/BZD) system and by multiple negative feedback loops. These include an ultrashort CRH-mediated loop, a short hypothalamic POMC gene-derived peptide loop and a long glucocorticoid-mediated negative feedback loop.

hypothalamic IR-rCRH or pituitary ACTH secretion in vitro. Mouse EGF stimulated IR-rCRH secretion in a dose-dependent fashion, but failed to cause pituitary ACTH secretion. These data suggest that EGF, and/or one of its naturally occurring analogs, may participate in the physiological activation of the HPA axis at times during which the concentrations of these factors are raised in the systemic circulation and/or locally in the hypothalamus. Such states may include trauma, surgery, and possibly emotional stress.

SUMMARY AND CONCLUSION

In summary, 5HT, ACh, NE, E and DA appear to stimulate hypothalamic CRH secretion whereas activation of the GABA/BZD system seems to decrease the responsivity of the CRH neuron to stimulatory neurotransmitters (Fig. 6). Hypothalamic CRH released from the hypothalamic neuron not only activates the HPA axis, but also stimulates the locus coeruleus-norepinephrine system (LC) and the central sympathetic system (CSS). CRH also induces secretion of hypothalamic POMC gene-derived peptides, such as ACTH, β-EP, α-MSH and CLIP. These peptides as well as CRH itself, decrease the responsivity of the CRH neuron to stimulatory inputs. In addition, glucocorticoids restrain the activity of both the CRH neuron and the locus coeruleus and may also inhibit the secretion of POMC gene-derived peptides by the POMC neurons of the arcuate nucleus.

Hypothalamic CRH secretion is regulated also by a number of mediators of the immune response, such as IL-1, IL-2, TNF-α and $PGF_{2\alpha}$, PAF and EGF. Although the physiologic significance of this regulation is largely unknown, it is tempting to speculate that cytokines and mediators of inflammation released in vivo may activate the HPA axis to trigger a glucocorticoid-mediated counter-regulatory mechanism to restrain the immune system (Fig. 7).

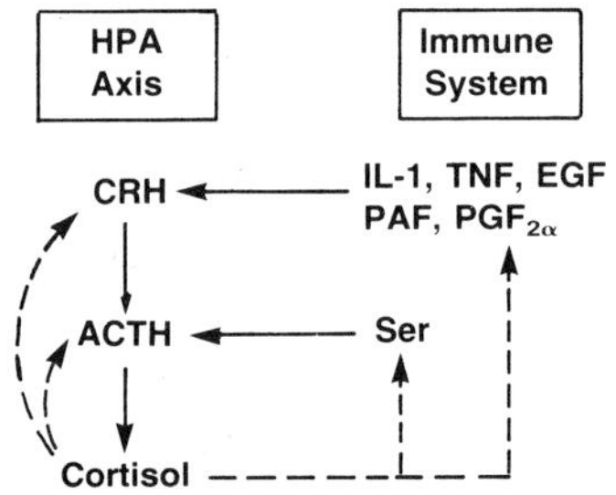

Fig. 7. Schematic representation of the interactions between the HPA axis and the immune system. Continuous lines represent stimulatory inputs and interrupted lines represent inhibitory inputs.

In conclusion, our in vitro hypothalamic organ culture system allowed us to examine the regulation of CRH secretion in a direct and specific manner. Some of our observations may help with better understanding of the role played by CRH in the complex symptomatology of stress. In making extrapolations and interpretations from the in vitro data, however, we should try to keep in mind the words of Claude Bernard, "... If we break up a living organism by isolating its different parts it is only for the sake of ease in analysis and by no means in order to consider them separately. Indeed when we wish to ascribe to a physiological quality its value and true significance we must always refer it to this whole and draw our final conclusions only in relation to the effects in the whole".

REFERENCES

1. Antoni FA, Palkovits M, Makara GB, Linton EA, Lowry PJ, Kiss JZ 1983 Immunoreactive corticotropin-releasing hormone in the hypothalamo-infundibular tract. Neuroendocrinology 36:415-423

2. Swanson LW, Sawchenko PE, Rivier J, Vale W 1983 Organization of ovine corticotropin-releasing factor immunoreactive cells and fibers in the rat brain: An immunohistochemical study. Neuroendocrinology 36:165-186

3. Vale W, Spiess J, Rivier C, Rivier J 1981 Characterization of a 41-residue ovine hypothalamic peptide that stimulates secretion of corticotropin and β-endorphin. Science (USA) 213:1394-1397

4. Brown MR, Fisher LA, Rivier J, Spiess J, Rivier C, Vale W 1982 Corticotropin-releasing factor: effects on the sympathetic nervous system and oxygen consumption. Life Sci 30:207-210

5. Britton DR, Koob GF, Rivier J, Vale W 1982 Intraventricular corticotropin-releasing factor enhances behavioral effects of novelty. Life Sci 31:363-367

6. Ehlers CL Henriksen SJ, Wang M, Rivier J, Vale W, Bloom FE 1983 Corticotropin-releasing factor produces increases in brain excitability and convulsive seizures in rats. Brain Res 278:332-336

7. Kalin NH, Shelton SE, Kraemer GW, McKinney WT 1983 Corticotropin-releasing factor administered intraventricularly to rhesus monkeys. Peptides (Fayetteville) 4:217-220

8. Morley JE, Levine AS 1982 Corticotropin-releasing factor, grooming and ingestive behavior. Life Sci 31:1459-1464

9. Sirinathsinghji DJS, Rees LH, Rivier J, Vale W 1983 Corticotropin-releasing factor is a potent inhibitor of sexual receptivity in the female rat. Nature (London) 27:232-235

10. Sutton RE, Koob GF, LeMoal M, Rivier J, Vale W 1982 Corticotropin-releasing factor produces behavioral activation in rats. Nature (London) 297:331-333

11. Britton DR, Britton KT 1981 A sensitive open field measure of anxiolytic drug activity. Pharmacol Biochem Behav 15:577-582

12. Jones MT, Gillham B, Di Renzo G, Beckford U, Holmes MC 1981 Neural controlof corticotrophin secretion. In: Front Horm Res Vol 8, TJB van Wimerma Greidanus, (ed) p 12-43 (Karger, Basel)

13. Telegdy G, Vermes I 1973 The role of serotonin in the regulation of the hypophysis-adrenal system. In: Brodish A, Redgate ES (eds) Brain-Pituitary-adrenal Interrelationships. Karger, Basel, p 332-338

14. Fehm HL, Voigt KH, Lang RE, Pfeiffer EF 1980 Effects of neurotransmitters on the release of corticotropin releasing hormone (CRH) by rat hypothalamic tissue _in vitro_. Exp Brain Res 39:229-234

15. Rose JL, Goldsmith PC, Holland FJ, Kaplan SL, Ganong WF 1976 Effect of

electrical stimulation of the canine brain stem on the secretion of ACTH and growth hormone (GH). Neuroendocrinology 22:352-362

16. Hillhouse EW, Burden J, Jones MT 1975 The effect of various putative neurotransmitters on the release of corticotropin-releasing hormone from the hypothalamus of the rat in vitro. I. The effect of acetylcholine and noradrenaline. Neuroendocrinology 17:1-11

17. Szafarczyk A, Alonso G, Ixart G, Malaval F, Assenmacher I 1985 Diurnal-stimulated and stress-induced ACTH release in rats is mediated by ventral noradrenergic bundle. Am J Physiol 249:E219-E226

18. Antoni FA 1986 Hypothalamic control of adrenocorticotropin secretion: advances since the discovery of 41-residue corticotropin-releasing factor. Endocr Rev 7:351-378

19. Calogero AE, Bernardini R, Margioris AN, Gallucci WT, Munson PJ, Tamarkin L, Tomai TP, Brady L, Gold PW, Chrousos GP 1988 Serotonin stimulates rat hypothalamic corticotropin-releasing hormone secretion in vitro. Submitted for publication

20. Schuermeyer TH, Avgerinos P, Gold PW, Gallucci WT, Tomai TP, Cutler GB, Loriaux DL, Chrousos GP 1984 Human corticotropin-releasing factor in man: Pharmacokinetic properties and dose-response of plasma adrenocorticotropin and cortisol secretion. J Clin Endocrinol Metab 59:1103-1108

21. Briaud B, Koch B, Lutz-Bucher B, Miolhe C 1979 In vitro regulation of ACTH release from neurointermediate lobe of rat hypophysis. II. Effect of neuro-transmitters. Neuroendocrinology 28:377-385

22. Calogero AE, Gallucci WT, Bernardini R, Saoutis C, Gold PW, Chrousos GP 1988 Effect of cholinergic agonists and antagonists on rat hypothalamic corticotropin-releasing hormone secretion in vitro. Neuroendocrinology 46:303-308

23. Hadjian AJ, Culty M, Chambaz EM 1984 Stimulation of phosphatidylinositol turnover by acetylcholine, angiotensin II and ACTH in bovine adrenal fasciculata cells. Biochem Biophys Acta 804:427-433

24. Hedge GA, Smelik PG 1968 Corticotropin release: Inhibition by intrahypothalamic implantation of atropine. Science (USA) 159:891-892

25. Hedge GA, de Wied D 1971 Corticotropin and vasopressin secretion after hypothalamic implantation of atropine. Endocrinology 88:1257-1259

26. Gilad GM, Mahon BD, Finkelstein Y, Koffler B, Gilad VH 1985 Stress-induced activation of the hippocampal cholinergic system and the pituitary-adrenocortical axis. Brain Res 347:404-408

27. Bruni JF, Hawkins RL, Yen SSC 1982 Serotonergic mechanism in the control of β-endorphin and ACTH release in male rats. Life Sci 30:1247-1254

28. Fuller RW 1981 Serotonergic stimualtion of pituitary-adrenocortical function in rats. Neuroendocrinology 32:118-127

29. Mueller GP, Twohy CP, Chen HT, Advis JP, Meites J 1976 Effects of L-tryptophan and restrain stress on hypothalamic and brain serotonin turnover, and pituitary TSH and prolactin release in rats. Life Sci 18:715-724

30. Spinedi E, Negro-Vilar A 1983 Serotonin and adrenocorticotropin (ACTH) release: direct effects at the anterior pituitary level and potentiation of arginine

vasopressin-induced ACTH release. Endocrinology 112:1217-1223

31. Gibbs DM, Vale W 1983 Effect of the serotonin reuptake inhibitor fluoxetine on corticotropin-releasing factor and vasopressin secretion into the hypophyseal portal blood. Brain Res 280:176-179

32. Axelrod J, Reisine TD 1984 Stress hormones: Their interaction and regulation. Science (USA) 224:452-459

33. Axelrod J, Muller RA, Henry JP, Stephens PM 1970 Changes in enzymes involved in the biosynthesis and metabolism of noradrenaline and adrenaline after psychosocial stimulation. Nature (London) 225:1059-1060

34. Udelsman R, Norton JA, Jelenich SE, Goldstein DS, Linehan WM, Loriaux DL, Chrousos GP 1987 Responses of the hypothalamic-pituitary-adrenal and renin-angiotensin axes and the sympathetic system during controlled surgical and anesthetic stress. J Clin Endocrinol Metab 64:986-994

35. Calogero AE, Gallucci WT, Chrousos GP, Gold PW 1988 Catecholamine effect upon rat hypothalamic corticotropin releasing hormone secretion in vitro. J Clin Invest (In Press)

36. Valentino RJ, Foote SL, Aston-Jones G 1983 Corticotropin-releasing factor activates noradrenergic neurons of the locus ceruleus. Brain Res 270:363-367

37. Gold PW, Goodwin FK, Chrousos GP 1988 Clinical and biochemical manifestations in depression. Relationship to neurobiology of stress. N Engl J Med. In Press

38. Van Loon GR, Scapagnini U, Moberg GP, Ganong WF 1971 Evidence for central adrenergic neural inhibition of ACTH secretion in the rat. Endocrinology 89:1464-1469

39. Benson H, Herd JA, Morse WH, Kelleher RT 1970. Hypothensive effects of chlordiazepoxide, amobarvital and chlorpromazine on behaviorally-induced elevated blood pressure in squirrel monkey. J Pharmacol Exp Ther 173:399-406

40. Dasgupta S, Mukherjee B 1967 Effect of chlordiazepoxide or stomach ulcers in rabbits induced by stress. Nature (London) 215:1183

41. Makara GB, Stark E 1974 Effect of gamma-aminobutyric acid (GABA) and GABA antagonist drugs or ACTH release. Neuroendocrinology 16:178-190

42. Britton WT, Morgan J, Rivier J, Vale W, Koob GF 1985 Chlorodiazepoxide attenuates response suppression induced by corticotropin-releasing factor in the conflict test. Psychopharmacology 86:170-174

43. DeLean A, Munson PJ, Rodbard D 1978 Simultaneous analysis of families of sigmoidal curves: Application to bioassay, radioligand assay, and physiological dose-response curves. Am J Physiol 235:E97-E102

44. Calogero AE, Gallucci WT, Chrousos GP, Gold PW 1988 Interaction between GABAergic neurotransmission and rat hypothalamic corticotropin releasing hormone secretion in vitro. Submitted for publication

45. Baxter CF 1970 The nature of gamma-aminobutyric acid. In: A Lajtha (Ed) Handbook of Neurochemistry, vol 2, Plenum Press, NY pp 289-353

46. Keller-Wood ME, Dallman MF 1984 Corticosteroid inhibition of ACTH secretion. Endocr Rev 5:1-24

47. Suda T, Yajima F, Tomori N, Sumitomo T, Nakagami Y, Ushiyama T, Demura H, Shizume K 1986 Inhibitory effect of adrenocorticotropin on corticotropin-releasing factor release from rat hypothalamus in vitro. Endocrinology 118:459-461

48. Suda T, Tomori N, Yajima F, Ushiyama T, Sumitomo T, Nakagami Y, Demura H, Shizume K 1987 A short negative feedback mechanism regulating corticotropin-releasing hormone release. J Clin Endocrinol Metab 64:909-913

49. Boscaro M, Sonino N, Paoletta A, Rampazzo A, Mantero F 1988 Evidence for ultra-short loop autoregulation of adrenocorticotropin secretion in man. J Clin Endocrinol Metab 66:255-257

50. Ono N, Bedran De Castro JC, McCann SM 1985 Ultrashort-loop positive feedback of corticotropin (ACTH)-releasing factor to enhance ACTH release in stress. Proc Natl Acad Sci (USA) 82:3528-3531

51. Calogero AE, Gallucci WT, Gold PW, Chrousos GP 1988 Multiple feedback regulatory loops upon rat hypothalamic corticotropin-releasing hormone secretion: Poential clinical implications. J Clin Invest (In Press)

52. Krieger DT, Liotta AS, Brownstein MJ 1977 Presence of corticotropin in Brain of normal and hypophysectomized rats. Proc Natl Acad Sci (USA) 74:648-652

53. Sawchenko PE, Swanson LW, Joseph SA 1982 The distribution and cells of origin of ACTH(1-39)-stained varicosities in the paraventricular and supraoptic nuclei. Brain Res 232:365-374

54. Nikolarakis KE, Almeida OFX, Herz A 1986 Stimulation of hypothalamic beta-endorphin and dynorphin release by corticotropin-releasing factor (in vitro). Brain Res 399:152-161

55. Uehara A, Gottscholl PE, Dahl RR, Arimura A 1987 Interleukin-1 stimulates ACTH release by an indirect action which requires endogenous corticotropin-releasing factor. Endocrinology 121:1580-1582

56. Sapolsky R, Rivier C, Yamamoto G, Plotsky P, Vale W 1987 Interleukin-1 stimulates the secretion of hypothalamic corticotropin-releasing factor. Science (USA) 238:522-524

57. Basedowsky H, Del Rey A, Sorkin E, Dinarello CA 1986 Immunoregulatory feedback between interleukin-1 and glucocorticoid hormones. Science (USA) 233:652-654

58. Calogero AE, Luger T, Gallucci WT, Gold PW, Chrousos GP 1987 Interleukin-1 and interleukin-2 stimulate hypothalamic corticotropin-releasing hormone but not pituitary ACTH secretion in vitro. Program of the 69th Annual Meeting of the Endocrine Society, Indianapolis, IN (Abstract 1001)

59. Healy DL, Hodgen GD, Schulte HM, Chrousos GP, Loriaux DL, Hall NR, Goldstein AL 1983 The thymus-adrenal connection: Thymosin has corticotropin releasing activity in primates. Science (USA) 222:1353-1355

60. Bernardini R, Calogero AE, Gold PW Chrousos GP 1988 Platelet activating factor stimulates adrenocorticotropin secretion in vivo. Clin Research 36:391A (Abstract)

61. Bernardini R, Listwak SJ, Calogero AE, Gold PW, Chrousos GP 1988 Interleukin-1 and -2 effects on hypothalamic corticotropin releasing hormone (CRH) secretion. Program of the UCLA Symposium on Therapeutic Application of Biologicals, Keystone, CO (Abstract) In Press

62. Bernardini R, Calogero AE, Gold PW, Chrousos GP 1988 Effect of arachidonic acid

metabolites on rat hypothalamic corticotropin releasing hormone secretion _in vitro_. Program of the 8th International Congress of Endocrinology, Kyoto, Japan (Abstract) In Press

63. Bernardini R, Calogero AE, Chrousos GP, Saoutis C, Gold PW 1987 Platelet activating factor stimulates hypothalamic corticotropin releasing hormone secretion _in vitro_. Program of the 17th Annual Meeting of the Society for Neuroscience, New Orleans. LA (Abstract 450.9)

64. Luger A, Calogero AE, Kalogeras K, Gallucci WT, Gold PW, Loriaux DL, Chrousos GP 1988 Interaction of epidermal growth factor with the hypothalamic-pituitary-adrenal axis: Potential physiologic relevance. J Clin Endocrinol Metab 66:334-337

IS THE BENZODIAZEPINE/GABA RECEPTOR CHLORIDE IONOPHORE COMPLEX INVOLVED IN PHYSICAL AND EMOTIONAL STRESS?

Ramon Trullas, Hratchia Havoundjian, and Phil Skolnick

Laboratory of Neuroscience, NIDDK
National Institutes of Health, Bethesda, MD 20892

I. Introduction

The response of an organism to physical and emotional stressors requires a complex integration of somatic, endocrine and behavioral systems. While causal relationships between specific receptor/effector systems in the central nervous system (CNS) and stress-induced somatic, endocrine and behavioral responses have not been firmly established, both physical and emotional stressors have been shown to affect a variety of neurotransmitters, neuromodulators, second messenger systems, and their associated recognition sites in the CNS (this volume and [1] for overviews). Thus, it is not surprising that most schemata describing stress-induced alterations in the CNS focus on the hypothalamus, although both the control and generation of stress must be initiated in higher centers.

It has been known since antiquity that the response to both physical and emotional stressors can be modified by drugs. Ethanol must be considered the oldest and perhaps most widely used anti-anxiety agent, while barbiturates and benzodiazepines can be considered synthetic means of coping with stress or anxiety. In experimental animals, it can be demonstrated that appropriate doses of such agents can blunt physically or emotionally provoked activation of the HPA axis, somatization (e.g. activation of the cardiovascular system), and improve social, operant, and non-operant "performance" in situations thought to model stress or anxiety in man (so-called "conflict tests") [2].

During the past decade, it has become apparent that many clinically useful anti-anxiety agents perturb one or more components of an oligomeric protein complex, often termed the benzodiazepine/GABA receptor chloride ionophore complex (supramolecular complex) [2-5]. Furthermore, it has been demonstrated that administration of certain high affinity ligands of the benzodiazepine receptor (best exemplified by C-3 substituted ß-carbolines such as 3-carboethoxy-ß-carboline) induce a syndrome that mimics the somatic, endocrine, and behavioral manifestations of stress or anxiety. Although most of the latter studies have been performed in rodents and non-human primates [6-10], administration of compounds such as N-methyl-ß-carboline-3-carboxamide (FG 7142) to a small number of normal volunteers produced not only a somatic and endocrine mimicry of stress, but an affective description of severe anxiety [11,12]. These pharmacological studies have implicated the supramolecular complex as a physiologically relevant neurochemical mediator of stress or anxiety. The supra-molecular complex is currently thought to consist of 50-60 kdalton subunits containing discrete recognition sites for both benzodiazepines (the benzodiazepine receptor, which also recognizes other chemically diverse compounds such as ß-carbolines, imidazobenzodiazepines, cyclopyrrolones, and triazolopyridazines) [13] and a subpopulation of receptors for τ-aminobutyric acid (GABA), the principal inhibitory neurotransmitter in the mammalian CNS. Electrophysiological and biochemical studies suggest that occupation of benzodiazepine and/or GABA receptors by appropriate ligands selectively affects the movement of Cl^- ions through specific channels in the cell membrane, resulting in an altered neuronal excitability [14,15]. The molecular arrangement of these channels with benzodiazepine/GABA receptor subunits is still a matter of intensive investigation [4,5]. Nonetheless, it is clear that the benzodiazepine/GABA receptor complex and its associated chloride channel (which contain multiple, allosteric recognition sites for compounds such as the barbiturates and pyrazolopyridines) [16,17] are functionally, if not physically associated. If the GABA-gated chloride channel is viewed as the effector component of the supramolecular complex, then by analogy with other signal transducing systems (e.g. the ß-adrenoceptor coupled adenylate cyclase system), it is the <u>effector</u> rather than recognition component which can undergo rapid and robust activation or inhibition. Thus, studies performed in our laboratory during the past two years have focussed on the impact of both physical and emotional stress on these benzodiazepine receptor coupled, GABA-gated chloride channels.

In this chapter, we summarize these findings which demonstrate that both physical and emotional stressors produce rapid and robust changes in the supramolecular complex which **precede** both a significant activation of the HPA axis and alterations in other neurotransmitter systems that have been associated with the stress response.

<u>II. Environmentally-induced modification of GABA-gated chloride channels.</u>

A. Effects of swim stress: In our initial studies, brief ambient temperature swim was selected as a model physical stressor, and Cl^- enhanced [^{3}H]flunitrazepam binding measured in well-washed preparations of rat cerebral cortex. The ability of permeant anions (such as Cl^-) to increase the apparent affinity of [^{3}H]benzodiazepines for their receptors (first described by Costa, et al. in 1979 [18]) is thought to reflect the coupling between chloride channels and benzodi-azepine receptors, and might thus indicate whether stress affected these chloride channels. Table 1 (left panel) illustrates the effects of this ambient temperature swim on tissues obtained from animals housed in a multi-user animal room containing mixed species (rats, guinea pigs) which also served as a procedure room (i.e. animals were handled, injected, and sacrificed in this area).

While Cl^- ions increased [^3H]flunitrazepam binding in a concentration dependent fashion in cortical membranes from swim-stressed and control animals, both the efficacy [the maximum increase in [^3H]flunitrazepam binding observed under these conditions (E_{max})] and potency of Cl^- [manifest as a lower EC_{50} (i.e. that concentration of Cl^- needed to produce a half-maximum increase in [^3H]flunitrazepam binding)] were significantly greater in the swim-stressed group. These findings are summarized in Table 1 (left panel) and in greater detail in [19-21]. A similar increase in the E_{max} (but not potency) of Br^- and I^- was observed in cortical membranes from stressed rats [20], but was not seen with ions such as acetate or phosphate which are relatively impermeable through chloride channels. These observations suggest a marked specificity of this effect, rather than merely increasing the ionic strength of the incubation medium. These observations were perhaps the first neurochemical evidence that stress could affect the chloride ionophore component of the supramolecular complex. Of considerable interest was the observation that no statistically significant differences in either [^3H]flunitrazepam binding or GABA-enhanced [^3H]flunitrazepam (which appears to reflect [15,22] that subpopulation GABA receptors coupled to benzodiazepine receptors) were observed between groups in the absence of Cl^-. These findings will be discussed in more detail in a subsequent section of this chapter.

These initial observations suggested that stress altered either the chloride ionophore or the coupling between the chloride ionophore and benzodiazepine receptor. However, since the physical significance of Cl^- enhanced [^3H]flunitrazepam binding is not well-defined, [^{35}S]t-butylbicylophosphorothionate (TBPS) [23] binding was employed as a more direct means of assessing stress-induced changes in this system. While a stoichiometric relationship between the E_{max} of Cl^- enhanced [^3H]flunitrazepam and the number of [^{35}S]TBPS binding sites has recently been evinced [24], it is evident that these parameters are not identical. The non-identity of these measures could reflect the differential regulation of an anion binding site [25,26] and the [^{35}S]TBPS binding site (see below). Alternatively, halide enhanced [^3H]benzodiazepine binding may measure only that population of chloride channels linked to benzodiazepine receptors and [^{35}S]TBPS binding <u>all</u> GABA-gated chloride channels [27,28], or halide-enhanced [^3H]benzodiazepine binding may reflect changes in the "coupling" (transduction of information) between chloride channels and benzodiazepine receptors. Squires, et al. [23] initially described the binding characteristics of [^{35}S]TBPS (a "cage" convulsant structurally related to picrotoxinin) and demonstrated that this compound labelled a population of sites at or near the GABA/benzodiazepine receptor linked chloride channel. More recent electrophysiological [29] and biochemical evidence [30] strongly suggests that such cage convulsants act directly at GABA-gated chloride channels. Furthermore, it has been shown that under a defined set of incubation conditions, an excellent correlation is observed between the permeability characteristics of a series of anions and the apparent affinity of this radioligand [30]. Other recent studies have demonstrated that incubation of brain tissue with the enzyme phospholipase A_2 produces a parallel loss in the number (B_{max}) of [^{35}S]TBPS binding sites and pentobarbital stimulated $^{36}Cl^-$ efflux from synaptoneurosomes [31,32]. These findings imply that the number of [^{35}S]TBPS binding sites may be related to the number of chloride channels in an "open" conformation. Thus, alterations in both the number of [^{35}S]TBPS binding sites and the apparent affinity of this radioligand under a defined set of conditions may be used to deduce information about the biophysical properties of GABA-gated chloride channels.

Table 1 (left panel) summarizes the effects of swim stress on [^{35}S]TBPS binding in a corresponding group of rats. A statistically significant increase in both the **number** (B_{max}) of binding sites (~34%) and **apparent affinity** (K_d reduced by ~23%) of [^{35}S]TBPS was observed in cortical membranes from the stressed group. Based on the relationships between [^{35}S]TBPS binding and GABA-gated chloride channels described in the previous paragraph, it may be inferred that swim-stress increased both the number and permeability of GABA-gated chloride channels in rat cerebral cortex.

B. Effects of housing conditions

These initial studies raised a number of important issues about the physiological relevance of the observed phenomena. For example, the observed changes might be attributed to physical exertion rather than to the "anxiety" produced by subjecting the animal to a novel (and most certainly stressful) environment, despite the finding that significant increases in Cl^- enhanced [^{3}H]flunitrazepam were observed after as little as five minutes ambient temperature swim [20]. These considerations prompted a series of experiments designed to determine whether the home-cage environment influenced GABA-gated chloride channels on either a tonic or acute (stress-induced) basis. In our initial studies, animals were housed in a colony room containing mixed species (rats, mice, guinea pigs), where laboratory procedures (e.g. injection, decapitation) were performed by a number of investigators throughout the day. Subsequent studies were performed in a "protected" environment, where animals were housed in metal cages inside a light-tight, sound attenuated cabinet with an exhaust fan and two indirect fluorescent lights providing a 12 h light/dark cycle. Food and water (and identical 12 h light/dark cycles) were available to both animals housed in the colony room and the "protected" environment. It was observed that in identically prepared cortical membranes from animals housed in this protected environment, the number of [^{35}S]TBPS binding sites and the E_{max} of Cl^- enhanced [^{3}H]flunitrazepam binding were significantly lower than in tissues prepared from rats housed in the "conventional" animal room. While swim-stress produced a robust increase in the B_{max} of [^{35}S]TBPS and E_{max} of Cl^--enhanced [^{3}H]flunitrazepam binding in both groups, the **absolute** values of these parameters were significantly lower in cortical tissues prepared from animals housed in the protected environment than those housed in the conventional facility. These findings are summarized in Table 1. Although stress-induced decreases in both the K_d of [^{35}S]TBPS and EC_{50}

Table 1: Effects of housing conditions and swim-stress on [^{35}S]TBPS and Cl^- enhanced [^{3}H]flunitrazepam binding to cerebral cortical membranes. Values represent X±SEM of at least 4 animals/group. Animals were housed in a conventional colony room or environmental chamber ("protected environment"). No more than two animals were removed from a single cage in each experiment. These data are from references [20] and [21]. EC_{50} refers to that concentration of NaCl producing a half-maximum increase in [^{3}H]flunitrazepam binding. E_{max} is the maximum increase in [^{3}H]flunitrazepam binding obtained in the presence of 400 mM NaCl (see references [20-21] for details). Symbols: *, Significantly different from the respective control group ($p<0.05$, Duncan´s test); ‡ Significantly different from the corresponding group of the other housing environment ($p<0.05$, Duncan´s test).

ENVIRONMENT	ANIMAL ROOM		"PROTECTED" ENVIRONMENT	
	CONTROL	STRESS	CONTROL	STRESS
<u>Cl^--Enhanced [^{3}H]flunitrazepam Binding</u>				
EC_{50} (mM)	92 ±5	59 ±4*	83 ±8	65 ±5*
E_{max} (%)	33 ±2‡	58 ±3*‡	20 ±3‡	39 ±1*‡
<u>[^{35}S]TBPS Binding</u>				
K_D (nM)	37.2 ±3	28.6 ±2*	35.3 ±2	24.9 ±2*
B_{max} (fmol/mg prot)	2030 ±56‡	2722 ±63*‡	1454 ±94‡	2131 ±59*‡

of Cl^- were also apparent in animals housed in the protected environment, neither the "basal" nor the stress-induced changes in these parameters were significantly different between the housing conditions. These findings suggest that the number of [^{35}S]TBPS binding sites (which presumably represents the number of channels in an "open" conformation) is under both tonic and acute regulation, while the K_d of [^{35}S]TBPS (which is presumably related to the permeability characteristics of GABA-gated chloride channels) [30] appears to be under acute, but not tonic regulation by the environment. These findings also imply that stress-induced perturbation of the K_d and B_{max} of [^{35}S]TBPS may be controlled by different mechanisms.

Both biochemical and pharmacological studies suggest that the swim-stress induced changes in [^{35}S]TBPS and Cl^- enhanced [^{3}H]flunitrazepam binding may be an attempt of the organism to compensate for stressful perturbation of the environment. Such evidence includes: 1) the addition of pentobarbital (200-400 μM) to cortical membranes of non-stressed animals results in an E_{max} of Cl^- enhanced [^{3}H]flunitrazepam binding that is not significantly different from values obtained in swim-stressed rats [20]; 2) the increase in apparent affinity of [^{35}S]TBPS elicited by swim stress is mimicked in control animals by addition of flunitrazepam (10-100 nM) to well-washed membranes, and a full mimicry of stress on [^{35}S]TBPS binding is observed by adding flunitrazepam to cortical homogenates of control animals prior to extensive washing [33]; 3) swim-stress increases muscimol stimulated $^{36}Cl^-$ flux in cortical synaptoneurosomes [34] in stressed animals; 4) a stress-induced increase in the threshold of convulsants that act through the supramolecular complex (e.g. picrotoxin [35], bicuculline [36], and the ß-carboline DMCM [in preparation], but not of compounds such as strychnine [35], which elicit convulsions through glycine-gated chloride channels.

We believe the results obtained by varying the housing conditions provide further support for this hypothesis since animals housed in a conventional facility have a significantly higer number of [^{35}S]TBPS binding sites than do rats housed in a protected environment. Conventional housing facilities are known to be more "stressful" to animals than a protected environment [37-39], and animals housed in our facility were additionally subject to random (intermittent) entry by investigators, presumably accompanied by distressful olfactory and visual cues. Further evidence that increases in Cl^- enhanced [^{3}H]flunitrazepam binding and the number of [^{35}S]TBPS binding sites represent an attempt to compensate for stressful changes in the environment was provided by data obtained from animals housed in this "protected" environment. During the course of these initial investigations, we sometimes observed an apparent effect of the order in which "naive", but **not** stressed animals were sacrificed. That is, the E_{max} of Cl^- to increase [^{3}H]flunitrazepam binding (in cortical membranes) was observed to progressively increase as the animals were sacrificed, despite a randomization of the order in which tissues were processed and assayed to eliminate the introduction of experimental bias [21]. In earlier studies we restricted the number of animals sacrificed to two "naive" and two "stressed" animals per experiment [19,20]. Nonetheless, these observations suggested that the benzodiazepine receptor complex could be even more sensitive to "stress" than our initial experiments with ambient temperature swim indicated. Analysis of these data revealed a significant increase in Cl^- enhanced [^{3}H]flunitrazepam binding in cortical and hippocampal (but **not** cerebellar) membranes from the second cohort removed from a common cage compared to the first. Thus, Cl^- enhanced [^{3}H]flunitrazepam binding in hippocampal and cortical

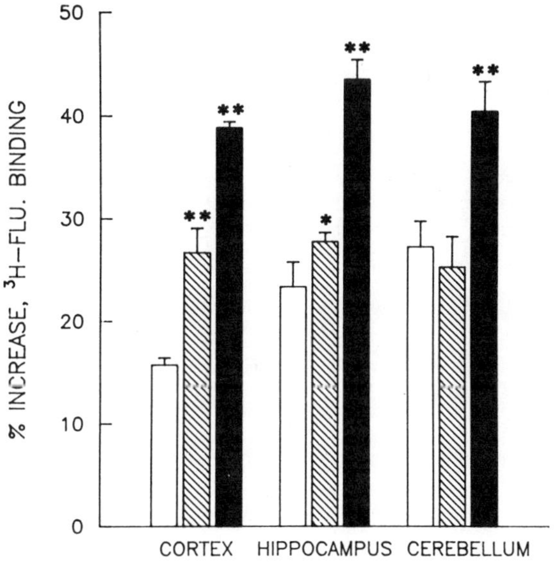

Figure 1: Effects of cohort removal and swim-stress on chloride-enhanced [^{3}H]flunitrazepam binding: regional specificity. Rats were group housed in an environmental chamber for at least seven days. The two "naive" animals were removed from the same cage. The time required to remove the first cohort from the cage, close the environmental chamber and decapitate the animal and repeat the procedure for the second cohort was ≤15 s. The stressed rats (10 min. ambient temperature swim) were removed from a different cage. Tissues were prepared and assayed for [^{3}H]flunitrazepam binding in the presence of 400 mM NaCl as described in references [19-21]. Symbols: Open bars, first cohort; hatched bars, second cohort; solid bars, swim-stressed (10 min., ambient temperature) animals. Cerebral cortex and hippocampus: Cl^- enhanced [^{3}H]flunitrazepam binding was significantly different in the second cohort removed from a common cage and swim-stressed rats compared to the first cohort (cortex: $p<0.01$ and $p<0.01$, respectively, paired t-test; hippocampus: $p<0.05$ and $p<0.01$, respectively paired t-test). Cerebellum: "Stress" significantly different from non-stressed ($p<0.01$); no differences between first and second cohorts removed from a common cage. A consistent "kill order" effect was not observed if the "second" animal killed was removed from a different cage in the same environmental chamber or a different chamber. The concentration of radioligand in these experiments was ~0.7 nM. From Trullas, et al. 1987 [21].

membranes of the second cohort removed from a common cage was intermediate to that of the first animal removed from that cage, and animals subjected to a 10 minute ambient temperature swim (Fig. 1).

A similar effect of cohort removal was not consistently observed if the second animal sacrificed was taken from a different cage in the same environmental chamber or a different environmental chamber. A parallel increase in the number and apparent affinity of [^{35}S]TBPS binding sites as a result of cohort removal has also been observed in cortical membranes [21]. Thus, it appears that changes in benzodiazepine/GABA receptor coupled chloride channels can occur in ≤15 s (the interval required to sequentially remove and decapitate two animals from the same environmental chamber), and can be provoked by a perturbation of the environment as "subtle" as that of an investigator removing a cohort from a common cage. Similar "minor" perturbation of the environment can also produce significant elevations in circulating levels of ACTH and corticosterone that are first observed 2.5 and 5 minutes, respectively, after application of the stressor [40,41]. These findings demonstrate that "stress"-induced alterations in the benzodiazepine receptor complex precede a significant activation of both the pituitary-adrenal axis and other brain transmitter/receptor systems (e.g catecholamines as measured by *in vivo* voltammetry [42]) that have been associated with stress. Further, the alterations in both [^{35}S]TBPS binding and halide-enhanced [^{3}H]flunitrazepam binding observed after swim-stress cannot be solely attributed to increases in motor activity since this effect of cohort removal was **not** observed in cerebellar membranes, despite the robust increase in halide-enhanced [^{3}H]flunitrazepam binding seen in this tissue after ambient temperature swim.

In our initial studies, no changes in [^{3}H]flunitrazepam binding (in the absence of permeant anions) were observed following ambient temperature swim. The literature on stress-induced changes in benzodiazepine receptors following application of stressors is conflicting, with reports of no change, small increases, or small decreases in either the K_d or B_{max} of [^{3}H]benzodiazepines following chronic or acute stressors [43, ref. 44 for review]. Part of the discrepant findings may be attributable to the method of tissue preparation, where inclusion of (e.g.) Cl^- could differentially effect radioligand binding to benzodiazepine receptors. Further, no differences in GABA-enhanced [^{3}H]flunitrazepam binding were observed following swim stress in the absence of Cl^- [20]. We used this measure to assess that population of GABA receptors coupled to benzodiazepine receptors since electrophysiological studies [15,22] suggest that the subpopulation of GABA receptors functionally linked to benzodiazepine receptors are most accurately assessed by measuring GABA-enhanced [^{3}H]benzodiazepine binding. Reports concerning stress induced changes in GABA receptors have been as conflicting as those concerned with stress-induced changes in benzodiazepine receptors. For example, Soubrie, et al. [35] reported no difference in [^{3}H]muscimol (a GABAmimetic) or GABA-enhanced [^{3}H]flunitrazepam binding in cortical tissue of rats following a brief, cold water swim stress, while Skerritt, et al. [45] reported rapid increases in [^{3}H]GABA binding in mouse brain following swim stress. In contrast,

Biggio and colleagues [46-49] have reported consistent reductions in the B_{max} of a population of low affinity GABA receptors after a brief exposure to footshock that is accompanied by an increase in the apparent affinity of [^{3}H]GABA for the remaining population of low affinity receptors. The effects of footshock were reported to be more robust in "handling-habituated" than "naive" animals [46]. Subsequently, these changes were attributed to an action at benzodiazepine receptors, since the _in vitro_ addition of diazepam (5 µM) reversed this effect, and the benzodiazepine antagonist Ro 15-1788 (100 nM) blocked the effect of diazepam [48,49]. These studies were performed in the presence of 50 mM Cl^-, which, by analogy to our findings, might reflect an alteration at the chloride ionophore rather than at GABA receptors. Further, measurement of [^{3}H]GABA or muscimol binding may not be identical to that subpopulation of low affinity GABA receptors linked to the benzodiazepine receptor complex. However, Maddison, et al. [50] have recently reported that consecutive, daily saline injections produced a highly significant increase in the number of low affinity GABA receptors (in whole brain) compared to uninjected animals. The latter findings are clearly consistent with the increase in [^{35}S]TBPS binding obtained both after acute swim or handling habituation [which will be described in detail in a subsequent portion of this chapter], which suggests that chronic (or acute) exposure to such stressors can produce a compensatory increase in "GABAergic tone". Nonetheless, an important aspect of all these studies concerns the different types of stressors applied. That is, aspects of controllability and intensity vary with the type of stressor applied, and these factors could account for the differences in "GABAergic tone" reported. _In toto_, our studies do not preclude changes in components of the supramolecular complex other than the chloride ionophore, but further studies are clearly needed to clarify these issues.

III. A putative relationship between "emotionality" and GABA-gated chloride channels

More recent studies from our laboratory have focussed on examining the hypothesis that the regulation of GABA-gated chloride channels may reflect a compensatory mechanism modulating the reactivity of the organism to stressors. According to this hypothesis, the adaptation of an organism to anxiety-producing stimuli after repeated exposure would be accompanied by an increase in both the affinity and number of GABA-gated chloride channels. To test this hypothesis, animals housed in the protected (low-stress) environment were handled twice daily for two minutes in an open field (a 75 x 50 cm wooden arena without walls) for 12 days. The defecation score (number of fecal boli, often considered as a measure of emotionality [51,52]) was recorded during these trials. At the end of this 12 day habituation period, the activity and defecation scores were measured in a different open field under a bright light. The animals were then returned to their home cage and [^{35}S]TBPS binding subsequently measured in both "naive" and swim-stressed animals. These studies revealed that, while the defecation score decreased significantly throughout the habituation period (2.7 ±0.4 vs. 0.8 ±0.3, $p<0.01$, on the first and last days of the habituation period,

Table 2. Effects of stress on [^{35}S]TBPS binding in cortical tissue from handling habituated animals.

	Handling Habituated Control	Handling Habituated Swim Stress
K_D	38 ±2	33 ±1 *
B_{max}	2013 ±222	2441 ±67 *

Animals were handled twice daily for 12 days. Stress consisted of a ten minutes ambient temperature swim. Values represent X±SEM of 4 animals in the control group and 16 animals in the stress group. * Significantly different from control group ($p<0.05$, Duncan´s test). The K_D (nM) and B_{max} (fmol/mg protein) values in non-handling habituated animals were respectively: 35.3 ±2, 1454 ±94 in controls and 24.9 ±2, 2131 ±59 in swim stressed animals (see Table 1). As a habituation index, the defecation score for each animal was measured after each handling. Defecation scores decreased from (X±SEM) 2.7 ±0.4 after the first handling period to 0.8 ±0.3 after the 12th day handling period.

respectively), handling habituation significantly **increased** the B_{max} of [^{35}S]TBPS in cortical tissue as compared to non-handled animals (compare Table 1, right and Table 2). Following swim stress, [^{35}S]TBPS binding was also increased in the handling habituated group, but (as a % of non-swim stress) this increase (21%) was not nearly as robust as that observed in the non-handled animals kept in the same protected environment (47%).

Furthermore, an analysis of [^{35}S]TBPS binding in different brain regions from these animals following swim stress revealed that animals with high defecation scores (total number of fecal boli during the period of handling habituation) (i.e. high emotionality) had significantly lower B_{max} values in hippocampal membranes than did the less emotional animals (Table 3). No significant differences were observed between high and low "emotionality" animals in cortical or cerebellar B_{max} values of [^{35}S]TBPS after swim-stress. However, a significantly higher affinity for [^{35}S]TBPS (lower K_D) was found in the cerebellum of low emotionality rats. Interestingly, the relationship between the emotionality score and the B_{MAX} values of [^{35}S]TBPS binding in cortical and cerebellar membranes appeared to follow an inverted **U**-shaped function. This observation is consistent with previous reports showing a similar **U**-shaped relationship between predictability of aversive stimuli, emotionality, and plasma corticosterone levels [53,54]. In summary, these data suggest that in this group of random bred Sprague-Dawley rats habituation to handling and the subsequent decrease in handling induced defecation are accompanied by an increase in basal levels or [^{35}S]TBPS binding. Furthermore, animals with a high degree of "emotionality" showed (in hippocampal membranes) a smaller

Table 3. [^{35}S]TBPS binding in different brain areas from handling habituated animals grouped according to "emotionality".

EMOTIONALITY SCORE	CORTEX K_D	CORTEX B_{MAX}	HIPPOCAMPUS K_D	HIPPOCAMPUS B_{MAX}	CEREBELLUM K_D	CEREBELLUM B_{MAX}
(0-20)	35 ±2	2205 ±155	33 ±2	2006 ±86	32 ±1*	1849 ±33
(21-47)	34 ±1	2628 ±52	33 ±2	2031 ±42	35 ±1	2058 ±39
(60-69)	31 ±2	2456 ±179	29 ±1	2149 ±38	34 ±1	1954 ±47
(70-90)	33 ±2	2477 ±22	33 ±1	1383 ±358*	36 ±1	1855 ±128
TOTAL	33 ±1	2441 ±67	32 ±1	1926 ±99	34 ±1	1929 ±39

Animals were handled for two minutes twice daily for 12 days in an open field. They were returned to their home cages and subjected to a ten minutes ambient temperature swim 24-96 hours after the final handling habituation period. The emotionality sccre represents the total number of fecal boli from each animal during the total period of handling. Animals with scores in the medium range (48-59) were not used for [^{35}S]TBPS binding assays. Values represent X±SEM of at least 3 animals per group. * Significantly different from all other groups ($p<0.05$, Duncan´s test).

Table 4. Pearson correlation coefficients between defecation scores and [^{35}S]TBPS binding in different brain regions after swim stress in handling habituated animals.

	CORTEX		HIPPOCAMPUS		CEREBELLUM	
	K_D	B_{MAX}	K_D	B_{MAX}	K_D	B_{MAX}
DEFTOT	-.252	.337	-.259	-.431*	.538*	.296
DEF IN NEW OPEN FIELD	-.017	.029	.106	-.468*	.187	-.065

Animals were subjected to a handling habituation protocol as described in the text. Correlations were obtained from a group of 16 animals. * $p<0.05$. Deftot refers to the cumulative defecation score from each animal over 12 days of handling habituation. After the habituation period, the defecation score was measured again for each animal in a novel open field. The correlation between both defecation scores was $r_{(14)}=0.47$ ($p<0.01$).

increase in [^{35}S]TBPS binding in response to swim stress than animals with a lower "emotionality" score. Significant negative correlations were obtained between both the cumulative defecation score over the 12 days period of handling habituation, the defecation score on the first exposure to a novel open field after the habituation period, and the B_{max} of [^{35}S]TBPS in the hippocampus (after swim stress). That is, the more "emotional" the animal, the lower the B_{max} of [^{35}S]TBPS binding in hippocampal membranes (Table 4). This finding is consistent with the proposed involvement of septo-hippocampal circuitry in anxiety [55], and supports the notion that a higher number [^{35}S]TBPS binding sites (in both the "basal" and stress situations) in specific brain regions may be related to the reactivity of an organism when confronted by stressors. These findings must also be considered in light of the fact that the "emotionality" range within a randomly bred colony of rats should be smaller than the range found among (for example) several inbred strains of mice. Examination of such inbred strains of mice may thus offer a more robust test of this proposed relationship.

Other studies examining [^{35}S]TBPS binding in hippocampal and cortical membranes of adrenalectomized ("ADX") and sham-adrenalectomized ("SHAM") rats further support the hypothesis that the number of GABA-gated chloride channels may be "up-regulated" in response to chronic stress or anxiety (Table 5). In these, as in previous studies, a clear effect of cohort removal was observed in both SHAM and ADX animals (Table 5). Values for [^{35}S]TBPS binding in brain areas from the first cohort removed from a common cage and the pooled first/second cohorts removed are included for comparison. In the SHAM animals, swim-stress elicited a significant increase in the number of [^{35}S]TBPS binding sites and apparent affinity of this radioligand in both cortical and hippocampal membranes. If the values of the first cohort are compared to the swim-stressed animals, the B_{max} increased ~61% and ~28% in cortex and hippocampus, respectively. In the ADX animals, B_{max} values of the first cohort removed were increased ~23% and ~13%, in cortical and hippocampal membranes respectively, compared to SHAM rats. While the B_{max} values of [^{35}S]TBPS in these tissues were not statistically significantly different between ADX and SHAM animals in the control situation, the relative **increase** in [^{35}S]TBPS values in the former group was much lower after stress (~28% and ~9% in cortical and hippocampal membranes, respectively) than in SHAM rats (61% and 28% respectively). These findings are consistent with a recent report that adrenalectomy abolishes stress-induced increases in muscimol-stimulated $^{36}Cl^-$ uptake in these brain areas [34]. The lack of glucocorticoids, mineralocorticoids, and\or medullary hormones (e.g. epinephrine) may be affecting the animal as a chronic stressor, resulting in the higher B_{max} of [^{35}S]TBPS in ADX compared to SHAM rats in the "non-stressed" situation. The

Table 5. Effects of adrenalectomy on stress-induced changes in [^{35}S]TBPS binding.

	CORTEX		HIPPOCAMPUS	
	K_D	B_{MAX}	K_D	B_{MAX}
<u>SHAM</u>				
First Cohort	43 ±2	1465 ±130	44 ±3	1362 ±81
First + Second	41 ±2	1898 ±155	39 ±3	1453 ±73
Stress	32 ±1**	2371 ±91**	35 ±2*	1744 ±98*
<u>ADRENALECTOMIZED</u>				
First Cohort	43 ±3	1803 ±150	46 ±4	1547 ±187
First + Second	41 ±2	2071 ±142	43 ±2	1642 ±121
Stress	32 ±1**	2312 ±76*	33 ±2**	1686 ±83

First + Second refers to a pool of control animals that were either the first or second cohorts removed from a common cage. See text and reference [21] for details. Stress refers to rats that were subjected to a 10 minutes ambient temperature swim.
* Significantly different from control first, ** Significantly different from both control groups, ($p<0.05$, Duncan´s test).

subsequent response to a different stressor in these animals might not be as robust (as a % of on non-stressed levels) as in the SHAM animals. Another possibility that would explain the high basal levels of [^{35}S]TBPS binding in ADX animals and, as a result of that, its lowered response to stress, is that the ADX animals (studied at least two peaks post surgery) lack circulating levels of progesterone metabolites such as 5α- and 5ß-pregnane-3α-ol-20-one. These compounds have been shown to inhibit [^{35}S]TBPS binding and possess pentobarbital like actions in electrophysiological paradigms [56]. While the tonic influence of circulating levels of these compounds on GABA-gated chloride channels is not known, it is possible that the changes in [^{35}S]TBPS binding found in ADX rats may be due to alterations in the levels of these steroids.

IV. Conclusions

These studies have documented rapid, stress induced changes in GABA-gated chloride channels of the central nervous system. Since the characteristics of [^{35}S]TBPS binding appear to reflect the biophysical properties of this hormone-gated ion channel, the speed at which changes occur (preceeding statistically significant increases in pituitary and adrenal hormones or other neurotransmitters which have been associated with stress) should not be surprising. The increase in [^{35}S]TBPS binding observed after chronic stress situations (housing and handling) supports the hypothesis that regulation of GABA-gated chloride channels modulates the reactivity of the organism to stressors. The neuroanatomical localization of the supramolecular complex taken together with pharmacologic evidence that many drugs which modify the components of emotional and physical stress are ligands of this system strongly suggests that the supramolecular complex plays a key role in the control of stress and anxiety.

Acknowledgements: R.T. acknowledges the financial support of the Comite Conjunto Hispano-Norteamericano and The Burroughs and Wellcome Fund. H.H. was a Howard Hughes Medical Research Scholar of the National Institutes of Health.

References

1. Usdin, E., Kvetnansky, R., and Axelrod, J. (Editors) (1984): Stress: The role of catecholamines and other neurotransmitters (2 volumes), Gordon and Breach Science Publishers, New York.
2. Usdin, E., Skolnick, P., Tallman, J., Greenblatt, D., and Paul, S. (Editors) (1982): Pharmacology of Benzodiazepines, MacMillan Press, London.
3. Tallman J., Paul S., Skolnick P., and Gallager D. (1980):Receptors for the age of anxiety: The pharmacology of benzodiazepines. Science 207:274-281.
4. Olsen, R. and Venter, J. (Editors) (1986): Benzodiazepine/ GABA receptors and chloride channels: structural and functional properties (Receptor Biochemistry & Methodology, Vol. 5). A.R. Liss, Inc., New York.

5. Squires, R. (Editor) (1987): GABA and benzodiazepine receptors. CRC Press, Boca Raton (in press).
6. Ninan, P., Insel, T., Cohen, R., Cook, J. Skolnick, P. and Paul, S. (1982): Benzodiazepine receptor-mediated experimental "anxiety" in primates. Science 218:1332-1334, 1982.
7. File, S.: Animal anxiety and the effects of benzodiazepines. In: Pharmacology of benzodiazepines. Usdin, E., Skolnick, P., Tallman, J., Greenblatt, D. and Paul, S. (Editors) (1982): MacMillan Press, London, pp. 355-363.
8. Corda, M., Blaker, W., Mendelson, W., Guidotti, A. and Costa, E. (1983): ß-Carbolines enhance shock-induced suppression of drinking in the rat. Proc. Natl. Acad. Sci. USA 80: 2072-2078.
9. Insel, T., Ninan, P., Aloi, J., Jimerson, D., Skolnick, P. and Paul, S. (1984): A benzodiazepine receptor-mediated model of anxiety. Studies in nonhuman primates and clinical implications. Arch. Gen. Psychiatry 41:741-750.
10. Crawley, J., Ninan, P., Pickar, D., Chrousos, G., Linnoila, M. Skolnick, P., and Paul, S. (1985): Neuropharmacological antagonism of the ß-carboline-induced "anxiety" response in rhesus monkeys. J. Neurosci. 5:477-485.
11. Dorow, R., Horowski, R., Paschelke, G., Amin, M., and Braestrup, C. (1983): Severe anxiety induced by FG 7142, a ß-carboline ligand for benzodiazepine receptors. Lancet i:98-99.
12. Dorow, R., Duka, T., Sauerbrey, N. and Holler, L. (1987): ß-Carbolines: new insights into the clinical pharmacology of benzodiazepine receptor ligands. In: Clinical Pharmacology in Psychiatry (Psychopharmacology Series 3). Dahl, S., Gram, L., Paul, S. and Potter, W. (Editors), Springer Verlag, Berlin, pp. 37-51.
13. Skolnick, P. and Paul, S. (1986): Benzodiazepines and non-benzodiazepines. In: Receptor binding in drug research. C. O´ Brien (Editor), Marcel Dekker, Inc., New York, pp. 53-76.
14.Study, R. and Barker, J. (1983): Cellular mechanisms of benzodiazepine action. J. Amer. Med. Assoc. 247:2147-2151.
15. Chan, C. and Farb, D. (1985): Modulation of neurotransmitter action: control of the τ-aminobutyric acid response through the benzodiazepine receptor. J. Neurosci. 5:2365-2372.
16. Paul, S., Marangos, P. and Skolnick, P. (1981): The benzodiazepine-GABA-chloride ionophore receptor complex: common site of minor tranquilizer action. Biol. Psychiat. 16:213-229.
17. Leeb-Lundberg, F., Snowman, A., and Olsen, R. (1981): Perturbation of benzodiazepine receptor binding by pyrazolopyridines involves picrotoxinin/barbiturate receptor sites. J. Neurosci. 1:471-477.
18. Costa, T., Rodbard D., and Pert, C. (1979): Is the benzodiazepine receptor coupled to a chloride anion channel? Nature 277:315-317.
19. Havoundjian, H., Paul, S., and Skolnick, P. (1986): Rapid, stress induced modification of the benzodiazepine receptor coupled chloride ionophore. Brain Res. 375:403-408.
20. Havoundjian, H., Paul, S., and Skolnick, P. (1986): Acute, stress-induced changes in the benzodiazepine/GABA receptor complex are confined to the chloride ionophore. J. Pharmacol. Exp. Ther. 237:787-793.

21. Trullas, R., Havoundjian, H., Zamir, N., Paul, S., and Skolnick, P. (1987): Environmentally-induced modification of the benzodiazepine/ GABA receptor coupled chloride ionophore. Psychopharmacol. (in press).
22. Gallager, D., Lakoski, J., Gonsalves, S., and Rauch, S. (1984): Chronic benzodiazepine treatment decreases post-synaptic GABA sensitivity. Nature 308:74-76.
23. Squires, R., Casida, J., Richardson, M., and Saederup, E. (1983): [^{35}S]t-Butylbicyclophosphorothionate binds with high affinity to brain specific sites coupled to τ-aminobutyric acid-A and ion recognition sites. Mol. Pharmacol. 23:326-336.
24. Havoundjian, H. and Skolnick, P. (1986): A quantitative relationship between Cl^- enhanced [^{3}H]flunitrazepam and [^{35}S] t-butylbicyclophosphorothionate binding to the benzodiazepine/ GABA receptor chloride ionophore complex. Molec. Brain Res. 1: 281-287.
25. Edwards, C. (1982): The selectivity of ion channels in nerve and muscle. Neuroscience 7:1335-1366.
26. Hille, B. (1984): Ionic Channels of Excitable Membranes. Sinauer Publishing Co., Sunderland, MA.
27. Wamsley, J., Gee, K., and Yamamura, H. (1983): Comparison of the distribution of convulsant/barbiturate and benzo-diazepine receptors using light microscopic autoradiography. Life Sci 33:2321-2329.
28. Wamsley, J., Gehlert, D. and Olsen, R. (1986): The benzo-diazepine/ barbiturate-sensitive convulsant/GABA receptor /chloride ionophore complex: autoradiographic localization of individual components. In: Benzodiazepine/GABA receptors and chloride channels: structural and functional properties (Receptor Biochemistry & Methodology, Vol. 5). Olsen, R. and Venter, J. (Editors), A.R. Liss, Inc., New York, pp.299-313.
29. Akaike, N., Hattori, K., Oomura, Y., Carpenter, D. (1985): Bicuculline and picrotoxin block τ-aminobutyric acid gated Cl^- conductance by different mechanisms. Experientia 41:70-71.
30. Havoundjian, H., Paul, S., and Skolnick, P. (1986): The permeability of τ-aminobutyric acid gated chloride channels is described by the binding of a cage convulsant, [^{35}S]t-butylbi-cyclophosphorotionate. Proc. Natl. Acad. Sci. (USA) 83:9241-9244.
31. Schwartz, R., Skolnick, P., Seale, T. and Paul, S. (1986): Demonstration of GABA/barbiturate-receptor-mediated chloride transport in rat brain synaptoneurosomes: a functional assay of GABA receptor-effector coupling. In: GABAergic Transmission and Anxiety, G. Biggio and E. Costa (Editors), Raven Press, New York, pp. 33-49.
32. Havoundjian, H., Cohen, R., Paul, S. and Skolnick, P. (1986): Differential sensitivity of "central" and "peripheral" benzodiazepine receptors to phospholipase A_2. J. Neurochem. 46:804-811.
33. Trullas, R., Havoundjian, H., and Skolnick, P. (1987): Stress-induced changes in [^{35}S]t-butylbicyclophosphorothionate binding to GABA gated chloride channels are mimicked by in vitro occupation of benzodiazepine receptors. J. Neurochem. (submitted).
34. Schwartz, R., Wess, M., Labarco, R., Skolnick, P., and Paul, S. (1987): Acute stress enhances the activity of the GABA-gated chloride channel in brain. Brain Res. (in press).
35. Soubrie, P., Thiebot, M., Jobert, A., Montasruc, J., Henry, F., and Hamon, M. (1980): Decreased convulsant potency

of picrotoxin and pentylenetetrazole and enhanced [^{3}H]flunitrazepam binding cortical binding following stressful manipulations in rats. Brain Res. 189:505-517.
36. Drugan, R., McIntyre, T., Alpern, H. and Maier, S. (1985): Coping and seizure susceptibility: control over shock protects against bicuculline-induced seizures. Brain Res. 342:9-17.
37. Mason, J., Maher, J., Hartley, L., Mougey, E., Perlow, M., and Jones, L. (1976): Selectivity of corticosteroid and catecholamine responses to various natural stimuli. In: Psychopathology of human adaptation. Serban, G. (Editor), Plenum Press, New York, pp. 147-171.
38. Riley, V., Fitzmaurice, M., and Spackman, D. (1981): Psychoneuroimmunologic factors in neoplasia: studies in animals. In: Psychoneuroimmunology. Ader, R. (Editor), Academic Press, New York, pp. 31-102.
39. Nemeth, S. (1984): Possible distorting of experimental results by stress. In: Stress: The role of catecholamines and other neurotransmitters (Vol. 1), Editors: E. Usdin, R. Kvetnansky, and J. Axelrod, Gordon and Breach Science Publishers, New York, 1984, pp. 59-68.
40. Cook, D., Kendall, J., Greer, M., and Kramer, R. (1977): The effect of acute or chronic ether stress on plasma ACTH concentration in the rat. Endocrinol. 93:1019-1024.
41. Guillemin, R., Vargo, T., Rossier, J., Minick, S., Ling, N., Rivier, C., Vale, W., and Bloom, F. (1977): ß-Endorphin and adrenocorticotropin are secreted concommitantly by the pituitary gland. Science 197:1367-1369.
42. Ikeda, M. and Nagatsu, T. (1985): Effect of short term swimming stress and diazepam on 3,4-dihydroxyphenylacetic acid (DOPAC) and 5-hydroxyindoleacetic acid (5-HIAA) levels in the caudate nucleus: an in vivo voltammetric study. Naunyn-Schmiedeberg´s Arch. Pharmacol. 331:23-26.
43. Braestrup, C., Nielsen, M., Nielsen, E. and Lyon, M. (1979): Benzodiazepine receptors in brain as affected by different experimental stresses: the changes are small and not unidirectional. Psychopharmacol. 65:273-277.
44. Havoundjian, H., Trullas, R., Paul, S. and Skolnick, P. (1987): A physiological role of the benzodiazepine receptor complex. In: Molecular mechanisms of neuronal responsiveness (Adv. In Exp. Med. and Biol.). Ehrlich, Y., Berry, W., and Lenox, R. (Editors). Plenum Press, New York, in press.
45. Skerritt, J., Triskidoon, P., and Johnston, G. (1981): Increased GABA binding to mouse brain following acute swim stress. Brain Res. 215:398-403.
46. Biggio, G., Corda, M., Concas, A., Demontis, G., Rosetti, Z., and Gessa, G. (1981): Rapid changes in GABA binding induced by stress in different areas of rat brain. Brain Res. 229:363-369.
47. Biggio, G. (1983): The action of stress, ß-carbolines, diazepam, and Ro 15-1788 on GABA receptors in the rat brain. In: Benzodiazepine recognition site ligands: Biochemistry and Pharmacology. Biggio, G. and Costa, E. (Editors), Raven Press, New York, pp. 105-119.
48. Concas, A., Corda, M., and Biggio, G. (1985): Involvement of benzodiazepine recognition sites in the footshock-induced decrease of low affinity GABA receptors in the rat cerebral cortex. Brain Res. 341:50-56.
49. Corda, M. and Biggio, G. (1986): Stress and GABAergic

transmission: biochemical and behavioural studies. In: GABA-ergic transmisstion and anxiety. Biggio, G. and Costa, E. (Editors), Raven Press, New York, pp. 121-136.
50. Maddison, J.E., Dodd, P.R., Jonhston, G.A.R., Farrell, G.C. (1987) Effect of handling and injection stress on rat brain GABA receptor binding. Neurosci. Lett., suppl. 27, S101
51. Broadhurst, P. (1957): Determinants of emotionality in the rat. I. situational factors. Brit. J. Psychol. 48:1-12.
52. Blizard, D. (1971): Autonomic reactivity in the rat: effects of genetic selection for emotionality. J. of Comp. Physiol. Psych. 76:282-289.
53. Hennessy, J.W., and Levine, S. (1979) Stress, Arousa53. the Pituitary-Adrenal System: A Psychoendocrine Progress in Psychobiol. & Physiol. Psychol., 8:133-178. Hypothes
54. Endröczi, E., and Fekete, T. (1971) Correlations between the pituitary-adrenal function and the exploratory activity, learning behavior and limbic functions. In: Hormones and Brain Function. Lissak, K. (Editor), Plenun Press, New York, pp. 399-408.52.
55. Gray, J. (1982): The neuropsychology of anxiety. Oxford University Press, New York.
56. Majewska, M., Harrison, N., Schwartz, R., Barker, J., and Paul, S. (1986) Steroid hormone metabolites are barbiturate-like modulators of the GABA receptor. Science 232:1004-1007.

STRESS AND IMMUNITY

CORTICOSTEROIDS AND THE IMMUNE RESPONSE

Henry N. Claman

University of Colorado School of Medicine
Denver, CO 80262

INTRODUCTION

Stress is a powerful stimulus to glucocorticoid ("steroid") production. Glucocorticoids, in turn, have important effects on the immune system. In principle, one could develop the theme; stress $\rightarrow$ steroid production $\rightarrow$ immune modulation. In practice, however, one should recognize that most of our knowledge of the effects of glucocorticosteroids on the immune response comes from pharmacological studies where exogenous steroids are given and where their effects on the immune system are then measured. Stress is omitted.

SPECIES DIFFERENCES

It was noted years ago that there are important species differences with respect to the lympholytic properties of glucocorticoids [1]. In certain species, i.e. mouse, rat and rabbit, steroids in vivo or in vitro quickly produce thymic atrophy, inhibition of antibody production and of cell-mediated immunity and lysis of lymphocytes. These species may be called "steroid sensitive" (Table I). Other species are more steroid resistant, in that these same effects are more difficult to obtain with steroids. As an example, mouse thymocytes can be lysed with 10^{-6} M hydrocortisone in vitro, while human thymocytes require 10^{-3} M hydrocortisone to get the same result [2]. Several points should be made. First, these differences have only been studied in detail with regard to lympholysis; species differences with regard to the effects of steroids on lymphocyte function or on other inflammatory cells have not been so well characterized. Second, the differences between functionally distinct classes of lymphocytes may not be a reflection of the number of steroid receptors [3]. Third, these differences are most likely quantitative rather than qualitiative and absolute. Nevertheless, results in our species should not be extrapolated to other species with out careful consideration. There are some data which show that human T lymphocytes become steroid sensitive after activation, but doses about 10^{-4} molar were needed to show this [4].

TABLE I. Species differences in the susceptibility of lymphocytes to lysis by glucocorticosteroids in vitro.

Sensitive	Resistant
Mouse	Guinea pig
Rat	Sheep
Rabbit	Monkey
	Man

STEROIDS AND CELL TRAFFIC

Large, acute doses of intravenous glucocorticosteroids given to normal human volunteers produce a neutrophilia and also a decrease in peripheral blood lymphocytes, eosinophils, basophils and monocytes [5]. These changes reach their maximum in 4-6 hours and the blood levels of the cells have generally returned to normal by 24 hours. Most of these effects reflect changes in patterns of cell traffic. Neutrophilia results both from increased release of immature cells from the bone marrow as well as from decreased margination of neutrophils from the endothelium, thus making them more accessible to the flow of blood which is sampled. Lymphopenia in man is mainly a reflection of a rapid movement of blood lymphocytes into the tissues [6]; it is not a result of lympholysis. The monocytopenia and eosinopenia may also reflect the redistribution of cells into the tissues, but recent evidence indicates that steroids in vitro inhibit the growth of eosinophil precursors [7]. The mechanism of the basopenia is not known.

It is also possible to examine the effects of corticosteroids on cellular traffic patterns in inflammatory responses. A well-studied example is the effect of steroids on the influx of cells into an inflammatory focus, represented by a Rebuck skin window [8]. Normally, neutrophils migrate into the window and are followed by monocytes. In patients on daily steroids, neutrophils and monocytes reach the coverslip in greatly diminished numbers. In patients on alternate morning corticosteroid therapy, neutrophil emigration is more impaired during the "on steroid" day as compared to the "off steroid" day. Monocyte emigration is impaired on both days. It should be emphasized that these studies really only count migrating cells; they do not measure the phagocytic or killing abilities of the cells which do reach the window. Nevertheless, they are significant additions to our knowledge of the effects of glucocorticoids on the inflammatory response.

GLUCOCORTICOIDS, CYTOKINES AND CELLULAR ENZYMES

Cytokines, including monokines, interleukins and interferons are powerful mediators of immune responses. In general, glucocorticoids inhibit cytokine production, including that of interleukin-1 (IL-1) [9, 10], interleukin-2 (IL-2) [11, 12], the IL-2 receptor [12], interferon-γ (IFN-γ) [12], and perhaps various colony-stimulating factors (CSF's) such as IL-3 [13]. In general, these cytokines are pro-inflammatory, and so their inhibition by corticosteroids represent an important anti-inflammatory mechanism. In addition, corticoids inhibit the production of a variety of more specialized pro-inflammatory mediators. Among these are the macrophage products collagenase, elastase and plasminogen activator [14].

The anti-inflammatory effects of steroids have long been attributed to their ability to stabilize lysosomal membranes and thereby to inhibit the release of pro-inflammatory molecules, particularly from neutrophils [15]. These experiments were done with lysosomes removed from rabbit liver and used high doses of steroids in vitro. Therefore, the applicability of these studies to man has been questioned. Indeed, some data have indicated that human neutrophil lysosomes are rather resistant to steroids [16] while other data indicate that they are sensitive at 10^{-6} to $5x10^{-4}$ molar concentrations of steroids [17, 18]. The subject requires further work.

STEROIDS AND CELL-SURFACE MOLECULES

Current models of immunoregulation indicate that antigen presentation to T cells occurs via cells (macrophages and/or B cells) bearing class II major histocompatibility complex (MHC) molecules on their surface. Therefore, the effects of corticosteroids on these molecules is of interest. Steroids do inhibit the expression of murine class II MHC antigens [9] but the effects on human cells are more complex. Steroids may inhibit antigen presentation by human monocytes but there seems to be no relation between class II expression and antigen presentation [19]. In fact, some experiments show that steroids increase class II expression in human monocytes [20].

Another cell surface molecule of interest is the Fc receptor (FcR) for immunoglobulins [21]. Corticosteroids inhibit the expression of FcR [22] and this is probably a prime mechanism which accounts for the prompt improvement seen when patients with autoimmune hemolytic anemia or autoimmune thrombocytopenia are given corticosteroids. That is, the steroids impair the clearance of antibody-coated (opsonized) red cells or platelets by inhibiting FcR (and C3 receptor) expression in the reticuloendothelial system.

GLUCOCORTICOIDS, FEVER, THE VASCULATURE AND PROSTAGLANDINS

A rise in body temperature is a prominent accompaniment of inflammation and may indeed represent part of the body's anti-inflammatory response. Steroids are anti-pyretic, probably because they impair the production of endogenous pyrogen [23], which appears to be the same as endogenous pyrogen. Furthermore, steroids decrease blood flow when they are applied locally [24] and also decrease the vasodilatation which accompanies inflammation [25]. However, the mechanism of this change is not yet understood.

Some of the above effects of corticosteroids may operate through their interactions with the arachidonate pathway. One of the more remarkable discoveries in our understanding of the action of steroids is that they stimulate production of materials which inhibit the conversion of membrane phospholipids to arachidonate by inhibiting phospholipase A_2 (PLA-2) [26]. These substances, variously known originally as lipomodulin [27] and macrocortin [28] have been renamed lipocortin. However, much remains to be learned in this area because, if steroids inhibited arachidonate production via PLA-2, one would expect a coordinate inhibition of products of both lipoxygenase and cyclooxygenase pathways, yet this is not always seen [29, 30].

CORTICOSTEROIDS AND ANTIBODIES

Glucocorticoids easily suppress antibody synthesis in steroid-sensitive species. In man, however, the situation is more complex. A 5-day course of methylprednisolone (96 mg/day) was followed by a20% lowering of serum immunoglobulin levels [31]. This fall was determined to be caused by both a decrease in immunoglobulin production and an increase in its catabolism. In more chronic experiments, however, specific antibody production was not less in steroid-treated patients [32]. In vitro studies further complicate the picture. Immunoglobulin production by cells of patients on high-dose steroids is decreased [33]. Paradoxically, however, more recent studies show that B cells from normal subjects given a bolus of steroids actually produce more immunoglobulin in vitro [34]. When the cells in question are exposed to steroids in vitro, IgE synthesis increases [35]. At the moment, there is no comprehensive schema which integrates all these findings, although one group believes that the effect on immunoglobulin production in vitro occurs by inhibition of leukotrienes [36]. It is likely that the net effect of steroids on antibody production will depend on the customary variables, including the steroid dose and duration, the type of antigen, and the influence of T helpers, T suppressors and B cells.

CORTICOSTEROIDS, T CELLS AND CELL-MEDIATED IMMUNITY

One of the earliest experiments on steroids and lymphocytes showed that steroids inhibited phytohemagglutinin-driven cell proliferation in vitro [37]. It has been known for years that corticosteroids inhibit the in vivo expression of delayed-type hypersensitivity reactions. A well-documented study showed that it required an average of 13.6 days for oral prednisone (40 mg/day) to inhibit the tuberulin test [38]. This effect may be largely due to inhibition of interleukins 1 and 2 production, as mentioned above. The actual destruction of T cells by corticosteroids is probably not important in man. As mentioned above, recent in vitro studies indicating lympholysis by steroids have used excessive doses, such as 1 mg/ml, which are neither physiological nor pharmacological [4]. Furthermore, the rapid recovery from steroid-induced lymphopenia in man [39] is more easily attributed to redistribution effects of nondividing T cells than on rapid regeneration of a population of recirculating cells which is known to have a long half-life. The important question — how long does steroid-induced anergy remain after steroids have been withdrawn? — is difficult to answer.

Information about the effects of steroids on T cell subsets requires more study. Hydrocortisone inhibited T suppressor activity in cultured human blood [40] yet there appears to be no difference in the number of steroid receptors when CD3 (all T cells), CD4 (helpers) and CD8 (suppressors) are studied [41]. With regard to cytotoxic T cell (CTL) function, it is clear that this can be inhibited by steroids in vitro, but (at least in the mouse) preincubation of primed CTL is needed [42].

CONCLUSION

The immune system is a complicated series of networks of cells and molecules. As all nucleated cells have glucocorticosteroid receptors, the effects of these hormones are many and varied. Furthermore, the interpretation of data is complicated by the fact that careful attention must be made to the species being studied, the steroid drug and its dosage, in vivo and in vitro parameters, etc. Much remains to be done to bring more order into our understanding of how corticosteroids affect the immune system.

REFERENCES

1. H.N. Claman, N. Engl. J. Med. 287, 388-397 (1972).
2. H.N. Claman, J.W. Moorhead, and W.H. Benner, J. Lab. Clin. Med. 78, 499-507 (1971).
3. A.S. Fauci, T. Murakami, D.D. Brandon, D.L. Lorianx, and M.B. Lippsett, Cell. Immunol. 49, 43-50 (1980).
4. N. Galili, U. Galili, E. Klein, L. Rosenthal, and B. Nordenskjold, Cell. Immunol. 50, 440-444 (1980).
5. H.N. Claman, Clinics in Immunol. & Allergy 4, 317-329 (1984).

6. A.S. Fauci, Immunology 28, 669-680 (1975).
7. F.T. Slovick, C.N. Abboud, J.K. Brennan, and M.A. Lichtman, Blood 66, 1072-1079, 1985.
8. D.C. Dale, A.S. Fauci, and S.M. Wolff, N. Engl. J. Med. 291, 1154-1158 (1974).
9. D.S. Snyder, and E.R.Unanue, J. Immunol. 129, 1803-1805 (1982).
10. C.A. Dinarello, Rev. Inf. Dis. 6, 51-95 (1984).
11. S. Gillis, G.R. Crabtree, and K.A. Smith, J. Immunol. 123, 1624-1631 (1979).
12. K. Grabstein, S. Dower, S. Gillis, D. Urdal, and A. Larsen, J. Immunol. 136, 4503-4508 (1986).
13. J.A. Culpepper, and F. Lee, J. Immunol. 135, 3191-3197 (1985).
14. Z. Werb, J. Exp. Med. 147, 1695-1712 (1978).
15. G. Weissmann, and L. Thomas, J. Clin. Invest. 42, 661-669 (1963).
16. R.H. Pevsellin, and L.C. Ku, J. Clin. Invest. 54, 919-925 (1974).
17. I.M. Goldstein, D. Roos, G. Weissmann, and H.B. Kaplan, Inflammation 1, 305-315 (1976).
18. L.J. Ignarro, Arth. Rheum. 19, 73-83 (1977).
19. T.L. Gerrard, T.R. Cupps, C.H. Jurgensen, and A.S. Fauci, Cell. Immunol. 85, 330-339 (1985).
20. T.L. Gerrard, T.R. Cupps, C.H. Jurgensen, and A.S. Fauci, Cell. Immunol. 84, 311-316 (1984).
21. A.D. Schreiber, and M.M. Frank, J .Clin. Invest. 51, 575-582 (1972).
22. G.R. Crabtree, A. Munck, and K.A. Smith, Nature 279, 338-339 (1979).
23. G.M. Dillard, and P. Bodel, J. Clin. Invest. 49, 2418-2426 (1970).
24. T.P. Greeson, N.E. Levan, R.I. Freedman, and W.H. Wong, J. Invest. Derm. 61, 242-244 (1973).
25. R.H. Ebert, and W.R. Barclay, Ann. Int. Med. 37, 506-518 (1952).
26. G.P. Lewis, and P.J. Piper, Nature 254, 308-311 (1975).
27. F. Hirata, Adv. Prostagl. Thromboxane Leukotriene Res. 11, 73-97 (1983).
28. R.J. Flower, and G.J. Blackwell, Nature 278, 456-459 (1979).
29. R.P. Schleimer, D.A. Davidson, L.M. Lichtenstein, and N.F. Adkinson, J. Immunol. 136, 3006-3011 (1986).
30. J.-L. Robin, D.C. Seldin, K.F. Austen, R.A. Lewis, J. Immunol. 135, 2719-2726 (1985).
31. W.T. Butler, and R.D. Rossen, J. Clin. Invest. 52, 2629-2640 (1973).
32. M. Tuchinda, R.W. Newcomb, and B.L. DeVald, Int. Arch. Allergy 42, 533-544 (1972).
33. R. McMillan, R. Longmire, and R. Yelenosky, J. Immunol. 116, 1592-1596 (1976).
34. T.R. Cupps, L.C. Edgar, C.A. Thomas, and A.S. Fauci, J. Immunol. 132, 170-175 (1984).
35. A. Ray, Z. Hemady, and R.E. Rocklin, N. Engl. Reg. Allergy Proc. 8, 81-84 (1987).
36. J.S. Goodwin, and D. Atluru, J. Immunol. 136, 3455-3460 (1986).
37. P.C. Nowell, Cancer Res. 21, 1518-1521 (1961).
38. S. Bovornkitti, P. Kangsadal, P. Sathirapat, and P. Oonsombatti, Dis. Chest 38, 51-55 (1960).
39. A.S. Fauci, D.C. Dale, and J.E. Balow, Ann. Int. Med. 84, 304-315 (1976).
40. P.E. Lipsky, W.W. Ginsburg, F.D. Finkelman, and M. Ziff, J. Immunol. 120, 902-910 (1978).
41. V.R. Martins, M.I.A. Birolli, A.J.S. Duarte, and M.M. Brentani, Cell. Immunol. 105, 443-446 (1987).
42. R.P. Schleimer, A. Jacques, H.S. Shin, L.M. Lichtenstein, and M. Plaut, J. Immunol. 132, 266-271 (1984).

LIPOCORTINS

Bernard Rothhut and Françoise Russo-Marie

Unité des Venins
Unité Associée Institut Pasteur / INSERM N° 285
25 rue du Dr Roux
75724 Paris Cedex 15, France

INTRODUCTION

The various membrane events occuring during the inflammatory reaction and leading to the formation of lipid-derived mediators (prostaglandins, leukotrienes and PAF-acether) always involve a membrane enzyme activation namely phospholipase A_2. This activation leads to the cleavage of the phospholipids present in the membrane and permits the release of an unsaturated fatty acid from the 2 position - usually, arachidonic acid - and to the simultaneous formation of a lysophospholipid. If the phospholipase A_2-target phospholipid is an ether phosphatidyl choline, the lyso derivative formed is lyso PAF, the precursor of PAF-acether (Platelet Activating Factor, 1-O- alkyl- 2 - acetyl - sn-glycero- 3 - phosphocholine) (Figure 1). In inflammatory cells, the signal induced by inflammatory stimuli is transduced by a succession of events leading to the phospholipase A_2 stimulation , allowing consequently the release in the cell of either arachidonic acid, precursor of prostaglandins and leukotrienes, or lyso-PAF, precursor of PAF-acether. These products, once formed, leave the cell and are able to stimulate neighbouring cell membranes, inducing again phospholipase A_2 stimulation through stimulus-coupling events. An oversimplified view of the inflammatory reaction can be presented as an event stimulating the membrane of inflammatory cells, event coupled to transducing systems allowing the formation of lipid-derived mediators. These mediators, being stimuli by themselves amplify the phenomenon, rendering the inflammatory reaction rather similar to a nuclear reaction where the pile is phospholipase A_2, and the combustible, phospholipids, always renewed by powerful acylating systems. Would phospholipase A_2 be alone, the reaction would be endless and could become harmful for the biological environment where it takes place. Consequently, there must be some naturally occuring biological systems capable of controlling such a reaction. The discovery of lipocortin, a natural endogenous phospholipase A_2 inhibitor, has focused the attention on the central role of phospholipase A_2 in the inflammatory reaction. Indeed, this discovery has been linked to the knowledge of the biochemical and cellular mode of action of glucocorticosteroids, which inhibit phospholipase A_2 activity through the induction of the synthesis of lipocortin (Figure 2).

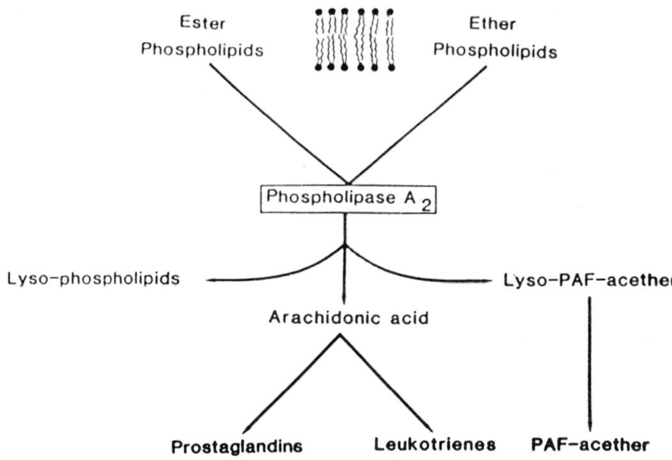

Fig.1. Central role of phospholipase A_2, the membrane enzyme responsible of the formation of lipid derived mediators

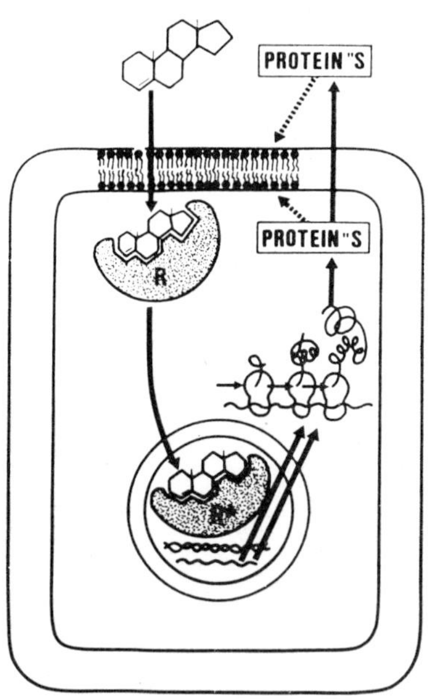

Fig.2. Scheme of glucocorticosteroid action on the lipocortins formation, involving receptor binding, transcription and traduction.

In this paper, we shall discuss how lipocortin has been discovered, and how, since its sequence has been known, the role of lipocortin and phospholipase A_2 can exceed their known functions in the inflammatory reaction. In addition, there is not only one single lipocortin but a family of lipocortins, probably derived from a common ancestor gene.

The lipocortin discovery

The initial discovery of lipocortin has been linked to the knowledge of the mechanism of action of anti-inflammatory steroids. The central role of arachidonic acid and its metabolites has been established since the discovery of the mode of action of aspirin-like drugs and all non steroidal anti-inflammatory drugs, anti-inflammatory because they were able to inhibit the prostaglandin forming enzyme, cyclooxygenase. Nevertheless, anti-inflammatory steroids were not able to inhibit cyclooxygenase although they were inhibiting also the prostaglandin formation. Preliminary studies had shown that steroids interfered with the arachidonic acid cascade, inhibiting its release from the endogenous membrane phospholipids by inducing the synthesis of a phospholipase A_2 proteic inhibitor (see figure 1). This antiphospholipase A_2 protein has been initially described in three different systems. Blackwell et al (1) have shown its presence in rat peritoneal macrophages treated with glucocorticosteroids and called it "macrocortin". Hirata et al (2) reported the same phenomenon in rabbit peritoneal neutrophils and called it "lipomodulin". Our group has also found a similar phenomenon in rat renomedullary interstitial cells in culture and has proposed the name of "renocortin" (3). A collaborative study has allowed to conlude that we were indeed dealing with the same protein, since these proteins were immunologically identical, possessed the same biological properties (i.e. inhibition of phospholipase A_2), the same biological regulation (induction by glucocorticosteroids) and the same biochemical characteristics (a molecular weight of 40 kDa and the property to be cleaved into active proteolytic fragments of 30 and 15 kDa). The common name of "lipocortin" was therefore adopted (4).

A constitutive expression

If lipocortin has been discovered through the fact that it is under the genomic control of glucocorticosteroids, a great number of questions remains unanswered, as for exemple : how is it inhibiting phospholipase A_2 ? Is there an endogenous lipocortin, and in such a case, how is it controlled by corticosteroids? Is the newly formed lipocortin inhibiting phospholipase A_2 from inside or outside the cell... This question is rather difficult to speculate on, since the cellular phospholipase A_2 is not well defined.

Preliminary studies done by the different groups have demonstrated the presence of lipocortin in the supernatant of cells or in the peritoneal fluid of elicited macrophages treated with dexamethasone, indicating that the protein was not only synthesized but also secreted outside of the cells. Hirata (5) had demonstrated that lipocortin was present within the cell. Our group (6), utilizing a monoclonal antibody against rat renal lipocortin, had reported its presence in control cells as well as in dexamethasone-treated cells, confirming the presence of an endogenous constitutive lipocortin. Using an immunoprecipitation after a ^{35}S-methionine labeling of the cells, we could not find any difference in the two types of cells, suggesting that if corticosteroids were indeed stimulating the neosynthesis of lipocortin, this synthesis occured on a large preexisting pool of lipocortin. It is therefore possible that corticosteroids induce the synthesis of a slightly different lipocortin possessing a very high antiphospholipase A_2 activity.

Regulation by phosphorylation

Hirata had first reported that the lipocortin activity could be regulated by phosphorylation ; once lipocortin is phosphorylated, it looses it ability to inhibit phospholipase A_2 (5). In addition, a regulation of lipocortin activity implicating protein kinase C has been reported in rabbit neutrophils (5) and tyrosine kinase in murine thymocytes (7). Our group has reported the presence of lipocortin in platelets. In these cells, the lipocortin activity is modified after stimulating platelets with either thrombin or phorbol ester, implicating protein kinase C (8). In addition, we have demonstrated that lipocortin phosphorylation after thrombin or phorbol ester treatment regulated the endogenous activity of phospholipase A_2. These various results have allowed us to propose a scheme of membrane activation (Figure 3) involving phospholipase A_2 together with lipocortin : the activation of a membrane receptor by a ligand induces the stimulation of phospholipase C allowing polyphosphoinositide hydrolysis into diacylglycerol (DAG) and inositol triphosphate (IP3). Diacylglycerol stimulates protein kinase C, which phosphorylates various proteins among which one is lipocortin. Once phosphorylated, lipocortin does not inhibit anymore phospholipase A_2. Nevertheless, this sole modification is not sufficient , a second event is required and is secondary to IP3 release, i.e. a calcium mobilization. Calcium stimulates then phospholipase A_2, which is now detached from its natural endogenous inhibitor, lipocortin. To confort this hypothesis, recent data obtained in our laboratory have shown that protein kinase C activation (mimicked by PMA) and calcium mobilization (mimicked by the calcium ionophore A23187) were necessary to obtain a full phospholipase A_2 activation in platelets.

Multiple phosphorylation sites

Lipocortin has been recently sequenced, cloned and expressed in E.Coli, by the group of Biogen (10). Using Northern blotting analysis, they could demonstrate the presence of lipocortin mRNA even in the absence of glucocorticosteroids in many cells as macrophages, in the lung, the spleen, the thymus, and to a lesser extent in the liver, the heart, the muscles and the kidney, confirming the presence of an endogenous level of lipocortin. More, the Biogen group has demonstrated the presence of lipocortin in some cells and its absence in other cells, as lymphoid cells, or certain nervous cell lines. As already proposed by Hirata et al (5, 7), and Touqui et al (8), lipocortin has various sites of phosphorylation on tyrosine and / or serine. In the A 431 cell line, lipocortin is phosphorylated by the tyrosine kinase activity of epidermal growth factor (EGF) receptor (11). This lipocortin possessing a real 35 kDa molecular weight has been called lipocortin I.

Other lipocortins

The same Biogen group, has not only cloned and sequenced lipocortin I, but has purified two different proteins, lipocortin I and II from human placenta (12). Both proteins exhibit the same antiphospholipase A_2 biological properties, but share 50% sequence homology. More, peptide analysis, immunological properties and the knowledge of the complete sequence have shown that lipocortin I, the EGF receptor tyrosine kinase substrate of the A 431 cells and calpactin II were similar proteins sharing more than 80% sequence homology (13,14). Similar studies have demontrated that lipocortin II was a human analogue of protein p 36, the substrate of p 60 ^{scr} from cellular and viral oncogenes c- and v-src. This p 36 protein had already been identified in chicken embryo fibroblasts and in bovine intestine brush border preparations, and had been shown to be identical to the calpactin I heavy chain (15,16). Calpactin I is a heterotetramer present in high amount in the enterocyte brush border and is formed by two subunits of p36 and two light chain subunits, each one of a molecular weight of 10 kDa, p10. In presence of calcium, the tetramer binds to actin and spectrin (two proteins of the cell cytoskeleton) but also to phospholipids. The calcium-calpactin-phospholipid complex possesses a higher affinity for actin than the calpactin-calcium complex alone. Spectrin is localized at the intracellular face of the plasmic membrane, and therefore, calpactin I is mostly found in the membrane. In addition, there exists a family of proteins binding calcium and biomembranes, proteins which have been highly conserved during evolution : antibodies directed against calelectrin, a calciprotein found in the fish "torpedo marmorata", recognize in mammalian tissues, a series of proteins, namely p70, p36 and p 32.5 (17). P36 seems therefore to be lipocortin II, i.e. the calpactin I heavy chain (18).

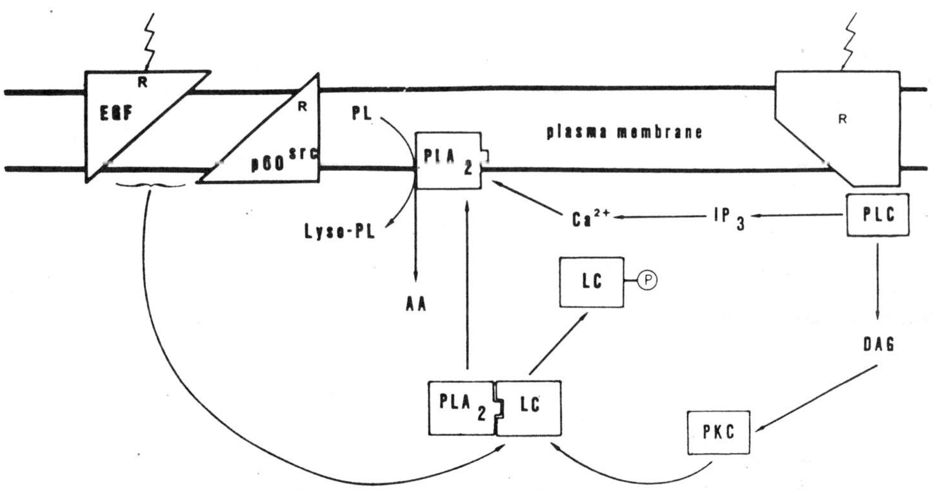

Fig. 3. Speculative scheme of the regulation of lipocortins during the cellular activation. R : Receptor. EGF : epidermal growth factor. LC : lipocortin. LC-P : phosphorylated lipocortin. PL : phospholipid. PLC : phospholipase C. DAG : diacylglycerol. PKC : protein kinase C. IP_3 : inositol triphosphate. On the right side : the pathway involving protein kinase C and serine or threonine phosphorylation, on the left side : the pathway involving tyrosine kinase activation and tyrosine phosphorylation.

Conclusions

These various results question what the real roles of lipocortins are : how can the different lipocortins interfere with the inflammatory reaction, is their antiphospholipase A_2 property the only one or do they possess other functions. More, how is lipocortin secreted and found outside of the cells since there is no sequence of signal peptide in the lipocortin gene.

We propose now that there exist one or more lipocortins whose major function would be to inhibit endocellular phospholipase A_2. Indeed, it is well known that under resting conditions, the enzyme is inactive, whereas when a specific inflammatory stimulus triggers the cell membrane, it induces a cascade of events through transducing systems, among which, the activation of protein kinase and tyrosine kinase, lipocortin phosphorylation, phospholipase A_2 activation and the production of lipid-derived mediators, necessary for the inflammatory reaction to take place. These latter agents reinitiate the whole stimulus-coupling event after interacting with specific receptors on other cell membranes. We propose that corticosteroids, by increasing the number of molecules of lipocortin, will slow down the whole process, by inhibiting the stimulus response coupling.

This very simple scheme is a very partial explanation of the role of lipocortin and does not allow to answer all the questions raised by the discovery of these proteins. Why are there so many different lipocortins, and if so, why are they all inhibiting phospholipase A_2, binding calcium and biomembranes. Is the antiphospholipase A_2 function the only important one or are there other cell functions, due to the possible interactions of lipocortins with phospholipids which might play a role in cell differentiation.

REFERENCES

1. Blackwell GJ, Carnuccio R, DiRosa M, Flower RJ, Parente L, Persico P. Macrocortin : a polypeptide causing the antiphospholipase like effects of glucocorticoids. Nature 1980; 287 : 147-9.

2. Hirata F, Schiffmann D, Venkatasubramanian K, Salomon D, Axelrod J. A phospholipase A_2 inhibitory protein in rabbit neutrophils induced by glucocorticosteroids. Proc Natl Acad Sci USA 1980; 77 : 2533-6.

3. Cloix JF, Colard O, Rothhut B, Russo-Marie F. Characterization and partial purification of "renocortins" : two polypeptides formed in renal cells causing the antiphospholipase like action of glucocorticoids. Br J Pharmacol 1983; 79 : 312-21.

4. DiRosa M, Flower RJ, Hirata F, Parente L, Russo-Marie F. Lipocortins, nomenclature announcement. Prostaglandins 1984; 28 : 441-4

5. Hirata F. The regulation of lipomodulin, a phospholipase inhibitory protein, in rabbit neutrophils by phosphorylation. J Biol Chem 1981; 256 : 7730-3.

6. Rothhut B, Russo-Marie F, Cousin M, Lando D. Renocortin : a phospholipase A_2 inhibitory protein induced by anti-inflammatory steroids : pharmacological and biological properties, development of monoclonal antibodies. In Proceedings of the 9th IUPHAR Congress of Pharmacology, P Turner, W Paton, JF Mitchell (eds), The Macmillan Press ltd. London, 1984 : 43-6

7. Hirata F, Matsuda K, Notsu Y, Hattori T, DelCarmine R. Phosphorylation of a tyrosine residue of lipomodulin in mitogen-stimulated murine thymocytes. Proc Natl Acad Sci USA 1984; 81 : 4717-21.

8. Touqui L, Rothhut B, Shaw A, Fradin A, Vargaftig BB, Russo-Marie F. Platelet activation - a role for a 40 K antiphospholipase A_2 protein indistinguishable from lipocortin. Nature 1986; 321 : 177-80.

9. Blake-Pepinsky R, Sinclair LK, Browning JL et al. Purification and

partial sequence analysis of a 37 KDa protein that inhibits phospholipase A2 activity from rat peritoneal exudates. J Biol Chem 1986; 261 : 4246-9.

10. Wallner BP, Mattaliano RJ, Hession C et al. : Cloning and expression of human lipocortin, a phospholipase A_2 inhibitor with potential anti-inflammatory activity. Nature 1986; 320 : 77-80.

11. Blake-Pepinsky R, Sinclair LK. Epidermal growth factor-dependent phosphorylation of lipocortin. Nature 1986; 321 : 81-84.

12. Huang K-S, Wallner BP, Mattaliano RJ, Browning JL et al. Two human 35 KD inhibitors of phospholipase A_2 are related to substrates of pp60 srcfr and of epidermal growth factor receptor kinase. Cell 1986; 46 : 191-9.

13. De BK, Misono KS, Lukas TJ, Mroczkowski B, Cohen S. A calcium-dependent 35-kilodalton substrate for epidermal growth factor receptor/kinase isolated from normal tissue. J Biol Chem 1986; 261 : 13784-92.

14. Glenney JR, Jr. Two related but different forms of the 36,000 Mr tyrosine kinase substrate (calpactins) which interact with phospholipid and actin in a Ca+fr-dependent manner. Proc Natl Acad Sci USA 1986; 83 : 4258-62.

15. Kristensen T, Saris CJM, Hunter T, Hicks LJ, Noonan DJ, Glenney JR, Tack B. Primary structure of bovine calpactin I heavy chain (p36), a major cellular substrate for retroviral protein-tyrosine kinases: homology with the human phospholipase A_2 inhibitor lipocortin. Biochemistry 1986; 25 : 4497-503.

16. Saris CJM, Tack BF, Kristensen T, Glenney JR, Hunter T. The cDNA sequence for the protein tyrosine kinase substrate p36 (Calpactin I heavy chain) reveals a multidomain protein with internal repeats. Cell 1986; 46 : 201-12.

17. Geisow MJ, Fritsche U, Hexham JM, Dash B, Johnson T. A consensus amino-acid sequence repeat in Torpedo and mammalian Ca+dependent membrane-binding proteins. Nature 1986; 320 : 636-8.

18. Geisow MJ. Common domain structures of Ca+fr and lipid-binding proteins. FEBS Lett 1986; 203 : 99-103.

IMMUNOLOGICALLY-MEDIATED PITUITARY-ADRENAL ACTIVATION

J. Edwin Blalock

Department of Physiology and Biophysics
The University of Alabama at Birmingham
Birmingham, AL

INTRODUCTION

Perhaps the most interesting recent observation concerning interactions between the immune and neuroendocrine systems is the finding that such communication is bidirectional. That is, while it is generally accepted that the neuroendocrine system can regulate immunologic functions, it is only now becoming apparent that the immune system can regulate neuroendocrine functions. In the latter instance, the best studied system seems to be immunologic activation of the pituitary adrenal axis. Initial data supporting such activation was the observation of an elevation of circulating glucocorticoid hormones which coincided with peak antibody titers during an immune response.[1] Subsequently, this effect was in part attributed to a factor(s) of unknown composition which was derived from activated leukocytes.[2] We now know that there are a number of well defined immunologically-derived molecules which are involved in a lymphoid pituitary adrenal axis and can account for the earlier observations. These molecules include lymphokines and monokines as well as leukocyte-derived neuroendocrine peptide hormones. This article will review the various pathways by which the immune system can activate the pituitary adrenal axis.

PRODUCTION AND ACTION OF LEUKOCYTE-DERIVED ACTH

The first indication that leukocytes might be an extrapituitary source of proopiomelanocortin (POMC)-derived peptides was the observation that virus-infected human peripheral blood cells contained a substance which was reactive with a monospecific antibody against a synthetic fragment of ACTH.[3] In the ensuing years, this antigen was shown to be bona fide ACTH. The shared characteristics of the immunologically and pituitary-derived ACTH included: related antigenicity as determined with monospecific antibodies against the synthetic hormone; identical retention time on reverse phase high performance liquid chromatography; identical molecular weights; coordinate production with endorphins; and shared biological activity.[3-8] Most recently, the mRNA for the ACTH precursor (proopiomelanocortin, POMC) has been identified in lymphocytes as well as macrophages[9-11] and the amino acid sequences of mouse splenic and pituitary ACTH were found to be identical (E. M. Smith and J. E. Blalock, unpublished observation).

While immunostimulants were the first described inducers of leukocyte POMC-derived peptides, it was subsequently found that cells of the immune system would respond to the classic hypothalamic regulator of pituitary ACTH. Corticotropin releasing factor (CRF) was observed to elicit the *de novo* synthesis and release of leukocyte-derived ACTH and β-endorphin. As with pituitary cells, arginine vasopressin (AVP) at 10-fold higher concentrations than CRF was observed to have intrinsic CRF activity. Further, CRF and AVP together acted in an additive fashion to induce leukocyte-derived POMC peptides and such induction was blocked by dexamethasone.[12] Thus cells of the immune system are not unlike corticotrophs with respect to control of the POMC gene by a positive hypothalamic signal and feedback inhibition by a synthetic adrenal hormone.

A striking difference between pituitary and leukocyte production of POMC-derived peptides concerns processing of the hormone precursor. For example, Newcastle disease virus or CRF induction leads to the production of peptides with the molecular weight of ACTH (1-39) and β-endorphin. Thus with regard to processing and with these two stimuli, cells of the immune system look like a composite of cells of the anterior and intermediate lobes of the pituitary where ACTH 1-39 and β-endorphin are primary processing products, respectively. In addition to having proteolytic processing enzymes, macrophages, at least, must also harbor acetylating enzymes since splenic macrophages produce acetylated derivatives of β-endorphin.[10] Unlike either anterior or intermediate lobe pituitary cells, immunocytes also possess what appears to be a unique processing pathway. Bacterial lipopolysaccharide (LPS) treatment results in the in vitro production of an ACTH with a molecular weight of 2,900.[6] This finding suggests that with LPS, ACTH (1-39) is further cleaved to a smaller form, possibly ACTH (1-23 to 26). In fact, an enzymatic activity is present in LPS-treated splenocytes which will convert ACTH (1-39) to a form with a molecular weight of 2,900 (D. V. Harbour, E. M. Smith, and J. E. Blalock, submitted for publication). The above findings point to cells of the immune system having processing pathways which are unique as well as in some instances composites of the anterior and intermediate lobe of the pituitary gland. At the present time, we do not know the relative contribution to alternate processing of the stimuli (i.e., CRF vs. LPS) as opposed to the possible different leukocyte types which respond to a particular stimulus.

The observation that cells of the immune system are a source of secreted ACTH suggested that in certain circumstances the immune system might replace the pituitary gland requirement for an ACTH mediated adrenal glucocorticoid response. This idea was supported by the observation that Newcastle disease virus (an inducer of leukocyte ACTH in vitro) infection of hypophysectomized mice caused a time dependent increase in corticosterone production which was inhibited by dexamethasone. The spleens of such infected mice were also positive for ACTH unless they were pretreated with dexamethasone.[5] A similar situation was observed in patients who suffer from pituitary ACTH deficiency. Such individuals who were undergoing pyrogen testing (typhoid vaccine, another in vitro inducer of leukocyte ACTH) of pituitary function showed an increase in the percentage of their circulating mononuclear leukocytes that were positive for ACTH. The response in hypopituitary children peaked at approximately six to eight hours after administration of typhoid vaccine.[13] Based on these observations, we have suggested that a lymphoid adrenal axis may exist, which in part accounts for the increases in corticosteroid levels which are generally observed during infections.

A possible in vivo corollary of the aforementioned ability of CRF and arginine vasopressin to elicit in vitro leukocyte ACTH has also been observed. We found an increase in the percentage of circulating immunoreactive ACTH positive leukocytes in normal short children who were undergoing assessment of pituitary function by insulin tolerance testing. In contrast to the previously described peak in the percentage of ACTH positive leukocytes which occurred at 6-8 hr post-typhoid vaccination, the peak was observed at 45 min to 1 hr after the administration of insulin. Insulin alone or in combination with low glucose levels did not induce in vitro production of leukocyte ACTH.[13] Considering both the in vivo and in vitro data, it is tempting to speculate that in centrally mediated stressful situations leukocytes may also produce ACTH in response to an encounter with CRF, which perhaps occurs as these cells pass through the portal circulation, between the hypothalamus and pituitary gland. A finding that is consistent with the possible in vivo action of CRF in such a setting is the observation that this hypothalamic releasing factor will elicit an ACTH response in individuals who are ACTH deficient secondary to removal of a pituitary adenoma (H. L. Fehm, personal communication). Collectively, these results suggest that the postulated lymphoid adrenal axis may also have a central nervous component and thus might be more appropriately termed a hypothalamic immuno-pituitary adrenal axis. Such an axis might also explain the earlier observation of bacterial polysaccharide (Piromen) induced cortisol responses in seven out of eight patients who underwent pituitary stalk sectioning.[14] These findings might result from leukocyte ACTH and/or by the induction of IL-1 or HSF by bacterial polysaccharide which in turn could cause pituitary ACTH release (see below). More recently, a patient presented with the clinical and laboratory features of the ectopic ACTH syndrome. While the absence of a gradient in ACTH concentration, between the inferior petrosal sinuses and the periphery, argued against pituitary overproduction, no ACTH-producing tumor was found and bilateral adrenalectomy was performed to correct the hypercortisolism. Six months later, a pseudotumor containing only normal fat and inflammatory tissue was observed in the patient. Upon removal of this inflammatory tissue, basal plasma ACTH levels immediately returned to normal and leukocytes within the inflammatory tissue stained positively for ACTH by an immunoperoxidase procedure.[15]

LYMPHOKINE AND MONOKINE ACTIVATION OF THE PITUITARY ADRENAL AXIS

Lymphokines and monokines seem to represent a second set of signals that can originate in the immune system and regulate the pituitary adrenal axis. Interferons (IFNs) were perhaps the first to be observed to have a hormonal function. Interferon was found to induce an in vitro ACTH-like steroidogenic response when placed on cultured adrenal cells.[16] A similar finding was observed during clinical testing of IFN-α where it was observed to cause elevations in circulating cortisol.[17] In this instance, the simplest explanation would be that IFN-α acted directly on the adrenal glands to elicit a steroidogenic response. While this has yet to be proven, it seems an attractive possibility considering the in vitro effects of this lymphokine.

Other lymphokines and monokines seem to regulate the pituitary-adrenocortical axis at sites other than that postulated for interferon. Thymosin-α-1, for instance, seems to cause in vivo ACTH release through an action on the central nervous system which in turn stimulates the pituitary gland.[18] Interleukin (IL) 2 was also recently shown to cause an elevation in circulating ACTH and cortisol levels during clinical testing.[19] In this case, the action of IL-2 may well be on the pituitary gland since this lymphokine can cause POMC production and release from pituitary tumor cells and corticotrophs (20 and our unpublished observation).

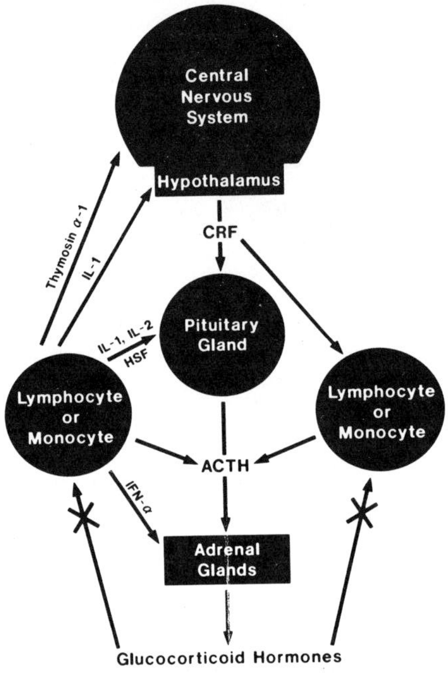

Fig. 1. Pathways for Immunologic Activation of the Pituitary Adrenal Axis. Please see text for description.

Some time ago, IL-1 was reported to function as the endogenous pyrogen in the fever response during endotoxemia, presumably by interacting with thermoregulatory centers of the hypothalamus (for review, see 21). Thus, an IL-1 action on the hypothalamus to subsequently activate pituitary ACTH seems plausible. Interestingly, IL-1 synthesis has been reported in astrocytes and glial cells and in the brain of endotoxin-treated mice, thus showing the possible local production of this monokine.[22] It has also been shown that IL-1 is equipotent with CRF and arginine vasopressin in its ability to stimulate ACTH release from pituitary cells.[23-25] A newly described monokine, hepatocyte stimulating factor (HSF), has also been shown to have CRF-like activity except that it is 3-fold more potent than either IL-1 or CRF-arginine vasopressin. It was concluded that since both IL-1 and HSF are released during inflammatory processes they may play a very important role in activation of the pituitary adrenal axis.[23] In fact, IL-1 was recently shown to elicit ACTH and glucocorticoid synthesis and release in vivo.[26] Though these studies are in their infancy, they nonetheless seem to clearly show that lymphokines and monokines can modulate the pituitary-adrenal axis.

POSSIBLE PATHWAYS FOR IMMUNOLOGIC ACTIVATION OF THE PITUITARY ADRENAL AXIS

Figure 1 shows a model for the various ways by which the immune system might participate in activation of the pituitary adrenal axis. This might ultimately originate from the recognition of two general classes of stimuli, cognitive (i.e., fear, pain, etc.) or noncognitive (i.e., immunostimulants such as viruses, tumor cells, bacteria, antigens, etc.). In the case of immunostimulants (left side of figure), recognition can lead to the production of either leukocyte ACTH or IFN-α that can directly activate adrenal steroidogenesis. Alternatively, though not mutually exclusively, IL-1, IL-2 or HSF can directly activate pituitary synthesis and release of ACTH. IL-1 and thymosin-α-1 can also potentially cause CRF release by an action on the hypothalamus or other parts of the central nervous system. In the case of cognitive stimuli (right side of figure), hypothalamically released CRF could elicit leukocyte ACTH production and subsequent adrenal steroidogenesis. This, of course, would represent one way in which the immune system could be aware of classical sensory perception. In both the cognitive and noncognitive pathways, a feedback loop involving adrenocortical steroids is present. This results in the inhibition of the further synthesis of leukocyte-derived ACTH, lymphokines, and monokines.

One consequence of the above view of the immune system is that it allows for the integration of this collection of organs with the nervous system. Further, it is our view that a primary function of the immune system is to serve as a sensory organ for stimuli that are not recognized by the central and peripheral nervous systems.[27] Immunologic recognition or sensing then results in signal transmission to the neuroendocrine system via leukocyte-derived hormones, lymphokines, and monokines. Ultimately, a physiologic change, such as stress, results. The immune system thus might be considered our "sixth sense."

ACKNOWLEDGEMENTS

This work was supported by NIH grant R01 DK38024. I thank Diane Weigent for typing this manuscript.

REFERENCES

1. Besedovsky, H.O., Sorkin, E., Keller, M., and Muller, J., 1975, Changes in blood hormone levels during the immune response, Proc. Soc. Exp. Biol. Med. 150:466-
2. Besedovsky, H.O., del Rey, A., and Sorkin, E., 1981, Lymphokine containing supernatants from Con A stimulated cells increase corticosterone blood levels, J. Immunol. 126:385-
3. Blalock, J.E., and Smith, E.M., 1980, Human leukocyte interferon: Structural and biological relatedness to adrenocorticotropic hormone and endorphins, Proc. Natl. Acad. Sci. U.S.A. 77:5972-5974.
4. Smith, E.M., and Blalock, J.E., 1981, Human lymphocyte production of ACTH and endorphin-like substances: Association with leukocyte interferon, Proc. Natl. Acad. Sci. U.S.A. 78:7530-7534.
5. Smith, E.M., Meyer, W.J., and Blalock, J.E., 1982, Virus-induced increases in corticosterone in hypophysectomized mice: A possible lymphoid adrenal axis. Science 218:1311-1313.
6. Harbour-McMenamin, D., Smith, E.M., and Blalock, J.E., 1985, Bacterial lipopolysaccharide induction of leukocyte-derived ACTH and endorphins, Infect. Immun. 48:813-817.
7. Lolait, S.J., Lim, A.T.W., Toh, B.H., and Funder, J.W., 1984, Immunoreactive β-endorphin in a subpopulation of mouse spleen macrophages, J. Clin. Invest. 73:277-280.
8. Blalock, J.E., and Smith, E.M., 1985, A complete regulatory loop between the immune and neuroendocrine systems, Fed. Proc. 44:108-111.
9. Westley, H.J., Kleiss, A.J., Kelley, K.W., Wong, P.K.Y., and Yuen, P.-H., 1986, Newcastle disease virus-infected splenocytes express the pro-opiomelanocortin gene, J. Exp. Med. 163:1589-1594.
10. Lolait, S.J., Clements, J.A., Markwick, A.J., Cheng, C., McNally, M., Smith, A.I., and Funder, J.W., 1986, Pro-opiomelanocortin messenger ribonucleic acid and posttranslational processing of beta endorphin in spleen macrophages, J. Clin. Invest. 77:1776-1779.
11. Endo, Y., Sakata, T., and Watanabe, S., 1985, Identification of proopiomelanocortin-producing cells in the rat pyloric antrum and duodenum by in situ mRNA-cDNA hybridization, Biomed. Res. 6:253-256.
12. Smith, E.M., Morrill, A.C., Meyer, W.J., and Blalock, J.E., 1986, Corticotropin releasing factor induction of leukocyte-derived immunoreactive ACTH and endorphins, Nature 322:881-882.
13. Meyer, W.J., III, Smith, E.M., Richards, G.E., Cava-lo, A., Morrill, A.C., and Blalock, J.E., In vivo immunoreactive ACTH production by human leukocytes from normal and ACTH-deficient individuals, J. Clin. Endocrinol. Metabol. in press.
14. Van Wyk, J.J., Dugger, G.S., Newsome, J.F., and Thomas, P.Z., 1960, The effect of pituitary stalk section in the adrenal function of women with cancer of the breast, J. Clin. Endocrinol. Metab. 20:157-172.
15. Dupont, A.G., Somers, G., Van Steviteghem, A.C., Warson, F., and Vanhaelst, L., 1984, Ectopic adrenocorticotropin production: Disappearance after removal of inflammatory tissue, J. Clin. Endocrinol. Metabol. 58:654-658.
16. Blalock, J.E., and Harp, C., 1981, Interferon and adrenocorticotropic hormone induction of steroidogenesis, melanogenesis and antiviral activity, Arch. Virol. 67:45-49.
17. Roosth, J., Pollard, R.B., Brown, S.L., and Meyer, W.J., III, 1986, Cortisol stimulation by recombinant interferon- 2, J. Neuroimmunol. 12:311-316.

18. Hall, N., and Goldstein, A., 1985, Neurotransmitters and host defense, in "Neural Modulation of Immunity," R. Guillemin, M. Cohn, and T. Melnechuk, eds., Raven Press, New York, pp. 143-156.
19. Lotze, M.T., Frana, L.W., Sharrow, S.O., Robb, R.J., and Rosenberg, S.A., 1985, In vivo administration of purified human interleukin 2. I. Half-life and immunologic effects of the Jurk at cell line-derived interleukin 2. J. Immunol. 134:157-166.
20. Farrar, W.L., 1984, Endorphin modulation of lymphokine activity, in "Opioid Peptides in the Periphery," F. Fraioli, A. Isidori, and M. Mazzetti, eds., Elsevier Publishers B.V., pp. 159-165.
21. Oppenheim, J.J., and Gery, I., 1982, Interleukin 1 is more than an interleukin, Immunol. Today 3:113-
22. Fontana, A., Weber, E., and Dayer, J.-M., 1984, Synthesis of interleukin 1/endogenous pyrogen in the brain of endotoxin treated mice: A step in fever induction?, J. Immunol. 133:1696-1698.
23. Woloski, B.M.R.N.J., Smith, E.M., Meyer, W.J., III, Fuller, G.M., and Blalock, J.E., 1985, Corticotropin-releasing activity of monokines, Science 230:1035-1037.
24. Beach, J.E., Bernton, E.W., Holaday, J.W., Smallridge, R.C., and Fein, H.G., 1986, Interleukin-1 modulates secretion by rat pituitary cells in monolayer culture, in "Abstracts of the 1st International Congress of Neuroendocrinology," #322, 106.
25. Kehrer, P.H., Gaillard, R.C., Dayer, J.-M., and Muller, A.F., 1986, Human interleukin 1 β stimulates ACTH release from rat anterior pituitary cells in a cAMP and prostaglandin E_2 independent manner, in "Abstracts of the 1st International Congress of Neuroendocrinology," #327, pp. 107.
26. Besedovsky, H., del Rey, A., Sorkin, E., and Dinarello, C., 1986, Immunoregulatory feedback between interleukin-1 and glucocorticoid hormones, Science 233:652-654.
27. Blalock, J.E., 1984, The immune system as a sensory organ, J. Immunol. 132:1067-1070.

GLUCOCORTICOID ANTAGONISTS AND THE ROLE OF GLUCOCORTICOIDS AT THE RESTING AND STRESS STATE

Louisa Laue, D. Lynn Loriaux, and George P. Chrousos

Developmental Endocrinology Branch, National Institute of Child Health and Human Development, Bethesda, MD

INTRODUCTION

Glucocorticoids act on almost all organs and tissues. Their levels are rarely constant, showing both pronounced circadian variations and/or a rise during stressful situations. It is thought that glucocorticoids evoke an adaptation to stress via a variety of mechanisms including mobilization of fuel for metabolism, enhancement of cardiovascular function, and production of euphoria or behavioral activation. Glucocorticoids also suppress the immune system. It has been argued that this effect helps prevent autoimmune activation following "fight or flight" situations that result in trauma and inflammation.

Glucocorticoid suppression of immune function has been thought to occur only during the administration of pharmacologic doses of steroids. These doses however, are frequently within the range of glucocorticoid production seen during stress (1,2). Studies in rodents, moreover, have suggested that glucocorticoids exert antiinflammatory effects even at basal, non-stress concentrations (3-6). The availability of a specific glucocorticoid receptor antagonist, RU 486, has allowed us to examine whether or not basal glucocorticoid secretion exerts anti-inflammatory effects and whether or not glucocorticoids suppress the inflammatory response through a receptor-mediated mechanism.

GLUCOCORTICOID EFFECTS ON INFLAMMATION

One of the most impressive clinical manifestations of glucocorticoid-induced immunosuppression is inhibition of the inflammatory response (7-10). Glucocorticoids cause a redistribution of circulating leukocytes which impedes the access of neutrophils and monocytes to an inflammatory site. Additionally, glucocorticoids inhibit the secretion and action of various cytokines (monokines and lymphokines) and also block phospholipid metabolism. Both initiation and potentiation of the inflammatory cascade is dependent upon various lipids secreted by both migrant and local cells. These mediators are derived from either polyunsaturated fatty acids, primarily arachidonic acid, or glycerolipids, primarily lyso-platelet activating factor (lyso-PAF)(11-13). These lipids are vasoactive agents as well as potent leukocyte stimulants and attractants serving as mediators of the inflammatory response (9,10,14).

Glucocorticoids exert their effects through binding to specific intracellular receptors (15-18). Once formed, the glucocorticoid receptor complex, considered to be a trans-regulating factor, modulates the transcription rate of specific genes. Glucocorticoid-induced phospholipase inhibitory proteins which presumably function as "second messengers" to mediate the antiinflammatory glucocorticoid effect have been demonstrated in a variety of *in vivo* and cellular systems (19-30). These proteins have collectively been termed "lipocortins." Two highly homologous but distinct proteins inhibiting PLA_2 activity *in vitro* have been isolated from various tissues, and designated as lipocortin I and II (25-30).

The enzyme phospholipase A_2 (PLA_2) is critical for the formation of arachidonic acid and lyso-PAF which are subsequently metabolized by cyclooxygenase to form prostaglandins and cyclic endoperoxides, by lipooxygenase to form leukotrienes, and by acetyl-CoA:1-alkyl-2-lyso-sn-glycero-3 phosphocholine acetyltransferase to form platelet activating factor (PAF).

Until recently, the antiinflammatory effects of glucocorticoids have been considered to be pharmacologic, since supraphysiologic doses of steroids are frequently administered to obtain full effect. This belief has been challenged by recent demonstrations that glucocorticoids exert antiinflammatory effects at basal concentrations in adrenalectomized rats and observations that glucocorticoid-induced immunosuppression occurs within the stress range of glucocorticoid production (2-6,31). We have reexamined the idea that glucocorticoids in normal physiologic concentrations modulate immune and inflammatory responses utilizing an animal model of carrageenin-induced inflammation in association with the glucocorticoid antagonist RU 486 (17-beta-hydroxy 11-beta-(4-dimethylaminophenyl) 17-alpha (propynyl)-estra 4.9 diene-3-one).

RU 486 binds with high affinity to the glucocorticoid receptor and exhibits strong, dose-dependent antiglucocorticoid action *in vitro* (32-43) and *in vivo* (37,44-53). Like glucocorticoids, RU 486 binds to the glucocorticoid receptor to form a complex which subsequently translocates to the nuclei of target cells. Despite its nuclear localization, however, the ability of this complex to bind to DNA is qualitatively inferior to that observed with glucocorticoid agonists (40,54). This results in failure of the RU 486-receptor complex to efficiently modulate gene transcription and to produce the glucocorticoid effect.

ANTIGLUCOCORTICOID EFFECTS ON INFLAMMATION IN A RODENT MODEL

Subcutaneous inflammation was induced with carrageenin according to the method Fukuhara and Tsurufuji (55), with some modification. 120-150 gram Harlan-Spraque-Dawley rats were divided into groups of 10 animals each, and injected subcutaneously with 8 ml of air forming a dorsal air pouch. Twenty-four hours later, 4 ml of 2% carrageenin solution was injected into this preformed pouch. Carrageenin produced an inflammatory exudate which could be aspirated from the pouch and the volume and concentration of leukocytes, prostaglandins and leukotrienes quantitated.

Time-course of acute inflammation

Vascular permeability in the acute exudative stage of this inflammatory model is most prominent 2-8 hours after carrageenin injection. The vascular response then declines and subsides by 24 hours (55). The inflammatory response between 1 and 7 hours is characterized in Fig 1. Exudate volume and concentration of leukocytes, prostaglandin E_2 and leukotriene B_4 were highest 7 hours following carrageenin administration.

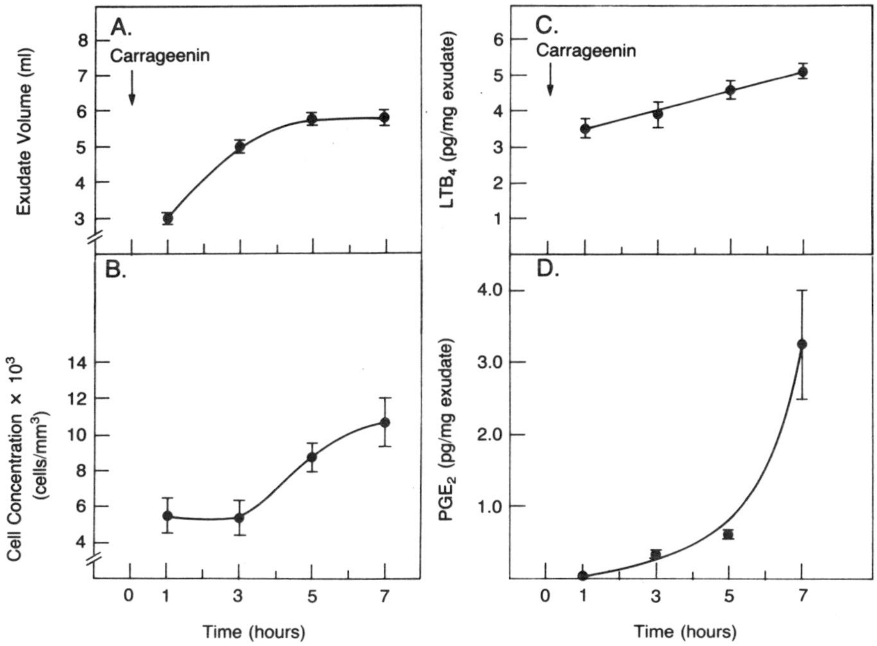

FIG 1. Time-course of the carrageenin-induced inflammatory response. Highest values were obtained after 7 hours of inflammation for exudate volume (panel A), leukocyte concentration (panel B), leukotriene B_4 (panel C) and prostaglandin E_2 (panel D). Each volume represents the mean ± SE from 10 animals (From reference 5).

Glucocorticoid effects on acute inflammation

To quantitate the antiinflammatory action of glucocorticoids in this model, dexamethasone at doses ranging from 0 to 1 mg per rat was administered by intraperitoneal injection one hour prior to carrageenin administration. The animals were sacrificed after 7 hours and the inflammatory response was evaluated. Exudate volume and concentration of leukocytes, prostaglandin E_2 and leukotriene B_4 were all suppressed by dexamethasone in a dose-dependent fashion (Fig. 2).

Antiglucocorticoid effects on acute inflammation

To examine the ability of a glucocorticoid antagonist to obliterate the antiinflammatory properties of exogenous glucocorticoid in this model, we next administered a fixed dose of dexamethasone concomitantly with increasing doses of RU 486. Using the ED_{80} for dexamethasone (approximately 0.1 mg/rat), we concurrently gave doses of RU 486 ranging from 0 to 10 mg/rat by intraperitoneal injection to intact rats one hour prior to carrageenin injection. RU 486 was able to antagonize the suppressive effects of dexamethasone on the inflammatory response in a dose-dependent fashion. The dose of 10 mg of RU 486 restored exudate volume and the concentration of leukocytes, prostaglandin E_2 and leukotriene B_4 to control values (Fig. 3).

We next examined whether or not endogenous basal levels of glucocorticoids exert a tonic level of antiinflammatory activity. Blocking the

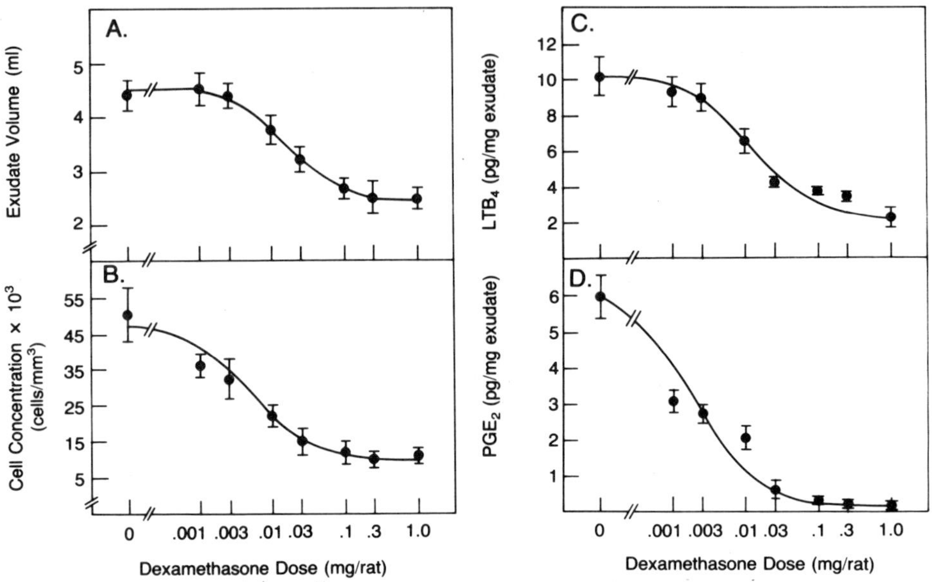

FIG 2. Dose-response relationships between dexamethasone administration and suppression of exudate volume (panel A), leukocyte concentration (panel B), leukotriene B_4 (panel C) and prostaglandin E_2 (panel D) levels. Each value represents the mean ± SE from 10 animals. (From reference 5).

effects of endogenous glucocorticoids in intact animals with RU 486 resulted in changes in the intensity of the inflammatory response to carrageenin. We administered increasing doses of RU 486 ranging from 0 to 10 mg/rat by intraperitoneal injection one hour prior to carrageenin administration. Exudate volume and the concentration of leukocytes, prostaglandin E_2 and leukotriene B_4 were maximally enhanced with 0.3 mg RU 486. Higher doses of RU 486 resulted in less enhancement (Fig. 4).

The "biphasic" shape of dose-response curves for the RU 486-induced enhancement of inflammation can be explained either by the presence of small intrinsic glucocorticoid agonist activity in RU 486 (or in its *in vivo* metabolites) or by the expected compensatory increase of glucocorticoids. The *in vivo* metabolic products from mono-or didemethylation of the dimethylamino group and/or 17-side chain hydroxylation of RU 486 have attenuated antagonist activity and possibly some agonist effect (56). RU 486 has been shown to suppress oCRH-stimulated pituitary ACTH secretion in patients with primary adrenal insufficiency, suggesting that this molecule possesses some agonist activity (57). RU 486 however, causes a rise in plasma ACTH, ß-lipotropin and ß-endorphin concentrations, suggesting its main effect is one of glucocorticoid antagonism. The possibility that the effects of RU 486 on inflammation are mediated by ß-endorphin, ß-lipotropin or ACTH remains to be evaluated in this model.

The antagonism of glucocorticoid-induced suppression of inflammation by RU 486 suggests that the inhibition of vascular permeability, leukocyte aggregation and prostnoid formation are all receptor-mediated effects. Similarly, RU 486 has been shown to antagonize dexamethasone-induced suppression of cotton-induced granuloma formation in adrenalectomized rats (45). In other cellular inflammation systems employing mouse, rat, or human reticuloendothelial cells (41-43), RU 486 has been shown to be an optimal glucocorticoid antagonist. This supports the conclusions of Tsurufusi *et al.* who employed weak, less specific glucocorticoid receptor antagonists, such

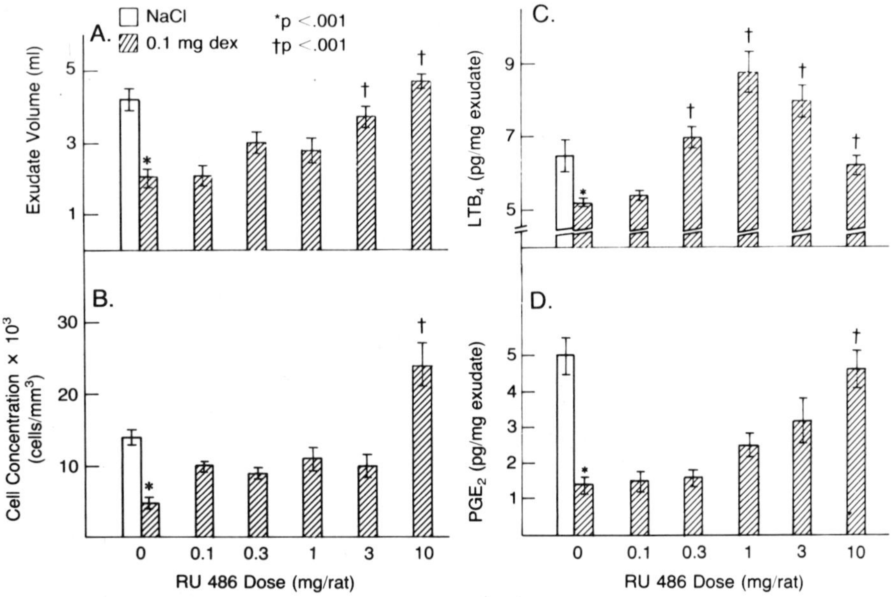

FIG 3. Antagonism of the suppressive effects of dexamethasone on the inflammatory response by the antiglucocorticoid RU 486. 0.1 mg dexamethasone (hatched bars) caused significant reduction of both exudate volume (panel A) and leukocyte concentration (panel B), compared to non-dexamethasone treated controls (open bar). Administration of 10 mg RU 486 restored both volume and leukocyte concentration (panels A and B) to control values. Dexamethasone-induced suppression of both leukotriene B_4 and prostaglandin E_2 levels (panels C & D) were also antagonized completely with 10 mg RU 486. Each value represents the mean ± SE from 10 animals. Stars represent statistical comparisons using Student t test between dexamethasone and saline-treated controls, not receiving RU 486. Daggers represent statistical comparisons, using multiple Students t tests with the Bonferroni correction between dexamethasone-treated controls and animals receiving varying doses of RU 486. (From reference 5).

as progesterone, 17α-methyltestosterone and fluoxymesterone to block the effect of dexamethasone on inflammation. Transcription and protein synthesis inhibitors such as actinomycin D and cycloheximide are also effective in blocking the anti-inflammatory effect of glucocorticoids, suggesting that glucocorticoid receptors and induction of protein synthesis are involved in the antiinflammatory action of dexamethasone (58-60).

The antiinflammatory effects of glucocorticoids have been examined predominately in the stress state (2,4,61-65) or following exogenous administration of pharmacologic doses of steroids (9,10,66). Our findings support the hypothesis of Munck _et al._ that in the rat, endogenous basal levels of glucocorticoids may harness the inflammatory cascade (1,67). A potentiation of this basal tonic antiinflammatory action of glucocorticoids occurs during stress when glucocorticoid secretion is markedly enhanced.

The concept that glucocorticoid antagonists may be able to enhance the immune response and thereby improve tumor surveillance was tested in a glucocorticoid receptor positive rat MCA fibrosarcoma tumor model. RU 486 treatment significantly retarded sarcoma growth compared to vehicle treated control rats (Fig. 5, p<0.05). Final tumor weight was significantly lower, while adrenal and splenic weights were greater in the RU 486-treated rats than in the vehicle-treated controls (p<0.05). Neither mean survival nor

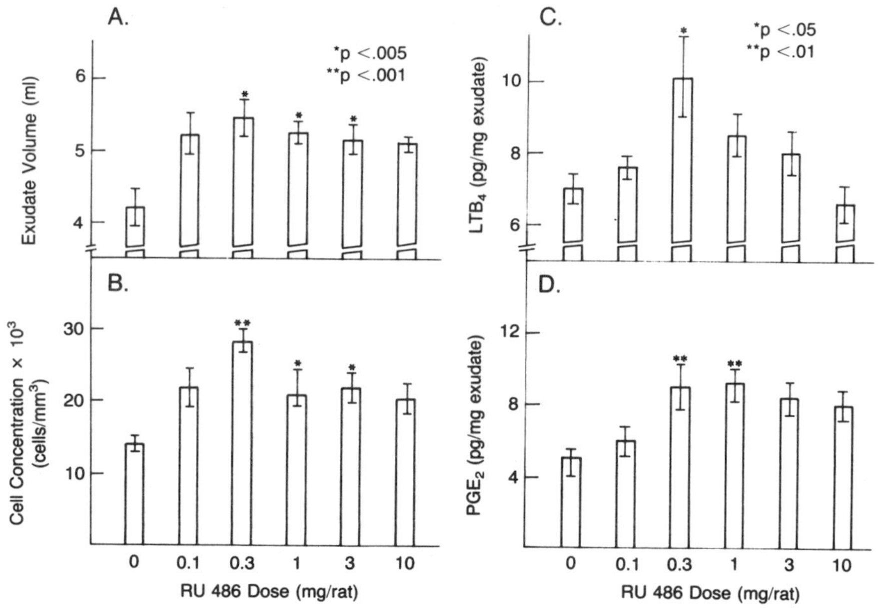

FIG 4. Enhancement of the inflammatory response following administration of the glucocorticoid antagonist RU 486 to intact rats. RU 486 at doses ranging from 0 to 10 mg per rat administered by i.p. injection resulted in increased exudate volume (panel A) and leukocyte concentration (panel B). Similarly, leukotriene B_4 and prostaglandin E_2 levels in the exudate (panels C&D) were maximally enhanced with 0.3 mg RU 486 per rat. Each value represents the mean ± SE from 10 animals. Stars represent statistical comparisons using multiple Students t tests with the Bonferroni correction between control and RU 486 treated animals (From reference 5).

change in host weight (an index of cachexia) was significantly altered by RU 486 treatment (68). The retardation of tumor growth may be a result of RU 486-induced immune enhancement or of relative dependency of the tumor on glucocorticoids for growth.

This lack of substantive therapeutic effect with rat fibrosarcoma does not exclude the possibility that other, less malignant tumors may respond to antiglucocorticoid therapy. Consequently, glucocorticoid antagonists may have some role in the treatment of glucocorticoid receptor positive solid tumors for which standard therapy has not been successful or is not available.

GLUCOCORTICOID AND GLUCOCORTICOID ANTAGONIST EFFECTS IN MAN

Acute administration of glucocorticoids results in neutrophilia and lymphopenia, presumably as a result of redistribution of these cells between the intravascular space and other lymphoid compartments (9). We have examined the effects of chronic changes in glucocorticoid action on various subpopulations of human leukocytes in patients with endogenous Cushing's syndrome and normal volunteers treated with RU 486 (10 mg/kg/d) for one week. Although neutrophilia was observed in the hypercortisolemic patients, no changes in absolute lymphocyte count and lymphocyte phenotypes were observed in either Cushing's patients or normal volunteers treated with RU 486 (69). This is in contrast to the known effects of acute glucocorticoid administration on T lymphocytes (9).

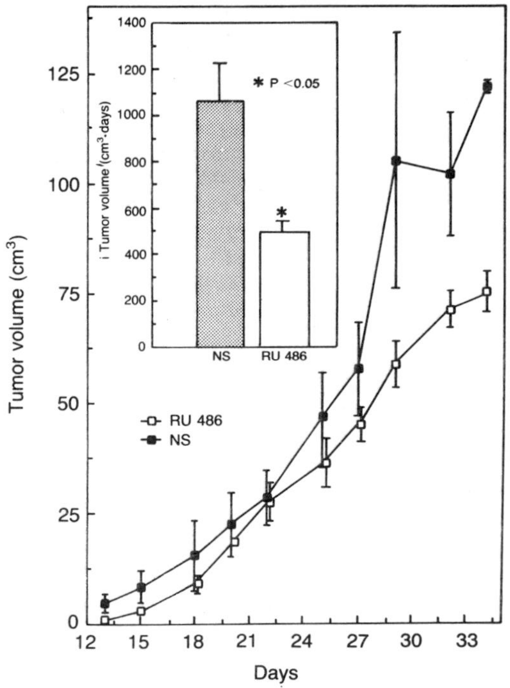

FIG 5. Mean serial tumor volumes determined by 3-dimensional caliper measurement and time-integrated tumor volumes (insert) for RU 486- and vehicle-treated tumor bearing rats. Sarcoma growth was significantly reduced in the RU 486-treated animals ($p<0.05$). (From reference 68).

It appears, thus, that chronic changes in glucocorticoid action associated with states of either hyper- or hypo-cortisolism in man do not alter total lymphocyte number or subpopulations. Adaptive changes may take place during chronic alterations of cortisol exposure to restore the normal distribution and number of circulating lymphocytes. Whether or not leukocyte function is affected, altering host immune competence, remains to be elucidated.

REFERENCES

1. A. S. Munck, P. M. Guyre, N. J. Holbrook, Physiological functions of glucocorticoids in stress and their relation to pharmacological actions. Endo. Rev., 5:25 (1984).
2. A. S. Munck, P. M. Guyre, Glucocorticoid physiology, pharmacology and stress, in: "Steroid Hormone Resistance", G. P. Chrousos, D. L. Loriaux, M. B. Lipsett, eds., Plenum Press, New York, 81 (1986).
3. R. J. Flower, The mediators of steroid action, Nature, 320:20 (1986).
4. L. Laue, S. Kawai, D. Brandon, D. L. Loriaux, G. P. Chrousos, Glucocorticoid and glucocorticoid antagonist effects on the inflammatory response, Ped. Res., 20:296A (1986).
5. L. Laue, S. Kawai, D. Brandon, D. Brightwell, K. Barnes, R. Knazek, D. L. Loriaux, G. P. Chrousos, Receptor-mediated effects of glucocorticoids on inflammation: Enhancement of the inflammatory response with a glucocorticoid antagonist, J. Steroid Biochem. (in press) (1988).
6. F. J. Flower, L. Parente, P. Persico, J. A. Salmon, A comparison of the acute inflammatory response in adrenalectomized and sham-operated rats, Br. J. Pharmac. 87:57 (1986).

7. M. DiRosa, Role in inflammation of glucocorticoid-induced phospholipase inhibitory proteins, Prog. Biochem. Pharmacol. 20:55 (1985).
8. B. Rothhut, F. Russo-Marie, Novel concepts in the mode of action of antiinflammatory steroids, Agents Actions Suppl. 14:171 (1984).
9. A. S. Fauci, Glucocorticosteroid therapy: Mechanisms of action and clinical considerations, Ann. Int. Med., 84:304 (1976).
10. H. C. Stevenson, A. S. Fauci, The effects of glucocorticoids and other hormones on inflammatory and immune responses, in: "Cellular Functions in Immunity and Inflammation", J. J. Oppenheim, D. L. Rosenstreich, M. Potter, eds., Elsevier Co., New York, 431 (1984).
11. G. Weissman, Pathways of arachiodonate oxidation to prostaglandins and leukotrienes, Seminars in Arthritis and Rheumatology, 13:123s (1983).
12. G. Weissman, Prostaglandins in acute inflammation, Current Concepts Monographs, Upjohn Co., Kalamazoo, Michigan, 1 (1980).
13. B. Samuelson, Mediators of immediate hypersensitivity reactions and inflammation, Science, 220:568 (1983).
14. J. J. Oppenheim, M. Potter, Immunity and inflammation, in: "Cellular Functions in Immunity and Inflammation", J. J. Oppenheim, D. L. Rosenstreich, M. Potter, eds., Elsevier Co., New York, 1 (1984).
15. D. Feldman, The role of hormone receptors in the action of adrenal steroids. Ann. Rev. Med., 26:83 (1975).
16. G. G. Rousseau, J. D. Baxter, Glucocorticoid hormone action, in: "Monographs in Endocrinology", J. D. Baxter, G. G. Rousseau, eds., Springer-Verlag, Germany, 12:49 (1979).
17. G. G. Rousseau, Structure and regulation of the glucocorticoid receptor, Mol Cell Endocrinol., 38:1 (1984).
18. B. W. O'Malley, U. T. Schrader, M. J. Tsai, Molecular actions of steroid hormones. in: "Steroid Hormone Resistance", G. P. Chrousos, D. L. Loriaux, M. B. Lipsett, eds., Plenum Press, New York, 1 (1986).
19. R. J. Flower, G. J. Blackwell, Antiinflammatory steroid induced biosynthesis of a phospholipase A2 inhibitor which prevents prostaglandin generation, Nature, 278:456 (1979).
20. R. Carnuccio, M. DiRosa, P. Persico, Hydrocortisone induced inhibition of prostaglandin biosynthesis in rat leukocytes, Br. J. Pharm., 68:14 (1980).
21. A. Danon, G. Assouline, Inhibition of prostaglandin biosynthesis by corticosteroids requires RNA and protein synthesis, Nature, 273:552 (1978).
22. M. DiRosa, P. Persico, Mechanism of inhibition of prostaglandin biosynthesis by hydrocortisone in rat leukocytes. Br. J. Pharm., 66:161 (1979).
23. G. J. Blackwell, R. Carnuccio, M. DiRosa, R. J. Flower, L. Parente, P. Persico, Macrocortin: A polypeptide causing the antiphospholipase effects of glucocorticoids. Nature, 287:147 (1980).
24. J. F. Cloix, O. Colard, B. Rothhut, F. Russo-Marie, Characterization and partial purification of "renocortins": two polypeptides formed in renal cells causing the anti-phospholipase-like action of glucocorticoids. Br. J. Pharm., 79:313 (1983).
25. B. P. Wallner, R. J. Mattaliano, C. Hession, R. L. Cate, R. Tizard, L. K. Sinclair, C. Fueller, E. P. Chow, J. L. Browning, K. L. Ramachandran, R. B. Pepinsky, Cloning and expression of human lipocortin, a phospholipase A2 inhibitor with potential antiinflammatory activity. Nature, 320:77 (1986).
26. G. J. Blackwell, R. Carnuccio, M. DiRosa, R. J. Flower, C. J. Langham, L. Parente, P. Persico, N. C. Russell-Smith, D. Stone, Glucocorticoids induce the formation and release of antiinflammatory and antiphospholipase proteins into the peritoneal cavity of the rat. Br. J. Pharm., 76:185 (1982).
27. F. Hirata, E. Schiffman, K. Venkatasubramanian, D. Salomon, J. Axelrod, A phospholipase A2 inhibitory protein in rabbit neutrophils induced by glucocorticoids. Proc. Nat. Acad. Sci. USA, 77:2533 (1980).

28. F. Hirata, M. Notsu, M. Iwata, L. Parente, M. DiRosa, R. J. Flower, Identification of several species of phospholipase inhibitory proteins by radioimmunoassay for lipomodulin. Biochem. Biophys. Res. Commun., 109:223 (1982).
29. B. Rothhut, F. Russo-Marie, J. Wood, M. DiRosa, R. J. Flower, Further characterization of the glucocorticoid-induced antiphospholipase protein "renocortin". Biochem. Biophys. Res. Commun., 117:878 (1983).
30. K. S. Huang, B. P. Wallner, R. J. Mattaliano, B. Tizard, C. Burne, A. Frey, C. Hession, P. McGray, K. L. Sinclair, E. P. Chow, J. L. Browning, K. L. Ramachandran, J. Tang, J. E. Smart, R. B. Pepinsky, Two human 35 kilodalton inhibitors of phospholipase A2 are related to substrates of $pp60^{v-src}$ and of the epidermal growth factor receptor/kinase, Cell, 46:191 (1986).
31. A. H. Tashjian, E. F. Coelkel, J. McDonough, L. Levine, Hydrocortisone inhibits prostaglandin production by mouse fibrosarcoma cells, Nature, 258:739 (1975).
32. I. Jung-Testas, E. E. Baulieu, Inhibition of glucocorticosteroid action in cultured L-929 mouse fibroblasts by RU 486, a new antiglucocorticosteroid receptor. Exp. Cell Res., 147:177 (1983).
33. N. Oliver, R. F. Newby, L. T. Furcht, S. Bourgeois, Regulation of fibronectin biosynthesis by glucocorticoids in human fibrosarcoma cells and normal fibroblasts, Cell, 33:287 (1983).
34. M. N. Chobert, R. Baronki, J. Finidori, M. Aggerbeck, J. Hanoume, D. Philibert, R. Deraedt, Antiglucocorticoid properties of RU 38486 in a differentiated hepatoma cell line, Biochem. Pharmacol., 43:3481 (1983).
35. S. Chasserot-Golaz, G. Beck, An approach to the mechanism of the potent antiglucocorticoid: 17 beta-hydroxy-11 beta-4-dimethylaminophenyl 17-alpha-propynyl-estra-4,9-diene-3-one, J. Steroid Biochem., 21:585 (1984).
36. E. Coezy, J. Bouhnik, E. Clauser, F. Pinet, M. Philippe, J. Menard, P. Carvol, Effects of glucocorticoids and antiglucocorticoids on angiotensinogen production by hepatoma cells in culture. In vitro, 20:528 (1984).
37. L. Proux-Ferland, J. Cote, D. Philibert, R. Deraedt, Potent antiglucocorticoid activity of RU 38486 on ACTH secretion in vitro and in vivo in the rat, 6th International Congress on Hormonal STeroids, J Steroid Biochem., 17:xvii, (Abstr), 80 (1982).
38. S. J. Lamberts, E. G. Bons, P. Uitterlinden, Studies on the glucocorticoid receptor blocking action of RU 38486 in cultured ACTH-secreting human pituitary tumor cells and normal rat pituitary cells. Acta. Endocrinological., 109:64 (1985).
39. B. Groner, H. Ponta, U. Rahmsclore, P. Herrlichl, M. Pfahll, N. E. Hynes, Glucocorticoid hormone interaction with cloned proviral DNA of mouse mammary tumor virus, J. Steroid Biochem., 20:95 (1984).
40. S. Bourgeois, M. Pfahl, E. E. Baullieu, DNA binding properties of glucocorticosteroid receptors bound to the steroid antagonist RU 486, Embo. J. 3:751 (1984).
41. D. Duval, S. Surant, F. Homo-Delarche, Effects of antiglucocorticoids on dexamethasone-induced inhibition of uridine incorporation and cell lysis in isolated mouse thymocytes, J. Steroid Biochem., 20:283 (1984).
42. D. Emile, P. Galanaud, E. E. Baulier, J. Dormont, Inhibition of in vitro immunosuppressive effects of glucocorticosteroids by a competitive antagonist RU 486, Immunol. Letters, 9:183 (1984).
43. D. A. Henderson, G. Neef, C. Bengston, M. Tobert, The antiprogestin and antiglucocorticoid activities of 11-beta substituted steroids can be separated, 7th International Congress of Endocrinology, Excerpta Medica International Congress Series 652, (Abstr), 1457 (1984).
44. D. Philibert, R. Deraedt, G. Teutsch, RU 38486: A potent antiglucocorticoid in vivo, Internat. Congress of Pharmacology, Tokyo, (Abst), 1463 (1981).

45. D. Philibert, RU 38486: An original multifaceted antihormone in vivo, in: "Adrenal Steroid Antagonism", M. K. Agarwal, ed., Walter de Gruyter and Co., New York, 77 (1984).
46. T. Gagne, M. Pons, D. Philibert, RU 486 a potent antiglucocorticoid in vivo and in vitro, J. Steroid Biochem., 23:247 (1985).
47. R. C. Gaillard, A. Riondel, W. Herrman, A. F. Muller, E. E. Baulieu, RU 486: A steroid with antiglucocorticosteroid activity that only disinhibits the human pituitary-adrenal system at a specific time of day, Proc. Nat. Acad. Sci., 81:3879 (1984).
48. R. D. Gaillard, D. Poffett, A. M. Riondel, J. H. Saurat, RU 486 inhibits peripheral effects of glucocorticoids in human. J. Clin. Endocrinol. Metab., 61:1009 (1985).
49. X. Bertagna, C. Bertagna, J. P. Luton, J. M. Husson, F. Girard, The new steroid analogue RU 486 inhibits glucocorticoid action in man, J. Clin. Endocrinol. Metabol., 59:25 (1984).
50. L. K. Nieman, E. E. Baulieu, D. Healy, R. Gaillard, G. Schaison, D. Shoupe, D. L. Loriaux, R. Udelsman, G. P. Chrousos, Clinical applications of a glucocorticoid and progesterone antagonist, Endocrine Reviews, (submitted) (1987).
51. J. P. Grunfeld, L. Eloy, A. M. Moura, D. Ganeval, B. Ramos-Grendo, M. Worcel, Effects of antiglucocorticoids on glucocorticoid hypertension in the rat, Hypertension, 7:292 (1984).
52. D. L. Healy, G. P. Chrousos, H. M. Schulte, R. F. Williams, P. W. Gold, E. E. Baulieu, G. D. Hodgen, Pituitary and adrenal responses to the antiprogesterone and antiglucocorticoid steroid RU 486 in primates. J. Clin. Endocrinol. Metabol., 57:863 (1983).
53. D. L. Healy, G. P. Chrousos, H. M. Schulte, P. W. Gold, G. P. Hodgen, Increased adrenocorticotropin, cortisol and arginine vasopressin secretion in primates after the antiglucocorticoid steroid RU 486: dose response relationships, J. Clin. Endocrinol. Metabol., 60:1 (1985).
54. M. Moguilewsky, D. Philibert, Biochemical profile of RU 486, in: "The Antiprogestin Steroid RU 486 and Human Fertility Control", E. E. Baulieu, S. J. Segal, eds., Plenum Press, New York, 87 (1985).
55. S. Tsurufuji, S. Hiroshi, K. R. Min, K. Ohuchi, Difference in the anti-inflammatory effects of indomethacin between acute and chronic states of carrageenin-induced inflammation, J. Pharm. Dyn., 1:8 (1978).
56. R. Deraedt, B. Vannier, R. Fournex, Toxicological study on RU 486, in: "The Antiprogestin Steroid RU 486 and Human Fertility Control", E. E. Baulieu, S. J. Segal, eds., Plenum Press, New York, 123 (1985).
57. L. Laue, G. P. Chrousos, W. Gallucci, P. Munson, K. Barnes, D. L. Loriaux, L. K. Nieman, G. Schaison, The antiprogestin and antiglucocorticoid steroid RU 486 inhibits the ACTH response to ovine corticotropin releasing hormone in man, J. Clin. Endo. Metabol., 66:290 (1988).
58. S. Tsurufuji, K. Sugio, F. Takemasa, The role of glucocorticoid receptor and gene expression in the anti-inflammatory action of dexamethasone, Nature, 280:408 (1979).
59. S. Tsurufuji, A. Kurihara, F. Ojima, Mechanisms of anti-inflammatory action of dexamethasone: Blockade by hydrocortisone mesylate and actinomycin D of the inhibitory effect of dexamethasone on leukocyte infiltration in inflammatory sites. J. Pharm. Exper. Ther., 229:237 (1983).
60. S. Tsurufuji, S. Kazuo, F. Takemasa, S. Yoshizawa, Blockade by antiglucocorticoids, actinomycin D and cyclohnexamide of anti-inflammatory action of dexamethasone against bradykinin, J. Pharm. Exper. Ther. 212:225 (1979).
61. M. Stein, S. Schleifer, S. Keller, The role of the brain and neuroendocrine system in immune regulation: Potential links to neoplastic disease. in: "Biological Mediators of Behavior and Disease", S. Levy, ed., Elsevier, Co, New York, 147 (1982).
62. J. Axelrod, T. D. Reisine, Stress hormones; their interaction and regulation, Science, 224:452 (1984).

63. M. Stein, S. Keller, S. Schleifer, Stress and immunomodulation: The role of depression and neuroendocrine function, J. Immunol., 135:827s (1985).
64. Y. Shavit, G. Terman, F. Martin, J. Lewis, J. Liebeskind, R. Gale, Stress, opioid peptides, the immune system and cancer, J. Immunol., 135:834s (1985).
65. A. Manjan, M. Collector, Stress-induced modulation of the immune response, Science, 196:307 (1977).
66. J. V. Fahey, P. M. Guyre, A. Munck, Mechanisms of antiinflammatory actions of glucocorticoids, Adv. Inflamm. Res., 2:21 (1981).
67. R. Carnuccio, M. DiRosa, R. J. Flower, A. Pinto, The inhibition by hydrocortisone of prostaglandin biosynthesis in rat peritoneal leukocytes is correlated with intracellular macrocortin levels, Br. J. Pharm., 74:322 (1981).
68. L. Laue, J. Peacock, D. Brandon, W. Gallucci, G. Cutler, D. L. Loriaux, G. P. Chrousos, J. Norton, Glucocorticoid receptor-mediated effects on rat fibrosarcoma growth, Cancer Res., 48:2703 (1988).
69. L. Laue, K. Barnes, T. Fleisher, Effects of chronic hypercortisolism and hypocortisolism on leukocyte subpopulations, Endo. Soc. 70:772A (1988).

PSYCHOLOGICAL INFLUENCES ON IMMUNITY: MAKING SENSE OF THE RELATIONSHIP BETWEEN STRESSFUL LIFE EVENTS AND HEALTH

Janice K. Kiecolt-Glaser
and Ronald Glaser

Department of Psychiatry
Department of Medical Microbiology and Immunology
Comprehensive Cancer Center
The Ohio State University College of Medicine
Columbus, Ohio 43210

INTRODUCTION

During the last two decades there has been ample documentation of the relationship between an accumulation of negative major life events and subsequent health impairments. Although there are a number of methodological problems with many of these studies, the consistency and pervasiveness of the effects are striking[1,2]. However, throughout most of this literature the size of the relationship between the number of negative life events and subsequent illness has been modest, with most studies showing correlations in the range of .30[1]. In fact, most individuals who experience major life events do not become ill, or they only experience relatively short illness episodes that are frequently quite comparable to those experienced by their population age-mates.

In this chapter we will present a conceptual framework for understanding the relationship between stressful events and an increased risk for infectious disease. We argue that an increase in psychological distress, sustained over time, leads to adverse immunological changes; these distress-related immunological changes provide one physiological pathway through which major and minor life changes could lead to an increased incidence of infectious disease and perhaps malignant disease.

While many distressed individuals may show immunological changes, actual organically-based illness episodes are a function of differential exposure to pathogens, as well as the prior health of the individual, particularly in regard to immune system function. Across those individuals who have equal exposure to an infectious agent, e.g., a virus, the probability of developing a clinical illness is likely to be at least partially dependent on the prior status of the individual's immune system, as is the intensity and duration of any clinical illness episode.

Extending this rationale, those individuals who are presumably most likely to show health changes in response to stressors are those whose immune system function is already compromised to some extent, either by an immunosuppressive disease like AIDS, or by a natural process like aging that is associated with immunological impairments. These individuals are more likely to be at risk because they are more likely to have poorer immunological defenses at the onset of a stressor; smaller stress-associated immunological decrements could have more important consequences. More speculatively, other at-risk groups could include chronically distressed individuals because of longer-term immunological changes.

While this conceptual framework is logical and consistent with considerable epidemiological data, a major caveat to be kept in mind is the paucity of certain kinds of critical supporting evidence. Very few studies have simultaneously shown a triad of heigthened distress, impaired immune function, and actual health changes in the same individuals; most studies have simply shown that two of the components may co-exist, e.g., distress and immunological alterations.

We will discuss possible connections among the heightened distress generated by major or minor stressful events, alterations in immune function, and actual health changes; evidence for each of the linkages will be examined separately. We will first describe evidence linking changes in immune function with illness, followed by discussions of the evidence linking greater distress (a reliable by-product of stressful life changes) with both immunological alterations and health changes. Finally, we will discuss the few studies that have included all three components and have shown a confluence among distress, immunity, and health changes.

IMMUNITY AND HEALTH

There is excellent evidence that individuals with significant impairments in immune function are at greater risk for infectious disease, at least among individuals who have major immunological impairments. For example, individuals with immunosuppressive diseases like AIDS, or individuals on immunosuppressive therapies like chemotherapy are at great risk for death from opportunistic infections[3].

Also relevant for our discussion is some of the work with older adults. Poorer immune function has been associated with higher rates of mortality in individuals over 80 years of age[4]. Pneumonia and influenza constitute the fourth leading cause of death among the elderly[5]. While the exact nature of immunological alterations associated with aging have been widely debated, there is good evidence that functional or qualitative aspects of immunity appear to be poorer in older adults[6,7].

There is also evidence that persistant nutritional problems are associated with immunological impairments and poorer health[8]. Taken together, there is consistent evidence that individuals with moderate to severe impairments in immune function are at greater risk for infectious disease. The links between mild or transient immunological impairments and health are not well-established.

IMMUNITY AND "STRESS"

Acute Stress

In a series of studies using medical student subjects, we have compared immunological and psychological data obtained during examinations with comparable data obtained one month previously when students were not taking examinations. Using this paradigm we have found differences in NK cell lysis using two different target cells, as well as differences in NK cell percentages using two different methods[9-11]. We have found poorer blastogenic responsiveness with two mitogens, concanavalin A (Con A) and phytohemagglutinin (PHA), lower percentages of total T-lymphocytes, lower percentages of helper and suppressor T-cells, lower percentages of total lymphocytes, and poorer gamma interferon production in lymphocytes stimulated with Con A[10-13].

We also found higher antibody titers to three latent herpesviruses, Epstein-Barr virus (EBV), herpes simplex virus using a Type 1 antigen (HSV-1), and cytomegalovirus during examinations[14]. Higher titers during examinations with lower levels during baseline periods are thought to reflect poorer cellular immune system control of virus latency[15].

Other investigators have found immunological changes associated with relatively commonplace stressful events. For example, Stone and colleagues[16] collected mood ratings over an eight-week period, with concurrent collection of parotid saliva in an intensive study of a small cohort of dental students. The antigen specific secretory immunoglobulin A response to a harmless protein (rabbit albumin) was lower on days with high negative mood relative to days with lower negative mood; similarly, the secretory immunoglobulin A respose was relatively higher on days with high positive mood, in contrast to days lower positive mood.

There is also evidence that bereavement and/or major depressive disorder may have adverse immunological consequences. Two studies of bereaved spouses have shown poorer blastogenic responsiveness following bereavement. In a cross-sectional study Bartrop and colleagues[17] found differences between bereaved spouses and nonbereaved comparison subjects. In a prospective study Schleifer and colleagues[18] found poorer mitogen responsiveness after the death of a wife than before bereavement among men whose wives had terminal breast cancer. Major depression has also been associated with poorer cellular immune function[19,29].

Chronic Stress

There is some evidence that exposure to chronic physical stressors (e.g., noise, shock, or rotation) may produce immunological adaptation or even enhancement of immune function in rodents[21,22]. In contrast, acute stresors are generally associated with immunological decrements[23].

Data from longitudinal studies of marital separation/divorce suggest that adaptation following separation occurs over a several-year period. Weiss[24] suggests that the establishment of a more resilient and stable identity may take 2-4 years. Data from a longitudinal study showed that it was 3.3 years after separation before the average woman's life assumed a sense of coherence and stability[25]. Cartwright[26] found depression-related REM sleep alterations that persisted two years after divorce in a sample of women. We were interested in the possibility of similarly persistant distress-related alterations in immune function following separation[27]. Self-report, immunological, and nutritional data were obtained from 38 separated or divorced women and 38 sociodemographically-matched married women. Women who had been separated one year or less had significantly poorer qualitative and quantitative immune function than their well-matched married counterparts. Moreover, among the separated/divorced women we found that shorter separation periods and greater continuing attachment to the (ex) husband were associated with poorer immune function and greater depression. Analyses of data from the married women showed that poorer marital quality was associated with both a poorer response on qualitative assays of immune function, as well as greater depression. These data are consistent with the epidemiological evidence that has linked marital disruption with increased mortality and morbidity, discussed shortly.

Other cross-sectional evidence also suggested that chronic stress was associated with more chronic immunological alterations. Immunological and psychological data obtained from 34 men and women were providing long-term care for a family member with Alzheimer's Disease[28]. Caregivers had significantly lower percentages of total T-lymphocytes and helper T-lymphocytes than sociodemographically-matched comparison subjects, as well as significantly lower helper/suppressor cell ratios. Caregivers also had significantly higher antibody titers to EBV than comparison subjects, suggesting that cellular immunity was less competent in controlling virus latency in the former. There were not significant differences between the groups in the percentages of natural killer cells or suppressor T-lymphocytes. Caregivers were significantly more distressed than comparison subjects.

Thus, evidence from two more chronically stressed samples of human subjects that suggests that prolonged or chronic psychological stress in humans does not lead to adaptation to the level of well-matched comparison subjects, in contrast to data from rodent studies. Moreover, these data suggest that chronically distressed individuals could be at greater risk for infectious illness. Epidemological data, discussed below, provide some indirect support for these assumptions.

EPIDEMIOLOGICAL EVIDENCE: HEALTH AND DISTRESS

There is a voluminous literature showing that higher

numbers of stressful life events are associated with both greater distress and increased health risks[1,2] that is too extensive to review here. Similarly, there are considerable epidemiological data consistent with higher rates of morbidity and mortality in more distressed populations. For example, the mortality rate among psychiatric patients is one and one-half to two times as high as among nonpatients, even when such high-risk groups as the chronically ill, aged, and alcoholic subpopulations are removed from consideration[29]. There is also evidence of greater cancer mortality among psychiatric patients[30], as well as among nonpsychiatric men who may be more depressed[31].

There is also evidence of significantly higher mortality among bereaved, separated, and divorced individuals[32-37]. Separated and divorced individuals are at much higher risk for both mental and physical illness than are their married age-mates[34]. While much of the relevant epidemiological literature does not provide data for infectious disease separately from other causes of morbidity and mortality, there is some evidence of significant risks for infectious disease associated with marital disruption. For example, separated/divorced adults have six times as many deaths from pneumonia as married adults[37]. Separated women have 30 percent more acute illnesses and physician visits than married women[35].

As noted earlier, poorer immune function has been associated with higher rates of mortality in individuals over 80 years of age[4], and pneumonia and influenza constitute the fourth leading cause of death among the elderly[5]. However, distress may have some important additive effects in this population: within the first year after psychiatric admission, there are <u>fifty</u> times more deaths from pneumonia among elderly psychiatric patients than found among their age-matched general population counterparts[38]. The ratio drops to twenty times that of their age-matched counterparts by the second year of hospitalization, suggesting that the hospital environment per se may not be as important a factor as the transition.

Moreover, while it is reasonable to assume that psychiatric inpatients are more distressed than community residents on the average, it should also be noted that depression is the modal reason for psychiatric hospitalization in the elderly[39].

If psychological resources like social support moderate distress and concurrently attenuate adverse immunological changes, then individuals with less interpersonal support should be at higher risk. A number of other investigators have also found evidence of lower morbidity and mortality associated with greater support[40-42].

CONVERGENT EVIDENCE: DISTRESS, IMMUNITY, AND HEALTH

There are few studies to date in which investigators have simultaneously found heightened distress, poorer immune function, and adverse health changes in the same individuals. The relative scarcity of such essential evidence may be related in large part to the low base rates for minor illness among relatively healthy populations; within a period of several months only a minority of individuals will actually fall ill. In addition, the accuracy or reliability of subjects' recall for short episodes of infectious disease is certainly better for more recent illnesses, so that differences in recall in studies demanding longer-term recall may obscure differences in illness rates as well.

Medical Students

We have not found significant differences in the health of our medical student subjects when we compared days of self-reported illness in baseline periods with illness during examinations, even though there were significant differences in immune function and distress[9-14]. However, in these earlier studies from our laboratory we obtained health information only for relatively brief time periods. In more recent work we have been following first-year medical students longitudinally, across the course of the academic year, and we are beginning to find differences in self-reported health using this strategy. We obtained serial blood samples from 40 students during the first, third, and fifth examination periods, as well as one month before each[13]. We found significant decrements in a variety of immunological measures during examinations as compared to baseline, including large changes in the production of gamma interferon by concanavalin-A stimulated lymphocytes, as well as changes in both plasma and intracellular levels of cyclic AMP. Moreover, changes in two different herpesvirus assays suggested reactivation of latent virus, and therefore poorer cellular immune system control of latent virus[15]: antibody titers to EBV fluctuated markedly during examinations, while T-cell killing by memory T-lymphocytes of EBV transformed autologous B-lymphocytes simultaneously declined during examinations.

Consistent with these immunological changes, our medical student subjects reported an increase in infectious disease symptoms, particularly upper respiratory tract infections, in the two weeks preceding and the two weeks following examination periods, compared to the comparable time periods around the baseline or low-stress samples. As in our past research, students also reliably reported greater distress during examinations.

Disclosure of Traumas and Immunity in Undergraduates

Psychotherapy is associated with a reduction in distress and positive behavioral changes. In addition, there is some evidence that even brief psychotherapy or psychological consultation is associated with fewer outpatient medical visits, fewer days of hospitalization, and lower medical costs[43,44]. Neither the mechanisms underlying these effects nor their generalizability is known.

In previous work from our laboratory we found that a distress-reducing intervention (relaxation training) was associated with improvements in some aspects of cellular immune function in older adults[45]. These data suggested that the reduction of distress might be associated with positive immunological changes.

In a subsequent study, it was hypothesized that actively confronting upsetting experiences would have positive physiological consequences[46]. Fifty healthy undergraduates were randomly assigned to write about either traumatic experiences or superficial topics for four consecutive days. Those subjects who wrote about traumatic events demonstrated an overall higher mitogen response following baseline in comparison to control subjects. Comparisons of the number of health center visits in the 15 weeks prior to the study with those during the six weeks of the study showed a significant drop in the number of visits for trauma subjects relative to those of controls. Although subjects in the trauma condition initially reported some negative feelings associated with the experiment, they were significantly happier than controls at the three-month follow-up. In addition, those participants who wrote about experiences that they had previously "actively held back" in discussions with others (n = 11) had a better proliferative response than those subjects who had not similarly held back (n = 14). Consistent with the results of our intervention study with older adults[45], these data suggest that psychotherapy or other distress-reducing therapies may have positive consequences for immune function and health.

West Point Cadets

In an innovative use of an opportunistic data set, Kasl and colleagues[47] examined psychological and immunological data from cadets who were seronegative for EBV (not latently infected) on entry into the West Point. Data collected over the four years the cohort was followed showed that a distress-related triad of risk factors (higher levels of motivation for a military career, poorer academic performance, and having a father who was an "overachiever") was associated with three illness indices: an increased risk for seroconversion, longer hospitalization in the infirmary following seroconversion, and higher antibody titers to EBV among those who seroconverted in the absence of clinical symptoms.

These data are consistent with the alterations in cellular immunity noted in the medical students during examinations[9-13], particularly those related to herpesvirus latency[14]. Taken together, these data also provide empirical support for the anecdotal speculation that has linked stress and the appearance, duration, and intensity of herpesvirus infections.

Other Considerations

There are a number of other factors that could contribute to poorer health in more distressed individuals. For example, Verbrugge[36] suggests that formerly individuals might have poorer health than their married counterparts because they may engage in other more risky behaviors, e.g., they may smoke, drink, have poorer diets, or sleep more poorly. In research from our laboratory we routinely include two biochemical nutritional assays with relatively longer and er half-lives; the data from these two plasma protein rs provide objective evidence of important aspects of ..tional status. We exclude data from subjects who are out of the normal range on these markers, or who report alcohol or drug abuse. While we find minor differences in sleep between more or less distressed subjects, they are not reliably correlated with immune function. Thus, although such factors may certainly play an important role in population differences in health, there is also evidence of physiological alterations in normal individuals that appear independent of these factors.

IMPLICATIONS

Reiterating our original proposal, we suggest that distress-related alterations in immune function may lead to greater health risks. We argued that behavioral influences on immunity were likely to have the most important health consequences for those individuals who were already vulnerable, either because their immune function is already compromised by a natural process like aging, or through more chronic stress-related alterations. While there may be some adaptation over time, prolonged or chronic psychological stress in humans does not appear to lead to adaptation to the level of well-matched comparison subjects. Epidemiological data indirectly support these assumptions, providing evidence of greater morbidity and mortality among more distressed populations. However, there are still very few studies to date in which investigators have simultaneously found heightened distress, poorer immune function, and adverse health changes in the same individuals; these data are essential for the demonstration of the relationship.

REFERENCES

1. P. A. Thoits, Dimensions of life events that influence distress: An evaluation and synthesis of the literature, in: "Psychosocial stress: Trends in theory and research," H. B. Kaplan, ed., Academic Press, New York (1983).
2. I. G. Sarason, B. R. Sarason, E. H. Potter, and M. H. Antoni, Life events, social support, and illness, Psychosom. Med. 47:156-163 (1985).
3. W. F. Batchelor, AIDS: A public health and psychological emergency, Am. Psychol. 39:1279-1284 (1984).
4. I. C. Roberts-Thomson, S. Whittingham, U. Youngchaiyud, and I. R. MacKay, Aging, immune response, and mortality, Lancet 2:368-370 (1974).
5. T. T. Yoshikawa, Geriatric infectious diseases: An emerging problem, J. Amer. Geriatrics Soc. 31:34-39 (1983).
6. S. J. Goodwin, R. P. Searles, and K. S. K. Tunk, Immunological responses of a healthy elderly population, Clin. Exper. Immunol. 48:403-410 (1982).
7. N. Braveman, Immunity and aging: Immunological and behavioral perspectives, in: "Perspectives on behavioral medicine," M. Riley, J. Matarazzo, and A. Baum, eds., New York, Academic Press (in press).

8. R. K. Chandra and P. M. Newberne, "Nutrition, immunity, and infection: Mechanisms of interactions," New York, Plenum Press (1977).
9. J. K. Kiecolt-Glaser, W. Garner, C. Speicher, G. M. Penn, J. E. Holliday, and R. Glaser, Psychosocial modifiers of immunocompetence in medical students, Psychosom. Med. 46:7-14 (1984).
10. J. K. Kiecolt-Glaser, R. Glaser, E. C. Strain, J. C. Stout, K. L. Tarr, J. E. Holliday, and C. E. Speicher, Modulation of cellular immunity in medical students, J. Beh. Med. 9: 5-21 (1986).
11. R. Glaser, J. Rice, C. E. Speicher, J. C. Stout, and J. K. Kiecolt-Glaser, Stress depresses interferon production concomitant with a decrease in natural killer cell activity, Beh. Neurosci. 100: 675-678 (1986).
12. R. Glaser, J. K. Kiecolt-Glaser, J. C. Stout, K. L. Tarr, C. E. Speicher, and J. E. Holliday, Stress-related impairments in cellular immunity, Psychiatric Res. 16: 233-239 (1985b).
13. R. Glaser, J. Rice, J. Sheridan, R. Fertel, J. Stout, C. E. Speicher, D. Pinsky, M.Kotur, A. Post, M. Beck, and J. K. Kiecolt-Glaser, Stress-related immune suppression: Health implications, Brain, Behavior, Immunity (in press).
14. R. Glaser, J. K. Kiecolt-Glaser, C. E. Speicher, and J. E. Holliday, Stress, loneliness, and changes in herpesvirus latency, J. Beh. Med. 8:249-260 (1985).
15. R. Glaser and T. Gotlieb-Stematsky, eds., "Human herpesvirus infections: Clinical aspects," Marcel Dekker, Inc., New York (1982).
16. A. A. Stone, D. S. Cox, H. Valdimarsdottir, L. Jandorf, J. M. Neale, Evidence that secretory IgA antibody is associated with daily mood. J. Pers. Soc. Psychol. (in press).
17. R. W. Bartrop, E. Luckhurst, L. Lazarus, L. G. Kiloh, and R. Penny, Depressed lymphocyte function after bereavement, Lancet 1:834-836 (1977).
18. S. J. Schleifer, S. E. Keller, M. Camerino, J. C. Thornton, and M. Stein, Suppresion of lymphocyte stimulation following bereavement, JAMA 250:374-377 (1983).
19. S. J. Schleifer, S. E. Keller, A. T. Meyerson, M. J. Raskin, K. L. Davis, and M. Stein, Lymphocyte function in major depressive disorder. Arch. Gen. Psychiatry 41:484-486 (1984).
20. S. J. Schliefer, S. E. Keller, S. G. Siris, K. L. Davis, and M. Stein, Depression and immunity, Arch. Gen. Psychiatry 42:129-133 (1985).
21. A. A. Monjan, and M. I.Collector, Stress-induced modulation of the immune response. Science 196:307-308 (1977).
22. L. S. Sklar and H. Anisman, Stress and coping factors influence tumor growth, Science, 205:513-515 (1979).
23. R. Ader, ed., "Psychoneuroimmunology", Academic Press, New York (1981).
24. R. S. Weiss, "Marital separation," Basic Books, New York (1975).
25. J. S. Wallerstein, J. B. Kelly, "Surviving the Breakup: How Children and Parents Cope With Divorce," Basic Books, New York (1980).

26. R. D. Cartwright, Rapid eye movement sleep charateristics during and after mood-disturbing events, Arch. Gen. Psychiat. 40:197-201 (1983).
27. J. K. Kiecolt-Glaser, L. Fisher, P. Ogrocki, J. C. Stout, C. E. Speicher, and R. Glaser, Marital quality, marital disruption, and immune function, Psychosom Med. 49:13-35 (1987).
28. J. K. Kiecolt-Glaser, R. Glaser, C. Dyer, E. C. Shuttleworth, P. Ogrocki, and C. E. Speicher, Chronic stress and immune function in family caregivers of Alzheimer's Disease victims, Psychosom. Med. (in press).
29. H. Babigian and C. L. Odoroff, The mortality experience of a population with psychiatric illness, Amer. J. Psychiat. 126:52-62 (1969).
30. B. H. Fox, Cancer death risk in hospitalized mental patients, Letters, Science 201: 966-967 (1978).
31. R. B. Shekelle, W. J. Raynor, A. M. Ostfeld, D. C. Garron, L. A. Bieliauskas, S. C. Liv, C. Maliza, and O. Paul, Psychological depression and the 17 year risk of cancer, Psychosom. Med. 43: 117-125 (1981).
32. V. L. Ernster, S. T. Sacks, S. Selvin, and N. L. Petrakis, Cancer incidence by marital status: U.S. Third National Cancer Survey, J. Natl. Can. Inst. 63:587-585 (1979).
33. W. D. Rees, and S. G. Lutkins, Mortality of bereavement, Brit. Med. J. 4:13-16 (1967).
34. B. L. Bloom, S. J. Asher, and S. W. White, Marital disruption as a stressor: A review and analysis, Psychol. Bull. 85:867-894 (1978).
35. A. R. Somers, Marital status, health, and use of health services. JAMA 241:1818-1822 (1979).
36. L. M. Verbrugge, Marital status and health. J. Marr. Fam. 41:267-285 (1979).
37. J. Lynch, "The broken heart," Basic Books, New York (1977).
38. T. J. Craig and S. P. Lin, Mortality among elderly psychiatric patients: Basis for preventive intervention, J. Am. Geriatr. Soc. 29:181-185 (1981).
39. K. Solomon, The depressed patient: Social antecedents of psychopathologic changes in the elderly,J. Am. Geriatr. Soc. 29:14-18 (1981).
40. D. Blazer, Social support and mortality in an elderly community population, Amer. J. Epidem., 115:684-694 (1982).
41. C. I. Cohen, J. Teresi, and D. Holmes, Social networks, stress, and physical health: A longitudinal study of an inner-city elderly population, J. Gerontol. 40: 478-486 (1985).
42. S. Cohen and S. L. Syme, eds., "Social support and health," Academic Press, New York (1985).
43. E. Mumford, H. J. Schlesinger, and G. V. Glass, Reducing medical costs through mental health treatment: Research problems and recommendations, in: "Linking health and mental health," A. Broskowski, E. Marks, and S. H. Budman, eds., Sage Publications, Beverly Hills (1981).
44. K. Jones and T. Vischi, Impact of alcohol, drug abuse, and mental health treatment on medical care utilization: A review of the literature, Medical Care 17:supplement (1980).

45. J. K. Kiecolt-Glaser, R. Glaser, D. Williger, J. Stout, G. Messick, S. Sheppard, D. Ricker, S. C. Romisher, W. Briner, G. Bonnell, and R. Donnerberg, Psychosocial enhancement of immunocompetence in a geriatric population, Health Psychol 4:25-41 (1985).
46. J. W. Pennebaker, J. K. Kiecolt-Glaser, R. Glaser, Disclosure of traumas and immune function: Health implications for psychotherapy, manuscript submitted for publication.
47. S. V. Kasl, A. S. Evans, and J. C. Niederman, Psychosocial risk factors in the development of infectious mononucleosis, Psychosom. Med. 41:445-466 (1979).

PHYSICAL STRESS

ENDOCRINE RESPONSES TO MULTIPLE SENSORY INPUT RELATED TO INJURY

D.A.Bereiter,E.J.DeMaria,W.C.Engeland and D.S.Gann*

Departments of Surgery and Neurobiology
Brown University/ Rhode Island Hospital
Providence, R.I. 02902 USA

Traumatic injury is a complex physiological stimulus that evokes a diverse array of autonomic and endocrine responses (1). Although a complete understanding of the compensatory responses to traumatic injury is not yet possible, a useful experimental approach has been to consider these responses as expressions of the efferent limb of a broad neuroendocrine reflex. The reflex derives its sensory input from multiple sources, requires integration by central neural pathways and is expressed as the autonomic or endocrine adjustments to the applied stimulus. An overview of this reflex is seen in Fig. 1. Several sources of afferent input that likely accompany traumatic injury are indicated as are several brain regions thought to be of major importance in neural integration and modulation of the afferent sensory input. The animal preparations used to explore the components of this reflex often require the use of anesthesia; however, valuable information may be obtained that is relevant for understanding the responses to traumatic injury as well as for understanding the compensatory responses to surgical stress in man.

To assess the influence of a given experimental stimulus on autonomic and endocrine systems, several aspects of the stimulus should be considered. First, the stimulus should be quantifiable and should be applied across a range of submaximal magnitudes. The threshold of stimulus magnitude necessary to evoke the activation of different endocrine or autonomic systems may permit the issue of causality versus parallel pathways for various effectors to be addressed. Second, the effect of repetition of the stimulus on the evoked responses should be assessed. Experimental designs often include repeated presentation of stimuli so that it is important to establish that responses evoked by subsequent stimulus periods are not altered by some feature of the initial stimulus. Similarly, the initial stress of traumatic injury is often followed by a second stressor, namely, surgical intervention. Considerable evidence indicates an apparent plasticity in the autonomic and endocrine responses to certain stressors, the influence of repeated stimuli being dependent on the magnitude

*To whom correspondence should be addressed

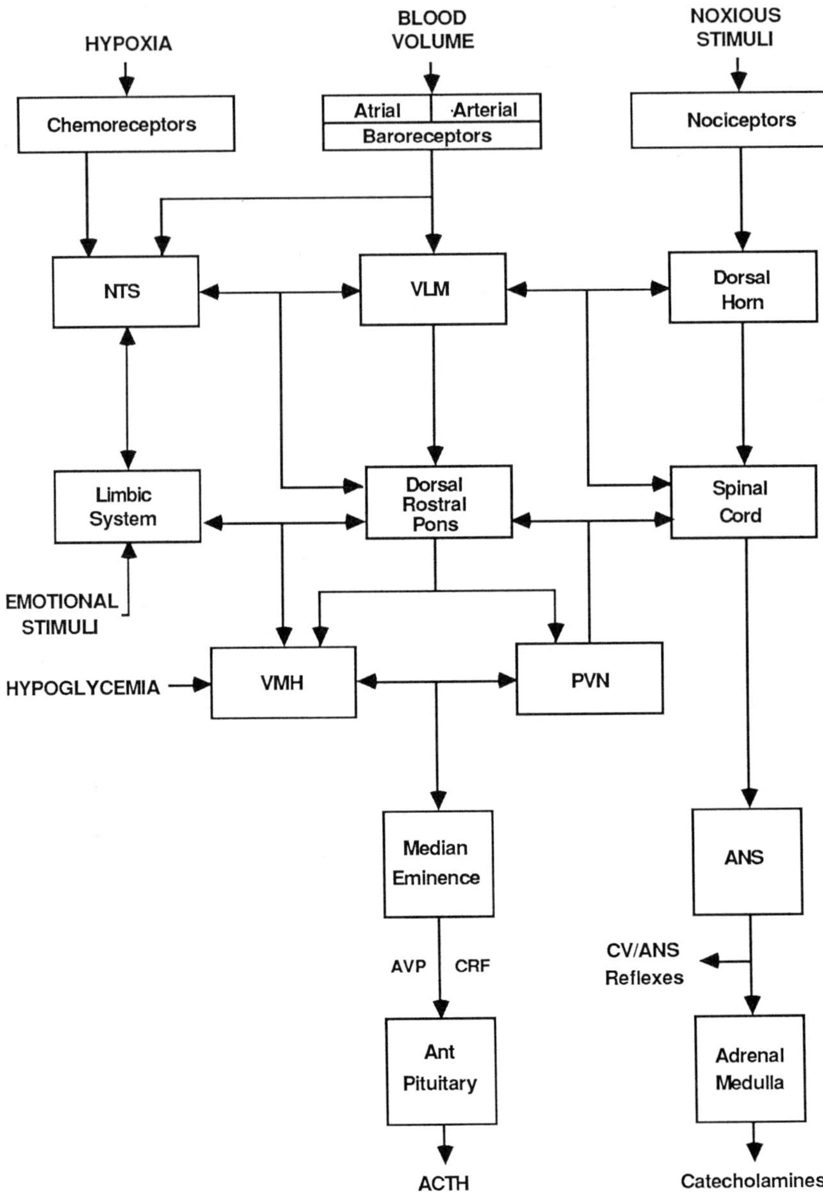

Figure 1. Schematic overview of the central neural structures that process sensory information relevant to the autonomic and endocrine responses to injury. Abbreviations: ACTH=adrenocorticotropin, ANS=autonomic nervous system, AVP=arginine vasopressin, CRF=corticotropin releasing factor, NTS=nucleus tractus solitarius, PVN=paraventricular nucleus of the hypothalamus, VLM=ventrolateral medulla, VMH=ventromedial nucleus of the hypothalamus.

and time separation of each stressor. Third, traumatic injury includes input from multiple sensory modalities and interactions between distinctly different sensory systems may evoke responses that could not be predicted on the basis of the responses to stimulation of individual sensory systems. Stimulus quantification, repeatability and interaction are three general features of the stimulus-response relationship that should be considered if the underlying central neural substrate for complex stressors such as traumatic injury is to be examined in any detail.

Significant shifts in body fluids, either through edema or through actual loss of blood, often accompany injury. To assess the influence of blood loss in autonomic and endocrine systems, we have made extensive use of acute hemorrhage in laboratory animals. Among the components of this stimulus that may alter hormone release are included magnitude of blood loss (2,3), rate of blood loss (4,5), and past history of the preparation (6,7,8). Quantification of hemorrhage is readily made by determination of blood volume using dye dilution. The majority of our experiments have used larger animals such as dogs or cats. This permits multiple samples of blood of sufficient volume to be assessed for several hormones without further compromising the preparation beyond the loss of blood due to the hemorrhage. By determining the plasma concentrations of multiple hormones from a given animal, possible interactions between hormonal systems may be examined. For example, the results shown in Fig. 2 demonstrate that an initial small hemorrhage of 10% blood volume (open bars) in awake dogs causes no significant change in plasma adrenocorticotropin (ACTH, Fig. 2A), whereas plasma vasopressin (Fig 2B) and plasma angiotensin II (Fig. 2C) are elevated. This indicates that the threshold magnitude of blood loss necessary for hormone release is different for these three hormone systems. Although many studies have suggested that vasopressin and angiotensin II (9) may modify the release of ACTH, these data suggest that such modification of ACTH release could not have been predicted by the plasma values seen after a 10% loss of blood volume. In contrast, a 20% loss of blood volume evoked a significant increase in ACTH, vasopressin and angiotensin II (data not shown) and is consistent with the hypothesis that release of ACTH may be modified by coincident release of vasopressin and angiotensin II. Also shown in Fig. 2 are the hormonal responses to a second hemorrhage of 10% blood volume (solid bars) sustained 5 hours after the initial hemorrhage and retransfusion. These data indicate that the plasma concentrations of ACTH, (Fig. 2A) vasopressin (Fig 2B), and angiotensin (Fig. 2C) are elevated significantly above prestimulus values. However, the data indicate further that repeated presentation of a 10% loss of blood volume does not potentiate the release of all hormones since plasma angiotensin II concentrations are elevated to similar values as those seen after the initial hemorrhage period. These results illustrate the differential effects of stimulus magnitude and repetition on various endocrine systems. Although the interpretation of these results alone cannot distinguish causality among endocrine responses from parallel control pathways, the importance of multiple hormone determinations in the same experimental preparation is apparent.

We have reported previously that nociceptor activation potentiated the release of ACTH after moderate blood loss (10,11), a result that may also apply to the release of

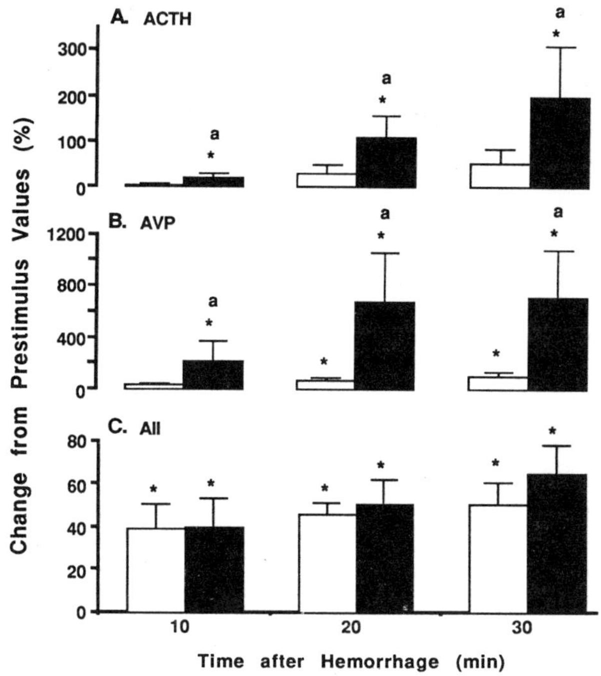

Figure 2. Percent change in: A. plasma concentration of adrenocorticotropin (ACTH), B. vasopressin (AVP), and C. angiotensin II (AII) after repeated loss of 10% blood volume in awake, chronically prepared dogs (n=8). Open bars: responses to initial hemorrhage (H1); solid bars: responses to second hemorrhage (H2) sustained 5 hours after H1 and retransfusion of shed blood. *P<.05 versus prestimulus value; a=P<.05 versus response to initial hemorrhage.

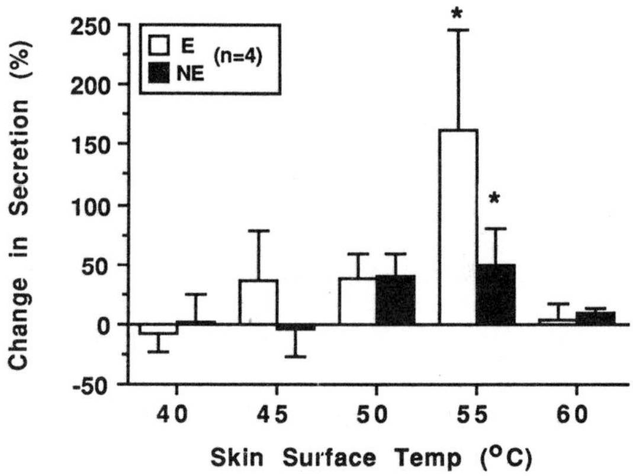

Figure 3. Effect of radiant heat stimulation of facial hairy skin on the adrenal secretion of catecholamines. Adult cats were anesthetized with chloralose and adrenal venous blood was collected directly from the left lumboadrenal vein as described previously (15). Each stimulus period lasted 20 seconds and at least 45 minutes elapsed between stimuli. The data are expressed as the peak change in secretion seen during the 10 minute sampling period. *P<.05 versus prestimulus value.

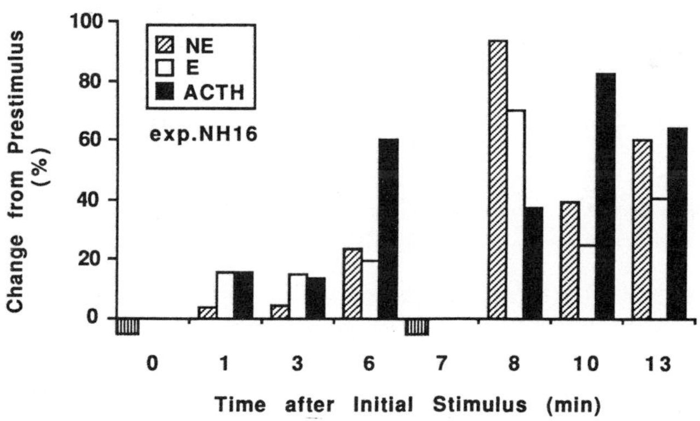

Figure 4. Effect of repeated stimulation of facial hairy skin by noxious radiant heat on the plasma concentration of ACTH and on the adrenal secretion of catacholamines in anesthetized cats. Stimulus parameters: 55 ^{0}C, 20 seconds, surface area of approx.100 mm^2.

vasopressin (12). Earlier studies (13,14) indicated that coincident stimulation of nerve or muscle influenced significantly the survival rate of dogs after acute hemorrhage. Although pain, via nociceptor activation, likely represents a major source of afferent neural information following traumatic injury, the role of nociceptor activity in the control of autonomic and endocrine function is not well defined. Unlike the chronic experimental preparations that may be used to assess the effects of blood loss on hormone release, nociceptor influence on endocrine function requires the use of anesthetized preparations. Recently, we have considered the effects of trigeminal stimuli on adrenal secretion of catecholamines and ACTH release in the anesthetized cat. Radiant heat stimulation of facial hairy skin evokes the adrenal secretion of catecholamines when the measured skin temperature approaches 55°C, a temperature which is perceived as noxious in awake cats (16). As seen in Fig. 3, a brief 20 sec exposure of facial skin (approx. surface area of 50mm^2) to radiant heat caused an increase in the adrenal secretion of epinephrine and of norepinephrine by 3 min with a maximum release of hormone seen after 55°C. However, increasing skin surface temperature to 60°C did not evoke hormone release and is consistent with reports of receptor inactivation at these higher temperatures (17). An additional feature of nociceptor activity is receptor sensitization to repeated stimuli (18,19). As seen in Fig. 4, repeated noxious radiant heat stimulation of facial hairy skin may also provoke an apparent potentiation of hormone release. These data indicate that, similar to the influence of blood loss on endocrine function, noxious stimuli evoke hormonal responses dependent on stimulus magnitude, and further, past history of receptor activity may influence the responses evoked by subsequent stimuli.

Although considerable detail exists concerning peripheral and central neural pathways involved in the sensory-discriminative aspects of nociception (20), the pathways concerned with autonomic or endocrine correlates of nociception remain ill-defined. Within the trigeminal system, primary afferent fibers that convey nociceptor information distribute within specific laminae of trigeminal nucleus caudalis (21). To determine if direct activation of neural elements within trigeminal nucleus caudalis influence hormone release, we have used electrical stimuli in the anesthetized cat. Since nucleus caudalis neurons that are responsive to noxious stimuli often exhibit a bursting firing pattern, stimuli were presented as a burst interpulse interval pattern or as a continuous interval pattern. As seen in Fig. 5, stimulation of lamina V-VI of nucleus caudalis caused a significant increase in secretion of epinephrine and of norepinephrine after burst interval pattern stimuli, whereas continuous interval pattern stimuli evoked only an increase in secretion of norepinephrine. Stimulation of brain stem areas ventral to trigeminal nucleus caudalis had no influence on adrenal secretion of epinephrine regardless of pattern of stimulation (22). In contrast, similar experiments that examined the effect of pattern of stimulation of trigeminal nucleus caudalis on release of ACTH indicated that burst and continuous stimulus interval patterns caused an equal increase in plasma ACTH concentration (23). To obviate the problem of activating fibers of passage by electrical stimuli, we have recently employed microinjections of the neuroexcitatory substance, L-glutamate, to examine the

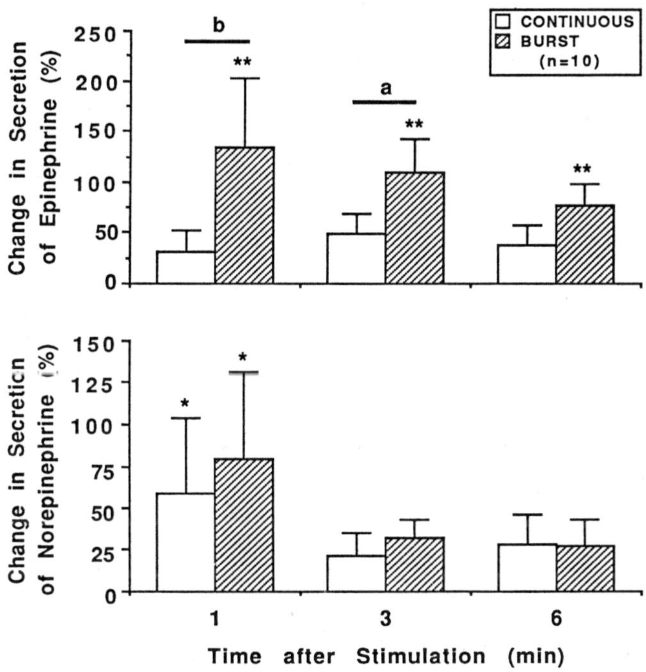

Figure 5. Adrenal secretion of catecholamines after electrical stimulation of lamina V/VI of trigeminal nucleus caudalis in the anesthetized cat. Lamina V/VI contains many neurons that process nociceptive sensory information and receives direct projections from primary nociceptive fibers that innervate orofacial regions. *P<.05, **P<.01 versus prestimulus value; a=P<.05, b=P<.01 between treatments. Stimulus parameters: continuous pattern=75uA, 0.2 msec duration, 20 Hz for 15 sec; burst pattern=75uA, 0.2 msec, 100 Hz trains, 200 msec train duration for 15 sec.

influence of local nucleus caudalis neural activity on hormone release. As shown in Fig. 6, glutamate activation of lamina V-VI neurons evoked an increase in adrenal secretion of epinephrine and in release of ACTH. These results indicate that localized stimulation of neurons in regions of trigeminal nucleus caudalis that process nociceptive sensory information causes a prompt increase in adrenal secretion of catecholamines and release of ACTH. The extent to which sensory-discriminative aspects of nociception and autonomic or endocrine correlates of nociception share common central neural pathways remains to be determined. However, our previous reports (10,11) that nociceptive and hemodynamic information interact to influence pituitary hormone release suggest that multiple parallel pathways are involved (24,25).

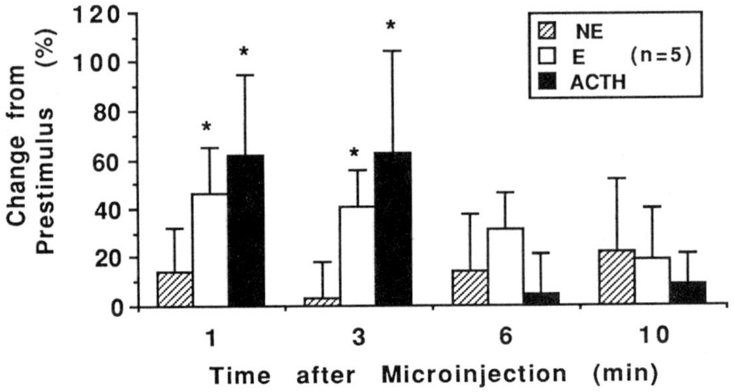

Figure 6. Adrenal secretion of catecholamines and plasma ACTH responses to chemical stimulation of lamina V/VI neurons in trigeminal nucleus caudalis of the anesthetized cat. L-glutamate (0.5 M, 160 nl)was microinjected directly into the brain stem via a 28 gauge cannula. *P<.05 versus prestimulus value.

In addition to physical stimuli, traumatic injury often elicits psychological stimuli that precipitate the emotions of fear and anxiety. Anxiety is a psychological state characterized by feelings of impending danger (26) and thus, the presence of anxiety is difficult to unequivocally demonstrate in animals. Experiments designed to elicit emotional responses usually involve classic Pavlovian conditioning paradigms that consist of pairing a neutral stimulus (a light or tone) with a noxious stimulus (electric shock). Following training, the animal associates the presentation of the neutral stimulus with the noxious or anxiety-arousing stimulus, and a physiological response is activated with presentation of the neutral stimulus alone.

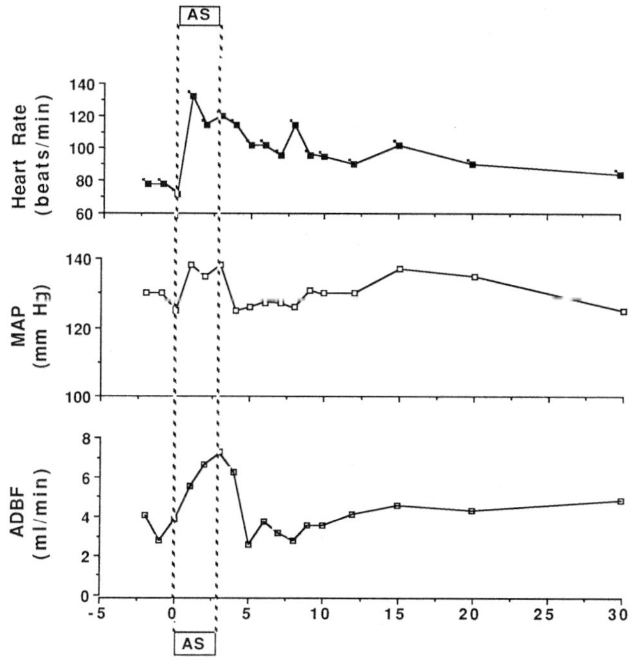

Figure 7. Hemodynamic responses to presentation of noise to an awake, chronically prepared dog. The acoustic stimulus (AS) was presented for 3 minutes and evoked a prompt change in heart rate, mean arterial pressure (MAP) and adrenal blood flow (ADBF).

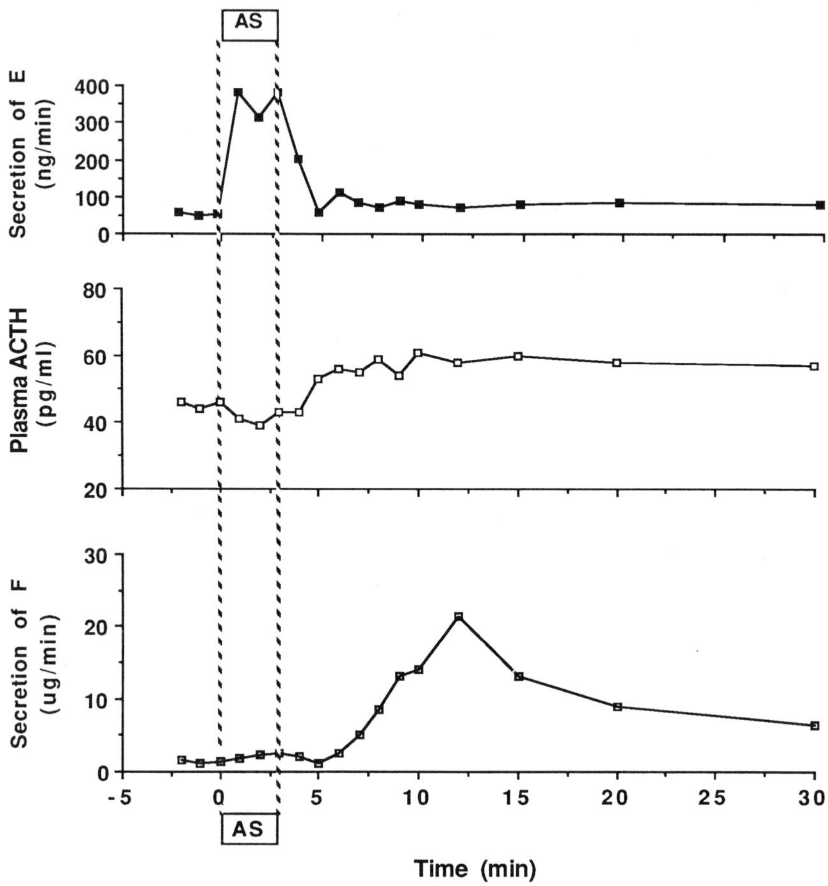

Figure 8. Pituitary-adrenal endocrine responses to presentation of noise to an awake, chronically prepared dog. The acoustic stimulus (AS) was presented for 3 minutes and evoked changes in the secretion of epinephrine (E), in the plasma concentration of ACTH and in the secretion of cortisol (F).

This "conditioned emotional response" consists of cardiovascular changes that include tachycardia, hypertension and renal vasocontriction (27) and endocrine changes that include increased circulating levels of corticosteroids, growth hormone and catecholamines (28). To assess the contribution of psychological stimuli to the hemodynamic and neuroendocrine responses that may accompany traumatic injury, awake dogs were exposed to a brief unexpected presentation of noise (75 dB, 0.25-8 kHz). The responses to acoustic stimuli presented with sudden onset have been described as innate, involuntary reactions that precede the emotions of fear and anxiety (29). The initial presentation of this acoustic stimulus results in the rapid onset of a sequence of events including hypertension, tachycardia, increased adrenal blood flow (Fig. 7) and increased adrenal secretion of catecholamines followed by increased adrenal secretion of cortisol (Fig. 8). Although hemorrhage and acoustic stimuli activate cardiovascular and endocrine responses, the dynamics of the responses are quite different. After acoustic stimuli, the epinephrine secretory response occurs during the presentation of the stimulus and ends prior to the onset of the cortisol secretory response. After hemorrhage, the epinephrine secretory response is coincident with the cortisol secretory response. Also, adrenal blood flow increases in response to acoustic stimuli but does not change or decreases after hemorrhage (30). The increase in adrenal blood flow may contribute to the cortisol response by increasing the delivery of ACTH (31). It is not yet determined if the responses to this acoustic stimulus actually evoke anxiety instead of an alerting response or a response to the novelty of the stimulus (32). However, the results indicate that psychological stimuli are potent activators of autonomic and neuroendocrine responses and that the adrenal medulla and the adrenal cortex can be affected differentially.

Experimental analyses of the mechanisms that underlie the endocrine and autonomic responses to traumatic injury remain difficult since traumatic injury activates multiple sensory components. It is apparent that the physiological responses to stimulation of individual sensory components such as nociceptor activation or loss of blood cannot predict accurately the responses evoked by injury. Interaction among sensory modalities and the past history of sensory receptor activity act to modulate the afferent signal or the central neural processing of sensory information. However, considerable progress in understanding the mechanisms underlying these responses has been made by experimental designs that address more accurately the complexity of the sensory components that constitute those of traumatic injury.

ACKNOWLEDGEMENTS

The authors thank A. Benetti, S. Carlson, and P. Miller for the excellent technical assistance. These studies were supported by NIH grants AM-26831 and GM-27946.

REFERENCES

1. D.S.Gann and M.P.Lilly, The neuroendocrine response to multiple trauma, World J Surg 7: 101-118 (1983).
2. D.A.Bereiter, A.M.Zaid, and D.S.Gann, Adrenocorticotropin responses to graded blood loss in the cat, Am J Physiol 247: E398-E404 (1984).

3. W.C.Engeland, D.P.Dempsher, G.J.Byrnes, K.Presnell and D.S.Gann, The adrenal medullary response to graded hemorrhage in awake dogs, Endocrinology 109: 1539-1544 (1981)
4. D.A.Bereiter, A.M.Zaid and D.S.Gann, The effect of rate of hemorrhage on sympathoadrenal catecholamine release in the cat, Am J Physiol 250: E69-E75 (1986).
5. D.A.Bereiter, A.M.Zaid and D.S.Gann, Effect of rate of hemorrhage on release of ACTH in cats, Am J Physiol 250: E76-E81 (1986).
6. M.P.Lilly, W.C.Engeland and D.S.Gann, Adrenal medullary responses to repeated hemorrhage in the anesthetized dog, Endocrinology 111: 1917-1924 (1982).
7. M.P.Lilly, W.C.Engeland and D.S.Gann, Responses of cortisol secretion to repeated hemorrhage in the anesthetized dog, Endocrinology 112: 681-688 (1983).
8. D.A.Bereiter and D.S.Gann, Potentiation of hemorrhage-evoked catecholamine release by prior blood loss in cats, Am J Physiol 250: E18-E23 (1986).
9. F.A.Antoni, Hypothalamic control of adrenocorticotropin secretion: advances since the discovery of 41-residue corticotropin-releasing factor, Endocrine Rev 7: 351-378 (1986).
10. D.A.Bereiter, P.M.Plotsky and D.S.Gann, Tooth pulp stimulation potentiates the adrenocorticotropin response to hemorrhage in cats, Endocrinology 111: 1127-1132 (1982).
11. D.A.Bereiter, P.M.Plotsky and D.S.Gann, Selective opiate modulation of the physiological responses to hemorrhage in the cat, Endocrinology 113: 1439-1446 (1983).
12. K.Shibuki, T.Onaka, M.Hamamura, K.Yagi, S.Ishikawa, T.Saito and S.Yoshida, Synergistic interactions between footshocks and non-osmotic hypovolemia on vasopressin secretion in rats, Brain Res 410: 140-142 (1987).
13. R.R.Overman and S.C.Wang, The contributory role of the afferent nervous factor in experimental shock: sublethal hemorrhage and sciatic nerve stimulation, Am J Physiol 148: 289-295 (1947).
14. S.C.Wang, The importance of the afferent nervous factor in experimental traumatic shock: the effect of chronic deafferentation, Am J Physiol 148: 547-556 (1947).
15. D.A.Bereiter, W.C.Engeland and D.S.Gann, Peripheral venous catecholamines versus adrenal secretory rates after brain stem stimulation in cats, Am J Physiol 251: E14-E20 (1986).
16. K.L.Casey and T.J.Morrow, Nocifensive responses to cutaneous thermal stimuli in the cat: stimulus-response profiles, latencies, and afferent activity, J Neurophysiol 50: 1497-1515 (1983).
17. P.Bessou and E.R.Perl, Response of cutaneous sensory units with unmyelinated fibers to noxious stimuli, J Neurophysiol 32: 1025-1043 (1969).
18. M.Fitzgerald and B.Lynn, The sensitization of high threshold mechanoreceptors with myelinated axons by repeated heating, J Physiol 265: 549-563 (1977).
19. D.R.Kenshalo Jr., R.B.Leonard, J.M.Chung and W.D.Willis, Responses of primate spinothalamic neurons to graded and repeated noxious heat stimuli, J Neurophysiol 42: 1370-1390 (1979).
20. D.D.Price and R.Dubner Neurones that subserve the sensori-discriminatve aspects of pain, Pain 3:307-338 (1977).
21. D.D.Price, R.Dubner and J.W.Hu, Trigeminothalamic neurons in nucleus caudalis responsive to tactile, thermal and

nociceptive stimulation of monkey's face, J Neurophysiol 39:936-954 (1976).

22. D.A.Bereiter, W.C.Engeland and D.S.Gann, Adrenal secretion of epinephrine after stimulation of trigeminal nucleus caudalis depends on stimulus pattern, Neuroendocrinology 45:54-61 (1987).
23. D.A.Bereiter and D.S.Gann, Caudolateral areas of medulla mediating release of ACTH in cats, Am J Physiol 251: R934-R940 (1986).
24. D.S.Gann, D.G.Ward and D.E.Carlson, Neural control of ACTH: a homeostatic reflex, Recent Prog Hormone Res 34: 357-396 (1978).
25. D.S.Gann, D.A.Bereiter, D.E.Carlson and K.V.Thrivikraman, Neural interaction in control of adrenocorticotropin, Fed Proc 44:161-167 (1985).
26. F.M.Berger, Effect of antianxiety drugs on fear and stress, Behav Sci 25: 315-323 (1980).
27. O.A.Smith, S.L.DeVito and C.A.Astley, Organization of central nervous system pathways influencing blood pressure responses during emotional behavior, Clin Exp Hypertension A6: 185-204 (1984).
28. J.W.Mason, A review of psychoendocrine research on the pituitary-adrenal cortical system, Psychosomatic Med 30:576-607 (1968).
29. C.Landis and W.A.Hunt, "The startle pattern," Rinehart and Co., New York,p. 12 (1939).
30. W.C.Engeland, M.P.Lilly and D.S.Gann, Sympathetic adrenal denervation decreases adrenal blood flow without altering the cortisol response to hemorrhage, Endocrinology 117: 1000-1010 (1985).
31. J.Urquhart, Adrenal blood flow and the adrenocortical response to corticotropin, Am J Physiol 209: 1162-1168 (1965).
32. S.W.Hennessey and S.Levine, Stress, arousal and the pituitary-adrenal system: a psychoendocrine hypothesis, in: "Prog psychobiol psychol," J.M.Sprague and A.M.Epstein, ed., Academic Press, New York, p.133-178 (1979).

HORMONAL RESPONSES TO SURGICAL STRESS

R. Udelsman[1,2] and G.P. Chrousos[1]

[1]Developmental Endocrinology Branch, National Institute of Child Health and Human Development, National Institutes of Health, Bldg. 10, Rm. 10N262, Bethesda, Maryland 20892 and [2]The Department of Surgical Sciences, The Johns Hopkins Hospital, 600 North Wolfe Street, Baltimore, Maryland 21205 USA

INTRODUCTION

Surgical stress encompasses a broad range of metabolic, traumatic, and psychologic stimuli which are imposed upon host homeostatic mechanisms during the perioperative period. Increased adrenal secretion of cortisol has been considered an essential component of the response to surgical stress (1). This notion formed the basis for supplementing stressed adrenal insufficient patients with high-dose glucocorticoids and has implications that extend widely into clinical practice. In 1971, over 5,000,000 patients in the United States received glucocorticoids at doses sufficient to cause adrenal suppression (2) and 10-20% of these patients were subsequently referred for a surgical procedure (3). The current standard of care for these patients includes perioperative high-dose glucocorticoids to allegedly prevent hemodynamic instability and cardiovascular collapse.

The role for increased secretion of glucocorticoids during the stress response is unclear (4). It has been suggested that excess glucocorticoids are required to prevent host responses from overreacting to stress (5). Glucocorticoid excess, however, can be associated with undesired immunosuppression, electrolyte imbalance, and delayed wound healing (6-9). A series of clinical studies have demonstrated that adrenal insufficient patients can tolerate surgical stress when they are replaced with relatively low doses of glucocorticoids (10-12).

We have recently described the perioperative stress-hormone responses in a homogeneous group of eleven patients who underwent neck explorations while receiving identical anesthetic agents (Forane/N_2O) (13). These patients had an intact hypothalamic-pituitary-adrenal (HPA) axis, renin-angiotensin axis, and sympathetic system. The data obtained from these patients represent the normal stress response in a well-defined population in which frequent blood samples were obtained before, during, and after surgery.

To better define the role of glucocorticoids during surgical stress in a controlled setting, we developed a primate model and examined the effects of glucocorticoids on the hemodynamic and metabolic adaptation to surgery (14). We studied the ability of adrenalectomized primates to tolerate surgical stress when given subphysiologic, physiologic, or pharmacologic doses of glucocorticoids. This model also allowed us to study the interactions of glucocorticoids and catecholamines during the perioperative period (15).

HORMONAL RESPONSES TO SURGICAL STRESS

We studied the responses of plasma CRH, ACTH, cortisol, norepinephrine, epinephrine and plasma renin activity in eleven patients undergoing parathyroid or thyroid surgery after identical preoperative sedation and under Forane, anesthesia. During surgical exploration plasma concentrations of ACTH, cortisol and epinephrine remained within their normal unstimulated morning ranges (Figures 1 and 2) (13).

The majority of the patients had no evidence of pulsatile episodes of ACTH secretion during the operation, but rather, secreted ACTH and cortisol continuously. There was a small elevation of plasma norepinephrine and plasma renin activity which was associated with a mild increase in heart rate and decrease in blood pressure. Anesthesia reversal, endotracheal extubation, and the early recovery period were associated with marked increases of plasma ACTH, cortisol and epinephrine and the return of plasma renin activity to basal levels. All hormones examined returned to basal levels by the first postoperative day.

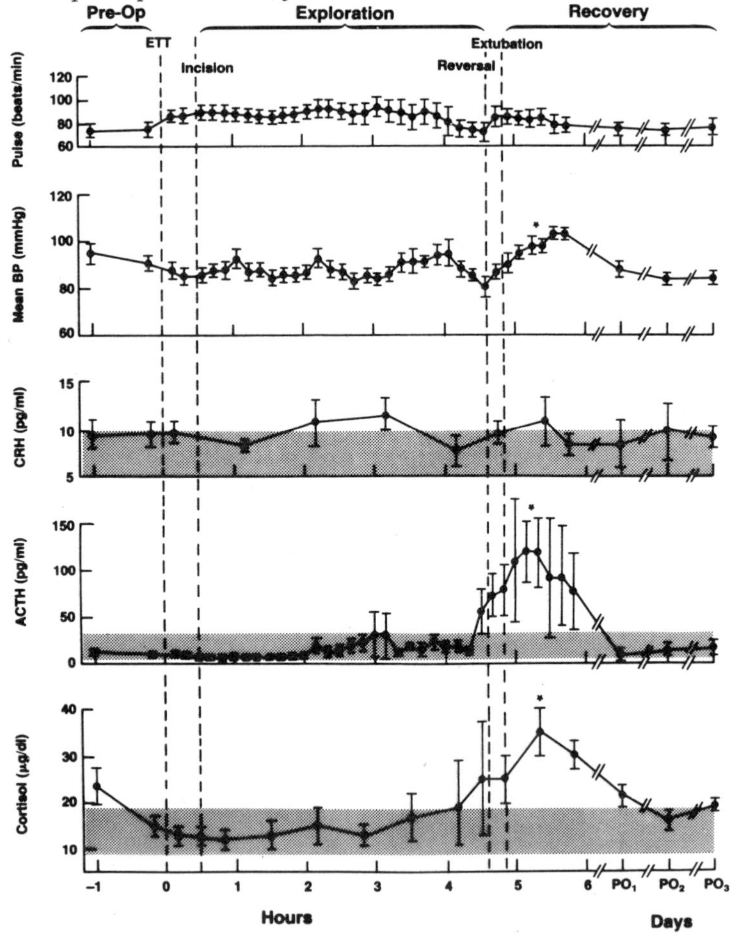

FIG. 1. Pulse rate, mean arterial blood pressure and plasma CRH, ACTH, and cortisol concentrations (mean ± SE in 11 patients before, during and after neck exploration under Forane,/N_2O anesthesia). The duration of surgery was normalized to 4.5 hours for the purpose of graphic representation. Preop = preoperative, Postop = postoperative, ETT = endotracheal tube. Shaded areas represent the normal unstressed morning ranges (Udelsman, et al., Ref: 13).

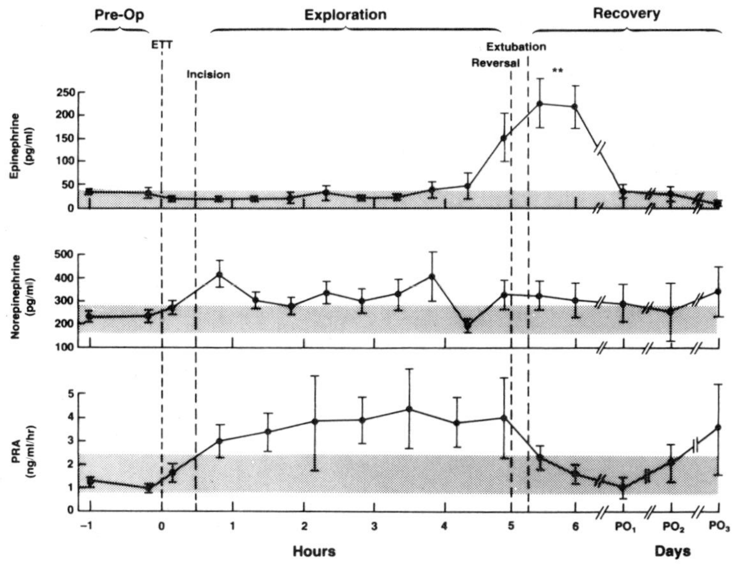

FIG. 2. Plasma concentrations of epinephrine, norepinephrine and plasma renin activity (mean $\pm$ SE) in 11 patients before, during, and after neck exploration as described in figure 1 (Udelsman, et al., Ref: 13).

The data suggest that with modern anesthetic techniques surgical trauma can be associated with mildly elevated levels of ACTH, cortisol and epinephrine. The levels of cortisol observed are of sufficient magnitude to allow maintenance of cardiovascular stability. Glucocorticoid secretion during the operation appears to be attained primarily by continuous rather than pulsatile ACTH secretion. The norepinephrine and the renin-angiotensin axis respond to maintain blood pressure during surgery and anesthesia. The immediate postoperative period is associated with profound elevations of the stress hormones ACTH, cortisol and epinephrine. The major determinant of ACTH, cortisol and epinephrine secretion appears to be anesthesia reversal and recovery rather than surgical trauma per se.

The mean intraoperative ACTH concentration was 11.5 $\pm$ 1.4 pg/ml, which is similar to the 24 hour ACTH mean level of 11.9 $\pm$ 4.4 pg/ml noted in normal unstressed subjects (1). Interestingly, the mean intraoperative cortisol level was 14.4 $\pm$ 1.5 mg/dl, approximately double the normal unstressed 24 hour plasma mean cortisol of 7.6 $\pm$ mg/dl (16). Thus, the ratio of mean ACTH/cortisol (0.8) during surgery was half of the ratio of the 24 hour mean ACTH/cortisol (1.6) in the unstressed state. This suggests that there might be an increase in the sensitivity of the adrenal glands to ACTH during surgery. Circulating or locally produced factors could alter the sensitivity of the adrenocortical cells directly or indirectly by modulating the blood supply to the gland.

Adequate ACTH secretion and moderate cortisol hypersecretion occurred during the perioperative period despite absence of an increase in circulating immunoreactive CRH. The most likely explanation is that the presumptive CRH elevation that occurs in the hypophyseal portal system is not detectable in the peripheral circulation because of dilution in the order of 1:5000 (17). To rule out the possibility that a greater surgical insult might result in an elevation of circulating CRH, we obtained additional perioperative samples from patients undergoing esophagectomies, pneumonectomies and hepatic lobectomies. In these patients with major surgical trauma and marked catabolism there were no significant rises in immunoreactive CRH (Udelsman and Chrousos, unpublished data).

One cannot rule out the presence of tissue factors with corticotropin releasing activities ("tissue-CRFs") that could be secreted locally by traumatized tissue and would not be detected by the hypothalamic CRH radioimmunoassay. Such a concept was first suggested by Witorsch and Brodish (18). Subsequently, Woloski et al. demonstrated that two macrophage products, interleukin-I and hepatocyte stimulating factor, have strong CRH activity (19). The aseptic inflammatory response that universally accompanies surgical trauma might initiate the local secretion of cytokines or other paracrine factors with CRH activity.

Plasma epinephrine and ACTH responded similarly to the stress of anesthesia reversal, endotracheal extubation and early recovery. In contrast, norepinephrine concentrations remained virtually unchanged during this period. This is in agreement with the concept that norepinephrine is primarily a cardiovascular regulator, whereas ACTH and epinephrine are reactive stress hormones secreted mainly in response to physical or emotional stressors. Cortisol, which appears to be a permissive hormone as far as circulation is concerned, may be an important stress hormone because it modulates the reactivity of the immune system via a large array of "homeostatic" mechanisms (20).

PRIMATE MODEL FOR THE STUDY OF SURGICAL STRESS

A surgical-stress hemodynamic model was developed in which chronically adrenalectomized monkeys could undergo a defined stress (cholecystectomy) while receiving replacement doses of mineralocorticoids and graded doses of glucocorticoids (21). Sham-adrenalectomized placebo-treated animals served as controls. The adrenalectomized macaques were randomly stratified to receive subphysiological (one-tenth the normal cortisol production rate; 1/10 X), physiological (1 X) or supraphysiological (ten-times the normal cortisol production rate; 10 X) glucocorticoid treatment for a period of four days prior to cholecystectomy.

The subphysiologically replaced adrenalectomized group (1/10 X) was hemodynamically unstable before, during and after surgery and had a significantly higher mortality rate compared to sham-operated controls (14). In this group systemic blood pressure was low and the cardiac index, systemic vascular resistance index and left ventricular stroke work index were all reduced indicating concurrent decreases in cardiac contractility and peripheral vascular tone (Figure 3). In contrast, the physiologically replaced group (1 X) was indistinguishable from either the supraphysiologically replaced group (10 X) or the sham-adrenalectomized controls (14).

These data suggest that physiologic glucocorticoid replacement, defined as the daily unstressed cortisol production rate, is both necessary and sufficient for primates to tolerate surgical stress. Supraphysiologic glucocorticoid treatment does not appear to offer either a hemodynamic or a survival advantage. On the other hand, stress induced cortisol hypersecretion does occur during the perioperative period. It is possible that hypercortisolemia is not in itself implicitly required for stress adaptation. Rather, stress related hypercortisolemia may be an epiphenomenon associated with an array of stress responses including increased secretion of proopiomelanocortin and beta-endorphin.

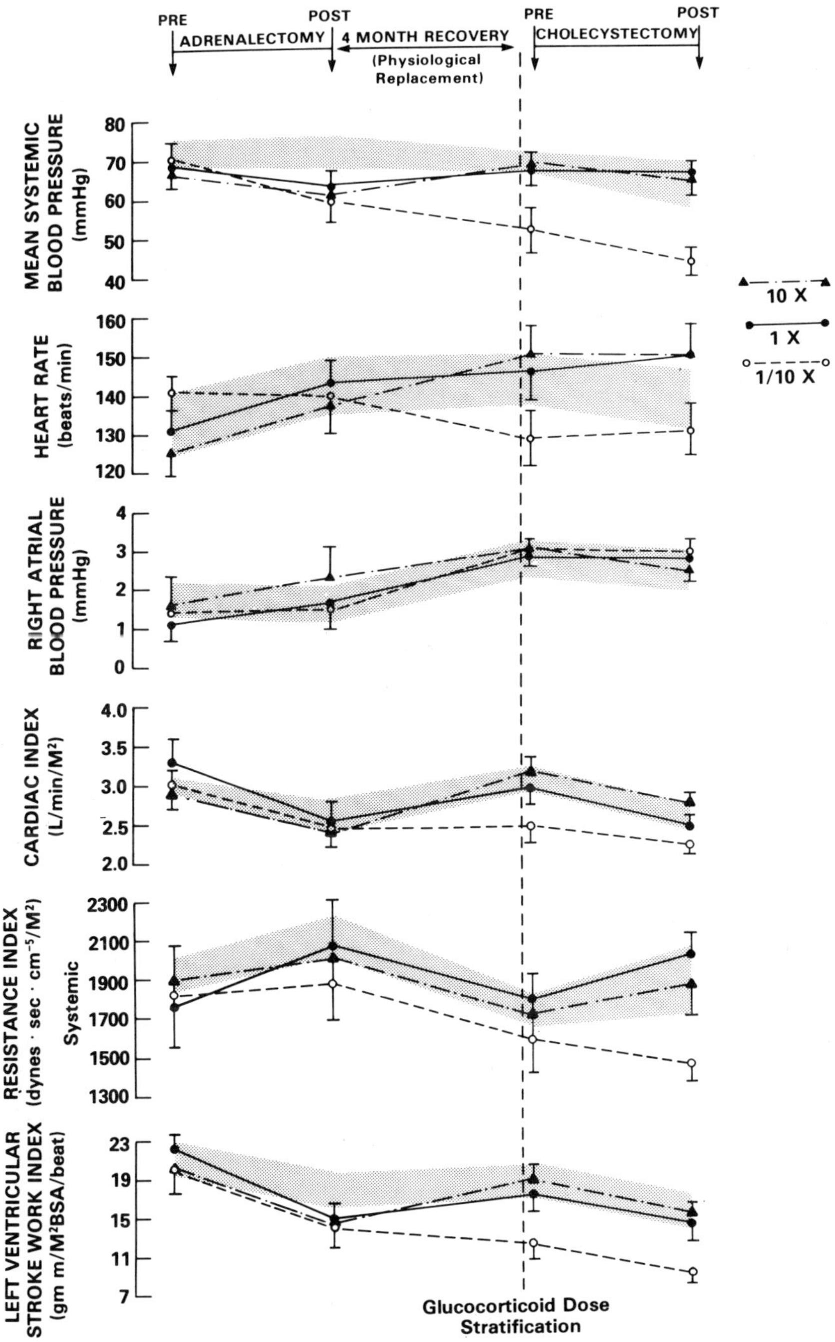

FIG. 3. Hemodynamic responses before and after adrenalectomy and cholecystectomy in subphysiologically (1/10 X), physiologically (1 X) and supraphysiologically (10 X) glucocorticoid-replaced primates. Sham-adrenalectomy placebo treatment group values are shown in gray. Results are expressed as mean ± SE (Udelsman, et al., Ref: 14).

CATECHOLAMINE-GLUCOCORTICOID INTERACTIONS

Since catecholamine-glucocorticoid interactions are known to affect vascular tone, we examined the effects of graded doses of glucocorticoid replacement upon extraadrenal catecholamine secretion in our stressed-adrenalectomized primates (15). Basal levels of epinephrine, norepinephrine and dopamine were similar in the sham-adrenalectomized-placebo treated controls, subphysiologically (1/10 X), physiologically (1 X), and supraphysiologically (10 X) replaced groups. During the period of surgical stress, however, there was a dissociation of catecholamine responses, as shown in Figure 4. The levels of extraadrenal, peripherally secreted catecholamines

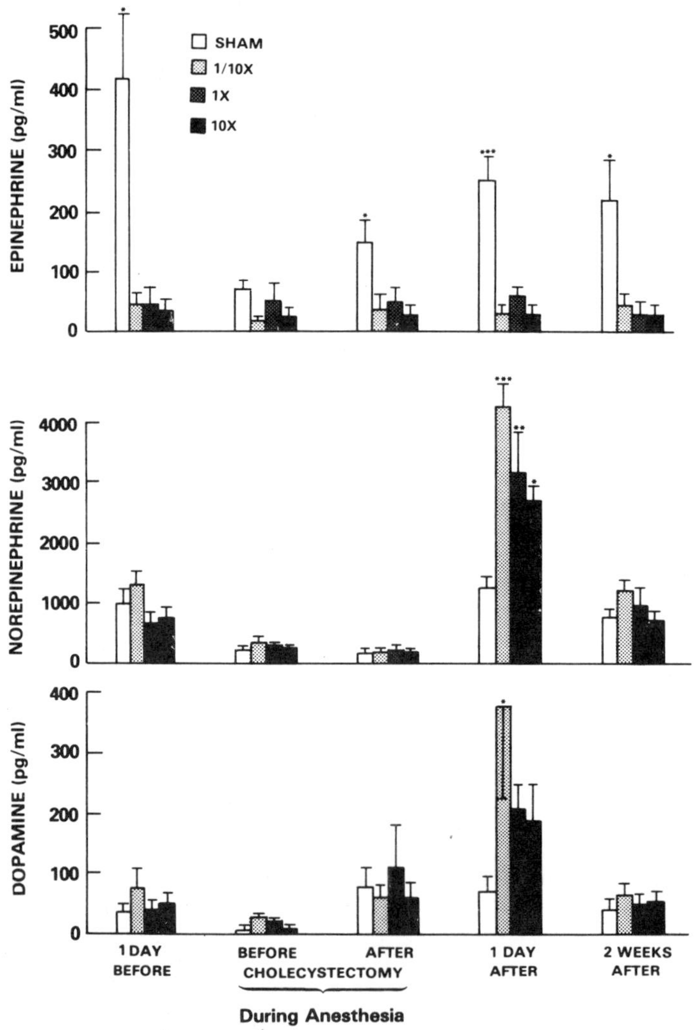

FIG. 4. Circulating levels of epinephrine, norepinephrine and dopamine (mean ± SE) obtained from sham- or bilaterally adrenalectomized monkeys one day before, during anesthesia (N_2O/Halothane), and one day, and two weeks after a cholecystectomy. The "during anesthesia" samples were obtained immediately before and after laparotomy. The dose of glucocorticoid administered to the adrenalectomized monkeys is indicated as a multiple of the physiologic cortisol production rate for cynomolgus macaques. *p<.01, **p<.005, ***p<.0001 (Udelsman, et al. Ref: 15).

(norepinephrine and dopamine) were inversely related to the dose of administered glucocorticoid. Thus, the subphysiologically replaced group (1/10 X) mounted the highest catecholamine response (Figure 4, "One Day After"). These data demonstrate that glucocorticoid insufficiency is associated with an exaggerated stress-induced catecholamine response. Thus, surgical stress in the setting of inadequate glucocorticoid (cortisol) replacement and epinephrine deficiency, stimulates additional sympathetic activity, probably as an adaptive mechanism to maintain homeostasis. Both norepinephrine and dopamine are recruited to compensate for the glucocorticoid and epinephrine deficiency.

SUMMARY AND CONCLUSIONS

Surgical stress is associated with activation of both the HPA axis and the sympathetic nervous system. Emergence from anesthesia appears to be a far more potent stimulus for both of these axes than the surgery itself. Intraoperative HPA activation can be moderate in degree and the mild cortisol elevation observed is compatible with an uneventful perioperative course. This observation was confirmed in adrenalectomized nonhuman primates where a physiologic replacement dose of cortisol was both necessary and sufficient to tolerate surgical stress. Taken together these human and nonhuman primate data suggest that the role of glucocorticoids during surgery is primarily permissive. Catecholamines appear to act in synergy with glucocorticoids during surgical stress.

REFERENCES

1. H. Selye, J. Clin. Endocrinol. Metab., 6:117-230 (1946).
2. N. P. Christy, Iatrogenic Cushing's Syndrome, in: "The Human Adrenal Cortex", N. P. Christy, ed., Harper and Row, New York 395-425 (1971).
3. J. C. Melby, Ann. Int. Med., 81: 505-512 (1974).
4. J. Ingle, Acta Endocrinol., 17:172-186 (1954).
5. A. Munck, P. M. Guyre, and N. J. Holbrook, Endocrine Reviews, 5:24-44 (1984).
6. C. G. Craddock, Ann. Int. Med., 88:564-566 (1978).
7. E. J. DeMaria, J. W. Reichman, P. R. Kenney, J. M. Armitage, and D. S. Gann, Ann. Surg., 202:248-252 (1985).
8. H. P. Ehrlich, and T. K. Hunt, Ann. Surg., 167:324-328 (1968).
9. A. S. Fauci, D. C. Dale, and J. E. Balow, Ann. Int. Med., 84:304-315 (1976).
10. H. Kehlet, Acta Anaesthesiol. Scand., 19:260-264 (1975).
11. H. Kehlet, and C. Binder, Br. J. Anaesth., 45:1043-1048 (1973).
12. F. S. Plumpton, G. M. Besser, and P. V. Cole, Anaesthesia, 24:3-18 (1969).
13. R. Udelsman, J. A. Norton, S. E. Jelenich, D. S. Goldstein, W. M. Linehan, D. L. Loriaux, and G. P. Chrousos, J. Clin. Endocrinol. Metab., 64:986-994 (1987).
14. R. Udelsman, J. Ramp, W. T. Gallucci, A. Gordon, E. Lipford, J. A. Norton, D. L. Loriaux, and F. P. Chrousos, J. Clin. Invest., 77:1377-1381 (1986).
15. R. Udelsman, D. S. Goldstein, D. L. Loriaux, and G. P. Chrousos, J. Surg. Res., 43 (1987).
16. G. P. Chrousos, A. Vingerhoeds, D. Brandon, M. Pugeat, C. Eil, D. L. Loriaux, and M. B. Lipsett, J. Clin. Invest., 69:1261-1269 (1982).
17. R. Y. Moore, The central nervous system and the neuroendodocrine regulation of reproduction. in: "Reproductive Endocrinology". S. S. C. Yen, and R. B. Jaff, eds., W. B. Saunders, Philadelphia 3-34 (1978).
18. R. J. Witorsch, and A. Brodish, Endocrinology, 90:1160-1167 (1972).

19. B. M. R. N. J. Woloski, E. M. Smith, W. J. Meyer III, G. M. Fuller, and J. E. Blalock, Science, 230:1035-1037 (1985).
20. C. E. Cooper, and D. H. Nelson, J. Clin. Invest., 41:1599-1605 (1962).
21. R. Udelsman, J. Bacher, W. T. Gallucci, P. W. Gold, M. L. Morin, D. Renquist, D. L. Loriaux, and G. P. Chrousos, J. Med. Primatol., 13:327-338 (1984).

HORMONAL RESPONSES TO THE STRESS OF EXERCISE

Anton Luger,[1] Patricia A. Deuster,[2] Philip W. Gold,[3] D. Lynn Loriaux,[1] and George P. Chrousos[1]

From the Developmental Endocrinology Branch, National Institute of Child Health and Human Development,[1] the Human Performance Laboratory, Department of Military Medicine, Uniformed Services University of the Health Sciences,[2] and the Biological Psychiatry Branch, National Institute of Mental Health,[3] Bethesda, Maryland

INTRODUCTION

Muscular activity is a component of the "fight or flight" response. Hormones which play an important role in the preservation of homeostasis are activated in response to acute exercise (1-9). Physical training appears to lead to a reduction in the "stress response" to a given workload (8,9). What role, this modulation of the "stress response" plays in the improved performance of trained athletes remains an intriguing question. We report in this brief review the response of several pituitary and adrenal hormones to various intensities of acute treadmill exercise stress. The effect of physical training on these hormonal responses was studied by comparing the responses of sedentary, untrained subjects to the responses of two groups of runners with different degrees of fitness (8).

SUBJECTS AND PROTOCOL

The first group consisted of seven untrained men. The runners were divided into two groups according to their level of weekly exercise: moderately trained runners (15-25 miles/ week, n=7) and highly trained runners (>45 miles/week, n=7). Characteristics of the three groups studied are shown in Table 1. All tests were performed in the evening, the subjects fasted for six hours before the exercise and the runners refrained from running on the day of the tests. Exercise was performed at three intensities (50, 70 and 90% $\dot{V}O_2$max: maximal oxygen uptake), and the three tests were separated from each other by one week. The running was intermittent (30 seconds on and 30 seconds off), and the slope and speed of the treadmill were adjusted to attain each workload. The duration of the exercise was 20 minutes at 50 and 70% $\dot{V}O_2$max and 10 minutes at 90% $\dot{V}O_2$max. This was preceded by a 5 minute warm-up period at all intensities. A cool-down period of 5 minutes was allowed after the exercise at 50 and 70% $\dot{V}O_2$max and a 10 minute cool-down took place after the exercise at 90% $\dot{V}O_2$ max.

An intravenous catheter was inserted before the test, and blood was collected before, during and after the exercise for measurement of hormones,

TABLE 1. Characteristics of Subjects Studied (From ref. 8).

Variable	Untrained Subjects	Moderately Trained Runners	Highly Trained Runners
Training Status	Sedentary	15-25 mi/wk	45 mi/wk
Age (yr)	35.7 ± 0.9	30.0 ± 3.4	31.6 ± 2.8
Height (cm)	177.3 ± 2.4	176.3 ± 3.3	174.9 ± 2.6
Weight (kg)	78.0 ± 3.2	73.8 ± 3.1	68.6 ± 1.9*
Body fat (%)	19.4 ± 2.1	9.8 ± 2.3§	8.1 ± 1.2§
$\dot{V}O_2max$ (ml/kg/min)	33.5 ± 1.5	47.3 ± 2.5+	57.4 ± 1.7+

* $p < 0.05$ as compared with untrained subjects

§ $p < 0.01$ as compared with untrained subjects

\+ $p < 0.001$ as compared with untrained subjects

lactate and potassium. Hormones were measured by radioimmunoassay, lactate by an enzymatic method, and potassium by flame photometry.

All data are expressed as mean ± SE. Differences among groups were evaluated by multivariate analysis of variance with repeated measures. If significant effects were detected, differences among groups were evaluated by t-tests with use of the Bonferroni correction. Correlations were calculated by the least-squares linear regression analysis.

HYPOTHALAMIC-PITUITARY-ADRENAL AXIS

Whereas treadmill exercise at 50% $\dot{V}O_2max$ had no effect on ACTH and cortisol secretion, and 70% $\dot{V}O_2max$ induced a moderate hormonal response, exercise at 90% $\dot{V}O_2max$ was accompanied by a steep increase of plasma ACTH and cortisol (Fig. 1). The peak levels were in the range seen with the insulin tolerance test and were comparable to those observed in response to surgery (10,11). The response of the hypothalamic-pituitary-adrenal axis was similar in the three groups of different training status at identical relative exercise intensities. When the hormonal responses were plotted against the absolute workload however, it became evident that highly trained runners required higher levels of oxygen consumption for the activation of the axis (Fig. 2).

An interesting finding with the evening exercise tests was the elevated basal plasma concentrations of ACTH and cortisol in the highly trained runners (Fig. 1). When the overall basal hormone values were compared, a significant difference emerged. This led us to investigate the hypothalamic-pituitary-adrenal axis in more detail using a CRH-stimulation test (Fig. 3). Ovine corticotropin releasing hormone (oCRH) at the dose of 1 μg/kg was administered as an i.v. bolus at 8 pm. Untrained and moderately trained subjects had similar responses to exogenous oCRH, the highly trained runners had a blunted response of plasma ACTH and cortisol. These findings are in

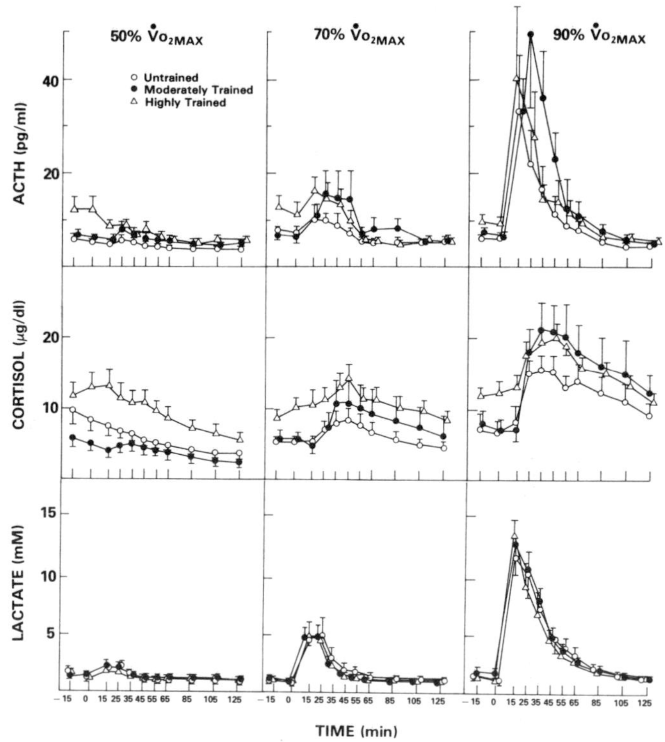

FIG 1: Mean (±SE) plasma ACTH, cortisol, and lactate responses to graded intensities of treadmill exercise (50, 70 and 90% $\dot{V}O_2$max). Untrained (o, n=7), moderately trained (●, n=7), and highly trained subjects (△, n=7) were studied. The tests began between 6 and 7 p.m. (From reference 8).

agreement with a recent study showing elevated urinary free cortisol excretion in highly trained female runners (12). The findings are also reminiscent of the alterations of the hypothalamic-pituitary-adrenal axis observed in patients with anorexia nervosa or depression (13,14). Whether this hypercortisolism is a consequence of the daily stress of exercise leading to a chronically activated hypothalamic-pituitary-adrenal axis or whether subjects with anorectic or depressive personality traits are overrepresented in the group of highly trained runners remains unresolved. The latter possibility has been suggested previously (15). Running has been used in the treatment of depression (16).

RENIN-ANGIOTENSIN-ALDOSTERONE AXIS

Basal and exercise-stimulated increases of plasma aldosterone concentrations were similar in the three groups of subjects tested at all three exercise intensities. The maximal aldosterone response was observed at 70% VO_2max with no further stimulation at 90% VO_2max. The known regulators of aldosterone release - plasma renin (PRA), potassium and ACTH - were tested simultaneously. Basal PRA and potassium concentrations were also similar in all three groups. Exercise induced only small increases in PRA at 50 and 70% VO_2max which failed to reach statistical significance. Increases of plasma potassium levels were observed at 50% VO_2max with only small additional responses at 70 and 90% VO_2max. No significant correlation between plasma aldosterone concentrations and PRA, ACTH or potassium levels could be demonstrated. Synergy between regulatory factors or participation of

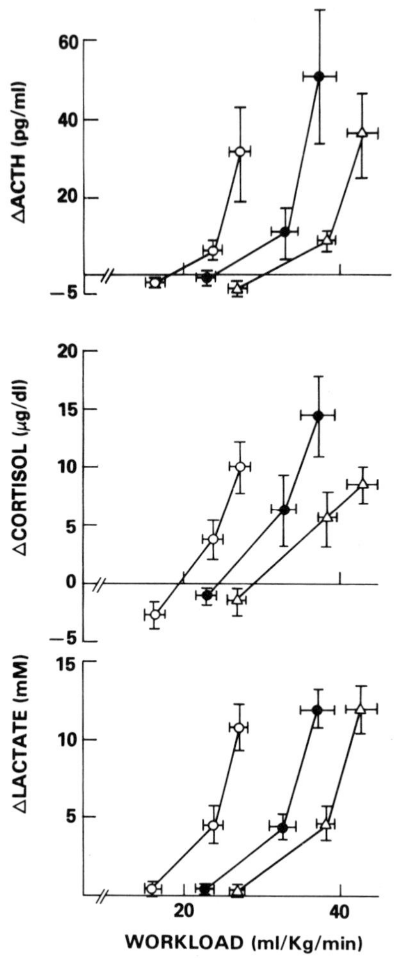

FIG 2: Mean (±SE) plasma ACTH, cortisol, and lactate responses to different absolute mean (±SE) workloads during treadmill exercise. (Symbols as in Fig. 1). (From reference 8).

other factors may explain this lack of correlation. These factors might include atrial natriuretic peptide(s) (17) or paracrine regulators of aldosterone secretion such as dopamine (18) or prostaglandins (19). A recent report has suggested that exercise is associated with increases in circulating atrial natriuretic peptide (20).

Aldosterone, PRA and potassium responses to treadmill exercise were dependent on relative exercise intensity, rather than on absolute workload, similar to the findings with the hypothalamic-pituitary-adrenal axis, suggesting that endurance training leads to adaptive changes that are associated with higher performance for the same hormonal response.

Core temperature did not vary among the three groups, neither under basal conditions, nor in response to exercise. Highly trained subjects however, perspired significantly more and at lower exercise intensities than untrained subjects. Thus, one would expect an increased aldosterone response to compensate for the increased salt loss through sweat in highly trained runners. The short duration of the exercise might have precluded

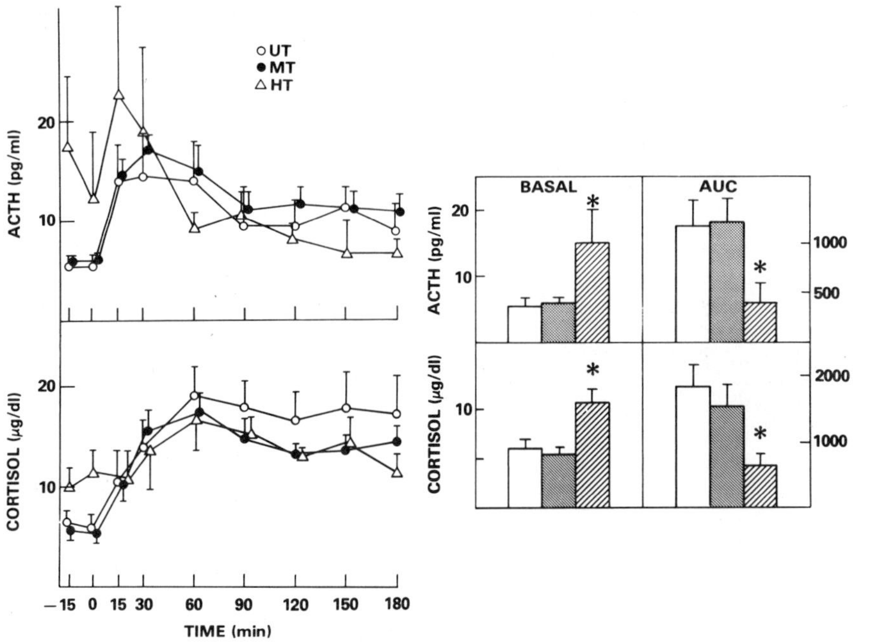

FIG 3: Mean (±SE) plasma ACTH and cortisol values before and after administration of ovine corticotropin-releasing hormone (1μg/kg). The left panel shows the results in untrained subjects (UT), moderately trained runners (MT), and highly trained runners (HT). The right panel shows the basal and oCRH-stimulated (time-integrated stimulation above baseline) values for plasma ACTH and cortisol in the three groups tested (UT, open bars; MT stippled bars; HT, hatched bars). * P<0.01 as compared with values in untrained subjects. AUC denotes area under the curve. (From reference 8).

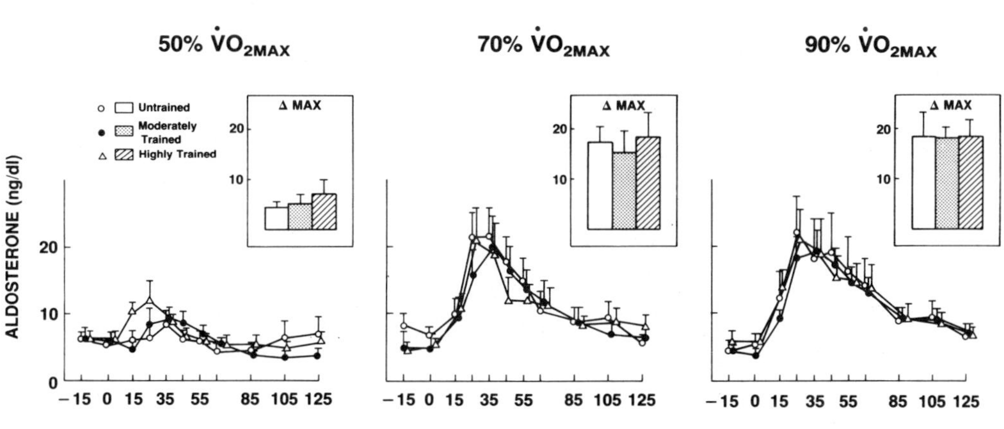

FIG 4: Mean (±SE) plasma aldosterone response to graded intensities of treadmill exercise (50, 70 and 90% $\dot{V}O_2$max). Inserts represent hormonal responses (peak minus baseline) (Symbols as in Fig. 3, unpublished information).

the manifestation of this effect which probably would be observed after longer exercise periods.

GROWTH HORMONE AND PROLACTIN

The response of growth hormone to treadmill exercise can be regarded as the most sensitive of all hormonal responses dealt with in this review. A significant increase was registered at the lowest intensity of 50% Vo_2 max. The maximal response was seen at 70% VO_2max with no further stimulation at 90%. Basal and exercise stimulated growth hormone levels were similar in the three groups of different training status (Fig. 5).

The prolactin response to treadmill exercise differed markedly from that of growth hormone. Exercise, at a relative intensity of 50% $\dot{V}O_2$max, was without effect on prolactin release. However, a clear but submaximal effect was seen at 70% $\dot{V}O_2$max which could be further stimulated at 90% $\dot{V}O_2$max. Thus prolactin stimulation through exercise parallels the activation of plasma ACTH and cortisol. Basal and exercise stimulated serum prolactin levels were similar in all three groups of subjects (Fig. 6).

SUMMARY

Exercise stimulates the hypothalamic-pituitary-adrenal and renin-angiotensin-aldosterone axes and causes growth hormone and prolactin release in an intensity-dependent fashion. Activation of the different endocrine responses examined occur at different exercise intensities. Subjects of differing training status respond similarly to matched exercise intensities. Highly trained subjects have elevated basal concentrations of ACTH and cortisol and a blunted response to exogenous corticotropin releasing hormone. These data are compatible with mild sustained hypercortisolism in highly trained runners.

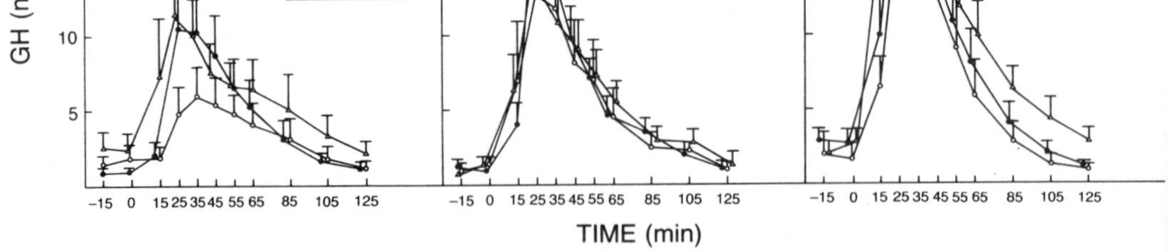

FIG 5: Mean (±SE) growth hormone response to graded intensities of treadmill exercise (50, 70 and 90% $\dot{V}O_2$max). (Symbols as in Fig. 3, unpublished information).

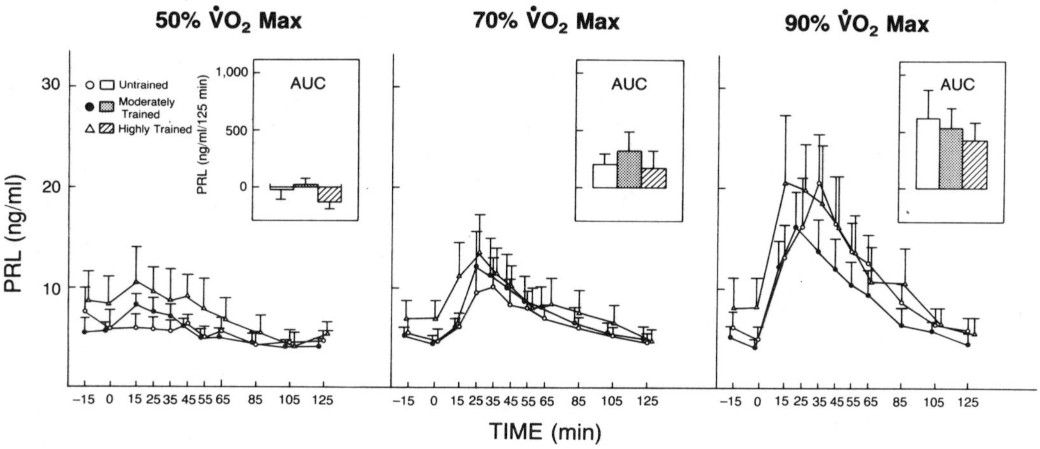

FIG 6: Mean (±SE) prolactin response to graded intensities of treadmill exercise (50, 70 and 90% $\dot{V}O_2$max). (Symbols as in Fig. 3, unpublished information).

CONCLUSIONS

Exercise is a quantifiable and reproducible stressor. It stimulates the hypothalamic-pituitary-adrenal and renin-angiotensin-aldosterone axes, and causes secretion of growth hormone and prolactin.

Physical training is associated with adaptive changes of hormone responses to exercise. A given workload is perceived as a smaller stressor in highly trained runners than in untrained subjects.

Running above a "threshold" mileage is associated with chronic changes of the hypothalamic-pituitary-adrenal axis reminiscent of those seen in patients with anorexia nervosa and depression.

REFERENCES

1. H. Selye, The general adaptation syndrome and the diseases of adaptation, J. Clin. Endocrinol., 6:117 (1946).
2. L. H. Hartley, J. W. Mason, R. P. Hogan, L. G. Jones, T. A. Kotchen, E. H. Mougey, F. E. Wherry, L. L. Pennington, P. T. Ricketts, Multiple hormonal responses to prolonged exercise in relation to physical training, J. Appl. Physiol., 33:607 (1972).
3. G. Brandenberger, and M. Follenius, Influence of timing and intensity of muscular exercise on temporal patterns of plasma cortisol levels, J. Clin. Endocrinol. Metab., 40:845 (1975).
4. H. Galbo, J. J. Holst, and N. J. Christensen, Glucagon and plasma catecholamine responses to graded and prolonged exercise in man, J. Appl. Physiol., 38:70 (1975).
5. J. R. Sutton, Hormonal and metabolic responses to exercise in subjects of high and low work capacities, Med. Sci. Sports, 10:1 (1978).
6. W. W. Winder, R. C. Hickson, J. M. Hagberg, A. A. Ehsani, and J. A. McLane, Training-induced changes in hormonal and metabolic responses to submaximal exercise, J. Appl. Physiol., 46:766 (1979).
7. H. Galbo, and P. D. Gollnick, Hormonal changes during and after exercise, Med. Sci. Sports, 17:97 (1984).
8. A. Luger, P. A. Deuster, S. B. Kyle, W. T. Gallucci, L. C. Montgomery, P. W. Gold, D. L. Loriaux, and G. P. Chrousos, Acute hypothalamic-pitui-

tary-adrenal responses to the stress of exercise. Physiologic adaptations to physical training, N. Engl. J. Med., 316:1309 (1987).

9. A. Viru, Hormones in muscular activity, "Vol 2: Adaptive effect of hormones in exercise", Boca Raton, Fla, CRC Press, 77 (1985).

10. P. C. Avgerinos, T. H. Schuermeyer, R. Udelsman, et al., Human corticotropin releasing factor as a tool for studying the pulsatile function of the hypothalamic-pituitary-adrenal axis, in: "Episodic hormone secretion: methods of analysis and normative date", W. Crowley, ed., John Wiley and Sons, Inc., New York (1987).

11. R. Udelsman, J. E. Norton, S. E. Jelenich, et al., Responses of the hypothalamic-pituitary-adrenal and renin angiotensin axes and the sympathetic system during controlled surgical and anesthetic stress, J. Clin. Endocrinol. Metab., 64:986 (1987).

12. A. L. Villanueva, C. Schlosser, B. Hopper, J. H. Liu, D. E. Hoffman, R. W. Rebar, Increased cortisol production in women runners, J. Clin. Endocrinol. Metab., 63:133 (1986).

13. P. W. Gold, H. Gwirtsman, P. Avgerinos, L. K. Nieman, W. T. Gallucci, W. Kaye, D. Jimerson, M. Ebert, R. Rittmaster, D. L. Loriaux, G. P. Chrousos, Hypothalamic-pituitary-adrenal function in anorexia nervosa: Pathophysiologic mechanisms in underweight and weight-corrected patients, N. Engl. J. Med., 314:1335 (1986).

14. P. W. Gold, D. L. Loriaux, A. Roy, M. A. Kling, J. R. Calabrese, C. H. Kellner, L. K. Nieman, R. M. Post, D. Pickar, W. T. Gallucci, P. Avgerinos, S. Paul, E. H. Oldfield, G. B. Cutler, and G. P. Chrousos, Responses to corticotropin releasing hormone on the hypercortisolism of depression and Cushing's disease: pathophysiologic and diagnostic implications, N. Engl. J. Med., 314:1329 (1986).

15. A. Yates, K. Leehey, and C. M. Shisslak, Running - an analogue of anorexia? N. Engl. J. Med., 308:251 (1983).

16. J. H. Greist, M. H. Klein, R. R. Eischens, J. Faris, A. S. Gurman, and W. P. Morgan, Running as treatment for depression, Comprehensive Psych., 20:41, 1979.

17. J. H. Laragh, Atrial natriuretic hormone, the renin-aldosterone axis, and blood pressure-electrolyte homeostasis, N. Engl. J. Med., 313:1330 (1985).

18. C. R. W. Edwards, E. A. S. Al-Dujaili, M. Boscaro, I. Gow, and B. C. Williams, Peptidergic and monoaminergic regulation of aldosterone secretion, in: Endocrinology of Hypertension, F. Mantero, E. G. Biglieri, C. R. W. Edwards, eds, Academic Press, London, New York, 11 (1982).

19. J. A. Oates, A. R. Whorton, J. F. Gerkens, R. A. Branch, J. W. Hollifield, and J. C. Frolich, The participation of prostaglandins in the control of renin release, Federation Proc., 38:72 (1979).

20. A. E. G. Raine, P. Erne, E. Buergisser, F. B. Mueller, P. Bolli, F. Burkart, and F. R. Buehler, Atrial natriuretic peptide and atrial pressure in patients with congestive heart failure, N. Engl. J. Med., 315:533 (1986).

STRESS, DEVELOPMENT AND REPRODUCTION

STRESS IN THE FETUS AND NEWBORN

Theodore A. Slotkin and Frederic J. Seidler

Department of Pharmacology
Duke University Medical Center
Durham, North Carolina 27710

INTRODUCTION

Birth and its attendant hypoxia, represent the first major stress to which all mammals are exposed. Even a normal, otherwise uneventful vaginal delivery is accompanied by a surge of stress hormones, including catecholamines and corticosteroids.[1-3] Indeed, the magnitude of the rise in plasma norepinephrine and epinephrine during parturition and in the immediate postnatal period far exceeds that seen in even the most extreme circumstances in adulthood: infants at birth experience typical catecholamine levels of 50 nmols/liter, compared to 5-7 nmols/liter in women during delivery or men during severe exercise, or 30 nmols/liter in patients with catecholamine-secreting tumors.[1] Where asphyxia occurs during delivery, the catecholamine levels may even exceed 200 nmols/liter, levels which in adulthood would cause a cerebral hemorrhage.

Recent work from a number of laboratories has shown that the catecholamine surge at birth is not, in fact, simply a stress response, but instead represents a major adaptive mechanism which prepares the organism for extrauterine life. Measurements of catecholamines in amniotic fluid indicate that the rise in adrenergic stimulation actually commences prior to birth, and that the catecholamine surge triggered during delivery has been prepared for in advance by the developing organism.[1,4] At least three classes of effects have been described which are dependent upon catecholamines in the perinatal period. The foremost adaptation is the sudden requirement for air-breathing, which necessitates the release of lung surfactant, the reabsorption of lung liquid, bronchiolar dilation and increased lung compliance; all of these are keyed to the catecholamine surge. Consequently, infants born by elective Caesarian section, where the surge has not occurred, are more likely to experience respiratory distress syndrome.[1,2] Because vaginal delivery causes significant hypoxia, responses must also be made to preserve blood flow to vital organs, such as heart, brain and adrenal medulla; again, these are keyed to catecholamines, which produce cardiovascular adjustments which are unique to the immature organism, and which resemble several aspects of the "diving reflex" seen in marine mammals. The third class of physiological adaptations keyed to the catecholamine surge relates to the metabolic status during the immediate postnatal period, in which mobilization of fuel from brown fat and glycogen stores takes place.[1,4]

Far from representing a non-specific stress response, then, the profound rise in catecholamines in the newborn is responsible for key physiological processes in the adaptation to extrauterine life. This paper summarizes recent progress identifying the mechanisms which evoke and control the release of catecholamines in the perinatal period, the role of catecholamines in neonatal physiological adaptations, and speculations about how abnormalities of development of these responses could contribute to vulnerability of the newborn to pathological situations.

ROLE OF THE ADRENAL MEDULLA

In the mature organism, the adrenal medulla acts in concert with the sympathetic nervous system to coordinate cardiovascular and metabolic responses to stressors such as hypotension, hypoglycemia and hypoxia. In the fetus and neonate, however, sympathetic efferent neural pathways are generally immature or non-functional, and consequently the adrenal medulla plays a predominating role in mediating adrenergic responses.[5] In some species, the adrenals are assisted by extra-adrenal chromaffin tissue, such as the para-aortic ganglia and organs of Zuckerkandl, which are prominent in the fetus and disappear with maturity.[1] The maturational differences in response patterns can best be illustrated in the rat, a species in which neuronal functional development occurs postnatally and in which extra-adrenal catecholamines are of less import. Sympathetic competence develops in the rat in three distinct stages:[3,4]

1. In the fetal stage and at birth, sympathetic target organs contain adrenergic receptors which are linked to functional responses. Nevertheless, reflex stimuli do not evoke responses due to immaturity of sympathetic efferent neurons.

2. By the end of the first week of postnatal life, sympathetic efferent neurotransmission becomes established, consequent to synaptogenesis at ganglionic and postganglionic sites.

3. During the second and third weeks after birth, the central nervous system becomes capable of relaying afferent sensory information from baroreceptors, to sympathetic efferents, resulting in appearance of fully-regulated sympathetic tone.

There is thus an extended period, including the time of birth, in which adrenergic responses can be elicited by circulating catecholamines (such as would be released by the adrenal medulla) acting without participation of neuronal stimulation. The first indication that the adrenal medulla of the newborn rat is capable of responding to its environment independently of neural input, was provided by studies in which pharmacological agents were used to trigger catecholamine release.[7,8] Subsequently, we found that physiological stimuli, such as hypoxia, could also evoke catecholamine secretion in the absence of functional innervation of the tissue by the splanchnic nerve.[3,9] Thus, exposing neonates to low oxygen conditions caused depletion of adrenal catecholamines which could not be prevented by administration of nicotinic blocking agents (Table 1). In contrast, older animals, in which transmission has become established, demonstrated a clear-cut interference of the blocking agent with secretory responses. An additional characteristic of the non-neurogenic response, was that there was no preferential release of either norepinephrine or epinephrine: both amines were secreted in exactly the same proportion as they were found within the adrenal. The responsiveness of the immature tissue was selective for certain stimuli, in

TABLE 1

EFFECTS OF HYPOXIA ON ADRENAL CATECHOLAMINES IN 1 AND 8 DAY OLD RATS

Age and Exposure	Pretreatment	
	Control	Chlorisondamine
1 Day Old	(catecholamine levels, ng/gland)	
Room air	294 ± 8	294 ± 6
Norepinephrine/ epinephrine	0.29 ± 0.01	
7% O_2, 120'	235 ± 4*	239 ± 8*
% depletion	20 ± 1	19 ± 3
Norepinephrine/ epinephrine	0.30 ± 0.01	
5% O_2, 75'	236 ± 12*	218 ± 14*
% depletion	20 ± 4	26 ± 5
8 Days Old		
Room air	1020 ± 40	1000 ± 20
Norepinephrine/ epinephrine	0.25 ± 0.02	
7% O_2, 120'	632 ± 44*	960 ± 48
% depletion	38 ± 4	4 ± 5
Norepinephrine/ epinephrine	0.32 ± 0.02*	
5% O_2, 50'	730 ± 80*	---
% depletion	28 ± 8	

Asterisks denote significant differences from comparably pretreated room air group. Chlorisondamine (5 mg/kg, s.c.) was administered 15 minutes before hypoxic exposure. Data show that depletion caused by hypoxia is not blocked by chlorisondamine in 1 day old rats but is blocked in 8 day olds. The ratio of norepinephrine/epinephrine does not change in the 1 day group, indicating no preferential release caused by hypoxia; in contrast, the ratio rises in 8 day old hypoxic animals, indicating preferential release of epinephrine. Data taken from Seidler and Slotkin.[3]

that hypoxia was a potent secretory signal, but hypoglycemia (which evokes purely neural reflex stimulation) was ineffective. Again, in the mature rat, the secretory patterns were completely different: specific secretion of norepinephrine or epinephrine was typical in response to physiological or pharmacological stimuli, due to the fact that, in adulthood, release occurs by selective activation of neurons innervating specific chromaffin cell subtypes;[10] furthermore, secretion in response to <u>either</u> hypoxia or hypoglycemia was a prominent feature of the adult pattern. Despite the fact that neural stimulation is not the trigger for catecholamine release in the neonate, the mechanism of non-neurogenic secretion is still via exocytosis, wherein the chromaffin granule membrane fuses with the cell membrane, leading to extrusion of the granule contents:[11,12] we were able to demonstrate clear-cut calcium-dependence of the secretory mechanism, loss of granule intergrity accompanying secretion, and translocation of the catecholamines into the extracellular space. Thus, regardless of the fact that secretion is non-neurogenic in the neonate, adrenergic target tissues have the same access to the released catecholamines as they do in the mature organism.

The fact that the non-neurogenic catecholamine release mechanism is unique to the fetus and neonate, raises the question of why this mechanism disappears with further development. Recent work from our laboratory has shown that the onset of neural competence is responsible for the loss of the specialized secretory mechanism.[3,9] First, by 8 days of age, when efferent sympathetic function is established, the immature secretory mechanism disappears and is replaced by a requirement for splanchnic nerve input. Thus, the response to hypoxia, which was independent of neural input at 1 day of age, was entirely blocked by a nicotinic antagonist (chlorisondamine) in the 8 day old (Table 1). Manipulations which accelerated the onset of functional innervation of the adrenal, such as neonatal hyperthyroidism or maternal stress during the last trimester, also caused premature loss of the non-neurogenic secretory capability. The definitive relationship between exposure of the adrenal to nerve input and the loss of specialized secretory responses was provided by studies in which the neonatal adrenal was denervated at 3 days of age, just prior to the onset of neural competence (Fig. 1). In these animals, the immature response pattern was maintained. Similarly, rats denervated in adulthood showed the reappearance of non-neurogenic response capabilities after 3 weeks without neural input to the adrenal. Thus, the development of neural stimulation itself appears to be responsible for the ontogenetic loss of the unique response pattern in the fetus and neonate.

PHYSIOLOGICAL RESPONSES TO CATECHOLAMINES IN THE NEWBORN

The finding that catecholamines released into the fetal and neonatal bloodstream are derived from the adrenal medulla, provides only one of the clues concerning the role of these substances in the perinatal period. Of equal import is the demonstration that the catecholamines are essential to survival. Studies from our laboratory[3,9] have shown that interference with adrenal catecholamine release, either through adrenalectomy (with or without replacement of corticosteroids) or through blockade of the effects of circulating catecholamines, leads to loss of the ability of the neonate to survive hypoxia (Figs. 2 and 3). In contrast, prevention of catecholamine release from sympathetic neurons (bretylium administration) has no discernable effect, a finding in concert with the relative immaturity of sympathetic neuronal function. Similarly, chlorisondamine, which blocks neural input to the adrenal, does not influence hypoxic survival time until the development of neural competence at 8 days of age. Two classes of catecholamine-mediated responses appear to be crucial to the adaptation of the immature organism to the hypoxia which accompanies birth. First, the respiratory adjustments are mediated predominantly via β_2-receptors. Thus, administration of ICI-118551, a β_2-antagonist, resulted in loss of the ability to survive hypoxia in the immediate neonatal period, but not at later stages of development (when respiratory adaptations to air-breathing were complete). The β_2-blocker did not influence the survival in total anoxia, as under that circumstance, respiratory adaptation is of no value. These results can be contrasted with the relative unimportance of β_1-mediated responses (primarily cardiac β-receptors); atenolol, for example, did not influence survival of the neonate during hypoxia.

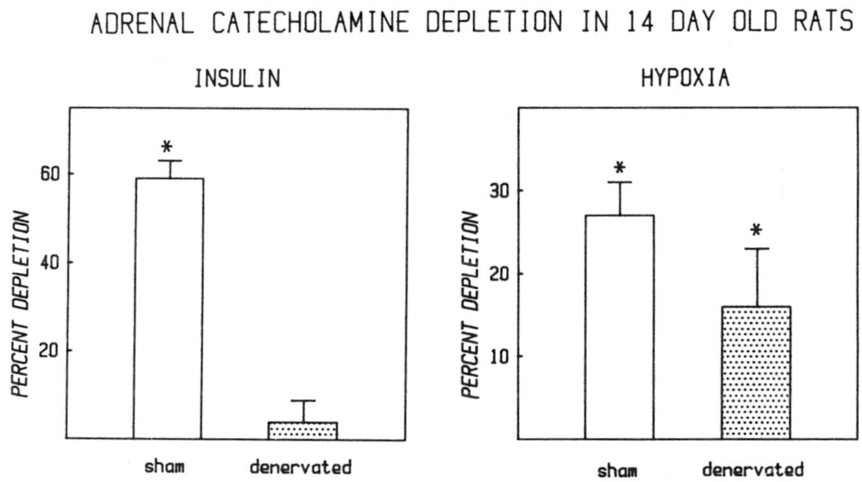

FIGURE 1

Adrenal catecholamine depletion caused by insulin or hypoxia in 14 day old rats after sham surgery or bilateral denervation at 3 days of age. Animals were killed 3 hr after administration of 20 IU/kg of insulin or after 2 hr at 7% oxygen. Asterisks denote significant depletions. The results indicate complete interference of denervation with development of a neurogenic response (insulin), but preservation of a non-neurogenic response (hypoxia). Data taken from Seidler and Slotkin.[9]

The second class of catecholamine-mediated responses which are essential to neonatal survival, are cardiovascular effects initiated through α-receptors.[3,13] These appear to be of more lasting import in that interference with α-adrenergic function (such as with phenoxybenzamine) causes loss of the ability to survive hypoxia not only during and after birth, but also well into the postnatal period. To some extent, these effects of catecholamines reflect differences between the neonate and the adult in receptor populations in cardiovascular tissues; in the heart, for example, α-receptors are present in concentrations exceeding those found in the adult,[14] and this population appears to be linked to some of the specialized needs of the newborn. Blockade of

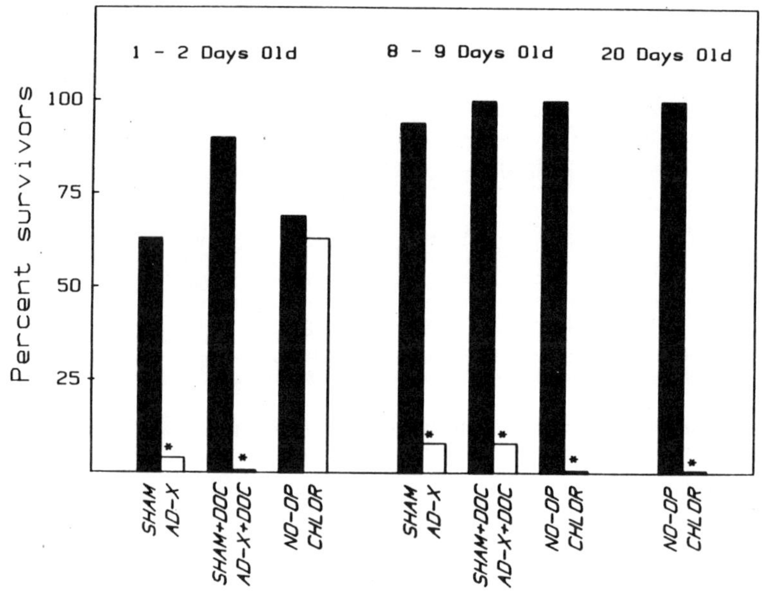

FIGURE 2

Survival during hypoxia (5% oxygen, 90 min at 1-2 days of age, 37.5 min at 8-9 days, 15 min at 20 days) in animals adrenalectomized with or without corticosteroid (DOC) replacement, or in animals pretreated with a nicotinic antagonist (CHLOR). Asterisks denote significant differences from comparable controls. The importance of adrenal catecholamines is shown by the failure to survive in the adrenalectomized groups, regardless of DOC administration. CHLOR results in mortality during hypoxia only in the 8-9 day and 20 day old groups, where neural input has become established. Data taken from Seidler and Slotkin.[3]

these receptors in 1 day old rats causes a failure of cardiac performance during hypoxia, resulting from a progressive decline in sinus rate and appearance of atrioventricular conduction defects (Fig. 4); again, this is not characteristic of more mature response patterns.[13]

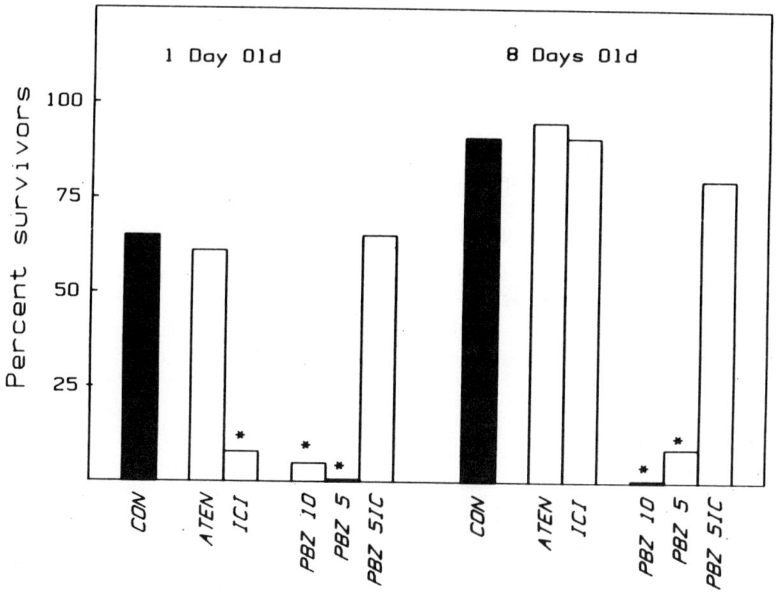

FIGURE 3

Effects of adrenergic blockade on survival during hypoxia. Data show that β_1 blockade with atenolol (ATEN) does not influence survival, but that β_2 blockade (ICI 118551) does in 1 day old rats. By 8 days of age, the effect of ICI is no longer present. α-Receptor blockade with phenoxybenzamine (PBZ) reduces hypoxic survival time at both 1 and 8 days of age, and the effect is exerted in the periphery, as intracisternal administration (i.c.) of PBZ has no effect. Data taken from Seidler and Slotkin.[3]

SYMPATHETIC NEURONAL HYPERACTIVITY IN THE NEONATE

As described above, the establishment of functional efferent sympathetic innervation of the adrenal medulla and other autonomic targets at the end of the postnatal week terminates the stage in which non-neurogenic responses can be elicited to stressors such as hypoxia. However, an additional phase of unusual responsiveness occurs during the third postnatal week, in which afferent sensory input first becomes capable of regulating sympathetic tone.[5,6] Measurements of catecholamine turnover in noradrenergic neurons, and of chromaffin granule utilization in the adrenal medulla, indicate a secondary surge of activity in this period;[11,15,16] similarly, direct measurement of sympathetic nerve impulse activity confirms that a two- to three-fold elevation in tone is present which subsides after weaning.[17] Sympathetic hyperactivity is also in evidence in responses to stressors, including hypoglycemia and hypoxia: increases in impulse activity are markedly higher in this phase than in adulthood.

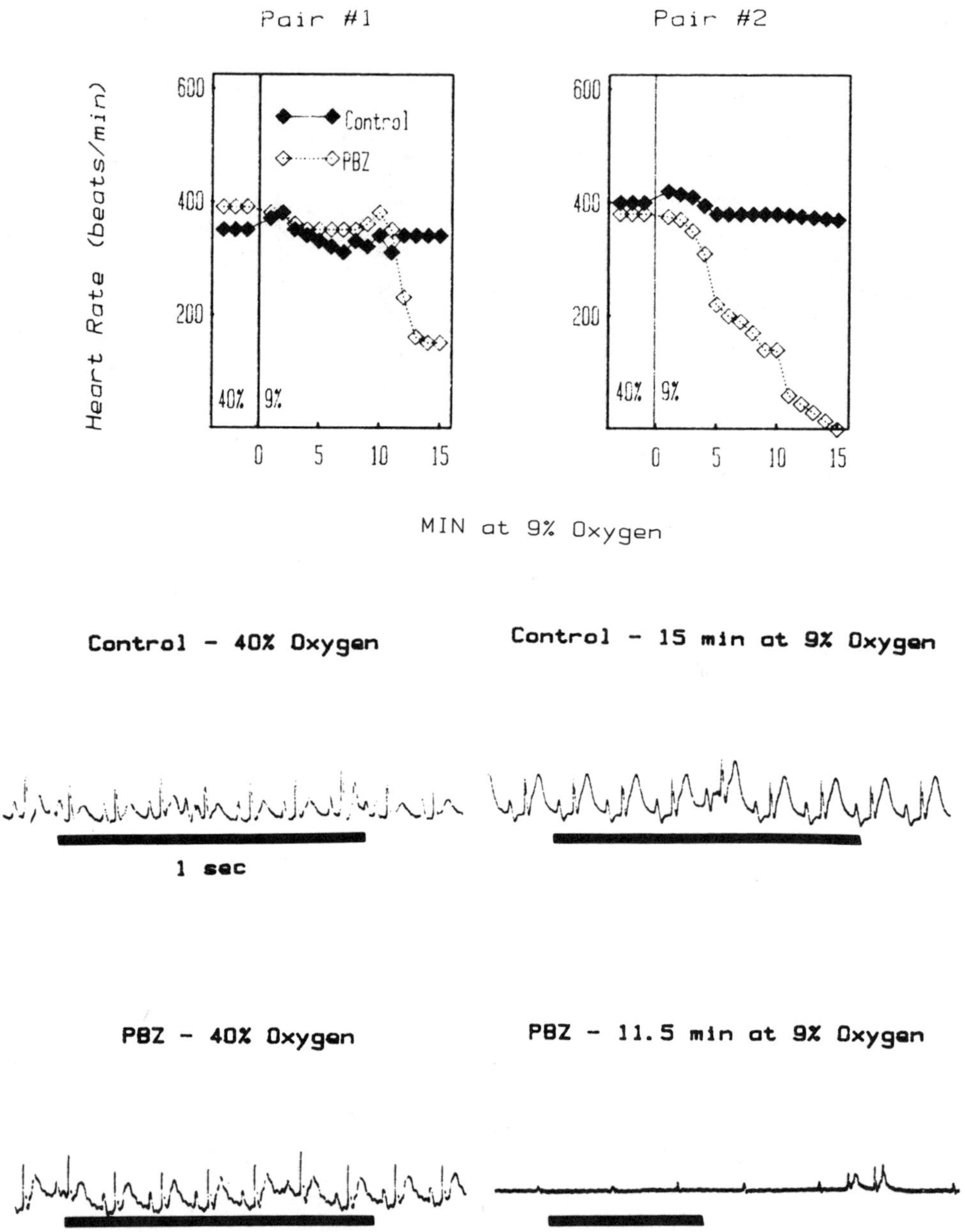

FIGURE 4

Effects of phenoxybenzamine (PBZ) on heart rates and EKG profiles of 1 day old rats during hypoxia (9% oxygen). Although PBZ does not affect heart rate or EKG under normoxic or hyperoxic conditions (40% oxygen), it impairs function during hypoxia. Control rats maintain their heart rates and conduction characteristics, whereas PBZ-treated rats show a marked decline in rate, slowing of sinus rhythm and atrioventricular block with junctional escape. Data taken from Seidler et al.[13]

Again, the question arises as to whether the apparent "abnormality" in sympathetic tone and regulation is maladaptive, or whether there is a functional role for catecholamines which is unique to this period of development. Recent evidence suggests that transient sympathetic hyperactivity may play a role in the final adjustments of end-organ sensitivity to sympathetic stimulation.[5,6] In the heart, the numbers of β-receptors increase substantially during the period in which sympathetic tone is elevated, accompanied by parallel maturation of the responsiveness of heart rate to adrenergic stimulation;[6] this ontogenetic shift

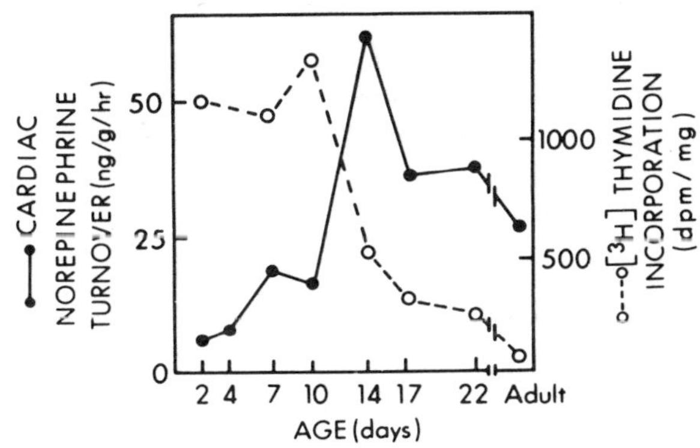

FIGURE 5

Turnover of norepinephrine in sympathetic nerve terminals of the developing rat heart (solid circles and lines) and comparison with DNA synthesis in the myocardium (open circles and broken lines). The results show the coincidence of the timing of sympathetic hyperactivity and the cessation of cell replication in the heart. Data taken from Slotkin.[6]

appears to be keyed to exposure of the tissue to catecholamines. A more intriguing possibility is that sympathetic hyperactivity regulates the pattern of macromolecule synthesis in adrenergic target tissues, terminating the phase of rapid cell replication and initiating growth by RNA-dependent cell enlargement.[6] Comparison of the time course of the surge in sympathetic tone (evaluated by the norepinephrine turnover rate) with the synthesis of cardiac DNA ([^{3}H]thymidine incorporation) indicates a precise correlation (Fig. 5). Either premature exposure to catecholamines, or delays in sympathetic neuronal development, cause a corresponding shift in the pattern of cell replication.[6]

DEVELOPMENTAL ABNORMALITIES

Because of the crucial role played by catecholamines in the period during and immediately after birth, perinatal factors which affect catecholamine actions, development of sympathetic nerves, or ontogeny of central regulation of sympathetic outflow all have adverse impact on development of the organism. The separation of neonatal adrenergic function into non-neurogenic and neurogenic phases produces corresponding differences in the vulnerability to insult by stress, drugs, hormonal imbalance or toxic chemicals. In general, situations which accelerate the onset of sympathetic nerve competence lead to the premature loss of non-neurogenic response mechanisms and its protective effect toward neonatal hypoxia. These include neonatal hyperthyroidism,[6] exposure to heavy metals,[18] maternal opiate or ethanol treatment[19] and maternal stress during the last trimester.[3,20] The latter factor is of particular importance in that even a repeated mild stress, such as daily subcutaneous injections of small volumes of slightly alkaline (pH 10) saline solutions, successfully alters sympathetic development. In addition to the short-term consequence of increasing the neonate's vulnerability to hypoxia, altering the development of sympathetic function during the period in which central control mechanisms are first set up can lead to long-term disruption of autonomic regulatory mechanisms, with consequent permanent abnormalities in cardiovascular status.[20] The acceleration of neural development is thus neither beneficial, nor innocuous.

The finding that subtle factors, such as maternal stress, can have a profound influence on neural development in the offspring, lends itself to a number of speculations concerning pathological states in the perinatal period. First, the incidence of respiratory distress syndrome in premature infants or infants delivered by elective Caesarian section would appear to reflect, in part, the absence of the catecholamine surge which normally accompanies delivery. To that extent, preventive measures could include delaying Caesarian sections until labor (and the surge) has begun, and avoidance of drugs (particularly β-adrenergic blocking agents) which interfere with the actions of the catecholamines on the lung; it is noteworthy that infants born to mothers treated with β-agonists (usually to arrest premature labor) are rarely found to have respiratory distress syndrome.[1,2] A second issue is the vulnerability of the newborn to hypoxic brain damage during difficult or breech deliveries. It should be emphasized that nearly <u>all</u> vaginally-delivered infants show clinical signs of hypoxic insult, but that the newborn is also relatively resistant to hypoxic damage.[1,2,21] The role of maternal stress and hormonal status could thus play a key role in determining whether a specific subset of the population is at greater risk, through two possible mechanisms:

1. Factors operating to accelerate sympathetic neuronal development could lead to loss of the specialized catecholamine secretory mechanism before birth, resulting in blunting of cardiovascular and respiratory adjustments to the stress of birth.

2. The same factors accelerate cell maturation and synaptogenesis in the central nervous system. Well-differentiated brain cells and synaptically-active neurons have a greater requirement for oxygen and are thus more vulnerable to hypoxia.

Finally, the potential participation of abnormalities of sympathetic and sympatho-adrenal function in Sudden Infant Death Syndrome is currently undergoing scrutiny. It is likely that similar factors influencing the timing of disappearance of specialized fetal and neonatal physiological mechanisms will be found to operate in infants at risk for sudden death.

Acknowledgements. Supported by USPHS HD-09713 and NS-06233.

REFERENCES

1. H. Lagercrantz and T.A. Slotkin, The "stress" of being born, Scientific American 254(4):100 (1986).
2. H. Lagercrantz, The development and functional role of the sympatho-adrenal system in the newborn, in: "Intensive Care in the Newborn, IV," L. Stern, H. Bard and B. Friis-Hansen, eds., Masson, New York (1983).
3. F.J. Seidler and T.A. Slotkin, Adrenomedullary function in the neonatal rat: Responses to acute hypoxia, J. Physiol. 358:1 (1985).
4. C.T. Jones, Circulating catecholamines in the fetus, their origin, actions and significance, in: "Biogenic Amines in Development," H. Parvez and S. Parvez, eds., Elsevier/North Holland, Amsterdam (1980).
5. T.A. Slotkin, Development of the sympatho-adrenal axis, in: "Developmental Neurobiology of the Autonomic Nervous System," P.M. Gootman, ed., Humana Press, New York (1986).
6. T.A. Slotkin, Endocrine control of synaptic development in the sympathetic nervous system: The cardiac-sympathetic axis, in: "Developmental Neurobiology of the Autonomic Nervous System," P.M. Gootman, eds., Humana Press, New York (1986).
7. C.J. Chantry, F.J. Seidler and T.A. Slotkin, Non-neurogenic mechanism for reserpine-induced release of catecholamines from the adrenal medulla of neonatal rats: Possible modulation by opiate receptors, Neuroscience 7:673 (1982).
8. J. Bartolome and T.A. Slotkin, Effects of postnatal reserpine administration on sympatho-adrenal development in the rat, Biochem. Pharmacol. 25:1513 (1976).
9. F.J. Seidler and T.A. Slotkin, Ontogeny of adrenomedullary responses to hypoxia and hypoglycemia: Role of splanchnic innervation, Brain Res. Bull. 16:11 (1986).
10. G.P. Lewis, Physiological mechanisms controlling secretory activity of adrenal medulla, Handb. Physiol., Sect. 7 6:309 (1975).
11. T.A. Slotkin, Maturation of the adrenal medulla. II. Content and properties of catecholamine storage vesicles of the rat, Biochem. Pharmacol. 22:2033 (1973).
12. F.J. Seidler and T.A. Slotkin, Non-neurogenic adrenal catecholamine release in the neonatal rat: Exocytosis or diffusion, Devl. Brain Res. 28:274 (1986).
13. F.J. Seidler, K.K. Brown, P.G. Smith and T.A. Slotkin, α-Adrenergic receptor stimulation mediates cardiovascular responses to hypoxia in the neonatal rat, Experientia, submitted.
14. T.A. Slotkin, R.J. Kavlock, T. Cowdery, L. Orband, M. Bartolome, W.L. Whitmore and J. Bartolome, Effects of neonatal methylmercury exposure on adrenergic receptor binding sites in peripheral tissues of the developing rat, Toxicology 41:95 (1986).
15. T.A. Slotkin, Maturation of the adrenal medulla. III. Practical and theoretical considerations of age-dependent alterations in kinetics of incorporation of catecholamines and non-catecholamines, Biochem. Pharmacol. 24:89 (1975).
16. F.J. Seidler and T.A. Slotkin, Development of central control of norepinephrine turnover and release in the rat heart: Responses to tyramine, 2-deoxyglucose and hydralazine, Neuroscience 6:2081 (1981).
17. P.G. Smith, T.A. Slotkin and E. Mills, Development of sympathetic ganglionic neurotransmission in the neonatal rat: Pre- and postganglionic nerve response to asphyxia and 2-deoxyglucose, Neuroscience 7:501 (1982).

18. T.A. Slotkin and J. Bartolome, Biochemical mechanisms of developmental neurotoxicity of methylmercury, Neurotoxicology 8:65 (1987).
19. T.A. Slotkin, P.G. Smith, C. Lau and D.L. Bareis, Functional aspects of development of catecholamine biosynthesis and release in the sympathetic nervous system, in: "Biogenic Amines in Development," S. Parvez and H. Parvez, eds., Elsevier/North Holland, Amsterdam (1980).
20. P.G. Smith and E. Mills, Abnormal development of blood pressure and growth in rats exposed to perinatal injection stress, Life Sci. 32:2497 (1983).
21. S.M. Rothman, Synaptic activity mediates death of hypoxic neurons, Science 220:536 (1983).

MATERNAL MODULATION OF PITUITARY-ADRENAL ACTIVITY DURING ONTOGENY

S. Levine, M. E. Stanton and Y. R. Gutierrez

Department of Psychiatry and Behavioral Sciences
Stanford University School of Medicine
Stanford, California 94305

The fundamental contribution of adrenal corticoid steroid hormones (CORT) in the regulation of a wide range of physiological and behavioral phenomena have now been well documented (Munck et al., 1984). Hormones secreted from the adrenal cortex have been shown to be essential for gene expression, metabolic processes, functions of the immune system and effect of brain function which concern not only the regulation of the adrenal cortex but also contribute to specific aspects of behavior, including learning and memory. In view of the profound role that these hormones play in the regulation of bodily functions, it is not surprising that the regulation of their secretion has been studied extensively. Studies of regulation have been conducted at every level of the regulatory system which includes the adrenal, pituitary, hypothalamus, and many other structures in the central nervous system. Based primarily on the classic contributions of Hans Selye, secretion of these hormones have generally been associated with stressful events. Early stress research emphasized that responses of the organism to a wide variety of stimuli were non-specific. These studies demonstrated increased activity in the adrenal cortex to a variety of "physical" stressors (Selye, 1950). However, in 1975, Mason suggested that much of the early research on stress, especially that involving exposure to a variety of physical stimuli, shared one important characteristic. One of the primary qualities of the stressful experience was exposure of the animal to a novel strange or unfamiliar environment. Therefore, a common element that may have accounted for the adrenal responses was the psychological relevance of the event rather than the particular physical stressors to which they were exposed. Mason was able to demonstrate that exposure of the animals to different physical stimuli, including heat, cold, reduction of caloric intake, was not always accompanied by activation of CORT (corticosterone-cortisol = CORT) particularly, when the stress was imposed in such a way that the animal could not readily detect changes in the environment.

Since increased adrenal secretion has been so closely associated with stress, it is not surprising that most of the research examining the influence of psychological factors on pituitary-adrenal (P-A) hormones have been concerned with the activation of this system. However, evidence has been accumulating which indicates that the P-A system is bi-directional (Levine, Weinberg, & Brett, 1979), i.e., psychological stimuli cannot only activate this system but can also inhibit it.

Inhibition of the P-A system is manifested by either reduced elevations of plasma CORT following events which would usually cause maximal elevation or by lower levels of CORT caused by an actual decrease of P-A output. Thus, if access to food and water is restricted to the morning, the normal daily rhythm of CORT hormones is reoriented to the period of the consummatory behavior (i.e., levels of plasma CORT are elevated in the morning, whereas they are usually low during that period in the nocturnal rodent) (Gray et al., 1978). Within 5 minutes of consuming food or water there is a marked decline of plasma CORT accompanied by concomitant drop in ACTH (Heybach & Vernikos-Danellis, 1979). These studies have shown that consummatory behavior initiates an inhibitory process which reduces both ACTH and CORT secretion at the level of the CNS, the pituitary and the adrenal. There is further evidence that indicates that when reinforcement occurs in an operant conditioning paradigm there is a comparable drop in CORT after consummatory behavior (Coe, Stanton, & Levine, 1983). It has been further demonstrated that the availability of a consummatory response can reduce an animals P-A response to a normally stressful situation (Levine, Weinberg, & Brett, 1979). Thus, when exposed to novelty animals usually show a significant elevation of CORT levels. However, if a consummatory response, such as drinking, is available in a novel environment this elevation is markedly reduced or absent. It has been argued that this inhibition of P-A activity is a consequence of the positive affective state which appetitive stimuli evoke (Coover, 1983). In summary, in the adult, psychological factors are an important determinant of P-A activity. A broad range of cognitive and affective processes both activate and inhibit the output of P-A hormones (Hennessy & Levine, 1979).

The wealth of knowledge on the question of how psychological factors regulate P-A activity in adults contrast sharply with the lack of knowledge concerning psychoendocrine regulation in developing animals. This may be due to the fact that only recently has there been an attempt to understand psychological development in animals. Over the past decade major advances have been made in the study of motivation, perception, learning and memory, and social behavior of developing animals (Aslin, Alberts, & Petersen, 1981; Kail & Spear, 1984; Spear & Campbell, 1979). There are several major reasons for addressing the issue of regulation of P-A activity in the developing organism. First, it appears that CORT in the developing organism acts on certain developmental processes which are not affected by these hormones in the adult organism. Thus, the normal transitions in development of gastrointestinal physiology appear to be mediated by CORT (Henning, 1978a,b; Martin & Henning, 1984). Second, abnormally high levels of CORT during the neonatal period appear to be cytotoxic and produce abnormalities in neural and behavioral development (Bohn, 1984). In addition, the adult response of the P-A system is also determined by stressful events occurring during the neonatal period. Thus, animals handled or shocked during sensitive periods of development show different patterns of CORT activity in adulthood than do animals who have been untreated during these periods. In view of the generally recognized pattern of individual differences in CORT activity to stress in the adult organism, it appears that one source of this variance can be attributed to early experience. (Levine & Mullins, 1966; Levine & Treiman, 1969; Weinberg, Smotherman, & Levine, 1980).

As in the adult, it appears that the major emphasis in examining the developing P-A system has been on those factors which activate the system during ontogeny. Almost no work has been presented on the inhibitory control of this system during development. There is some evidence (Goldman, Winget, Hollingshead, & Levine, 1973) which indicates that there is a deficit in the CORT feedback mechanism which inhibits the system at the physiological level during ontogeny. Studies on inhibitory processes as a consequence of psychological factors, however, are non-existent.

Maternal Contact and Inhibition of Pituitary-Adrenal System

One of the purposes of the series of studies that we are about to present was to examine the question of whether psychological factors can inhibit P-A activity in the developing rat. To accomplish these studies, we initially exposed rat pups, aged 12, 16, and 20 days postpartem, to novelty in order to determine if the resulting CORT response could be inhibited by consummatory behaviors. The ages we chose in the initial experiments were based upon the pattern of P-A activity described in the developing rodent. Elevations of CORT are small and often unreliable prior to day 12. Thus we focused on the period of development during which time we could only elicit a small CORT response (day 12) or a reliable P-A response and thus could examine inhibition. All pups tested were deprived for 24 hours before the testing. The deprivation procedure involved separating the pups from their mothers and placing them individually without food or water into separate compartments of a heated Plexiglas incubator. Pups were then immediately fitted with posterior oral cannulas following the technique of Hall and Rosenblatt (1977). Twenty-four hours later, each of the pups in each age group was assigned randomly to one of the following four treatment conditions: (1) Dam/Milk (D+M): pups suckled an anesthetized lactating dam and milk was infused; (2) Dam/NO Milk (D+ M-): pups suckled an anesthetized lactating dam, but no milk was infused; (3) No Dam/Milk (D- M+): no dam was present, but milk was infused; (4) No Dam/No Milk (D- M-): no dam present and no milk was infused. Hormone levels from these conditions were compared with two basal conditions: (1) Not Deprived (ND): pups remained undisturbed in their nest with a lactating dam, and (2) Not Tested (NT): pups were cannulated and deprived, but not exposed to the test situation.

For each of these conditions, trunk blood was collected from each individual pup by decapitation, and plasma was extracted and frozen until subsequent CORT assay. Further details of the procedure can be found in Stanton, Wallstrom, and Levine (1987).

The results of this experiment (Figure 1), show clearly that the novelty induced CORT response can be actively inhibited. However, it is also clear that this inhibition is not due to the consumption of infused milk. The critical variable that produced a suppression of the novelty induced CORT response was the presence of the anesthesized lactating female. At each of the ages tested, the developing rat pups placed in a novel environment in the presence of an anesthesized lactating dam showed no elevation of CORT as contrasted with those infants in the absence of the dam who all showed an elevation of plasma CORT.

The behavior of infant rat pups deprived of maternal care for 24 hrs and then exposed to an anesthetized female is to engage in active suckling. With the procedures involved, that is, presenting an anesthetized dam with only one infant, this suckling is non-nutritive (Lincoln, Hill, & Wakerley, 1973). However, many aspects of the mother-infant relationship, such as the dam's thermotactile contact with her pups -- which appears by behavioral indices to have a direct effect on the pup's level of arousal and thus, perhaps, on its P-A system -- do not appear dependent on suckling, per se (Hofer, 1978; Leon, 1979; Stanton et al., 1987). Thus, it is possible that the influence of the mother on suppressing infant's CORT responses to novelty may not be attributed at all to suckling but rather to contact between mother and infant.

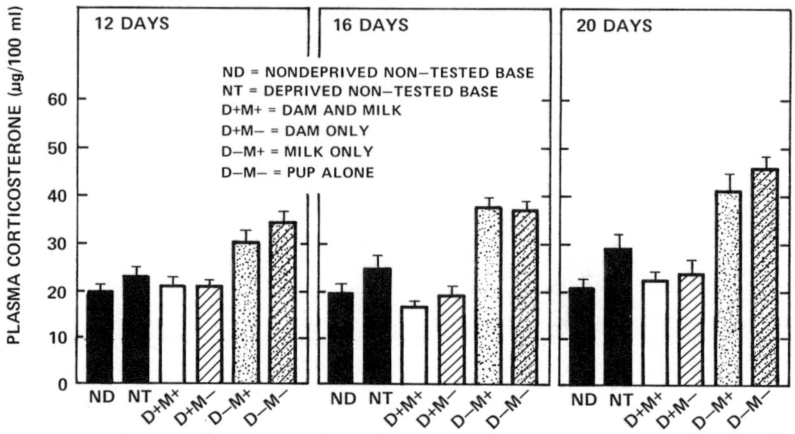

Fig. 1. Plasma corticosterone levels (µg/100 ml) for rat pups at 12, 16, and 20 days of age under the six sampling conditions (Mean +SEM).

A second experiment was therefore conducted to separate the effects of nipple attachment from the effects of contact with the dam, per se, under conditions which were otherwise the same as in the initial experiment. Two groups of infants were tested with a dam, one was allowed to suckle, non-nutritively, and the other group was not permitted to suckle (see Stanton et al., 1987, for further details). The results both replicate and extend the findings of the initial experiment. Suckling an anesthetized dam resulted, once again, in inhibition of the CORT response which occurs when pups are exposed to novelty. However, contact with the dam in the absence of suckling was equally as effective in inhibiting this CORT response.

In the previous experiments we demonstrated that maternal contact is apparently a necessary condition to inhibit the CORT response of preweanling rats when exposed to novelty. In the next experiment we asked two questions. First, whether contact is necessary for this effect. Second, we examined the importance of lactation, per se. In order to answer this question we exposed infant rats 12, 16, and 20 days of age both to lactating females and virgin females. The purpose of the second question was to investigate the possible role of olfactory cues. Leon and Moltz (1972) have shown that preweanling rat pups are attracted to the odor of maternal pheromone which is present in the anal excretia of lactating but not virgin females. Since olfaction plays a very important role in the developing rat, the possibility exists that the presence of maternal odors can inhibit the CORT response to novelty. To answer this question, infants were placed in the novel environment but contact was prevented by a wire mesh partition, approximately 2-3 cm from the surface of the dam's ventral. Thus, the pups in this condition could smell but not contact the stimulus dam. Non-tested and "pup-alone" conditions were also included for comparison.

The results clearly indicated that contact was, at all ages, necessary for inhibition of the stress response. However, the effect of the contact stimulus was somewhat dependent upon age. At 12 and 16 days of age, contact with the virgin female was as effective in inhibiting the CORT response as was contact with the lactating dam. However, at 20 days of age, contact with the virgin female was not nearly as effective as contact with the lactating dam. There thus appear to be age-related changes with regards to the nature of the stimulus which is capable of inhibiting the CORT response to stress. These data emphasize the importance of contact and do not seem to indicate that there is an important olfactory mechanism which, in the absence of contact, is responsible for this inhibition.

Since we have demonstrated that passive contact with an anesthesized lactating dam eliminates or suppresses the CORT response to novelty in 12, 16, and 20 day old pups, it was pertinent to determine whether an inhibition of the CORT response is related exclusively to contact with the maternal stimulus (i.e., lactating dam) as opposed to other social stimuli. Two experiments were conducted to examine this question (Stanton & Levine, 1985). One compared the effectiveness of contact with an anesthesized dam to that of contact with an adult male. The other compared an anesthetized dam with an anesthetized sibling pup. In the former experiment, contact with the male was partially effective in reducing stress response at 12 and 16 days of age, but nowhere as potent as contact with the lactating dam. At 20 days of age, contact with the adult male was entirely ineffective. The latter experiment showed that at no age was contact with an anesthetized pup in any way an effective inhibitor of the CORT stress response.

In summary, if we were to rank order the effectiveness of stimuli capable of suppressing the CORT response to novelty, clearly the anesthetized lactating dam is far and away the most effective stimulus at all ages tested. The virgin female appears to be equally effective in the younger ages, 12 and 16, but only moderately capable of reducing the stress response in day 20 pups. Contact with the adult male at day 12 and 16 results in only a moderate inhibition of the P-A response to novelty and is totally ineffective in 20 day old pups. Contact with the anesthetized sibling is not effective at any age.

This latter finding that sibling contact is incapable of suppressing the CORT response is of interest insofar as Hofer and Shair (1978) have reported that ultrasonic vocalization by pups in a novel environment is inhibited when a sibling is present in that environment. That vocalization and CORT responses are not concordant in the rat is consistent with other findings which appear extensively in the analysis of primate vocalization following separation (Levine, 1987). Thus, in the primate, low emissions of isolation calls are frequently associated with elevated CORT and more extensive vocalizations are often accompanied by reduced levels of circulating CORT.

The previous experiments reported in this series have now demonstrated that contact with a lactating dam, but not an adult male or sibling pup, attenuates the CORT distress response in preweanling rats. These data suggest that the phenomena of CORT suppression is under selective control of cues which are related to the maternal status of the animal which is contacted. Thus, if the cues related to maternal status are a critical variable in determining the degree to which the mother can inhibit the CORT response, it can be assumed that as the saliency of these cues is diminished, the ability of the mother to inhibit the CORT response to stress would also be diminished. This hypothesis was tested in a further experiment (Stanton & Levine, 1985). Weaning in the developing rat begins somewhere around 18 days of age and is a progressive process during which time the infant becomes less dependent upon the mother for thermoregulation and nutrition. The infant becomes more independent, spends less time in contact with the mother, and begins to eat solid food. In general, it is assumed, that weaning is concluded at about 28 days of age. In the next experiment, 20, 24, and 28 day old rat pups were assigned to three conditions. For the dam conditions, pups were placed in a novel test chamber for 30 minutes in the presence of an anesthesized lactating dam which they could contact or they were placed alone in a novel environment with no dam present. A comparison group was deprived for 24 hours but not tested. The data indicated that , as the animals achieved greater independence as a consequence of weaning, the capacity of the anesthetized lactating dam to inhibit the CORT stress response is reduced. At 20 days of age the dam is completely effective in reducing this response, at 24 days of age she is less effective and at 28 days of age, even though the presence of the dam yields a significantly lower CORT value, clearly these values are elevated over those seen in pups permitted contact at younger ages.

It should be noted that when one compares all of the data thus presented, the dam becomes less effective at approximately 20 days of age. Whereas the earlier ages, 12 and 16, both the anesthesized dam and virgin female are capable of completely inhibiting these stress responses to novelty. The rat does not appear to possess the capacity for specific recognition either by the mother or the pup of its own offspring or mother. Thus, for the infant rat early in development, the virgin female may be sufficiently similar to the lactating dam to be effective as an inhibiting stimulus. At 20 days of age, in some of the experiments, the anesthetized dam seems to perform the same function as seen in the earlier ages, whereas in other experiments, at 20 days of age, the presence of a lactating female did not completely eliminate the CORT response following exposure to the novel environment. However, at no time was the CORT of 20 day old animals with the lactating dam elevated to the level of the animals tested with either other social stimuli or alone. Some of the differences between the results of the different experiments can be accounted for by minor differences in procedure. However, in spite of these procedural differences, it does appear as though the 20 day old period is a transitional one. It represents an age in which the animal is showing less reliance on maternal cues and the maternal regulation of the inhibitory process appears to be diminishing at that age, and diminishes further as the animal becomes older and less reliant on the maternal cues.

In all the experiments we have thus far reported, some aspect of maternal contact appears to modulate the P-A response to stress. Contact, as an important dimension in reducing arousal, has received considerable attention in the primate literature. Harlow, 1959, was the first to show the importance of "contact-comfort" as a determinant of affiliation of infant rhesus monkeys toward their mothers. It has been argued that "contact-comfort" during infancy is necessary for normal, social-affective development in the monkey (Mineka & Suomi, 1978; Suomi, in press). Further, it has been demonstrated in squirrel monkeys that contact immediately reduces the P-A response following brief periods of separation (Coe, Wiener, Rosenberg, & Levine, 1985). Contact cues also play an important role in the formation of affiliative responses in infant rats (Alberts, 1981). Thus, there is ample evidence that in the infant rat, as in the infant primate, contact enhances maternal and, more generally, social affiliation. The findings of the present studies further indicate that an affiliative process of this kind can play a fundamental role in regulating endocrine responses of the developing organism. The findings of these studies are consistent with a growing body of evidence which indicates that physiological processes in developing infants are regulated by the mother in many subtle ways. The concept that elements of the mother-infant interaction are involved in regulating specific physiological processes of the infant has been elaborated by Hofer (1978). The evidence indicates that different physiological processes in the infant are controlled by specific aspects of the mother-infant interaction (Hofer, 1984, Hofer, & Shair, 1982).

The influence of maternal regulation is usually made manifest by separating the infant from the mother. Thus, certain physiological changes in the infant which develop during a long separation have been shown to be related to specific features of the mother-infant interaction which are absent during the period of maternal separation. One such example involves the decreasing cardiac rate in 2 week old rat pups after several hours of maternal separation. Regulation of cardiac physiology through the mother can be specifically linked to gastric fill of the mother's milk. Stomach loading milk reinstates normal cardiac function in otherwise deprived pups. This can be contrasted to the mother's influence on the infant's sleep patterns which can be traced more to the rhythmicity of milk delivery than the gastric fill of the milk, per se (Hofer & Shair, 1982). Another specific maternal influence is represented by the work of Schanberg, Evoniuk, and Kuhn (1984), who have analyzed maternal stimuli which regulate the activity of growth hormone and of a tissue enzyme related to growth, ornithinedecarboxylase (ODC). Depriving the infant of the mother causes a decline in growth hormone and ODC activity. Unlike the regulation of cardiac physiology, this decline does not reflect deprivation of milk. Rather, it can be traced precisely to the lack of vigorous tactile stimulation, active licking and stroking, which when applied in the absence of the mother restores growth hormone and ODC activity to normal. The present studies indicate that another physiological system of the infant, the P-A axis, may be regulated by a different but perhaps equally specific maternal variable, thermotactile contact. All of the research discussed thus far emphasizes the importance of maternal regulation in the physiological development of the infant. This process is a subtle and complex one which cannot be fully understood without detailed behavioral analysis of the mother-infant interaction. This research also represents an example of the importance of pursuing an integrated approach to understanding biological phenomena. It is illustrative of the fundamental contribution of psychobiology which emphasizes such an integrated approach.

Activational Properties of Maternal Deprivation on the Infant's P-A System

Thus far, in all the experiments we have discussed, the experimental paradigm involved separating infants from their mothers for 24 hours at specified ages. The primary purpose of this deprivation procedure was to increase the infant's motivation to engage in consummatory behaviors. However, this deprivation procedure could also have the consequence of increasing the relevancy of maternal cues. Therefore, we conducted an experiment comparing deprived and nondeprived animals placed in a novel environment with or without anesthetized dams. We hypothesized that the nondeprived pup would presumably be less motivated to seek the mother and thus the mother would be less salient to the pup. Thus we would predict that nondeprived pups that did not seek out the dam would not be affected by maternal presence and would display equivalent elevations of glucocorticoids following novelty with or without the mother (Figure 2). The results of this experiment not only totally contradicted our hypothesis but revealed a most surprising phenomena. That is, maternal deprivation resulted in a marked augmentation of the response to novelty. Thus, maternally deprived pups tested at days 12 and 16 displayed significant CORT elevations when exposed to novelty without an anesthetized dam. In contrast, age-matched nondeprived pups similarly tested did not show a CORT response. Although by 20 days of age, nondeprived pups displayed a CORT response to novelty, the deprived pups showed a significantly greater response. This augmentation of the response to novelty in the maternally deprived pups was of considerable magnitude. These data suggested that maternal factors were involved in the developing P-A system and that one of the functions of the mother is

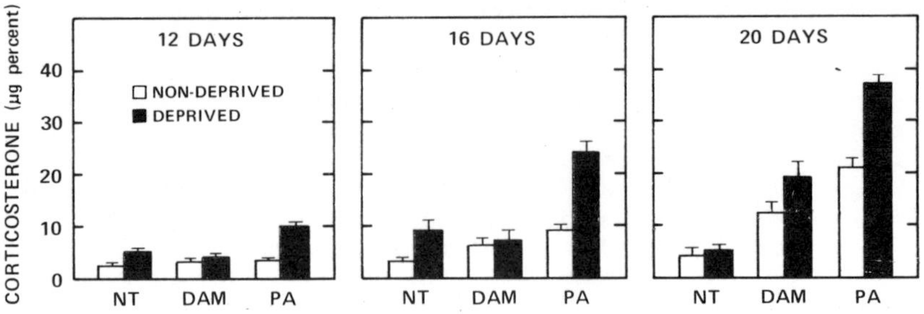

Fig. 2. Plasma corticosterone levels (μg %) for rat pups aged 12, 16, and 20 days under the six sampling conditions (Mean +SEM): NT denotes the nontested basal condition; DAM ("pup with dam") maternal-contact-plus-novelty-stress; PA ("pup alone") novelty-stress-only; Non-Deprived, taken directly from the home nest prior to the experiment; and Deprived, deprived for 24 hr prior to the experiment.

to exert an inhibitory influence on this system during development.

In all of the experiments we have thus far reported, the youngest animals we have used to investigate the effects of maternal deprivation on the regulation of P-A activity have been 12 days of age. Animals younger than 12 days were not used because of the large body of research indicating that prior to this age CORT stress responses either do not occur or are so minimal that they can not be studied reliably (Schapiro, 1968; Levine, 1970; Henning, 1978a). The hypotheses which have been presented to account for the absence of the stress response are numerous (DeKloet et al., in press). It has been postulated that the SHRP is due to diminished adrenal response to ACTH, reduced ACTH concentration in the pituitary, an absence of appropriate CRF releasing mechanisms in the hypothalamus. More recently, an elaborate theory has been presented, based on the current knowledge of CORT receptor systems in both the pituitary and the central nervous system, which postulates that the SHRP is a consequence of enhanced CORT negative feedback (Sapolsky & Meaney, 1986). However, all of this research was based on non-deprived rat pups, i.e., pups taken directly from the home nest and exposed to stress or ACTH injections which the present studies have demonstrated are less responsive to stress. However, since our experiments appear to demonstrate maternal regulation of the developing P-A system, it is possible that one of the major determinants of a SHRP may be maternal inhibition of P-A activity as a function of the mother-infant interaction in the home environment.

In order to study the influence of maternal regulation during an earlier neonatal period, we examined pups aged 4, 8, 12, 16, and 20 days either following exposure to novelty or exposure to novelty plus ACTH injection (8 IU/100 g). In order to control for the effects of injection, the novelty exposed animals were also given an injection of saline immediately prior to being placed in the novel arena. The results of this experiment (Figure 3) indicates, once again, that maternal deprivation potentiates the CORT response to novelty in pre-weanling rats. Potentiation occurs despite a general increase in stress responsiveness that occurs over the course of ontogeny. There are some interesting developmental differences. At 4 days of age, novelty did not activate the P-A system in either deprived or non-deprived animals. However, in both conditions there was a significant elevation of plasma CORT following ACTH injections. This elevation is augmented in the

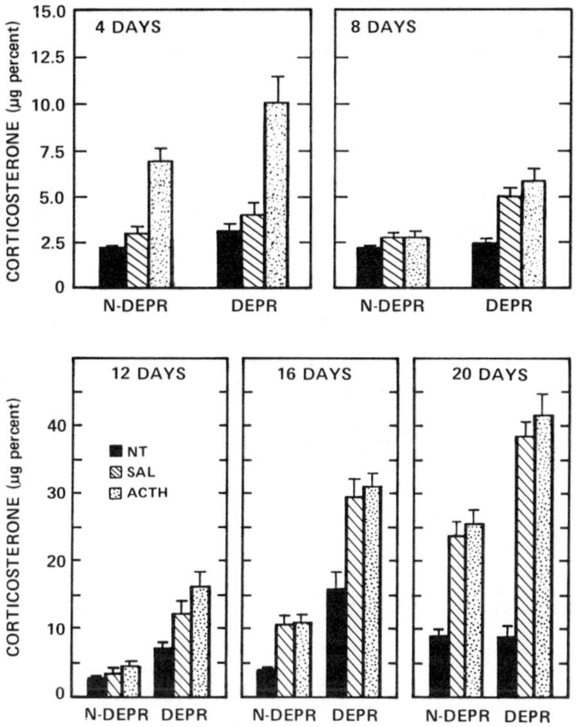

Fig. 3. Plasma corticosterone levels (μg %) for rat pups aged 4, 8, 12, 16, and 20 days under the six sampling conditions (Mean +SEM): NT, nontested basal group; SAL, novelty exposed following saline injection; ACTH, novelty exposed following ACTH injection; N-DEPR, nondeprived, i.e., taken directly from the home nest prior to novelty testing; DEPR, deprived for 24 hr prior to novelty testing. Note change in scale of vertical axis between upper and lower panels.

deprived animals. At 8 and 12 days of age there appears to be no response to ACTH or novelty in the non-deprived animals. The deprived animals, however, do show a response to novelty at 8 and 12 days of age albeit the response is significantly dimished relative to later ages. ACTH is capable of further augmenting the CORT response at 8 and 12 days, whereas at the older ages, the ACTH response is not significantly different from the response to novelty, although there is an augmentation of both responses in the deprived animals. In this study, as has been previously reported, at the earliest age the infant animal does have the capacity to respond to ACTH. However, during the SHRP between 8 and 16 days of age only the deprived animals are capable of showing a CORT response. At the older ages deprivation substantially increases the magnitude of the stress response which was nonetheless apparant under non-deprived conditions. At all ages, non-deprived pups showed a lower CORT elevation to exogenous ACTH administration than did deprived pups. These data suggest that increased adrenal sensitivity to ACTH in part accounts for the effects of deprivation on the CORT response to stress. In the non-deprived animals, in addition to all other potential reasons for a diminished stress response, one cannot avoid concluding that there is also a diminished sensitivity to ACTH (Stanton, Gutierrez, & Levine, in press). These data are in contrast to recent conclusions by Walker et al. (1986) that diminished adrenal sensitivity does not appear to be an important mechanism in the appearance of a SHRP.

A subsequent experiment was undertaken to both extend and replicate these results. There were a number of problems with the initial study. First, the response was only evaluated at 30 minutes, and it was thus difficult to make a clear statement as to whether non-deprived animals are indeed refractory to ACTH insofar as they may show an altered time course of the adrenal response to ACTH. Second, due to the fact that this was an initial probe, we used supermaximal doses of ACTH. Third, interpretation of the response to ACTH was confounded by the fact that the animals were also exposed to novelty during the 30 minute period following injection of ACTH.

A large study was undertaken in which we examined the ontogeny of the P-A response to two doses of ACTH over a period of 60 minutes in deprived and non-deprived animals between the ages of 4 and 16 days of age. The two doses of ACTH used were 1 IU/100g and 4 IU/100g. The results presented in Figure 4 are consistent with our previous experiment. The differences that occur as a consequence of maternal deprivation following ACTH injection are both unambiguous and striking. The pattern is most consistent. For all ages, injection of ACTH leads to an elevation of circulating levels of plasma CORT in deprived animals. Even during those periods of time when non-deprived animals are responding to ACTH, the response is not comparable to that observed in deprived animals. Clearly, the ACTH response in both deprived and non-deprived animals is biphasic. Thus, at the earliest age, 4 days of age, both deprived and non-deprived animals respond to ACTH with the deprived animals showing a significantly greater elevation of plasma CORT following the injection. At 8 days of age there is a marked dimunition of the response. However, once again, deprivation does result in an increase in the output of CORT. The non-deprived animals, however, do show a slight but significant increase at 1 hour following injection. All of these results are consistent throughout the older ages. There appears to be little question that maternal deprivation has a profound consequence on the developing P-A system and that there are important elements in the maternal behavior that appear capable of not only suppressing the system during the SHRP during which time there are minimal responses to ACTH injections and stress in the non-deprived animal. Moreover, even at the older ages, when an ACTH response is elicited, the response is still diminished in non-deprived animals when compared to deprived animals.

The concept of maternal regulation as it applies to the P-A system has one fundamental difference when compared with the maternal regulation of cardiac responses and growth hormone. For these two physiological systems the absence of maternal care results in suppression of these biological systems. The specific role of maternal behavior, nutrition and vigorous tactile stimulation, is to activate these responses. In contrast, the role of the mother appears to be to inhibit the P-A system. Maternal deprivation accentuates the CORT response in older pups. In younger animals a vigorous response to novelty and ACTH is present only in maternally deprived pups.

It has been hypothesized that the SHRP which ensues shortly after birth in the developing rat is a highly adaptive mechanism. Exposures to high levels of CORT during this critical period in development results in numerous detrimental effects on the developing nervous system (Bohn, 1984). Therefore, mechanisms that prevent elevated levels of circulating adrenal steroids would appear to be essential for normal CNS development.

One such mechanism could be the immaturity of various components of

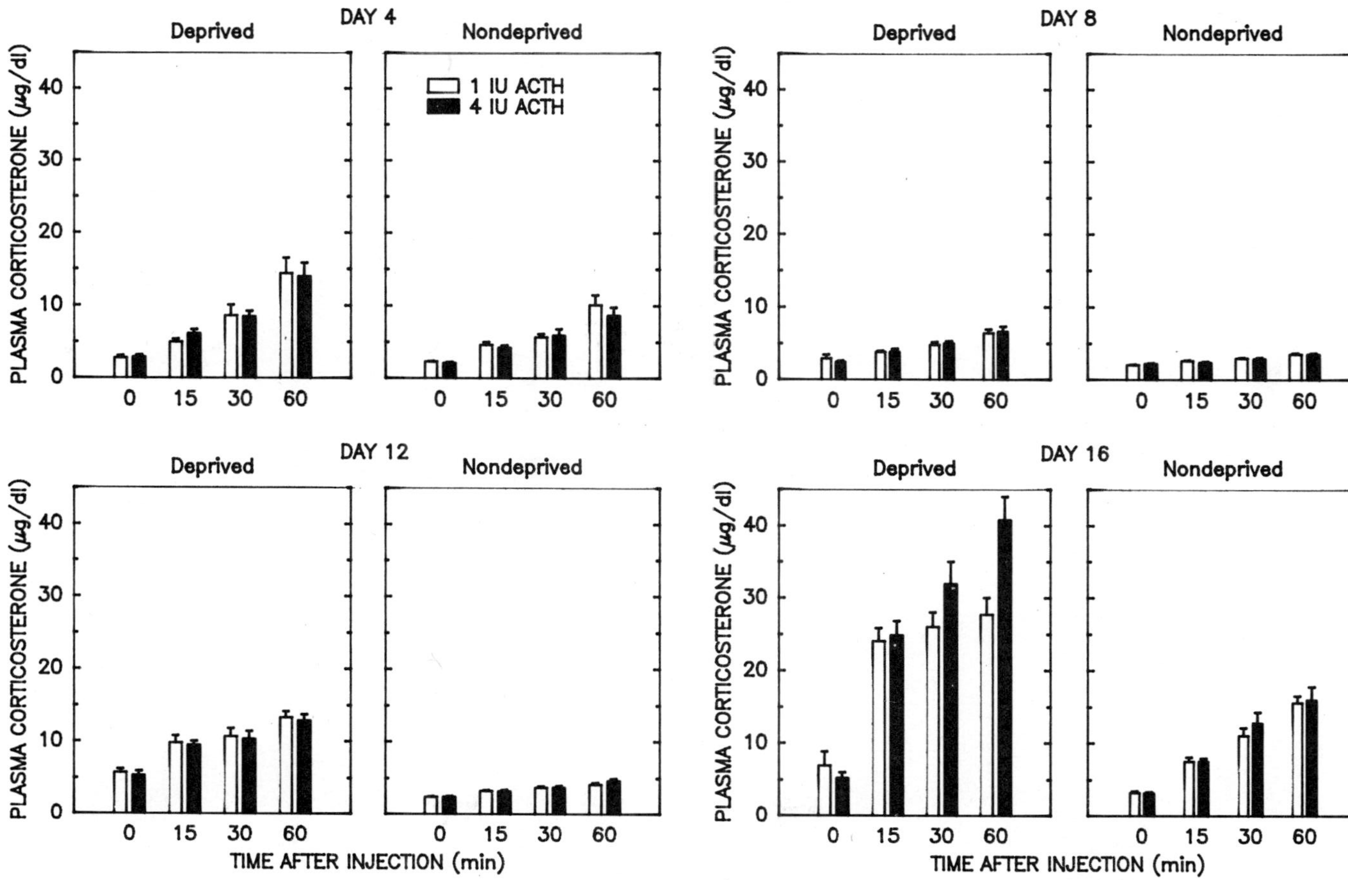

Fig. 4. Plasma corticosterone levels (ug/dl) for rat pups at 4, 8, 12, and 16 days of age under eight sampling conditions (Mean +SEM).

the P-A system. Thus even when the system does respond, as in the case of maternally deprived 8 and 12-day-old pups, the response is still significantly lower than is seen with equivalent doses of ACTH or exposure to novelty at the later ages. However, the data presented in this paper indicates that some aspect of the mother-infant interaction is yet another important mechanism which further inhibits the infant's capacity to elevate adrenocortical hormones. It is certainly not possible at this time to specify which specific aspect or aspects of the regulatory mechanisms involved in the P-A system are being altered as a function of maternal deprivation. They do indicate a change in the sensitivity of the adrenal to ACTH following 24 hours of deprivation. However, it is unlikely that the sensitivity of the adrenal is altered without altering some other component of the regulatory mechanisms. One aspect of endocrine function is that if a target organ, such as the adrenal, is deprived of its trophic hormone for a period of time that system becomes refractory to the trophic hormone. Then a considerable period of priming is required before it once again becomes responsive. One could postulate that ACTH is increased during the period of deprivation, so as to prime the adrenal and thus make it responsive to further ACTH stimulation by exogenous administration of this peptide. However, this begs the question further since one would now be required to postulate an additional mechanism that would trigger ACTH secretion. Certainly the most obvious mechanism would be increases in CRF activity. Thus, what would be required eventually to understand the complexity of the events which result in deprivation-induced increases in P-A activity is a more detailed analysis of the multiple components of this system which ultimately regulate the output of CORT from the adrenal.

In the adult, corticosteroids influence corticotropic activity by exerting feedback inhibition at both the hypothalamic and pituitary levels. During development, an enhanced glucocorticoid feedback signal at the pituitary level, due to the particular characteristics of the CORT-binding systems at this time, has been proposed as the mechanism responsible for the SHRP (Sapolski & Meaney, 1986).

There are three binding systems for glucocorticoids in the pituitary; these show a remarkably divergent developmental pattern. Thus, the Type 2 glucocorticoid receptor (GR) in the pituitary is present in approximately adult concentrations and is physiologically active (i.e., capable of inhibiting ACTH release) at the time of birth, and shows few changes thereafter. The Type 1 (MR) receptor is undetectable until the end of the first week of life; concentrations then rise rapidly and achieve adult values at approximately one month of age (Sakly & Koch, 1987). Finally, in the adult, CORT circulates partly bound to a plasma protein, corticosteroid-binding globulin (CBG). The free fraction (i.e., the faction not bound to CBG) is assumed to be available for interaction with the receptors. CGB-like proteins are also found in the pituitary corticotrophs, and are thought to diminish the feedback signal of CORT at this level. During development, concentrations of both plasma CGB and CBG-like molecules decrease precipitously after birth and are virtually undetectable between days 4 and 10 of life (Sakly & Koch, 1983). This decrease in CBG leads to an increase in the percentage of unbound CORT which can therefore interact freely with the receptors. During early development, the combination of high receptor levels (presumably of Type 2 receptors, as Type 1 receptors are not present until midway throught the SHRP) (Rosenfeld et al., in press) and low concentrations of plasma CBG and pituitary CBG-like molecules, would result in an augmented feedback signal at the pituitary level, such that even low circulating CORT levels would be sufficient to inhibit ACTH secretion. This, together with a diminished CRF signal, it has been argued, would account for the SHRP.

Although this may very well be the case, the fact still remains that the infant's responsivity (or lack thereof) is governed to a large extent by the mother. The mechanism whereby this maternal effect is 'transduced' into a diminished responsivity of the HPA axis remains unknown. Although a neural mechanism (i.e., suppression of neural input to the CRF-secreting cells) seems the more likely hypothesis, we certainly cannot discount, at this stage, the possibility that maternal regulation may act at the receptor level, by inducing changes in receptor affinity, capacity, or functional state.

Finally, an important property of the concept of maternal regulation is the time course of the separation-related phenomena. Hofer (1983) has distinguished between the immediate response of an infant to maternal separation which he believes is a non-specific aversive emotional response to the disruption of "the attachment bond" from the more slowly developing response to prolonged periods of separation which reveals the process of maternal regulation (by removing the regulatory influence). Thus, in order for us to invoke the concept of maternal regulation in relation to the P-A response of pups to stressful stimuli, it was necessary to determine whether this response varies with the duration of maternal deprivation. To accomplish this we investigated the effects of maternal deprivation on the pups' CORT levels following novelty exposure as a function of the length of the deprivation period. We examined the CORT response to novelty in rat pups aged 12, 16, and 20 days after 0, 1, 8, or 24 hours of deprivation. The results of this experiment as presented in Figure 5 indicate that maternal separation potentiates the pups CORT stress response. Exposure to novelty after 24 hours of deprivation at all ages raises CORT levels significantly above both basal (NT) and the non-deprived novelty stress conditions. It is also clear that this potentiation of this stress response is not an acute response to separation rather it appears to be a positively accelerating function of the deprivation period and appears only after relatively prolonged (8 to 24 hours) of separation. This property of the stress-potentiation effect is consistent with the concept of maternal regulation which postulates that regulatory influences of the mother are manifested only after relatively long periods of maternal separation (Hofer, 1983).

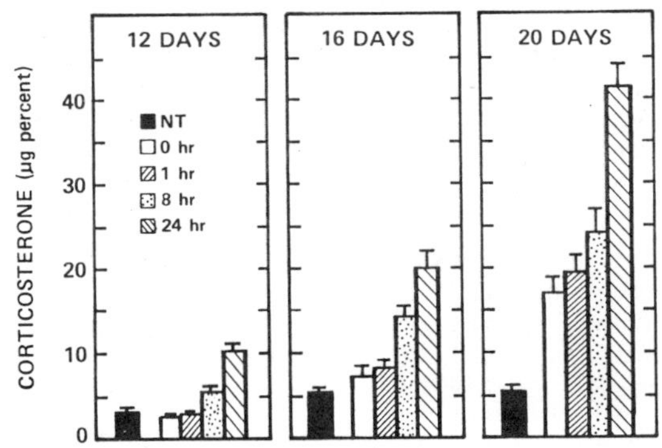

Fig. 5. Plasma corticosterone levels (μg %) for rat pups aged 12, 16, and 20 days under the five sampling conditions (Mean +SEM): NT denotes deprived nontested basal group; 0 hr, 1 hr, 8 hr, and 24 hr denote the deprivation periods of remaining four groups. These latter four groups were exposed to novelty for 30 min prior to blood sampling.

In conclusion, we believe that the extensive series of studies which we have reported all demonstrate that the orderly and adaptive pattern of the ontogeny of the P-A activity is in part regulated by some as yet unspecified aspect of maternal behavior and tends to support the hypothesis that mother-infant interaction is an important mechanism in inhibiting the infants capacity to elevate plasma levels of CORT during critical periods in development. Little in evolution is capricious and clearly what is emerging from these studies and other studies on maternal regulation of physiological responses is that the developing infant is only part of an equisite integrated system which includes specific genetic programming, the infant's behavior and the mother's behavior.

ACKNOWLEDGEMENTS

This research was supported by a grant from the National Institute of Mental Health HD-02881 and by a Research Scientist Award MH-19963 to Seymour Levine.

REFERENCES

Alberts, J. R., 1981, Ontogeny of olfaction: reciprocal roles of sensation and behavior in the development of perception, in: "The development of perception: Psychobiological perspectives," R. N. Aslin, J. R. Alberts, and M. R. Petersen, eds., Academic Press, New York.

Aslin, R. N., Alberts, J. R., and Petersen, M. R., eds., 1981, "The development of perception: Psychobiological perspectives," Academic Press, New York.

Bohn, M. C., 1984, Glucocorticoid induced teratologies of the nervous system, in: "Neurobehavioral teratology," J. Yanai, ed., Elsevier, New York.

Coe, C. L., Stanton, M. E., and Levine S., 1983, Adrenal responses to reinforcement and extinction: Role of expectancy versus instrumental responding, Behav. Neurosc., 97:654.

Coe, C. L., Wiener, S. G., Rosenberg, L. T., and Levine, S., 1985, Physiological consequences of maternal separation and loss in the squirrel monkey, in: "Handbook of squirrel monkey research," L. A. Rosenblum and C. L. Coe, eds., Plenum, New York.

Coover, G. D., 1983, Positive and negative expectancies: The rat's reward environment and pituitary-adrenal activity, in: "Biological and psychological basis of psychosomatic disease," H. Ursin and R. Murison, eds., Pergamon Press, Oxford.

De Kloet, R., Rosenfeld, P., Van Eekelen, J. A. M., Sutanto, W., & Levine, S., 1988, Stress, glucocorticoids and development, in: "Progress in Brain Research," vol. 73, G. J. Boer, M.G. P. Feenstra, M. Mirmiran, D. F. Swaab, and F. Van Haaren, eds., Elsevier, New York.

Goldman, L., Winget, C., Hollingshead, G. W., and Levine, S., 1973, Postweaning development of negative feedback in the pituitary-adrenal system of the rat, Neuroendocrinology, 12:199.

Gray, G. D., Bergfors, A. M., Levin, R., and Levine S., 1978, Comparison of the effects of restricted morning or evening water intake on adrenocortical activity in female rats, Neuroendocrinology, 25:236.

Hall, W.G. and Rosenblatt, J.S. (1977). Suckling behavior and intake control in the developing rat pup. J. Comp. Physiol. Psychol., 91: 1232.

Harlow, H. F., 1959, Love in infant monkeys, Scien. Amer., 200:68.

Hennessy, J. W. and Levine, S., 1979, Stress, arousal and the pituitary-adrenal system: A psychoendocrine model, in: "Progress in psychobiology and physiological psychology," vol. 8, J. M. Sprague and A. N. Epstein, eds., Academic Press, New York.

Henning, S., 1978a, Plasma concentrations of total and free corticosterone during development in the rat, Amer. J. Physiol., 235:E451.

Henning, S. J., 1978b, Postnatal development: Coordination of feeding, digestion, and metabolism. Amer. J. Physiol., 235:G199.

Heybach, J. P. and Vernikos-Danellis, J., 1979, Inhibition of adrenocorticotrophin secretion during deprivation-induced eating and drinking in rats, Neuroendocrinology, 28:329.

Hofer, M. A., 1978, Hidden regulatory processes in early social relationships, in: "Perspectives in ethology: Social behavior," vol. 3, P. P. G. Bateson and P. H. Klopfer, eds., Plenum, New York.

Hofer, M. A., 1983, On the relation between attachment and separation processes in infancy, in: "Emotion: Theory, research and experience, vol. 2: Early development," R. Plutnik, ed., Academic Press, New York.

Hofer, M. A., 1984, Early stages in the organization of cardiovascular control, Proc. Soc. Exp. Biol. Med., 175:147.

Hofer, M. A. and Shair, H., 1978, Ultrasonic vocalization during social interaction and isolation in two week old rats, Devel. Psychobiol., 11: 495.

Hofer, M. A. and Shair, H., 1982, Control of sleep-wake states in the infant rat by features of the mother-infant relationship, Devel. Psychobiol., 15:229

Kail, R. V. and Spear, N. E., 1984, "Memory development: Comparative perspectives," Erlbaum, Hillsdale, N.J.

Leon, M., 1979, Mother-young reunions, in: "Progress in psychobiology and physiological psychology," vol. 8, J. M. Sprague and A. N. Epstein, eds., Academic Press, New York.

Leon, M. and Moltz, M., 1972, The development of the pheromonal bond in the albino rat, Physiol. & Behav., 8:683.

Levine, S., 1970, The pituitary-adrenal system and the developing brain, Progr.in Brain Res., 32:79.

Levine, S., 1987, Psychobiologic consequences of disruption in mother-infant relationships, in: "Perinatal development: A psychobiological perspective," N. Krasnegor, E. Blass, M. Hofer, and W. Smotherman, eds., Academic Press, New York.

Levine, S. and Mullins, R. F., Jr., 1966, Hormonal influences on brain organization in infant rats, Science, 152:1585.

Levine, S. and Treiman, D. M., 1969, Determinants of individual differences in the steroid response to stress, in: "Physiology and pathology of adaptation mechanisms," E. Bajusz, ed., Pergamon, Oxford.

Levine, S., Weinberg, J., and Brett, L. P., 1979, Inhibition of pituitary-adrenal activity as a consequence of consummatory behavior, Psychoneuroendocrinology, 4:275.

Lincoln, D. W., Hill, A., and Wakerley, J. B., 1973, The milk ejection reflex of the rat: An intermittent function not abolished by surgical levels of anesthesia, J. Endocr., 57:459.

Martin, G. R. and Henning, S. J., 1984, Enzymic development of the small intestine: Are glucocorticoids necessary? Amer. J. Physiol., 246:G695.

Mason, J. W., 1975, A historical view of the stress field, J. Hum. Stress, 1:6.

Mineka, S. and Suomi, S. J., 1978, Social separation in monkeys, Psychol. Bull., 85:1376.

Munck, A., Guyre, P. M., Holbrook, N. J., 1984, Physiological functions of glucocorticoids in stress and their relation to pharmacological actions. Endocr. Rev., 5(1):25.

Rosenfeld, P., Sutanto, W., Levine, S., and De Kloet, E. R., in press, Ontogeny of Type I and II corticosteroid receptors in the rat hippocampus, Devl. Brain Res.

Sakly, M., and Koch, B., 1983, Ontogenetical variations of transcortin modulate glucocorticoid receptor function and corticotropic activity in the pituitary gland, Hormone & Metabolic Res., 15:92.

Sapolsky, R. M. and Meaney, M. J., 1986, Maturation of the adrenocortical stress response: Neuroendocrine control mechanisms and the stress hyporesponsive period, Brain Res. Rev., 11:65.

Schanberg, S. M., Evoniuk, G., and Kuhn, C. M., 1984, Tactile and nutritional aspects of maternal care: Specific regulators of neuroendocrine function and cellular development. Proc. Soc. Exp. Biol. Med., 175:135.

Selye, H., 1950, "Stress," Acta, Montreal.

Schapiro, S., 1968, Maturation of the neuroendocrine response to stress in the rat, in: "Early experience and behavior," G. Newton and S. Levine, eds., Thomas, Springfield, IL.

Spear, N. E. and Campbell, B. A., eds., 1979, "Ontogeny of learning and memory, " Erlbaum, Hillsdale, NJ.

Stanton, M. E., Gutierrez, Y. R., and Levine, S., in press, Maternal deprivation potentiates pituitary-adrenal stress responses in infant rats. Beh. Neurosc.

Stanton, M. E., and Levine, S., 1985, Maternal modulation of corticosteroid stress responses in infant rats. Paper presented at the annual meeting of the Intern. Soc. Devl. Psychobiol., October, Dallas, Texas.

Stanton, M. E., Wallstrom, J., and Levine, S., 1987, Maternal contact inhibits pituitary-adrenal stress responses in preweanling rats. Devl. Psychobiol., 20:131.

Suomi, S. J., in press, The role of touch in rhesus monkey social development, in: "The many facets of touch," C.C. Brown (Ed.), Johnson & Johnson, Skillman, NJ.

Walker, C. D., Meaney, M. J., Sapolsky, R. M., and Rivier, C., 1986, Increased sensitivity to glucocorticoid feedback during the stress non-responsive period in the neonatal rat, Endocrinology, 119:1816.

Weinberg, J., Smotherman, W. P., and Levine, S., 1980, Early handling effects on the intake of novel substances: Differential behavioral and adrenocortical responses, Behav. Neur. Biol., 29:446.

Weiss, J. M., 1972, Influence of psychological variables on stress-induced pathology, in: "Physiology, emotion & psychosomatic illness," R. Porter and J. Knight, eds., Elsevier, Amsterdam.

MATERNAL ALCOHOL CONSUMPTION AND STRESS RESPONSIVENESS IN OFFSPRING

Anna Newman Taylor, Berrilyn J. Branch,
Judith E. Van Zuylen and Eva Redei

Department of Anatomy and Laboratory of
Neuroendocrinology, Brain Research Institute
UCLA School of Medicine and
West Los Angeles VAMC, Brentwood Division
Los Angeles, CA 90024

INTRODUCTION

It is now generally acknowledged that pre- and/or postnatal environmental factors, such as stress and drugs, can exert long-lasting neurobehavioral effects in the individual. These factors may influence the development of central regulatory systems by acting directly on the fetal brain or indirectly through changes in the internal environment of the mother and fetus. The Fetal Alcohol Syndrome which has been recognized as a clinical entity for the past 15 years (1) is a case in point. Adverse effects of maternal alcohol consumption during pregnancy on growth and development of the offspring have been well documented in humans and laboratory animals (2). Mental retardation and behavioral deficits, such as hyperactivity, jitteriness, irritability and marked attentional and learning problems, are consistently observed in children who were exposed to alcohol *in utero* (3).

We have been struck by the apparent similarities between the neurobehavioral disturbances associated with fetal alcohol exposure and those ascribed to perinatal exposure to corticosteroids (4). On the basis of these similarities we hypothesized (5) that alcohol-induced activation of the brain-pituitary-adrenal (BPA) axis may contribute to the actions of alcohol on the developing organism. To validate this hypothesis we embarked on an investigation of the developmental and long-term effects of prenatal alcohol exposure on the responsiveness of the BPA axis in an animal model. With recent evidence for impaired feedback regulation of BPA responsiveness to stress in adult rats exposed to alcohol *in utero* (6), in this presentation we review and update our studies of the long-term effects. Our developmental data have recently been reviewed (7).

EXPERIMENTAL MODEL

In our studies ethanol is administered to pregnant rat dams via a liquid diet. Timed-pregnant Sprague-Dawley dams are received from Charles River Breeding Labs on day 6 of gestation and are divided into three weight-matched groups. The normal (N) group is allowed access to standard lab chow and water *ad libitum* throughout pregnancy. The ethanol (E) group receives a high protein liquid diet (BioServ, Inc., Frenchtown, NJ) containing 5.0% (w/v) ethanol during the last two weeks of gestation. To control for the effects of the liquid diet and nutritional intake, another group of dams is pair-fed (P) an amount of isocaloric liquid diet without ethanol equal to that ingested by dams in the E group. Fresh diet is presented each morning from day 8 of gestation until parturition. Intake of ethanol by E dams averages 11.6 g/kg/day over the two week period, which produces blood alcohol levels of 79 mg% at 0900 hr and 127 mg% at 2000 hr (8). At birth, the pups from each treatment group are pooled, sexed, randomly placed in groups of 10, and fostered to dams fed the standard lab chow diet. The pups and dams are weighed at weekly intervals until weaning on postnatal day 21, when pups are again randomized within treatment conditions and group-housed by sex. These pup randomization procedures minimize the potential contributions of litter effects when offspring are used in later experiments.

This regimen of prenatal ethanol administration does not usually result in any visible dysmorphologies in rats. Generally, at birth, E pups weigh less than P pups which, in turn, weigh less than N pups, although the weight differences are not consistently significant (7, 9). This trend can persist to adulthood. At birth, brain (9, 10) and plasma (7, 9) corticosterone (CORT) titers are significantly higher in E neonates than in the N or P controls. These differences in basal BPA function disappear by the third day of life (9) and we consistently find similar basal, nonstressed serum titers of CORT or ACTH in all three prenatal treatment groups (6, 11).

EXPERIMENTAL STUDIES OF THE LONG-TERM EFFECTS OF PRENATAL ETHANOL ON STRESS RESPONSIVENESS

In our first studies with this model we sought to determine whether prenatal ethanol exposure would produce a long-term effect on sensitivity to ethanol in adulthood. As an end-point, we used pituitary-adrenal activation by a challenge dose of ethanol which is a well-known biological response to the drug (12-14). Ethanol produced significantly higher plasma CORT levels in adult female E rats than in the P or N control rats (8). Thus, the BPA axis of adult E rats is hypersensitive to ethanol; but, is this a specific response to ethanol or is there a generalized increase in BPA responsiveness to other stressors? We found that certain neurogenically mediated stressors, such as cardiac puncture or noise and shake, also elicited greater plasma CORT responses in adult female E rats than in controls . In contrast, responses to metabolic stressors, such as cold or fasting, did not differentiate between E and control rats (11). We have since extended this list to include enhanced plasma CORT responses in E rats to an acute injection of morphine and to intermittent footshock (15). Our results demonstrate that prenatal ethanol exposure produces long-lasting effects on the responsiveness of the BPA axis to a broad range of stressors.

We have also identified persistent effects on behavioral responses to stress. Adult E rats show significantly greater stress-induced analgesia in response to an opioid-mediated form of footshock stress (16). E rats of both sexes are more sensitive to the analgesic effects of morphine (17). Adult female E rats consume more ethanol when stressed daily for at least one week with intermittent footshock than controls (18). E rats do not adapt like controls to a stressful swimming paradigm designed to test "behavioral despair" (19). Additionally, we have also found that prenatal ethanol exposure appears to cause alterations in thermoregulatory mechanisms since E rats show a greater degree of hypothermia following administration of ethanol (8), pentobarbital, diazepam (20,21), or morphine (22).

All the neuroendocrine and behavioral responses that we have studied share one common feature: E rats do not appear to differ from N or P controls when tested in adulthood under basal conditions. Instead, alterations in these functions are revealed only when the E animals are challenged by a stressor or pharmacological agent. It, therefore, appears that prenatal exposure to ethanol affects normal processes for reestablishing homeostasis after challenges with stress and drugs. How ethanol exposure during fetal life produces these long-lasting effects remains essentially unknown. We have begun experiments designed to elucidate possible mechanisms for the prenatally ethanol-induced malfunction that leads to enhanced responsiveness of the BPA axis in stress, as outlined below.

Having previously determined that adrenal sensitivity to ACTH is normal in E rats (11), we recently examined the possibility that the enhanced BPA response of E rats may be due to greater sensitivity of the pituitary to ACTH secretagogues. For these studies adult female rats from each prenatal treatment group were pretreated with chlorpromazine, morphine and pentobarbital to suppress endogenous release of ACTH secretagogues (23). Rats were then injected via the tail vein with 750 ng/100 g body weight (bwt) CRF, 600 ng/100 g bwt arginine vasopressin (AVP) or 0.25 ml of the phosphate buffer/BSA/ascorbic acid vehicle. Ten min later blood samples were drawn by cardiac puncture. All rats responded to AVP and CRF with a marked increase in serum ACTH levels and a corresponding increase in serum CORT (both hormones measured by radioimmunoassay). Responses of the E rats were not greater than those of controls; if anything, E responses were somewhat attenuated compared to controls. Thus, since neither the adrenal nor the pituitary of E rats is hyperresponsive to its secretagogues, it appears that the defect in the regulation of BPA function must lie within the CNS.

In a preliminary assessment of the integrity of the feedback loop in E rats, we administered the synthetic glucocorticoid, dexamethasone (DEX), and measured plasma CORT levels 4 hr later (24). At several doses of DEX (12.5, 25 and 50 µg/kg), plasma CORT levels were suppressed to a comparable degree in offspring of all prenatal treatment groups. At the lowest dose of DEX (3.125 µg/kg), there was no suppression of CORT in the controls, but E rats had a paradoxical elevation of CORT over basal levels. These results suggest that the temporal profiles of either suppression or sensitivity to DEX may be altered in E rats.

In our most recent studies we obtained further evidence for impaired glucocorticoid feedback in E rats (6). In these experiments we measured serum ACTH levels by RIA in adult female rats at 5, 10 or 15 min after a 1-min episode of intermittent footshock stress (1 mA, 0.5 sec/sec). Confirming our results for plasma CORT, basal prestress ACTH levels were unaffected by the prenatal treatment. ACTH was markedly elevated at 5 min after footshock and to the same extent in E and control groups. At 10 min postshock, ACTH declined toward basal levels in the N and P controls. In contrast, in the E rats at the 10 min time-point, ACTH remained elevated at a level which was significantly higher than that of the N and P controls. At 15 min poststress, ACTH in all three groups declined toward basal levels. These results not only confirmed our previous findings of enhanced pituitary-adrenal responses to stress in the E rats, but the differences in the time-course of the responses suggested that fast feedback inhibition may be impaired by prenatal ethanol exposure.

To further test this hypothesis we presented rats with two episodes of the 1-min intermittent footshock stress, separated by a 5-min interval. Trunk blood was collected at 5 or 10 min after the second shock episode. The peak ACTH response occurred at 5 min after the second shock in all three prenatal treatment groups. At this time, ACTH levels in the N and P control groups did not differ from each other, but the ACTH level in the E rats was significantly higher than the level of the two control groups combined. At 10 min after the second shock, ACTH levels in all three groups had declined and no longer differed from each other. Thus, the ACTH response of E rats to multiple stress presented in the time domain of fast feedback is enhanced. Since the primary site of action of fast feedback is the brain (25), these results indicate that glucocorticoid feedback inhibition at the brain level is either impaired or delayed in E rats. Indeed, we now have evidence that two weeks after adrenalectomy, the ACTH response of E rats to 1-min intermittent footshock stress is no longer higher than the response of control rats.

DISCUSSION

Our results demonstrate a consistent pattern of enhanced challenge-induced neuroendocrine and behavioral responses in adult rats that are exposed to ethanol *in utero*. Other investigators have also observed BPA hyperresponsiveness in rats (26) and a lack of inhibition of behavioral responses in rats (27) and children (3) exposed to ethanol prenatally. The central sites and mechanisms underlying the long-term actions of maternal ethanol consumption on these diverse neurobehavioral functions in offspring remain to be elaborated. On the basis of our initial findings of augmented plasma CORT responses to various stress and drug challenges in E rats, we postulated that BPA hyperresponsiveness could be due to dysfunction of central inhibitory components and/or enhancement of stimulatory components of the BPA regulatory system (11, 28, 29). Our recent studies showing impaired inhibition of the ACTH response to a brief stimulus and to multiple stress presented in the fast feedback time domain provide evidence for the persistent effects of prenatal ethanol exposure on one of the central inhibitory systems involved in BPA regulation.

In a series of elegant studies which Dr. McEwen reviewed at this conference, Sapolsky et al (30) demonstrated that glucocorticoid receptors in the hippocampus are involved in terminating the adrenocortical response to stress, albeit in a different time frame than in our paradigm. Whether prenatal ethanol exposure depletes glucocorticoid receptors in the hippocampus or, for that matter, in other CNS sites, remains to be determined. However, there is good reason to believe this may occur. Prenatal exposure to ethanol results in a permanent loss of hippocampal pyramidal cells (31) along with other alterations in hippocampal morphology (32). Since glucocorticoid receptors are known to be concentrated in hippocampal pyramidal cells (33, 34), their permanent loss following prenatal ethanol exposure could account for a corresponding receptor depletion and consequent impairment of feedback inhibition.

How ethanol exerts its effects on the ontogeny of central glucocorticoid receptors remains to be determined. Ethanol may have a direct cytotoxic action on neurogenesis, the region of vulnerability being dependent on when during site-specific developmental periods exposure to the drug occurs. Alternatively, the elevated maternal titers of CORT during gestation (7) and similarly elevated CORT titers at birth in brain (9, 10) and plasma (7, 9) of neonates exposed to ethanol during gestation may influence central glucocorticoid receptor ontogeny, since it has been reported that neonatal corticosteroid administration selectively decreases nuclear binding of the steroid in adult hippocampus (35). Likewise, it is possible that the glucocorticoid enhancement of cytotoxicity which has been observed in adult rats (36) may also occur in near-term fetuses being exposed to ethanol. These and other possibilities require examination. Determination of the mechanisms of action of ethanol on CNS development is essential for understanding and eventually treating the multiple deleterious and persistent effects of prenatal ethanol exposure.

ACKNOWLEDGEMENTS

The authors' research was supported by the Veterans Administration Medical Research Service.

REFERENCES

1. K. L. Jones, D. W. Smith, C. N. Ulleland and A. P. Streissguth, Lancet 1: 267-1271 (1973).
2. J. R. West, "Alcohol and Brain Development," Oxford Univ. Press, London and New York, 440 pp (1986).
3. A. P. Streissguth, The behavioral teratology of alcohol: Performance, behavioral, and intellectual deficits in prenatally exposed children, in "Alcohol and Brain Development," J. R. West, ed., Oxford Univ. Press, London and New York, pp. 3-44 (1986).
4. M. E. Weichsel, Jr., The therapeutic use of glucocorticoid hormones in the perinatal period: Potential neurological hazards, Ann. Neurol. 2: 364-366 (1977).
5. A. N. Taylor, B. J. Branch and N. Kokka, Neuroendocrine effects of fetal alcohol exposure, Prog. biochem. Pharmacol. 18: 99-110 (1981).
6. A. N. Taylor, B. J. Branch, J. E. Van Zuylen and E. Redei, Prenatal ethanol exposure alters ACTH stress response in adult rats, Alcoholism 10: 120 (1986).
7. J. Weinberg, L. R. Nelson and A. N. Taylor, Hormonal effects of fetal alcohol exposure, in "Alcohol and Brain Development," J. R. West, ed., Oxford Univ. Press, London and New York, pp. 310-342 (1986).
8. A .N. Taylor, B. J. Branch, S. H. Liu, A. F. Weichmann, M. A. Hill and N. Kokka, Fetal exposure to alcohol enhances pituitary-adrenal and temperature responses to ethanol in adult rats, Alcoholism 5: 237-246 (1981).
9. A. N. Taylor, B. J. Branch, B. Cooley-Matthews and R. E. Poland, Effects of maternal ethanol consumption on basal and rhythmic pituitary-adrenal function in neonatal offspring, Psychoneuroendocrinology 7: 49-58 (1982).
10. R. Kakihana, J. C. Butte and J. A. Moore, Endocrine effects of maternal alcoholization: Plasma and brain testosterone, dihydrotestosterone, estradiol and corticosterone, Alcoholism 4: 57-60 (1980).
11. A. N. Taylor, B. J. Branch, S. H. Liu and N. Kokka, Long-term effects of fetal ethanol exposure on pituitary-adrenal response to stress, Pharmac. Biochem. Behav. 16: 585-589 (1982).
12. F. W. Ellis, Effect of ethanol on plasma corticosterone levels, J. Pharmac. exp. Ther. 153: 121-127 (1966).
13. L. A. Pohorecky and L. S. Jaffe, Noradrenergic involvement in the acute effects of ethanol, Res. Commun. chem. Pathol. Pharmac. 12: 433-447 (1975).
14. J. H. Mendelson, M. Ogata and N. K. Mello, Adrenal function and alcoholism. 1. Serum cortisol, Psychosom. Med. 33: 145-157 (1971).
15. L. R. Nelson, A. N. Taylor, J. W. Lewis, R. E. Poland, E. Redei and B. J. Branch, Pituitary-adrenal responses to morphine and footshock stress are enhanced following prenatal alcohol exposure, Alcoholism 10: 397-402 (1986).
16. L. R. Nelson, A. N. Taylor, J. W. Lewis, B. J. Branch and J. C. Liebeskind, Opioid but not nonopioid stress-induced analgesia is enhanced following prenatal exposure to ethanol, Psychopharmacology 85: 92-96 (1985).
17. L. R. Nelson, A. N. Taylor, J. W. Lewis, B. J. Branch and J. C. Liebeskind, Morphine analgesia is potentiated in adult rats

prenatally exposed to ethanol, Brain Res. 372: 234-240 (1986).
18. L. R. Nelson, J. W. Lewis, J. C. Liebeskind, B. J. Branch and A. N. Taylor, Stress induced changes in ethanol consumption in adult rats exposed to ethanol *in utero*, Proc. West. Pharmacol. Soc. 26: 205-209 (1983).
19. L. R. Nelson, A. N. Taylor, J. W. Lewis, B. J. Branch and J. C. Liebeskind, Prenatal exposure to ethanol alters responding in a "behavioral despair" paradigm, Proc. West. Pharmacol. Soc. 27: 583-586 (1984).
20. A. N. Taylor, B. J. Branch, N. Kokka and R. E. Poland, Neonatal and long-term neuroendocrine effects of fetal alcohol exposure, Monogr. Neural Sci. 9: 140-152 (1983).
21. A. N. Taylor, B. J. Branch, M. A. Hill, D. Randolph and N. Kokka, Fetal exposure to ethanol enhances hypothermic responses to diazepam and pentobarbital in adult rats, Alcoholism (in press).
22. L. R. Nelson, J. W. Lewis, N. Kokka, B. J. Branch and A. N. Taylor, Prenatal exposure to ethanol potentiates morphine-induced hypothermia in adult rats, Neurobehav. Toxicol. Teratol. 8: 469-474 (1984).
23. A. Arimura, T. Saito and A. V. Schally, Assays for corticotropin-releasing factor (CRF) using rats treated with morphine, chlorpromazine, dexamethasone and nembutal, Endocrinology 81: 235-245 (1967).
24. L. R. Nelson, A. N. Taylor, E. Redei, B. J. Branch and J. C. Liebeskind, Corticosterone response to dexamethasone in fetal ethanol-exposed rats, Proc. West. Pharmacol. Soc. 28: 299-302 (1985).
25. M. E. Keller-Wood and M. F. Dallman, Corticosteroid inhibition of ACTH secretion, Endocrine Rev. 5: 1-24 (1984).
26. J. Weinberg and P. V. Gallo, Prenatal ethanol exposure: Pituitary-adrenal activity in pregnant dams and offspring, Neurobehav. Toxicol. Teratol. 4: 515-520 (1982).
27. E. P. Riley, S. Barron and J. H. Hanningan, Response inhibition deficits following prenatal alcohol exposure: A comparison to the effects of hippocampal lesions in rats, in "Alcohol and Brain Development," J. R. West, ed., Oxford Univ. Press, London and New York, pp. 71-104 (1986).
28. A. N. Taylor, L. R. Nelson, B. J. Branch, N. Kokka and R. E. Poland, Altered stress responsiveness in adult rats exposed to ethanol in utero: Neuroendocrine mechanisms, in "Mechanisms of Alcohol Damage in Utero," Ciba Found. Symp. 105: 47-59 (1984).
29. L. R. Nelson and A. N. Taylor, Long-term behavioral and neuroendocrine effects of prenatal ethanol exposure, in "Genetic and Perinatal Effects of Abused Substances," M. C. Braude and A. M. Zimmerman, eds., Academic Press, Inc., Orlando, FL, pp. 177-203 (1987).
30. R. M. Sapolsky, L. C. Krey and B. S. McEwen, Glucocorticoid-sensitive hippocampal neurons are involved in terminating the adrenocortical stress response, Proc. Natl. Acad. Sci. 81: 6174-6177 (1984).
31. D. E. Barnes and D. W. Walker, Prenatal ethanol exposure permanently reduces the number of pyramidal neurons in rat hippocampus, Dev. Brain Res. 1: 333-340 (1981).
32. J. R. West, S. L. Dewey, D. R. Pierce and A. C. Black, Jr., Prenatal and early postnatal exposure to ethanol permanently alters the rat hippocampus, in "Mechanisms of Alcohol Damage in Utero," Ciba Found. Symp. 105: 8-20 (1984).
33. J. M. H. M. Reul and E. R. DeKloet, Two receptor systems for corticosterone in the rat brain: Microdistribution and differential occupation, Endocrinology 117: 2505-2511 (1985).
34. K. Fuxe, A.-C. Wikstrom, S. Okret, L. F.Agnati, A. Harfstrand, Z.-Y. Yu, L. Granholm, M. Zoli, W. Vale and J.-A. Gustafsson, Mapping of glucocorticoid receptor immunoreactive neurons in the rat tel- and

diencephalon using a monoclonal antibody against rat liver glucocorticoid receptor, Endocrinology 117: 1803-1812 (1985).

35. C. Nyakas, G. Szabo and E. Endroczi, Corticosterone binding to hippocampal receptors following neonatal corticosterone treatment, in "Cellular and Molecular Bases of Neuroendocrine Processes," E. Endroczi, ed., Akademiai Kiado, Budapest, pp. 565-573 (1976).

36. R. M. Sapolsky, A mechanism for glucocorticoid toxicity in the hippocampus: Increased neuronal vulnerability to metabolic insults, J. Neuroscience 5: 1228-1232 (1985).

TEMPERAMENTAL INFLUENCES ON REACTIONS TO UNFAMILIARITY AND CHALLENGE[1]

Jerome Kagan, J. Steven Reznick and Nancy Snidman

Department of Psychology
Harvard University
Cambridge, MA

Introduction

Recent interest in the temperamental characteristics of young children, which has enjoyed a renaissance during the last decade, is correcting the theoretical imbalance created by a half-century of extreme environmental interpretations of behavioral variations among children. Some of this variation is to be viewed as the product of an interaction of tonic physiological states, genetic or prenatal in origin, and rearing conditions that facilitate or constrain the psychological biases created by these endogenous states (Goldsmith & Campos, 1986; Thomas & Chess, 1977). Many of these dispositions are transient during the first year of life; the few that show some stability are called temperamental.

The empirical record suggests that the temperamental quality defined by an initial tendency to withdraw, to seek a parent, and to inhibit play and vocalization following encounter with unfamiliar people and events persists over the childhood years, is part of a coherent network of behavioral and physiological characteristics, and has analogues in many mammalian species (Blizard, 1981; Bronson, 1981; Brush et.al. 1985; Goldsmith & Gottesman, 1981; Kagan & Moss, 1962; Kagan, Reznick, Clarke, Snidman & Garcia-Coll, 1984; McDonald, 1983). The tendency to withdraw from or approach an unfamiliar event emerges as a reliable individual characteristic at about one year in humans, two months in monkeys, and one month in cats and dogs. Scott and Fuller (1965) found that variation in reaction to novelty was one of the two most differentiating characteristics among five breeds of dogs, and both Soumi (1984) as well as Stevenson-Hinde, Stillwell-Barnes and Zunz (1980) have reported similar variation in laboratory reared macaque monkeys. This dimension is also central in Jung's theory of psychological types, which made introversion-extraversion the major differentiating characteristic among humans (Jung, 1924). It is also of interest that fearfulness is a highly heritable characteristic in infants under one year of age (Goldsmith, 1983; Goldsmith & Gottesman, 1981; Goldsmith & Campos, 1986), and introversion-extraversion is the most heritable and best preserved personality quality in adults (Floderus-Myrhed, Pedersen & Rasmuson, 1980; Conley, 1985; Eysenck, 1982).

A Longitudinal Study

Our laboratory has been following two independent groups of Caucasian children from intact middle and working class families who were selected from large samples when they were either 21 or 31 months of age because the child was either consistently shy, timid and fearful (behaviorally inhibited) or sociable, bold and fearless (uninhibited) when exposed to unfamiliar rooms, people, and objects. We had to screen over 400 children by telephone and laboratory observation to find groups of approximately 60 inhibited and 60 uninhibited children with equal numbers of boys and girls in each group (see Garcia-Coll, Kagan & Reznick, 1984; Snidman, 1984 for details).

The children in Cohort 1 were chosen at 21 months of age from a group of 117 who came to our laboratory where their behavior was taped and subsequently quantified. The situations were: initial meeting with an unfamiliar examiner in an unfamiliar room, and subsequent encounters with a second unfamiliar room containing toys, an unfamiliar woman displaying three acts with toys that were difficult to remember and to imitate, a third unfamiliar woman, a large metal robot, and, finally, a brief separation from the mother. Behavioral signs of inhibition were: Clinging to or remaining proximal to the mother; fretting or crying, interruption of play, cessation of vocalization, distress after the model's display of the three acts, and reluctance to approach the unfamiliar objects or people. The children who displayed these behaviors consistently across most of the incentives, as well as those who did not, were selected to form groups of 28 extremely inhibited and 30 extremely uninhibited children. These 58 children returned to the laboratory one month later to be observed in the same room with the same incentives -- the reliability of a continuous index of behavioral inhibition was 0.6.

The original index of inhibited behavior in Cohort 2 at 31 months was based primarily on behavior with an unfamiliar peer of the same sex and age, with both mothers present, and a brief episode in which the child encountered an unfamiliar woman. The behavioral indexes of inhibition, which were similar to those used with Cohort 1, included long latencies to interact with the unfamiliar child, adult or toys, withdrawal from the unfamiliar events, and long periods of time proximal to the mother while not playing with an object. Figure 1 shows the differences between the two groups in Cohort 1 for duration proximal to the mother across all the episodes and Figure 2 the percent of children who displayed crying to separation, remained in physical contact with the mother for more than ten seconds, and fretted or cried to a woman modeling the three acts. Figures 3 and 4 show the differences between the two groups of Cohort 2 on two of the selection criteria: total time proximal to the mother and latency to approach the unfamiliar peer. Over two thirds of the children classified as inhibited in Cohorts 1 and 2, but no uninhibited child, were proximal to their mother for more than six-and-a-half minutes. In Cohort 2 over 80 percent of the inhibited children, but no uninhibited child, took longer than ten minutes to make their first approach to the unfamiliar child.

Each of these samples has been seen on several additional occasions. Cohort 1 has been seen at four years, five-and-a-half years and seven-and-a-half years of age; Cohort 2 at 43 months and five-and-a-half years, with fifteen percent attrition in each cohort.

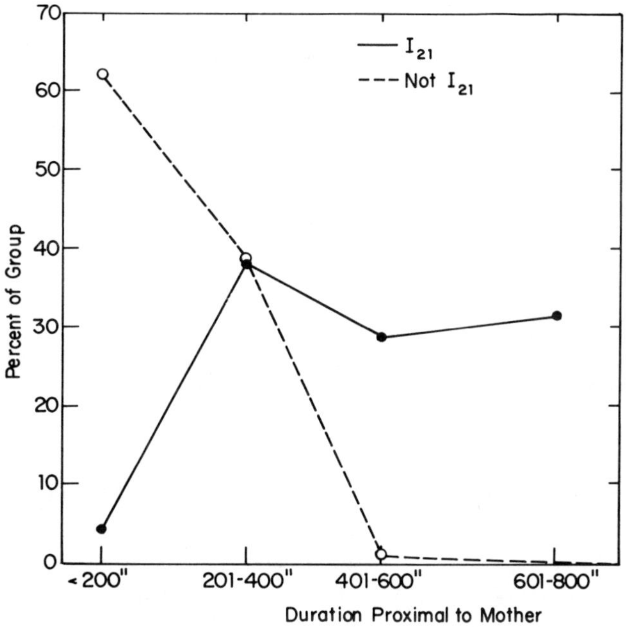

1. Differences between Cohort 1 inhibited and uninhibited children at 21 months on duration proximal to the mother.

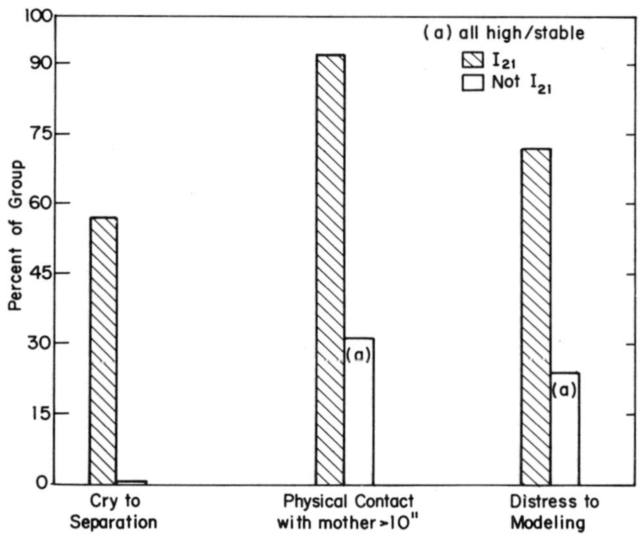

2. Differences between Cohort 1 inhibited and uninhibited children in separation fear, physical contact with the mother, and distress to modeling.

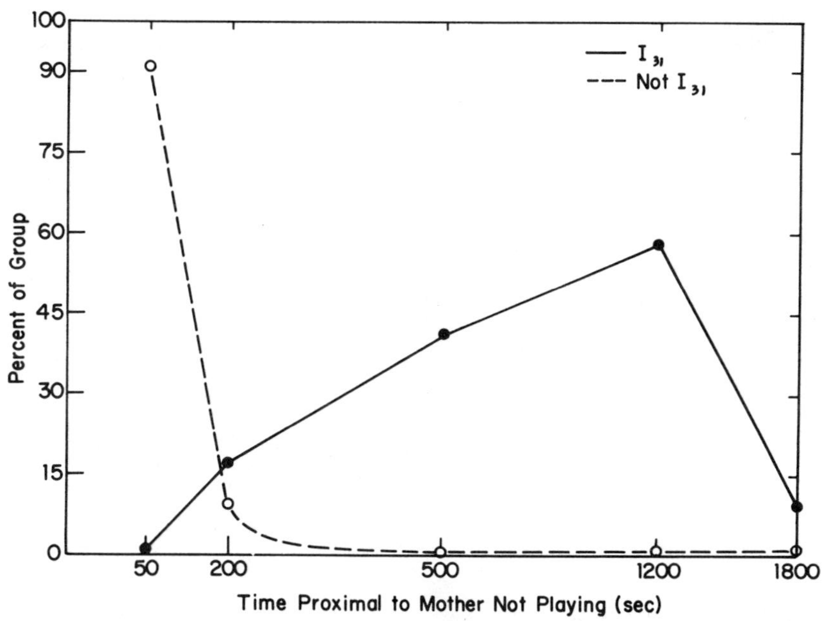

3. Differences between Cohort 2 inhibited and uninhibited children in time proximal to the mother at 31 months.

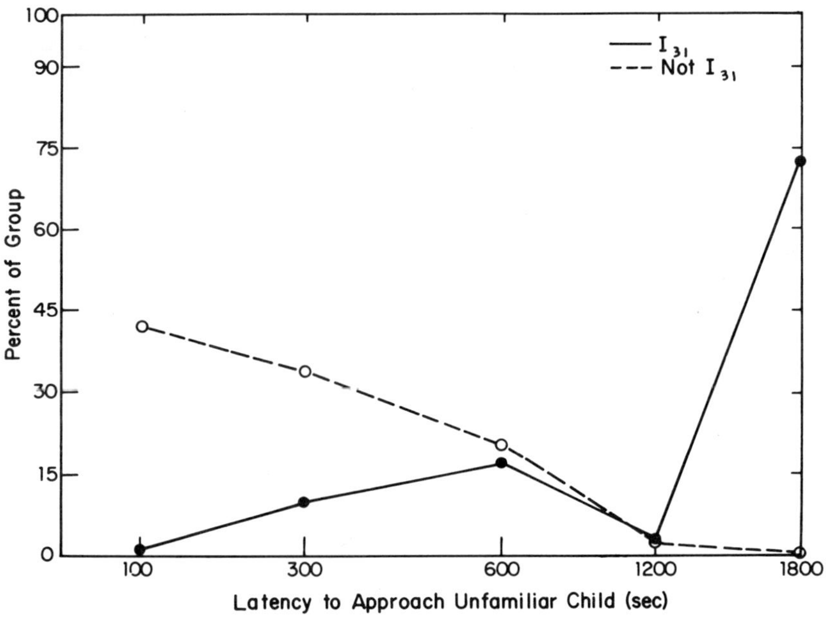

4. Differences between Cohort 2 inhibited and uninhibited children at 31 months for latency to approach an unfamiliar child.

On the second visit of Cohort 1 at four years of age and Cohort 2 at three-and-a-half years, the primary index of behavioral inhibition was derived from behavior with an unfamiliar child of the same sex and age in a laboratory playroom with both mothers present. At five-and-a-half years of age several indexes of inhibition derived from varied situations were aggregated to form a composite index for Cohort 1. The separate indexes were based on behavior with (1) an unfamiliar child in a laboratory setting, (2) classmates in a school setting, (3) a female examiner administering tests over a 90 minute session, and (4) unfamiliar objects suggestive of risk (a balance beam, a black box with a hole) in an unfamiliar laboratory room. The data gathered on Cohort 1 at seven-and-a-half years are still being analyzed. However, one of the indexes of inhibition was based on the child's behavior in a 90 minute laboratory play situation with eight or nine unfamiliar children of the same age and sex. The two major signs of behavioral inhibition were: infrequent spontaneous comments and large proportion of time spent sitting or standing apart from any other child (defined as greater than arm's length from a peer). These two variables were coded from a combination of a video tape of the entire group and a taped narration of each child's behavior dictated by one of nine coders who was blind to the child's behavioral assignment.

Preservation of Behavior

The behaviors that characterize inhibited and uninhibited children were moderately preserved from the original to the later assessments. The adjectives inhibited and uninhibited refer to the original classification at 21 or 31 months, unless stated otherwise. (See Table 1).

TABLE 1

Preservation of inhibited and uninhibited behavior across three assessments in the two cohorts (numbers refer to correlation coefficients).

COHORT 1	1	2	3	COHORT 2	1	2	3
Inhibition at:							
1. 21 months	-	.51 XX	.52 XX	1. 31 months	-	.59 XX	.55 XX
2. 4 years	-	-	.67 XXX	2. 43 months	-	-	.45 XX
3. 5 1/2 yrs	-	-	-	3. 5 1/2 yrs	-	-	-

XX $p < .01$
XXX $p < .001$

Inhibited behaviors were preserved from the earliest to the latest assessment for both cohorts. In Cohort 1, the correlation between the index of inhibition at 21 months and five-and-a-half years was 0.52; the comparable correlation for Cohort 2 from 31 months to five-and-a-half years was 0.55 (Reznick et. al., 1986). The preservation of each class of behavior was generalized to the school context. In her doctoral research, Gersten (1986) trained observers who were unaware of the child's classification to code the child's behavior in his or her kindergarten class during the first week of school in

September, as well as during a day in the Spring of the same academic year. The observers noted every fifteen seconds whether the target child was displaying one or more of a small number of responses. The children in Cohort 1 classified as inhibited at 21 months were more often alone and less often in social interaction with another child or adult ($r = 0.34$, $p < 0.05$).

One of the most sensitive indexes of inhibition in a laboratory situation, especially after four or five years of age, is a reluctance to talk spontaneously to an examiner during the individual testing session. This relation was robust for Cohort 2 at five-and-a-half years and Cohort 1 at both five-and-a-half and seven-and-a-half years. The two most sensitive variables are latency to utter the first few spontaneous comments to the examiner (i.e. an unprovoked statement that is not a reply to an examiner's probe or question), and the total number of spontaneous comments over the entire session. Figures 5 and 6 illustrate scatter plots of these two conversation variables for the children classified originally as inhibited or uninhibited in Cohorts 1 and 2. (Note: As of this writing we have not yet coded latency to the first six spontaneous comments for Cohort 2; hence, the use of slightly different variables for the two cohorts. However, the correlation between the latency to the second and the sixth comment is 0.90 in Cohort 1). Not one inhibited five-and-a-half year old in Cohort 2, but 14 uninhibited children, uttered their second spontaneous comment within three minutes of meeting the examiner, and, additionally, made 40 or more comments during the session. By contrast, 13 inhibited, but only two uninhibited, children had not made their second spontaneous comment after being in the room ten minutes and spoke less than ten times (chi square = 21.0 $p < 0.01$). Among Cohort 1 children at seven-and-a-half years, 12 uninhibited, but only five inhibited, children made their sixth spontaneous comment within five minutes and made more than 40 spontaneous comments, while 12 inhibited, but only two uninhibited, children took more than ten minutes before making their sixth comment and made fewer than 30 spontaneous comments (chi square = 9.6 $p < .001$). Figure 7 shows the relation between total spontaneous comments and total time in which the child remained apart from any peer in the group play situation for Cohort 1 at seven-and-a-half years. Ten uninhibited and two inhibited children spoke more than 40 times; 14 inhibited but six uninhibited children spoke less than 20 times (chi square = 8.4, $p < .001$).

There was more obvious preservation of uninhibited than inhibited behavior in both cohorts and we believe this is a result of socialization pressures. American parents regard outgoing social behavior as more desirable and adaptive than shy, cautious, timid behavior. About 40 percent of the original groups of inhibited children became less inhibited at five-and-a-half years, while less than ten percent of uninhibited children became more timid. Further, more boys than girls changed from an inhibited to an uninhibited behavioral style. Maternal interviews suggest that the mother's of the children who had become less inhibited over time self-consciously helped their children to overcome their shy behavior by introducing peers into the home and encouraging the child to cope independently with mildly stressful situations.

There are also more inhibited children who are later than first born -- about two-thirds -- and many more uninhibited children who are first born. This may be due to the fact that first born children are encouraged to be more independent of the mother when the next

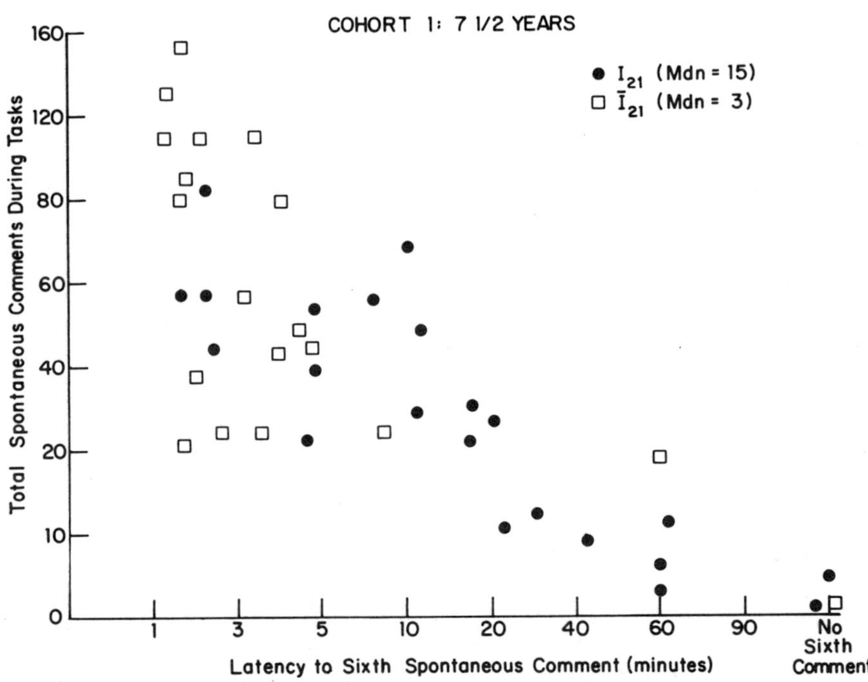

5. Relation between total spontaneous comments and latency to the sixth spontaneous comment during testing session with Cohort 1 children at seven-and-a-half years.

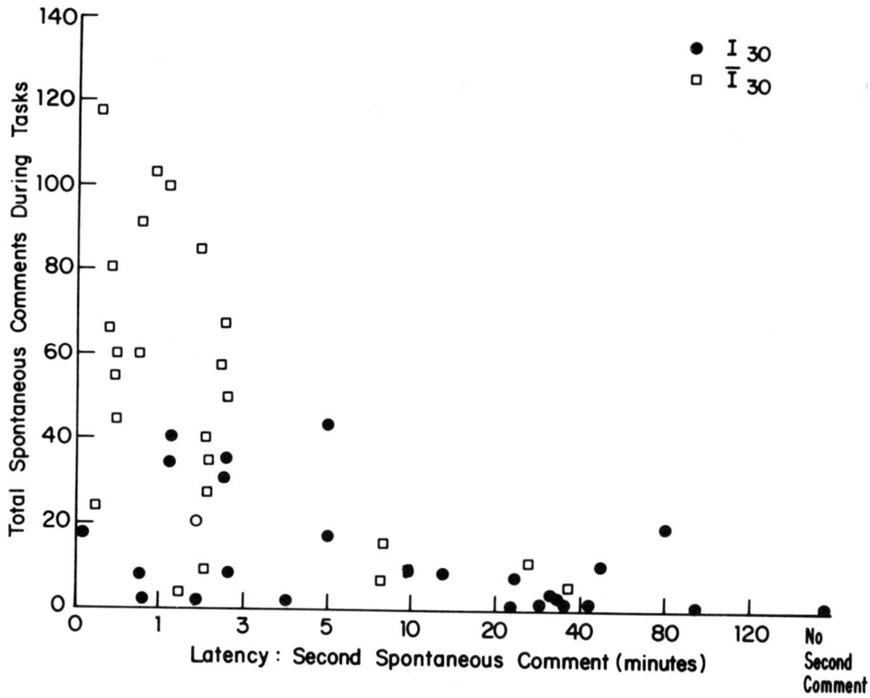

6. Relation between total spontaneous comments and latency to the second spontaneous comment during testing session with Cohort 2 children at five-and-a-half years.

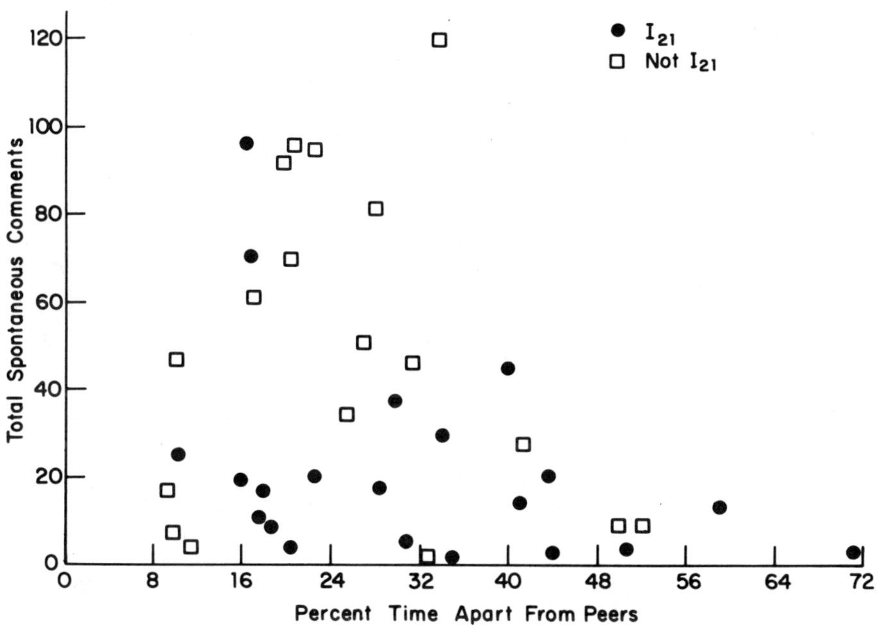

7. Relation between total spontaneous comments and percent time apart from peers for Cohort 1 children at seven-and-a-half years.

child arrives. A second possible interpretation assumes that fetal stress is less likely during the first than during subsequent pregnancies. A third interpretation is that later born status is associated with a more stressful set of experiences for the small group of infants who are born with a biological disposition to become uncertain to challenge and unfamiliarity. An infant with such a temperamental vulnerability might react with autonomic arousal to the mild, but unexpected, intrusions of an older four year old sibling who seizes a toy or pushes the infant from a chair.

Physiological Correlates

We believe that the consistently inhibited children have lower thresholds of excitability in those limbic structures that serve the related psychological states social scientists call uncertainty or anxiety. As a result, they are likely to show enhanced activity in the pituitary-adrenal axis, reticular activating system, and the sympathetic arm of the autonomic nervous system, as well as inhibition of action to unfamiliar events that cannot be assimilated and to threats in which they have no coping response. We now present data revealing a correlation between signs of reactivity in one or more of these systems and behavioral inhibition.

Heart Rate and Variability. Because we measured the child's heart rate variability to baseline and cognitive tasks on every assessment we can make the firmest statements about these two parameters. Heart period variability was defined as the average standard deviation of all the interbeat intervals during a particular episode. The mean heart period and variability for a multi-trial episode was the average of the values for the separate trials in that episode. Although we shall use the terms heart rate and heart rate variability in the text, all statistical analyses were performed on the heart period values. Heart period and heart period variability were always positively correlated under both relaxed conditions as well as under mild cognitive stress (correlations between 0.6 and 0.7). More important, more inhibited than uninhibited children had high and stable heart rates at every age of evaluation, with the magnitude of the correlation higher at five-and-a-half and seven-and-a-half years than at 21 months. We believe this is due to more effective sampling of heart rate under cognitive stress at the older ages. Recent research by Fox (personal communication) has shown that a high and stable heart rate in infancy is associated with the later development of inhibited behavior.

Table 2 shows the degree of preservation of differences in heart rate and heart rate variability for the two cohorts. Heart rate variability was preserved across all assessments in both cohorts. The preservation of differences in heart rate was significant at all ages for Cohort 2; for Cohort 1 the differences were only preserved from four to five-and-a-half years. The magnitudes of preservation for heart rate and variability matched those reported for behavior. Moreover, the inhibited children with the highest and most stable heart rates over the first two assessments were more likely to remain behaviorally inhibited than the inhibited children who had lower and more variable heart rates. Figure 8 shows the mean heart period for each of four assessments for four different groups in Cohort 1: The inhibited children who remained inhibited, the inhibited children who became less inhibited over time, the uninhibited children who remained uninhibited and, finally, the small number of uninhibited children who became somewhat more timid. Note that the consistently

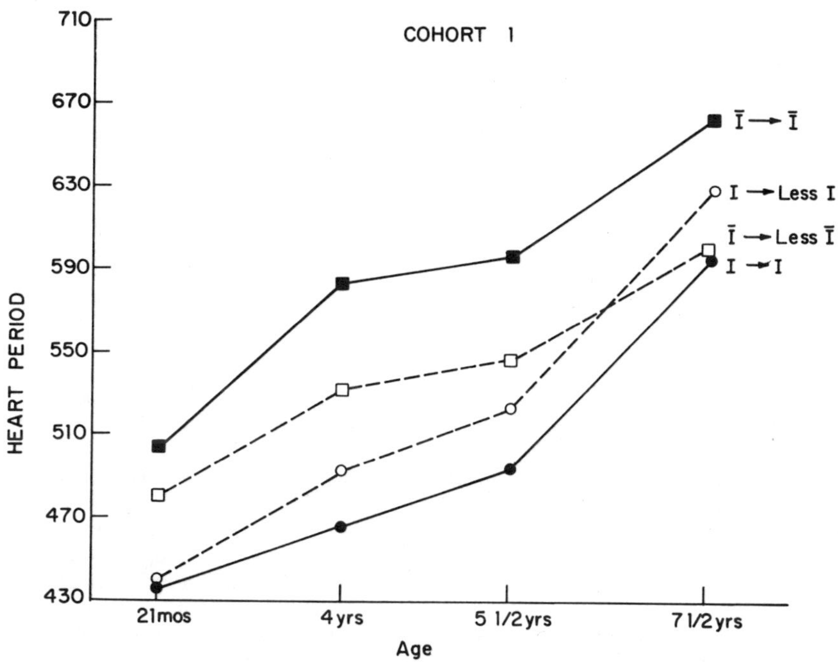

8. Mean heart period across four assessments for Cohort 1 children as a function of changes in behavior over time.

inhibited children had the highest heart rates and the consistently uninhibited children had the lowest heart rates. In addition, the children with consistently high heart rates had the largest number of unusual fears, and were most likely to have had symptoms during the first year of life suggestive of arousal of the limbic system, particularly chronic constipation, allergy and sleeplessness.

TABLE 2

Preservation of individual differences in heart rate and variability across three assessments in the two cohorts (numbers refer to correlations).

COHORT 1	21m.	4yr.	5 1/2yr.	**COHORT 2**	31m.	43m.	5 1/2
Heart rate:							
1. 21 months	-	.23	.15	1. 31 months	-	.53 XX	.59 XX
2. 4 years	-	-	.58 XX	2. 43 months	-	-	.40 X
3. 5 1/2 yrs.	-	-	-	3. 5 1/2 yrs.	-	-	-
Heart rate variability:							
1. 21 months	-	.49 XX	.39 XX	1. 31 months	-	.53 XXX	.61 XXX
2. 4 years	-	-	.64 XXX	2. 43 months	-	-	.49 XX
3. 5 1/2 yrs.	-	-	-	3. 5 1/2 yrs.	-	-	-

X $p < .05$
XX $p < .01$
XXX $p < .001$

Spectral analysis of the heart rate expresses the complex beat to beat variation as a power spectrum of the separate rhythms with peaks at characteristic frequencies. Peaks at low frequencies reflect sympathetic processes, especially blood pressure and temperature regulation. In human subjects a peak frequency between 0.2 and 0.5 Hz is due primarily to respiration and is mediated by vagal activity. Peaks at lower frequencies are due primarily, but not exclusively, to sympathetic activity. Analysis of the changes in the heart rate power spectrum of the children from Cohort 2 at 43 months of age reveal that more inhibited than uninhibited children shifted to greater power at the lower frequencies from a baseline period prior to cognitive testing to a subsequent baseline after the stress of several difficult cognitive procedures. Moreover, this index of sympathetic activity derived from the spectral analysis at 43 months predicted inhibited behavior at five-and-a-half years of age ($r = .47$, $p < 0.01$).

Pupillary dilation is a second potential index of sympathetic activity and the inhibited children from both cohorts showed significantly larger pupillary dilation to cognitive tasks than uninhibited children (recall memory for words and digits, a mental comparison of the relative size of objects, and inferring an object from its features). However, the pupil data were not as discriminating of the two groups as heart rate or heart rate variability.

The projections from limbic structures to the skeletal muscles of the larynx and vocal cords also seem to be at a higher level of excitability in inhibited children. Increased tension in these muscles is usually accompanied by a decrease in the variability of the pitch periods of vocal utterances (Stevens & Hirano, 1981). The increased muscle tension can be due not only to discharge of the nucleus ambiguus, but also, indirectly, to sympathetic activity that constricts arterioles serving the muscles of the larynx and vocal folds. Because the vocal cords do not maintain a steady rate as they open and close, the perturbations in the rate at which they open and close is a consequence of many factors, one of which is the tension in the laryngeal muscles (Lieberman, 1961). In her doctoral research, Coster (1986) has found that the inhibited children in Cohort 1 at five-and-a-half years of age showed less variability in the pitch periods of single word utterances spoken under psychological stress than uninhibited children.

Because norepinephrine is a primary neurotransmitter of the sympathetic nervous system, a urine sample collected from each child in Cohort 1 at the end of the test battery at five-and-a-half years was assayed for norepinephrine and its derivatives using mass fragmentography (Karoum, 1983). The assays yielded values for norepinephrine, normetanephrine, MHPG and VMA. Concentrations of each compound were transformed into moles per gram of creatinine and an index of total norepinephrine activity was computed. This index reflects primarily peripheral norepinephrine activity. There was a modest correlation between the index of norepinephrine activity at five-and-a-half years and inhibited behavior at four and five-and-a-half years (r = 0.34 with the index at age four, r = 0.31 with the index at age five-and-a-half).

In order to assess activity in the pituitary-adrenal axis, samples of saliva were gathered from both cohorts at five-and-a-half years of age in the laboratory as well as at home on three mornings before breakfast and, therefore, before the stress of the day had begun. These saliva samples were analyzed for unbound cortisol using a modification of the standard RIA method (Walker, 1984). In both groups, the average cortisol level across the three home samples was significantly correlated with the original index of inhibition. Cortisol levels in Cohort 1 were more discriminating of the two groups than any other physiological variables, for when the cortisol values were split at the median, a dichotomy of high versus low levels correctly predicted the original 21 month classification for 78 percent of the children. Moreover, inhibited children with high levels of cortisol in both home and laboratory samples differed from the inhibited children with lower cortisol values; the former were more fearful and had more psychosomatic symptoms (See Levine et. al. 1978; Tennes, Downey & Vernadakis, 1977).

Aggregate Index of Physiological Responsivity. Although the inhibited children in Cohort 1 showed the peripheral physiological consequences predicted from the hypothesis of lower thresholds of reactivity in the limbic system, only about one-third of the inhibited children showed signs of higher activity in all of the target systems, and most of the correlations among the individual variables were low and nonsignificant, with the single exception of a correlation of 0.84 between heart period and heart period variability. However, the statistical independence of the separate physiological variables is not atypical (See Nesse et. al., 1985). In order to assess the predictive power of an aggregate index of responsivity across the separate indexes we computed a mean standard score for eight physiological variables quantified on Cohort 1 at five-and-a-half years of age. The eight variables were: (1) cortisol level in the morning at home (2) cortisol level in the laboratory (3) heart period during cognitive tasks (4) heart period variability during cognitive tasks (5) pupillary dilation during cognitive tasks (6) variability of pitch periods of vocal utterances under cognitive stress (7) standard deviation of all the fundamental frequency values and, finally, (8) total norepinephrine activity in the laboratory urine sample. We reversed the values for variables 3,4, 6, and 7 so that a high standard score reflected greater limbic arousal. The correlations between the mean of the eight physiological variables and each of its component scores ranged from 0.36 to 0.56. The relation between the aggregate of the eight physiological variables and the index of behavioral inhibition at each age for 22 inhibited and 21 uninhibited children was highest with the original index of inhibition at 21 months ($r = 0.70$, $p < 0.001$), but significant at the two later ages ($r = 0.66$ with the index at age four; $r = 0.58$ with the index at five-and-a-half years). It is of interest that two of the three formerly inhibited children with the lowest physiological indexes had become increasingly less inhibited across the period of study while two of the three formerly uninhibited children with the highest indexes had become more inhibited at five-and-a-half years. Thus, the direction of change in behavior over the four years of study is accompanied by the expected level of physiological reactivity at five-and-a-half years.

Discussion

The temperamental qualities of inhibition and lack of inhibition are moderately stable over time and have reasonable correlates with peripheral physiological responses, each of which implies a higher level of excitability in parts of the limbic system. However, the stability correlations are in the range of 0.5; hence, some children change their behavior over the period of study. A few older children who had been behaviorally inhibited and had a high and stable heart rate at 21 months changed both their behavior and their cardiac pattern at five-and-a-half years, suggesting that both behavior and peripheral physiology are malleable following experience.

The group play data gathered at seven-and-a-half years, illustrated in Figure 7, are relevant to this discussion of change. Of the 21 children who had been classified as inhibited at 21 months, the scores of two boys on total spontaneous comments and total time apart from another child were similar to the mean score of the uninhibited children. However, these two boys resembled the inhibited children on other significant variables associated with inhibition. Thus, even though some aspects of their behavior had changed, other components of the coherence that defines inhibition did not change.

Psychology is more reluctant than biology, chemistry or physics to invent terms for transitional states. When a child who originally meets all of the defining criteria for a category loses some, but not all, of the defining qualities at a later age it is not clear what category term should be used for that subject. Physicians say a person is "in remission from a disease" even if the person still harbors the pathogen. We suggest it is useful to invent a special category for a child that had been inhibited and shy at two years of age and was sociable and uninhibited at five-and-a-half and seven-and-a-half years, but retained some of the important behavioral or biological characteristics of his original temperamental category.

Mechanisms. There are several possible explanations of the variation in inhibited behavior. During the forty years when behaviorism and psychoanalytic theory shared popularity, many American and European psychologists viewed fearfulness as a learned reaction. Two related explanations that are gaining initial consensus, while not denying the influence of learning, both implicate inherent differences in central nervous system functioning, especially the activity of limbic structures (Gray, 1982). One hypothesis holds that some infants are born with low thresholds of reactivity in the amygdala and hypothalamus when unfamiliarity or challenge occurs. A related hypothesis emphasizes the effectiveness of processes that inhibit limbic activity. Pavlov chose to emphasize the inhibition function, suggesting that organisms differed in the "strength of the central nervous system", where strength meant the ability to inhibit the build-up of excitation. Pavlov would have regarded his fearful dogs and our shy children as possessing a "weak nervous system".

The environment presents young children with at least three classes of events that invite cognitive work or overt behavior. One class contains unfamiliar events that provoke attempts at assimilation; a second includes events, often the actions of people, to which the child must issue some response; finally, the environment presents problems to which the child must generate solutions. Most of the time children assimilate unfamiliar events easily, issue socially effective actions and generate correct cognitive solutions with minimal delay. As a result there is minimal limbic arousal. But when there is a delay in the production of a coping response the child is simultaneously alerted psychologically and aroused physiologically. We believe that this state should be treated as a special psychological/physiological condition mediated, in part, by limbic structures. We choose to call this state uncertainty; others may prefer a different word. (See Schneiderman & McCabe, 1985).

We suggest that more inhibited than uninhibited children show the physiological reactions we quantified (tension in the laryngeal muscles, high and stable heart rate, pupillary dilation, high level of cortisol) because they have a lower threshold of excitability in the amygdala and hypothalamus. This hypothesis is in accord with the suggestions of many investigators (Aggleton & Passingham, 1981; Turner, Mishkin & Knapp, 1980; Sarter, & Markowitsch, 1985; Gloor, 1978; Tsuda & Tanaka, 1985). This hypothesis finds strong support in studies of domestic cats that differ in degree of defensive, non-aggressive behavior with rats (Adamec & Stark-Adamec, 1986). The descriptions of the defensive cats resemble closely our descriptions of inhibited children. The defensive cats are less likely to attack rats and are more inhibited in novel environments, as well as with humans. These behavioral qualities are stable over periods as long as three years. Further, stimulation of the basal amygdala evokes stronger multiple unit activity in the limbic areas of defensive compared with non-defensive animals. When the amygdala is stimulated with single pulses of increasing intensity, the peak height of the evoked potential in the ventromedial hypothalamus is significantly greater in defensive than in non-defensive cats. These data, together with evidence indicating the amygdala mediates the state of conditioned fear in the rat (Hitchcock & Davis, 1986), renders more credible the idea that differences in responsivity of limbic structures make an important contribution to the contrasting behaviors of inhibited and uninhibited children.

The reasons for the lower thresholds of reactivity in the limbic system are unclear; one possible contributing factor is high levels of central norepinephrine (Aston-Jones & Bloom, 1981; Bandura et. al., 1985; Reiser, 1984; Charney, Heninger, & Breier, 1984; Charney & Redmond, 1983). Assays of a large number of autonomic and hormonal variables in phobic women exposed to the sources of their fear revealed that blood levels of norepinephrine had the largest number of correlations with the other physiological indexes (Nesse, Curtis, Thyer, McCann, Huber-Smith, & Knopf, 1985). Thus, one hypothesis, albeit speculative, is that inhibited children have tonically higher levels of norepinephrine than uninhibited children (Ciarenello, 1983).

The lack of high correlation among the various physiological indexes that we regard as indicators of limbic arousal requires comment. This result, which we have noted is common, argues against the utility of a concept of general arousal. But that is a negative conclusion. A more constructive way to view this fact is to suggest that each behavioral and physiological system serves a hierarchy of incentive conditions, as a retinal cone responds to a band of wave lengths around the one to which it is most sensitive. Although the adrenal cortex and motor system seem to be preferentially sensitive to increases in uncertainty that cannot be coped with effectively, changes in heart rate and heart rate variability are less sensitive to unassimiliable uncertainty. One reason is that the primary function of the heart's activity is metabolic; heart rate and blood pressure must meet bodily needs first. Although unfamiliarity and challenge often lead to an increase in heart rate and a decrease in variability, this association is less reliable than the inhibition of motor behavior or the increase in cortisol. Further, the heart seems to be sensitive to the amount of effort, whether physical or mental, invested in a task (Kagan and Reznick, 1984). The relation between cognitive effort and heart rate and variability is more reliable than the relation between behavioral inhibition and these two cardiac variables. The pupil, too, appears to be maximally responsive to amount of cognitive effort and a less sensitive index of uncertainty. Increased pupil size to tasks was far less differentiating of inhibited and uninhibited children than either cortisol or heart rate. These facts, which have been replicated by other investigators, imply that we must differentiate between the state of uncertainty to unfamiliarity and the state generated by cognitive effort.

Finally, we believe it is necessary to differentiate between children who are endogenously prone to uncertainty and limbic arousal and those who become prone to uncertainty later in life as a result of stressful experiences. Some children who were classified as temperamentally uninhibited originally, but became inhibited later had experienced the stresses of divorce, marital quarreling, death of a grandparent, or severe mental illness in a family member after the original evaluation. A small proportion of these children also developed a high and stable heart rate; and some showed high cortisol levels. It is useful to classify these children in a different way because their histories and the origins of their biological characteristics differ from those of the inhibited children, and because the likelihood of reversibility may be greater for this group.

REFERENCES

Adamec, R.E., & Stark-Adamec, C. (1986). Limbic hyperfunction, limbic epilepsy, and interictal behavior. In B.K. Doane and K.E. Livingston (Eds.), The Limbic System. New York: Raven, 129-145.

Aggleton, J.P. & Passingham, R.E. (1981). Syndrome produced by lesions of amygdala in monkeys (Macaca mulatta). Journal of Comparative and Physiological Psychology, 95, 961-977.

Aston-Jones, G. & Bloom, F.E. (1981). Norepinephrine-containing locus coeruleus neurons in behaving rats exhibit pronounced responses to non-noxious environmental stimuli. The Journal of Neuroscience, 1, 887-900.

Bandura, A., Taylor, C.B., Williams, S.L., Mefford, I.N., & Barchas, J.D. (1985). Catecholamine secretion as a function of perceived coping self-efficiency. Journal Consulting Clinical Psychology, 53, 406-414.

Blizard, D.A. The Maudsley reactive and nonreactive strains. Behavioral Genetics, 1981, 11, 469-489.

Bronson, W.C. Toddlers' Behaviors with Age Mates. Norwood, New Jersey: Ablex, 1981.

Brush, F.R., Baron, S., Froehlich, J.C., Ison, J.R., Pellegrino, L.J., Phillips, D.S., Sakellaris, R.C., and Williams, V.N. Genetic differences in avoidance learning. In Rattus Norvegicus Journal of Comparative Psychology, 1985, 99, 60-73.

Charney, D.S., Heninger, G.R., & Breier, A. (1984). Noradrenergic function in panic anxiety. Archives of General Psychiatry, 41, 751-763.

Charney, D.S. & Redmond, D.E. (1983). Neurobiological mechanisms in human anxiety. Neuropharmacology, 22, 1531-1536.

Ciarenello, R.D. (1983). Neurochemical aspects of stress. In N. Garmezy and M. Rutter (Eds.), Stress, Coping, and Development in Children. New York: McGraw-Hill, 85-105.

Conley, J.J. (1985). Longitudinal stability of personality traits: A multitrait-multimethod-multioccassion analysis. Journal of Personality and Social Psychology, 49, 1266-1282.

Coster, W. (1986). Aspects of voice and conversation in behaviorally inhibited and uninhibited children. Unpublished doctoral dissertation. Harvard University.

Eysenck, H.J. (1982). Personality, Genetics, and Behavior. New York: Praeger.

Floderus-Myrhed, B., Pedersen, N & Rasmuson, I. (1980). Assessment of heritability for personality based on a short form of the Eysenck personality inventory. Behavioral Genetics, 10, 153-162.

Garcia-Coll, C., Kagan, J., & Reznick, J.S. (1984). Behavioral inhibition in young children. Child Development, 55, 1005-1019.

Gersten, M. (1986). The contribution of temperament to behavior in natural contexts. Unpublished doctoral dissertation, Harvard Graduate School of Education.

Gloor, P. (1978). Inputs and outputs of the amygdala: what the amygdala is trying to tell the rest of the brain. In K.E. Livingston and O. Hoznykiewicz (Eds.), Limbic Mechanisims. New York: Plenum, 189-209.

Goldsmith, H.H. (1983). Genetic influences on personality from infancy to adulthood. Child Development, 54, 331-355.

Goldsmith, H.H. & Campos, J.J. (1986). Fundamental issues in the study of early temperament: The Denver twin temperament study. In M.H. Lamb & A. Brown (Eds.), Advances in Developmental Psychology. Hillsdale, New Jersey: L. Erlbaum, 281-283.

Goldsmith, H.H. & Gottesman, I.I. (1981). Origins of variation in behavioral style. Child Development, 52, 91-103.

Gray, J.A. (1982). The Neuropsychology of Anxiety. Oxford: Clarendon Press.

Hitchcock, J., & Davis, M. (1986). Lesions of the amygdala, but not of the cerebellum or red nucleus block conditioned fear as measured with the potentiated startle paradigm. Behavioral Neuroscience, 100, 11-22.

Jung, C.G. (1924). Psychological Types. New York: Harcourt Brace and Company.

Kagan, J. and Moss, H.A. Birth to Maturity. New York: John Wiley, 1962.

Kagan, J. & Reznick, J.S. (1984). Task involvement and cardiac response in young children. Australian Journal of Psychology, 36, 135-147.

Kagan, J., Reznick, J.S., Clarke, C., Snidman, N., & Garcia-Coll, C. (1984). Behavioral inhibition to the unfamiliar. Child Development, 55, 2212-2225.

Karoum, F. (1983). Mass fragmentography in the analysis of biogenic amines: A clinical, physiological, and pharmacological evaluation. In S. Parvez, T. Nagatsu, I. Nagatsu, and H. Parvez (Eds.), Methods in Biogenic Amine Research. Amsterdam: Elsevier, 237-255.

Levine, S., Coe, C.L., Smotherman, W.P., & Kaplan, J.N. (1978). Prolonged cortisol elevation in the infant squirrel monkey after reunion with mother. Physiology and Behavior, 20, 7-10.

Lieberman, P. (1961). Perturbations in vocal pitch. Journal of Acoustical Society of America, 33, 597-603.

McDonald, K. (1983). Stability of individual differences in behavior in a litter of wolf cubs. Journal of Comparative Psychology, 97, 99-106.

Nesse, R.M., Curtis, G.C., Thyer, B.A., McCann, D.S., Huber-Smith, M., & Knopf, R.F. (1985). Endocrine and cardiovascular responses during phobic anxiety. Psychosomatic Medicine, 47, 320-332.

Reiser, M.F. (1984). Mind, Brain, Body. New York: Basic Books.
Reznick, J.S., Kagan, J., Snidman, N., Gersten, M., Baak, K., & Rosenberg, A. (1986). Inhibited and uninhibited behavior: A follow-up study, Child Development, 51, 660-680.

Sarter, M., & Markowitsch, H.J. (1985). Involvement of the amygdala in learning and memory. Behavioral Neuroscience, 99, 342-380.

Schneiderman, N. & McCabe, P.M. (1985). Bio-behavioral responses to stressors. In T.M. Field, P.M. McCabe & N. Schneiderman (Eds.), Stress and Coping. Hillsdale, New Jersey: L. Erlbaum, 13-61.

Scott, J.P. and Fuller, J.L. Dog Behavior: The Genetic Basis. Chicago: University of Chicago Press, Phoenix Edition, 1974 (originally published 1965).

Snidman, N. (1984). Behavioral restraint and the central nervous system. Unpublished doctoral dissertation. University of California, Los Angeles.

Stevens, K.N. & Hirano, M. (Eds). (1981). Vocal Fold Physiology. Tokyo: University of Tokyo Press.

Stevenson-Hinde, J., Stillwell-Barnes, R. & Zunz, M. (1980). Subjective assessment of rhesus monkeys over four successive years. Primates, 21, 66-82.

Suomi, S.J. The development of affect in rhesus monkeys. In N.A. Fox and R.J. Davidson (Eds.), The Psychology of Affective Development. Erlbaum, 1984, 119-159.

Tennes, K., Downey, K., & Vernadakis, A. (1977). Urinary cortisol excretion rates and anxiety in normal one-year-old infants. Psychosomatic Medicine, 39, 178-187.

Thomas, A. and Chess, S. Temperament and Development. New York: Brunner/Mazel, 1977.

Tsuda, A., & Tanaka, M. (1985). Differential changes in noradrenaline turnover in specific regions of rat brain produced by controllable and uncontrollable shocks. Behavioral Neuroscience, 99, 802-817.

Turner, B.H., Mishkin, M., & Knapp, M. (1980). Organization of the amygdalopetal projections from modality-specific cortical association areas in the monkey. Journal of Comparative Neurology, 19, 515-543.

Walker, R.F. (1984). Salivary cortisol determinations in the assessment of adrenal activity. In D.B. Ferguson (Ed.), Steriod Hormones in Saliva. Basel:Karger, 33-50.

FOOTNOTES

1. This research was supported in part by a grant from the John D. and Catherine T. MacArthur Foundation and the National Institute of Mental Health.

PHYSIOLOGICAL AND BEHAVIORAL ASPECTS OF STRESS IN ADOLESCENCE

Elizabeth J. Susman
The Pennsylvania State University

Editha D. Nottelmann
National Institute of Mental Health

Lorah D. Dorn
The Pennsylvania State University

Gale Inoff-Germain
National Institute of Mental Health

George P. Chrousos
National Institute of Child Health & Human Development

Adolescence is considered an inherently stressful period of development. The stressors that adolescents experience are thought to arise from both social-environmental and physiological sources. The interaction between these sources of stress and adolescent development is the focus of this chapter. Two issues will be addressed. The first is concerned with puberty related-hormone and physical growth changes that may precipitate or be affected by heightened stress reactivity of adolescents, in general. The second is concerned with individual differences in physiological and behavioral reactivity to specific environmental stressors. The theoretical rationale for the importance of these issues in the psychological development of adolescents also is discussed.

STRESS, PUBERTY-RELATED HORMONE CHANGES AND ADOLESCENT DEVELOPMENT

Adolescence is characterized by many socio-environmental stressors and many behavioral and psychiatric disorders. Other subtle mood and behavior changes also occur during adolescence. The mechanisms involved in these disorders and behavior changes, particularly during early adolescence, are thought to be linked to the physiological changes that occur at puberty. Behavior at puberty may be altered by two related processes. First, behavior may be altered directly through hormones acting on the central nervous system. Second, behavior may be altered indirectly by physical maturation changes. Changes in height, weight and body proportions are a stimulus for adolescents themselves and others of the adolescent's more mature status. This new status means that adolescents are exposed to increasing demands for more independent and responsible behavior. In brief, at puberty, changes in hormone levels may alter the brain biochemical milieu such that behavior is altered. Changes in hormone levels also are responsible for physical changes which, in turn, may affect behavior and adjustment.

At a folk wisdom level, hormones as influences on adolescent behavior have been considered for decades. At an empirical level, hormonal influences on adolescent behavior have come under study only recently. When hormones were considered, pituitary-gonadal axis hormones, and their manifestations in physical maturation and reproductive function, were the hormones presumed to influence behavior. Specifically, gonadal steroids have been linked to changes in spatial abilities[1] and aggression[2] at puberty. Similarly, the tempo of physical maturation, the timing and rate of maturation, has been linked to many aspects of the social and emotional development of adolescents[3]. However, none of these studies have measured actual levels of hormones.

While the stressfulness of adolescence is emphasized in the literature, the effects of social stressors on the hypothalamic-pituitary-adrenal axis hormones generally have been ignored in nonclinical samples of adolescents. Adolescence is an especially important period of development for examining physiological aspects of stress because environmental stressors and physiological stressors may be considered integral to the experience of adolescents. Adrenal activation as a consequence of environmental and physiological stressors may have enduring influences on puberty-related, hormonal processes during adolescence.

Stress and gonadal functioning

A major physiological effect of stress is the suppression of gonadal functioning in both animals and humans[4]. Specifically, in animals, environmental stressors, which are accompanied by adrenal activation, has been shown to alter gonadal function[5,6]. For instance, in mice, immobilization or handling stress delayed puberty[7]. In male rats, immobilization stress for a six hour period for one day resulted in decreased testosterone levels[5]. However, seven or 30 hours of immobilization stress did not affect physical maturation of male rats. These latter findings also indicate that the effects of stress on gonadal function depend on its acute or chronic nature. Other studies show that continuous stimulation of the pituitary-adrenal axis by peripherally administered corticotropin-releasing factor (CRF) was accompanied by persistent elevation of circulating levels of adrenocorticotropin hormone (ACTH) and corticosteroids and decreased levels of luteinizing hormone, testosterone and prolactin[8].

In human adults, elevated cortisol levels also are related to suppressed gonadal functioning. Specifically, higher levels of cortisol are related to lower levels of luteinizing hormone in males and females[9,10] and lower levels of testosterone in males[11,12]. In human adolescents, evidence for the effects of environmental stressors on gonadal function can be gleaned from three different types of stress-related problems: (1) delayed menarche as a potential consequence of rigorous exercise stress[13], (2) psychosocial dwarfism as a potential consequence of family and environmental stressors[14] and, (3) anorexia nervosa[15] and accompanying alterations in the menstrual cycle as a potential consequence of interpersonal stress. Delayed menarche as a sequela of rigorous exercise and dieting has been observed in ballet dancers[13]. More dance than nondance students were late maturers[16]. Furthermore, the dance students had more eating concerns and lower scores on family relationship and impulse control scales than the comparison group. Psychosocial dwarfism is characterized by arrestation in growth in children who experience extreme emotional and psychological deprivation[14]. In these children, bone age is retarded and weight for height is low. Growth catch- up is rapid when the children are moved to a less stressful environment. Anorexia nervosa is an eating disorder that is accompanied by a number of aberrations in hormone levels. A manifestation of these abnormalities is the absence of regular menstrual cycles indicating suppression of the gonadal axis. Cortisol levels also are high in patients with anorexia nervosa[15]. Higher levels of cortisol may reflect partially the emotional stress that accompanies the disorder.

In all three of these conditions, there is a delay or arrest of gonadal hormone-regulated processes. Furthermore, in these conditions, environmental or emotional stressors appear to precipitate or co-occur with altered hormonal status. However, the mechanisms involved in linking environmental stressors, adrenocortical activation and gonadal function in these and other conditions are not well understood. The implications of these effects of rather extreme stressors and hormonal alterations for normal adolescents are that even less extreme environmental stressors similarly may alter gonadal functioning and, in turn, the tempo of pubertal development.

HORMONES AND SOCIAL AND EMOTIONAL DEVELOPMENT AT PUBERTY

In a longitudinal study, we examined the interrelations among hormones of gonadal and adrenal origin, physical maturation and social-emotional development in young adolescents. The study was conducted by investigators in the Laboratory of Developmental Psychology, National Institute of Mental Health, and the Developmental Endocrinology Branch, National Institute of Child Health and Human Development.

Normal pubertal development consists of adrenarche, occurring between ages 6 and 8, and gonadarche, occurring between ages 8 and 13. Based on the interdependencies of most physiological systems as well as the centrality of adrenal functioning to other physiological processes, we believed that hormones of adrenal origin, as well as hormones of gonadal origin, were important to consider as possible influences on the behavioral changes that occur in young adolescents.

We studied a sample of 108 healthy adolescents: 56 boys, ages 10-14 years, and 52 girls, ages 9-14 years. There were three assessments, at 6-month intervals, of physical maturity, endocrine status, and psychological and behavioral functioning. The adolescents and their parents were seen during a 4-hour evening visit in a home-like research laboratory where the adolescents were assessed on psychological dimensions hypothesized to be related to hormone changes at adolescence. These dimensions included aggression, self-image, moods, cognition, behavior problems and qualitative aspects of parent-adolescent interactions. The adolescents also were seen during a 2-hour morning visit at an outpatient clinic. During the clinic visit, a physical examination was done to assess stage of pubertal development (Tanner criteria): breast stage and pubic hair stage for girls, genital stage and pubic hair stage for boys[17,18] height and weight, as well as head circumference for both sexes and testicular volume (Prader orchidometer) for boys. For the hormone level assessment, blood samples were collected during the second visit, which took place between 8 and 10 A. M. to minimize the effects of diurnal variations in the levels of the hormones that were measured. For each adolescent, a sample was drawn at 0, 20, and 40 minutes. The mean for the three samples was used in all subsequent analyses.

Radioimmunoassays were used to evaluate serum hormone levels of gonadotropins: luteinizing hormone (LH) and follicle stimulating hormone (FSH); gonadal steroids: testosterone (T), estradiol (E2) and the computed testosterone to estradiol ratio (T/E2), adrenal androgens: androstenedione (delta 4-A), dehydroepiandrosterone (DHEA) and dehydroepiandrosterone sulphate (DHEAS) and a glucocorticoid, cortisol. We also evaluated levels of testosterone-estradiol binding globulin (TeBG) and cortisol binding globulin (CBG) using competitive protein binding assays. Additional information on the radioimmunoassays and binding assays and characteristics of the sample are described elsewhere[19].

Our findings suggest links between behavior in young adolescents and hormones of gonadal and adrenal origin. We had hypothesized specific hormone-behavior relations (e.g. testosterone and aggressive behavior), but found these relations to be far more complex than the one-to-one relations that we had anticipated. Behavior tended to be related to a profile of asynchronies in hormone levels, that is, lower levels of some hormones and higher levels of other hormones. Relations between hormone levels and aggressive behavior and psychosocial adjustment will be used as two examples to illustrate the general pattern of results. The findings discussed are based on cross-sectional data from the first time of assessment.

Hormones and aggressive behavior

Aggressive behavior increases at puberty in many species[20] when there is a concomitant increase in androgen levels. The link between androgen levels and aggressive behavior in animals consistently is reported for males and sometimes for females[21,22]. In male human adolescents, relations between testosterone and aggressive behavior have been reported by Olweus and colleagues[2]. Testosterone appeared to have a direct causal influence on provoked (versus unprovoked) aggressive behavior. Testosterone also appeared to have an indirect causal influence on unprovoked aggression, with low frustration tolerance as a mediating factor. Olweus[23] suggests that a high level of testosterone makes a boy impatient and irritable which, in turn, may lead to unprovoked and destructive aggression.

We expected that higher testosterone levels would be related to higher levels of aggressive behavior in our sample of young adolescents[24]. However, the relations between testosterone, as well as other hormones, and aggressive behavior may be moderated by emotions, such as feelings of irritability, anxiety, and sadness. Thus, we hypothesized that higher levels of hormones, primarily androgens, would be related to higher levels of aggressive behavior, particularly if higher levels of androgens were accompanied by negative emotions.

The measures of emotional dispositions were adolescent self-ratings of their feelings (e.g. anxious and angry) and emotional tone (depressive affect) and impulse control, subscales of the Offer Self-Image Questionnaire for Adolescents[25]. The aggressive attributes were mother reports of delinquent (acting out) and aggressive behavior problems, subscales from the Child Behavior Checklist[26]. Rebellious and nasty behavior were based on mother ratings of the adolescent's moods and behaviors.

Our general pattern of findings for boys was that a profile of lower gonadal steroid levels and higher androstenedione levels was related to acting out and rebellious behavior. For instance, based on regression analyses, we found that lower E2, T/E2 ratio, TeBG and DHEAS were associated with more acting out behavior problems. Higher levels of androstenedione also were associated with more acting out behavior problems. Similarly, a profile of lower T/E2 ratio, lower TeBG and higher androstenedione levels was associated with more sad affect. Feeling anxious also was related to acting out. We found no relations for girls between hormones, emotions and aggressive behavior.

Hormones and adjustment

In another project based on the same study, Nottelmann and colleagues[27] examined the pattern of findings between psychosocial adjustment problems and indices of physical maturation; chronological age, pubertal status, and serum hormone levels. The behavioral measures used were the Offer Self-Image Questionnaire for Adolescents and the Child Behavior Checklist. For boys, in particular, Nottelmann and colleagues reported that adjustment problems were associated with a hormone profile that is characteristic of later maturation: relatively low sex steroid levels, or lower pubertal stage, and relatively higher androstenedione levels, frequently in conjunction with higher chronological age. That is, adjustment problems were higher in males with later maturation, lower pubertal stage for chronological age and lower gonadal steroid levels for chronological age. "Later maturation" is relatively later only in comparison to the other adolescents in the sample. Age of physical maturation for the adolescents in the study was well within the normal range.

There were fewer findings for girls than for boys. However, adjustment problems for girls were associated with higher levels of gonadotropins, lower levels of DHEAS and higher levels of androstenedione. This hormone profile sometimes was accompanied by lower pubertal stage.

STRESS AND ADRENAL AND GONADAL FUNCTIONING IN ADOLESCENTS

We found that emotions, aggressive behavior and adjustment problems were associated with a hormone profile of lower gonadal steroid levels frequently in combination with higher chronological age. This hormone and chronological age profile is typical of later maturers. We offer some tentative explanations for these findings.

A profile of lower gonadal steroid levels and negative emotions and problems in adjustment is consistent with the literature cited earlier reporting links between lower gonadal steroid levels and stress in both animals and humans. In the conditions discussed above (e.g. psychosocial dwarfism), the types of stress were more extreme than the undoubtedly relatively minor stressors that were experienced by the majority of adolescents in our sample. However, even minor stressors that are experienced chronically potentially may affect the tempo of puberty. Two possible pathways involving similar physiological mechanisms may be involved in linking stress and the tempo of puberty[28,29]. First, environmental stressors may stimulate adrenal activation, thereby suppressing the gonadal axis, in turn, producing relatively later physical maturation. The ongoing daily environmental stressors and adjustment problems experienced by adolescents may result in chronic adrenal activation. The resultant increases in CRH, ACTH and cortisol may suppress the gonadal axis at the gonadotropin or gonadal steroid levels. Suppression of the gonadal axis, in turn, may lead to delayed physical maturation and reproductive function. Second, later maturation and delayed reproductive function may act as a stressor, stimulating the adrenal axis, and, in turn, further suppressing the gonadal axis. Our cross-sectional findings do not allow us to make causal statement about the direction of influences between the different types of stressors. What appears most likely is that a vicious cycle between environment and physiological stressors may be characteristic of the development of adolescents. Hypotheses regarding the cycle of stressors, adrenocortical activation and gonadal functioning now are being examined in relation to nutritional stress (i.e. iron deficiency) in adolescents[30].

We also found that higher levels of androstenedione, particularly in boys, were associated with aggressiveness and other problems of adjustment. Consistently, higher levels of androstenedione were related to negative aspects of behavior. The mechanisms whereby androstenedione may affect behavior are unknown. Androstenedione generally is considered a weak androgenic substance with low binding affinity for the androgenic receptor. However, since androstenedione is an androgen, it is conceivable that it may have behavioral effects similar to those described for testosterone and behavior. That is, higher levels of androstenedione may be related to irritability, impulsivity or other emotions which, in turn, predispose adolescents to problem behavior.

Alternatively, higher androstenedione levels and negative aspects of behavior may reflect adrenal activation in response to environmental stressors, to which the poorly adjusted adolescent is likely to be highly sensitive. While androstenedione generally is not considered a "stress hormone", its secretion is influenced by ACTH in a fashion parallel to that of cortisol. Androstenedione may provide a window on degree of neuroendocrine arousal to chronic or acute environmental stressors similar to that of cortisol. In our study, poorly adjusted adolescents may have been more reactive physiologically to the acutely challenging situation, the clinic visit and the phlebotomy procedure, than the well adjusted child. Thus, relations between androstenedione levels and aggressive behavior and adjustment problems may indicate higher degrees of arousal in poorly adjusted adolescents in response to the challenging situation.

While our interpretation of relations between stress, hormones and adjustment may be appropriate for boys, it appears to be less appropriate for girls. In general, we are finding fewer and less consistent hormone-behavior relations for girls than for boys. The most obvious interpretation for lack of findings for girls is that menstrual cycle variations in hormone levels introduce error into the data. While this interpretation may be partially true, it seems overly simplistic. Even in toddlers, hormone-behavior relations were more frequent in boys than in girls in the Stanford Longitudinal Study[31]. Thus, it appears that gender differences in behavior emerge well before the onset of puberty and menstrual cycle influences in hormones. Perhaps pre- and perinatal organizational influences of hormones and socialization practices in the early childhood years together contribute to these sex differences.

INDIVIDUAL DIFFERENCES IN REACTIVITY TO STRESS

Adolescents potentially experience multiple social stressors: rejection by peers, forced independence from parents and pressure to achieve in increasingly competitive academic endeavors. Since the majority of adolescents are exposed to these stressors, but only a minority of adolescents actually exhibit serious moods and behavior problems, one would expect large individual differences in how adolescents react to and cope with these social stressors In general, individual differences in physiological and behavioral reactivity to stressful events are important for three reasons[32]. Heightened reactivity may be involved in the development of disease (e.g. cardiovascular and gastrointestinal disorders). Heightened physiological and behavioral reactivity also may identify individuals who are at risk for the development of anxiety, depression, or other affective disorders. Finally, heightened reactivity may reflect and influence psychosocial adjustment, including the quality of relationships with peers and family. Thus, adolescents' behavioral and physiological reactions to environmental stressors may reflect their sensitivity or resilience to stress, in general. In highly stress-reactive adolescents, adjustment problems or more serious psychiatric problems may ensue. Our earlier discussion of stress and gonadal functioning included the notion that environmental stressors may affect adrenocortical activation, which in turn may affect the onset and tempo of puberty and therefore, behavior. In addition, rapidly changing hormone levels also were considered as influences on behavior at puberty. While the problem of the interplay between gonadal hormone changes and adrenal activation is a fundamental one for understanding stress in adolescents, there are several other questions to be asked in addition to examining this larger problem. For instance, for adolescents in general, we know neither what situations are particularly stressful for them nor do we know their physiological and behavioral reactions to these stressors. For example, is there an association between physiological and behavioral reactivity to a known environmental stressor?

We examined the physiological and behavioral reactivity of adolescents to what may be considered an emotional and physical stressor, the visit to our outpatient clinic and the phlebotomy procedure. Cortisol levels and behavioral distress were used as indices of physiological and behavioral reactivity, respectively. The association between cortisol levels and behavior in challenging and uncertain environmental circumstances has been documented in primates[32], human infants[33,34,35] and young children[36]. Thus, we expected that cortisol levels and distress behavior also should be related in young adolescents.

To assess physiological reactivity, cortisol levels were obtained at three, 6-month intervals, as part of the larger study described earlier. Levels were obtained from three samples drawn at 20-minute intervals between the hours of eight and ten A. M. We inserted a heparin lock after the first stick so as to avoid the trauma of additional sticks.

To assess behavioral reactivity, we observed adolescents for distress behaviors during the clinic visit. The observations consisted of recording each occurrence of 16 different distress behaviors (e.g. physical resistance, verbal support seeking, whining and crying). Figure 1 shows the design for assessing physiological and behavioral reactivity. The clinic visit was divided into five Distress Behavior

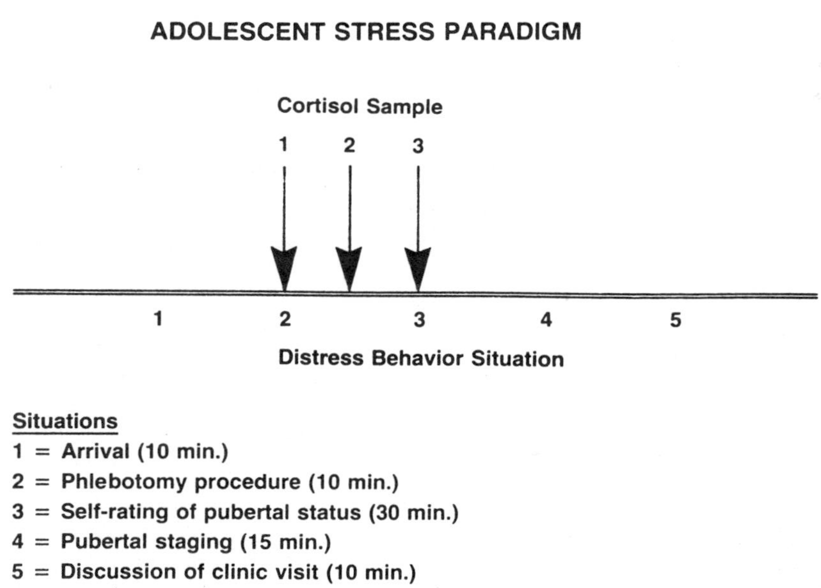

Figure 1. Distress Behavior Situations and timing of cortisol samples during outpatient clinic visit.

Situations: situation 1, arrival at the clinic; situation 2, a phlebotomy procedure and drawing blood sample 1; situation 3, drawing samples 2 and 3 and explanation of pubertal development; situation 4, physical examination; and situation 5, a feedback session for parent and adolescent.

There were significant correlations between cortisol levels and the number of distress behaviors during the phlebotomy procedure at the first clinic visit. A higher number of distress behaviors in situation 2, the phlebotomy procedure, was related significantly to higher cortisol levels in sample 1, $\underline{r} = .39$ $p < .05$, which was obtained concurrently. A higher number of distress behaviors in situation 2 also was related significantly to higher cortisol levels in sample 2, $\underline{r} = .47$, $p < .05$, which was obtained 20 minutes after the initial phlebotomy procedure. Number of distress behaviors was not related to cortisol level in sample 3, which was obtained 40 minutes later.

As in prior studies with humans and animals, the associations between distress behavior and cortisol levels were found in the most novel and challenging situation, the first clinic visit. As was also true in prior studies[37], we found dissociations between cortisol levels and distress behavior at the second and third time of assessment. One interpretation of these findings is that organisms may need novel environmental stimuli for the association between physiological and behavioral reactivity[38]. After the initial novel clinic visit, adolescents may have learned strategies for coping with the initially challenging situation.

Coping implies a strategy used by individuals to deal with threatening situations that enables them to behave more effectively[39]. Success of a coping strategy has been evaluated in two ways[38]: (a) At the behavioral level - does the behavior allow the individual to tolerate a stressful situation without debilitating anxiety or depression? (b) At the physiological level - do internal states (e.g. hormone levels) change? Based on these evaluation criteria, the adolescents' coping was successful following initial exposure to the phlebotomy procedure at the behavioral level but unsuccessful at the physiological level. There were statistically significant decreases in distress behaviors, but no significant decreases in cortisol levels over the one-year period.[40].

The importance of behavioral coping in adolescents in linked to sociocultural expectations of the behavior of adolescents. A developmental task of adolescence is to develop independent and mature strategies for dealing with challenging situations. Thus, even young adolescents are expected to exhibit stoic and adult-like behaviors, rather than the crying and resistant behavior exhibited by younger children, when confronted with medical procedures. Adolescents tend to show that they are aware of such expectations by acting embarrassed when they exhibit child-like behaviors. Therefore, after the initial clinic visit, adolescents may have adopted more mature behavior consistent with the environmental pressure to act in accordance with their newly acquired mature status. However, the lack of change in cortisol levels indicates that the clinic visits continued to be emotionally distressing.

The pattern of behavioral but not physiological coping has implications for understanding the reactions of adolescents to stressors in other situations as well as in our clinic situation. Reoccurring emotional arousal in the absence of behavioral distress may have negative long-term implications for adolescents. Adolescents may experience anxiety, but suppress distress behavior, because they feel they must. Parents or others may not spontaneously initiate nurturing and other supportive behaviors because they are unaware of the real feelings of the adolescents. Thus, parents and others may not do what may help to reduce and eventually alleviate anxiety in similar threatening situations. The cumulative effect of acute anxiety in challenging situations may lead to the development of chronic anxiety or depression.

SUMMARY AND CONCLUSIONS

Gonadal and adrenal hormone levels appear to be linked to the stressors experienced by young adolescents. Adjustment problems were accompanied by a profile of lower gonadal steroids and higher adrenal androgen levels, primarily androstenedione. Later gonadal maturation may be a result of stress suppressing the reproductive axis. Higher levels of androstenedione may be indicative of chronic levels of stress. However, the findings for androstenedione are complicated by the fact that androstenedione was related to cortisol only in males. Furthermore, androstenedione as a weak androgenic has low potential for affecting behavior, directly.

Cortisol levels were related to the frequency of distress behavior in a challenging situation. The relations disappeared with experience in the setting. These findings are consistent with prior animal and human studies. However, while distress behavior in a challenging situation decreased over the one-year period, cortisol levels did not. Sustained physiological arousal in a challenging situation may have long-term implications for the mental health of adolescents. A question for further exploration is whether individual differences in reactivity in one challenging situation, like the clinic visit, is predictive of reactivity in other settings.

Adolescence appears to be an ideal period of development in which to examine the interactions between environmental and physiological causes and sequelae of stress. It is characterized by measurable changes in hormonal status and physical maturation and behavior. Studying the intricate interactions between these two sets of changes has only just begun. A larger question yet to be examined is whether interactions between hormones and behavior are unique to adolescence or whether they are indicative of hormone-behavior processes characteristic of the entire life-span.

Adjustment and social stressors, adrenal activation, and reproductive maturation may constitute a "vicious" cycle of interrelated factors during adolescence. Adjustment problems could cause activation of the adrenal glands which would cause gonadal suppression and later maturation. The latter could constitute an added stressor reentering the cycle and potentiating the "abnormality". In our population of normal adolescents, this cycle is obviously active within a normal range. In these adolescents stress arising from either endogenous or exogenous sources was not a debilitating force. Rather, they fell within a normal range with various degrees of adjustment, adrenal activation and gonadal maturation. The relations that were uncovered among adjustment, stress, adrenal function and gonadal functioning can be explained on the basis of known interactions.

References

1. D. P. Waber, M. B. Mann, J. Merola, P. M. Moylan, Physical maturation rate and cognitive performance in early adolescence: A longitudinal examination, Dev Psych, 21:666 (1985).
2. D. Olweus, A. Mattsson, D. Schalling and H. Low, Testosterone, aggression, physical and personality dimensions in normal adolescent males, Psychosom Med. 42:253 (1980).

3. J. Brooks-Gunn, A. C. Petersen and D. Eichorn, The study of maturational timing effects in adolescence. Special issue on "Pubertal Timing," J Youth Adolesc. 14 (1985).
4. H. Selye, "Stress in Health and Disease," Butterworths, Reading, MA (1976).
5. R. Collu, W. Gibb and J. R. Ducharme, Effects of stress on the gonadal function, J Endocrinol Invest. 7:529 (1984).
6. G. Charpenet, Y. Tache, M. G. Forest, F. Haour, J. M. Saez, M. Bernier, J. R. Ducharme and R. Collu, Effects of chronic intermittent immobilization stress on rat testicular androgenic function, Endocrinol., 109:1254 (1981).
7. G. Sieck, J. A. Ramaley, Effects of early handling upon puberty: Correlations with adrenal stress responsiveness. Physiol Behav., 15:487 (1975).
8. C. Rivier and W. Vale, Effect of the long-term administration of corticotropin-releasing factor on the pituitary-adrenal and pituitary-gonadal axis in the male rat. J Clin Invest. 75:689 (1985).
9. M. Sakakura, K. Takebe, and S. Nakagawa, Inhibition of luteinizing hormone secretion induced by synthetic LRH by long-term treatment with glucocorticoids in human subjects. J Clin Endocrinol Metab. 40:774 (1975).
10. J. R. Sowers, B. F. Rice, and S. Blanchard, Effect of dexamethasone on luteinizing hormone and follicle stimulating hormone responses to LHRH and to clomiphene in the follicular phase of women with normal menstrual cycles. Horm Metab Res. 11:478 (1979).
11. D. C. Cummings, M. E. Quigley and S. S. C. Yen, Acute suppression of circulating testosterone levels by cortisol in men, J Clin Endocrinol Metab. 57:671 (1983).
12. G. Schaison, F. Durand and I. Mowszowicz, I. Effect of glucocorticoids on plasma testosterone in men. ACTA Endocrinologica. 89:126 (1978).
13. M. P. Warren, The effects of exercise on pubertal progression and reproductive function in girls, J Clin Endocrinol Metab. 51:1150 (1980).
14. G. F. Powell, J. A. Brasel, A. Raiti and R. M. Blizzard, Emotional deprivation and growth retardation simulating idiopathic hypopituitarism: II Endocrinologic evaluation of the syndrome, N Engl J Med. 276:1279 (1967).
15. P. Gold, H. Gwirtzman, P. C. Avgerinos, L. K. Nieman, W. T. Gallucci, W. Kaye, D. Jimerson, M. Ebert, R. Rittmaster, D. L. Loriaux and G. P. Chrousos, Abnormal hypothalamic-pituitary-adrenal function in anorexia nervosa. N Engl J Med. 314:1335 (1986).
16. J. Brooks-Gunn, Pubertal processes and girls' psychological adaptation, in: "Biological-psychological interactions in early adolescence," R. M. Lerner, T. T Fochs, eds., Lawrence Erlbaum Associates, Hillsdale, NJ (1987).
17. W. A. Marshall and J. M. Tanner, Variations in the pattern of pubertal changes in girls, Arch Dis Child. 44:291 (1969).
18. W. A. Marshall and J. M. Tanner. Variations in the pattern of pubertal changes in boys, Arch Dis Child. 45:13-23 (1970).
19. E. D. Nottelmann, E. J. Susman, L. D. Dorn, G. Inoff-Germain, G. F. Cutler, Jr., D. L. Loriaux and G. P. Chrousos, Developmental processes in early adolescence: Relations among chronologic age, pubertal stage, height, weight, and hormone levels, J Adolesc Health Care. 8:492-504 (1987).
20. G. E. Weisfeld and J. M. Berger, Some features of human adolescence viewed in evolutionary perspective, Hum Develop. 26:121 (1983).

21. B. E. Eleftheriou, R. L. Sprott, "Hormonal correlates of behavior: Vol. 1, A lifespan view," Plenum Press, New York (1975).
22. M. F. Bouissou, Androgens, aggressive behavior and social relationships in higher mammals, Horm Res. 18:43 (1983).
23. D. Olweus, Aggression and hormones: Behavioral relationships with testosterone and adrenaline, in: "Development of antisocial and prosocial behavior: Research, theories, and issues," D. Olweus, J. Block, M. Radke-Yarrow, eds., Academic Press, Orlando (1986).
24. E. J. Susman, G. Inoff-Germain, E. D. Nottelmann, D. L. Loriaux, G. B. Cutler, Jr. and G. P. Chrousos, Hormones, emotional dispositions, and aggressive attributes in young adolescents, Child Develop. 58:1114-1134 (1987).
25. D. Offer, E. Ostrov and K. I. Howard, "The Offer Self-Image Questionnaire for Adolescents. A Manual." Michael Reese Hospital, Chicago (1977).
26. T. M. Achenbach and C. Edelbrock, "Manual for the Child Behavior Checklist and Revised Child Behavior Profile," Department of Psychiatry, University of Vermont, Burlington (1983).
27. E. D. Nottelmann, E. J. Susman, G. Inoff-Germain, G. B. Cutler, Jr., D. L. Loriaux and G. P. Chrousos, Developmental processes in early adolescence: Relations between adolescent adjustment problems and chronologic age, pubertal stage and puberty-related serum hormone levels, J. Pediatr. 110:473-478 (1987).
28. G. P. Chrousos, Hypothalamic-pituitary-adrenal stress: Is there a vicious cycle? in "Coping with uncertainty: Biological, behavioral, and developmental perspectives," D. S. Palermo, ed., Lawrence Erlbaum Associates, Hillsdale, NJ. In press.
29. E. D. Nottelmann, E. J. Susman and G. P. Chrousos, Hormones and adjustment in puberty, in "Adolescence and puberty," J. Bancroft, ed., Kinsey Institute Symposia Series (in preparation).
30. A. C. Petersen, E. J. Susman and J. L. Beard, The development of coping responses during adolescence: Endocrine and behavioral aspects, in "Coping with uncertainty: Biological, behavioral, and developmental perspectives," D. S. Palermo, ed., Lawrence Erlbaum Associates, Hillsdale, NJ. In press.
31. H. M. Marcus, E. E. Maccoby, E. N. Jacklin and C. H. Doering, Individual differences in mood in early childhood: Their relation to gender and neonatal sex steriods, Dev. Psychobio. 18:327 (1985).
32. S. J. Suomi, New studies of social development in rhesus monkeys, Verh Dtsch Zool Ges. 77:37 (1984).
33. M. Gunnar, R. Fisch, S. Korsvik and J. Donhowe, The effects of circumcision on serum cortisol and behavior, Psychoneuroendocrinology. 6:269 (1981).
34. K. Tennes and D. Carter, Plasma cortisol levels and behavioral states in early infancy, Psychosom Med. 35:121 (1973).
35. K. Tennes, K. Downey and A. Vernadakis, Urinary cortisol excretion rates and anxiety in normal 1-year-old infants, Psychosom Med. 47:451 (1977).
36. K. Tennes and M. Kreye, Children's adrenocortical responses to classroom activities in elementary school, Psychosom Med. 47:451 (1985).
37. M. Gunnar, D. Marvinney, J. Isensee, Coping with uncertainty: New models of the relations between hormonal, behavioral and cognitive processes, in: "Coping with uncertainty: Biological, behavioral, and developmental perspectives," D. S. Palermo, ed., Lawrence Erlbaum Associates, Hillsdale, NJ. In press.

38. L. Levine, A psychobiological approach to the ontogeny of coping, _in_: "Stress, coping and development in children," N. Garmezy and M. Rutter, eds., McGraw-Hill, New York (1984).
39. D. A. Hamburg, B. Hamburg and S. DeGoza, Adaptive problems and mechanisms in severely burned patients, _Psychiatry_. 16:1 (1953).
40. E. J. Susman, L. D. Dorn, E. D. Nottelmann, G. Inoff-Germain and G. P. Chrousos, Cortisol and behavioral reactivity in young adolescents (in preparation).

PROLACTIN AND GROWTH HORMONE SECRETION IN STRESS

Seymour Reichlin

Professor of Medicine, Tufts University School of Medicine; Chief, Endocrine Division, New England Medical Center, Boston, MA 02111

Prolactin (PRL) has long been recognized to be an essential hormone for the regulation of milk production in many species of animals. Even the earliest bioassays of this hormone were sensitive enough to demonstrate that secretion was increased after suckling, and that this effect was mediated through a specific neurogenic reflex (1,2). Reflex suckling-induced PRL release is readily illustrated in normal post partum women (Fig.1.).

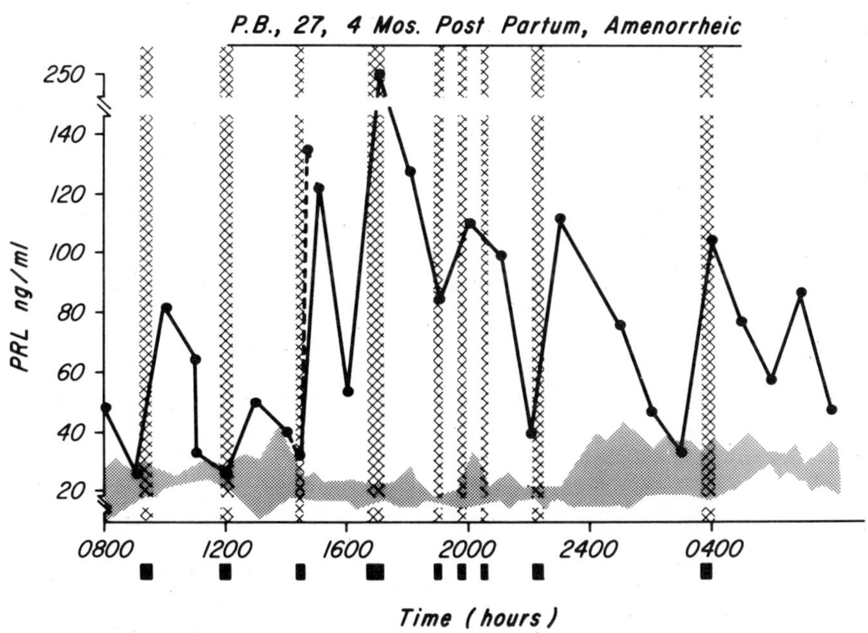

Fig. 1: The chief physiological function of prolactin is the stimulation of milk production which requires suitable hormonally mediated breast development. Prolactin-stimulated milk secretion is maintained by the suckling reflex initiated by nipple stimulation. This figure illustrates a 24-hour PRL secretory pattern in a post partum woman. The vertical bars indicate a suckling episode. The shaded area is the range of values in 5 non-pregnant women.

But PRL is not the only hormone released after nipple stimulation. Oxytocin is also released, a hormone responsible for milk "let down", a phenomenon due to the constriction of breast acinar myoepithelial cells which propel milk toward the nipple. The oxytocin response can be conditioned, and triggered by the sound of a hungry baby's crying (Fig. 2); curiously, PRL responses are not similarly conditioned (3).

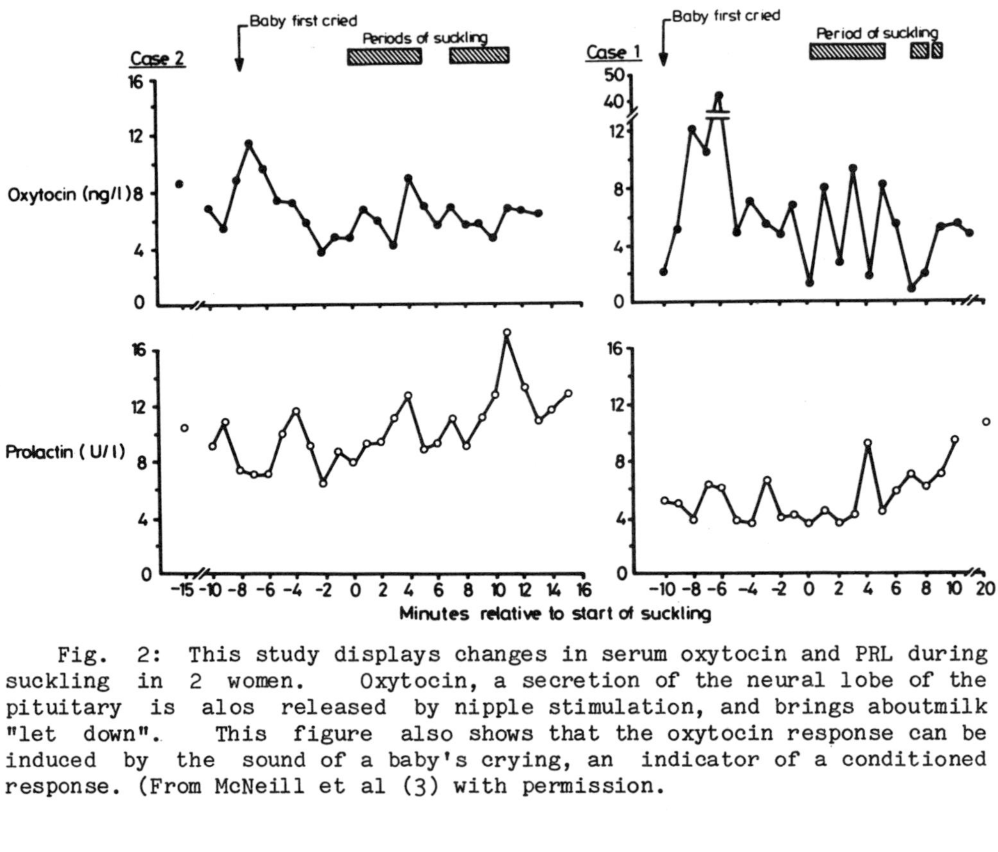

Fig. 2: This study displays changes in serum oxytocin and PRL during suckling in 2 women. Oxytocin, a secretion of the neural lobe of the pituitary is alos released by nipple stimulation, and brings aboutmilk "let down". This figure also shows that the oxytocin response can be induced by the sound of a baby's crying, an indicator of a conditioned response. (From McNeill et al (3) with permission.

These kinds of observations were completely predictable on the basis of earlier physiological and clinical work. Not predictable however was the finding that the release of PRL in man and rodents can be provoked by many kinds of stress. This realization did not come until a specific radioimmunoassay for human prolactin has been developed (4). By monitoring plasma levels of PRL it was soon recognized that PRL secretion was increased by exercise, general surgery, certain emotional states, myocardial infarction and hypoglycemia (Fig. 3), (1,2,5,6). Stress-PRL response is not entirely non-specific. Nesse and colleagues carried out an important study of this question (7). They assembled a group of patients with intense phobias focused on small animals such as snakes and rodents, and conducted therapeutic sessions with "flooding therapy". a procedure in which the individual was obliged to come into close contact with the phobic object. Despite the fact that the patients

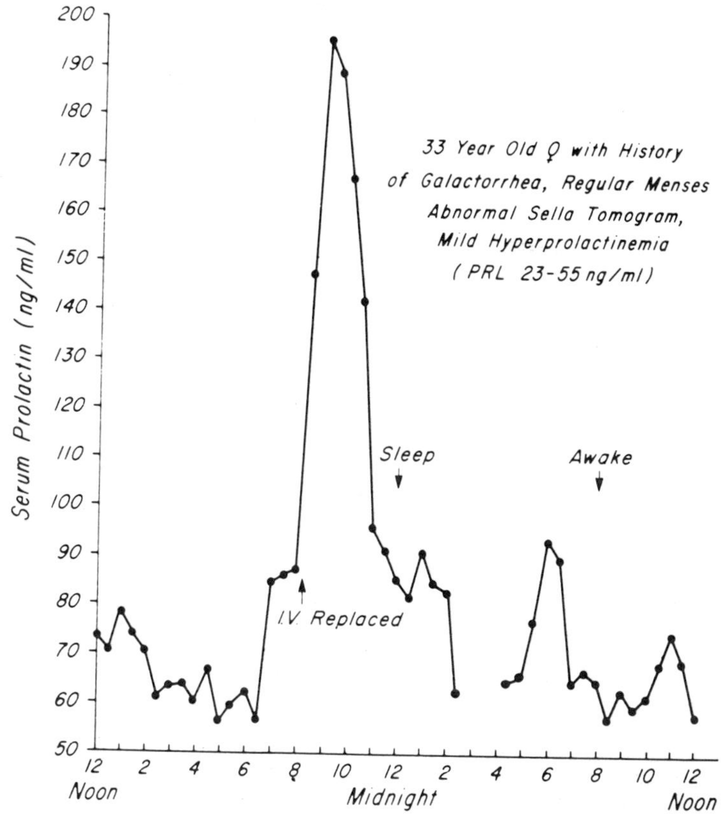

Fig. 3: Illustrated is the PRL secretory response to stress occuring incidental to a study of 24-hour PRL secretory patterns. The patient who later was shown to be in early pregnancy required the replacement of an indwelling venous catheter. Her angry response to the pain of a second venipuncture led to a marked increase in PRL level.

developed intense, often unbearable anxiety, they did not show an increase in PRL (Fig. 4). Indeed, the blood levels of this hormone fell progressively during both the control and the deconditioning session. I interpret this to mean that the anticipation of the session was a more specific stress than was the session itself.

The element of unpreparedness for stress (failure to be in coping "set") may be an important element in the PRL response to stress as has been reported for the pituitary-adrenal response (8). This was illustrated inadvertently in one of our patients who was undergoing 24 hour sampling of blood PRL as part of a study of galactorrhea-amenorrhea syndrome (Kapcala, Molitch & Reichlin, unpublished). At one point the catheter clogged, and it was necessary to replace it. The patient expressed intense anger with the staff and berated them; later analysis of plasma PRL revealed that an abrupt and dramatic rise in PRL levels had immediately followed this experience (Fig.3).

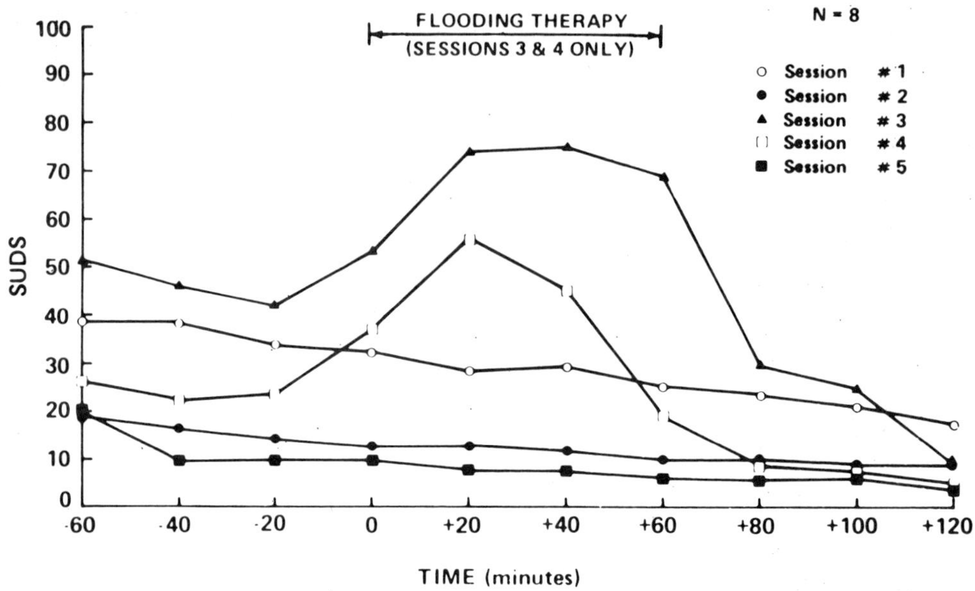

Fig. 4A: This figure illustrates part of a study of the effect of induced anxiety on PRL secretion. In this experiment, volunteers with known severe phobias were exposed in sessions 3 and 4 to the phobic object during the course of "flooding" therapy. In comparison with three control sessions, the flooding sessions produced very high SUDS scores, an index of severe anxiety. PRL response, sampled serially during this time are shown in Fig. 4B. (From Nesse et al, 7, with permission).

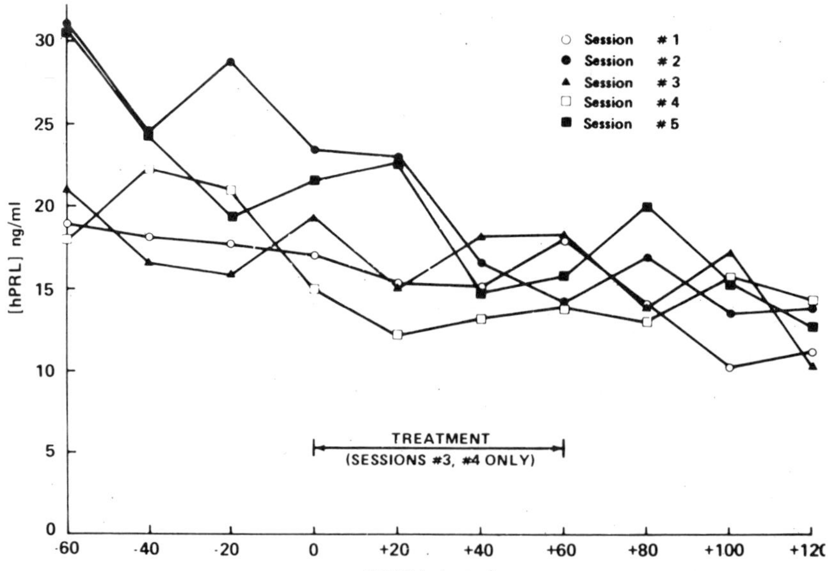

Fig. 4B: PRL levels were followed during the course of flooding therapy. Despite the occurence of severe, almost intolerable anxiety, PRL secretion was not enhanced. Initially mildly elevated levels fell.

In the rat, several stressful stimuli predictably increase blood PRL levels (9,10) these include forceful restraint, and ether anesthesia (Fig. 5, 6). This stress model has been used to analyze the neuroendocrine mechanisms that bring about PRL release (Fig. 7). The earliest hypothesis to be considered was that stress disinhibited the secretion of a normally acting prolactin inhibitory factor responsible for the normal tonic influence of the hypothalamus. The principal prolactin inhibitory factor is dopamine, a secretion of the arcuate-tuberoinfundibular complex (11,12), but GABA (13), and still other peptide PIF's have been described (14,15).

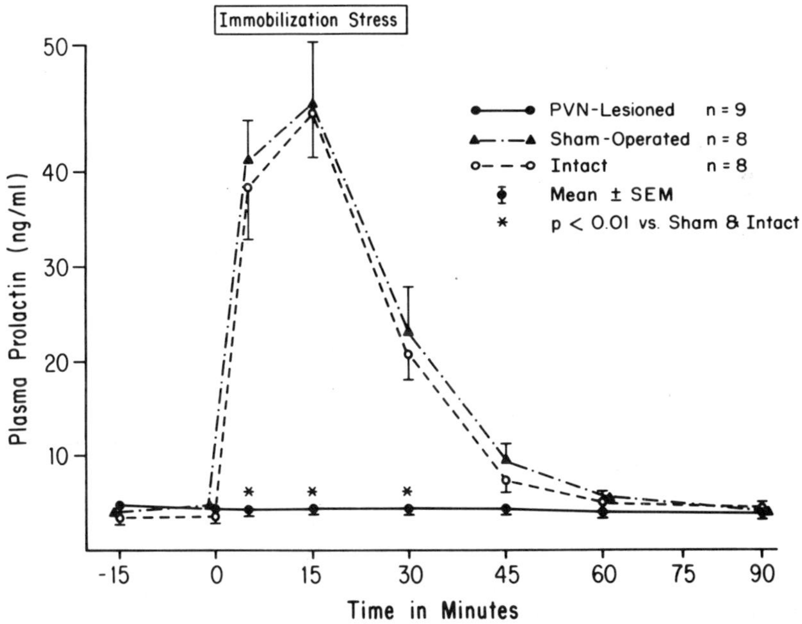

Fig. 5: Illustrated is the effect of forced restraint on serum PRL levels in rats with lesions of the paraventricular nucleus as compared with responses in sham operated and intact animals. The striking increase in intact animals is blocked completely by the lesions. Basal levels are unaltered. Since the lesions spared the dopaminergic tuberoinfundibular pathways, it appears that basal levels are determined by the latter system while the acute release in mediated by the paraventricular nucleus. (From Minamitani et al, 29, with permission).

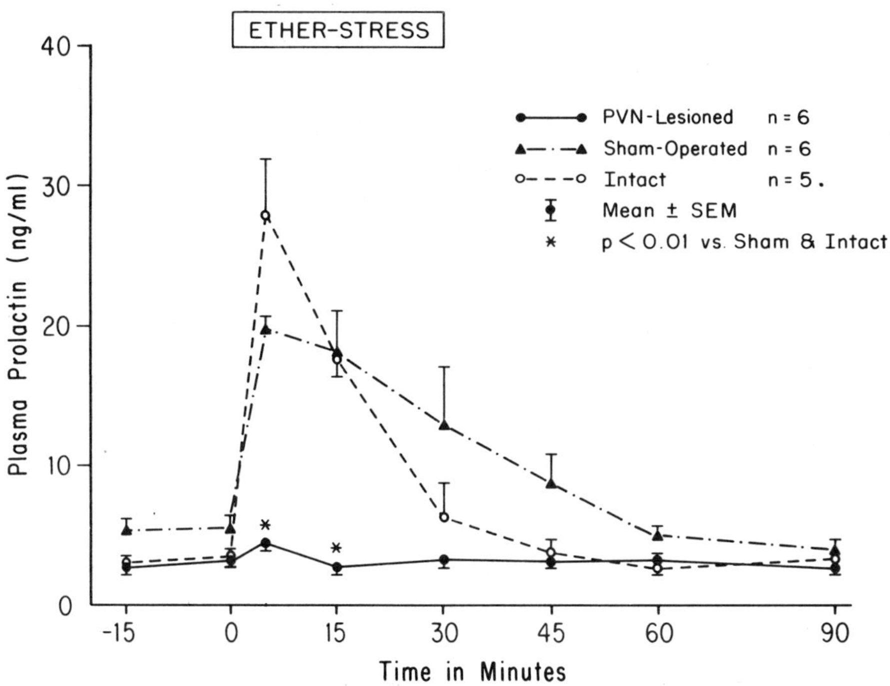

Fig. 6: Illustrated is the effect of exposure to ether vapor on serum PRL and the blocking effect of lesions of the paraventricular nucleus. (From Minamitani et al, 29, with permission).

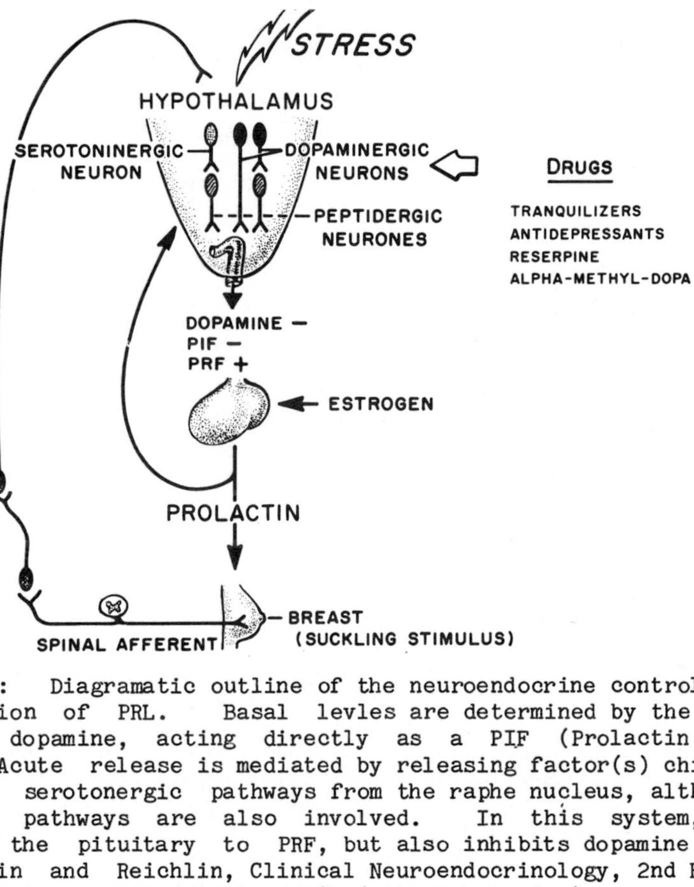

Fig. 7: Diagramatic outline of the neuroendocrine control system of the secretion of PRL. Basal levles are determined by the inhibitory action of dopamine, acting directly as a PIF (Prolactin inhibitory factor) Acute release is mediated by releasing factor(s) chiefly under control of serotonergic pathways from the raphe nucleus, although other peptidergic pathways are also involved. In this system, estrogen sensitizes the pituitary to PRF, but also inhibits dopamine secretion. (From Martin and Reichlin, Clinical Neuroendocrinology, 2nd Edition, FA Davis, Publishers, Philadelphia, 1987 (with Permission).

An alternative view is that stress-induced PRL release is triggered by the release of a prolactin releasing factor. One point in favor of this view is the finding that in rats in whom dopamine receptors have been blocked or all catecholamines depleted, (16,17) the elevated PRL levels are raised even further by stress. Crude extracts of hypothalamic tissue stimulate PRL release (18). Based on bioassays of known hypothalamic factors, there are several candidate PRF's. These include TRH, (19,20), VIP (21,23), PHI (24,25) and oxytocin (26). PHI and VIP are analog peptides co-synthesized as part of the VIP prohormone (27). The tuberoinfundibular TRH, PHI and VIP containing fibers arise in distinct populations of parvocellular neurons located in subdivisions of the paraventricular nucleus (28,29). We have demonstrated that destruction of the paraventricular nucleus by electrolytic techniques blocks PRL release in response to ether and restraint stress (Fig. 5,6) (29). Because these lesions had no effect on GH secretion, (known to be under control of the arcuate nucleus which is also the nucleus of origin of the tuberoinfundibular dopaminergic tract) we concluded that acute response cannot be due to disinhibition of tonic dopamine secretion.

As first pointed out by Hokfelt and colleagues, the paraventricular nucleus is the site of origin of many of the neurons that subserve the acute PRL and ACTH response to stress (30). The paraventricular nucleus contains cells of origin of the CRH neuronal projection to the median eminence (31,32). Destruction of this nucleus markedly reduces the ACTH and/or corticosterone response to a variety of stresses (Fig 8, 33-36). The paraventricular nucleus can be looked upon as the final common neurohumoral pathway through which stress acts on the anterior pituitary. It is innervated by a large number of fibers projecting from other parts of the brain, mediated by an extraordinary variety of neurotransmitters (34,35 Table 1). It appears likely that activation of these neurons is mediated by serotoninergic afferents. Treatment of rats with serotonin blockers abolishes many forms of stress induced PRL release in the rat (17), and the serotonin precursor 5-hydroxytryptophan stimulates PRL release (17,29); this precursor is ineffective in animals whose paraventricular nuclei have been destroyed (Fig. 9)(29).

There is some uncertainty as to which of the several candidate PRF's localized in this region is responsible for the effects of the lesions. One approach, followed by our group (10) and others (37,38,39) is to administer antisera directed against one of the peptides. Anti VIP antisera abolishes (Fig 10), or reduces (37-39) several stress induced PRL responses, and markedly reduces the PRL response to suckling (Fig. 11). The concentration of VIP in the hypophysial portal vessel system is increased by injection of 5-hydroxytryptophan (40). (29, with permission).

TABLE 1

NEUROACTIVE SUBSTANCES IDENTIFIED IN PARAVENTRICULAR NUCLEUS
(modified from Swanson and Sawchenko, Ann Rev. Neurosci.
1983 6:269-324)

CELL BODIES

oxytocin vasopressin somatostatin dopamine met-enkephalin
leu-enkephalin neuroptensin dynorphin substance P glucagon
VIP TRH renin CRH

FIBERS

Norepinephrine Epinephrine Serotonin Acetylcholine ACTH(1-39)
b-endorphin a-MSH GABA LHRH Pancreatic polypeptide bradykinin
prolactin angiotensin II

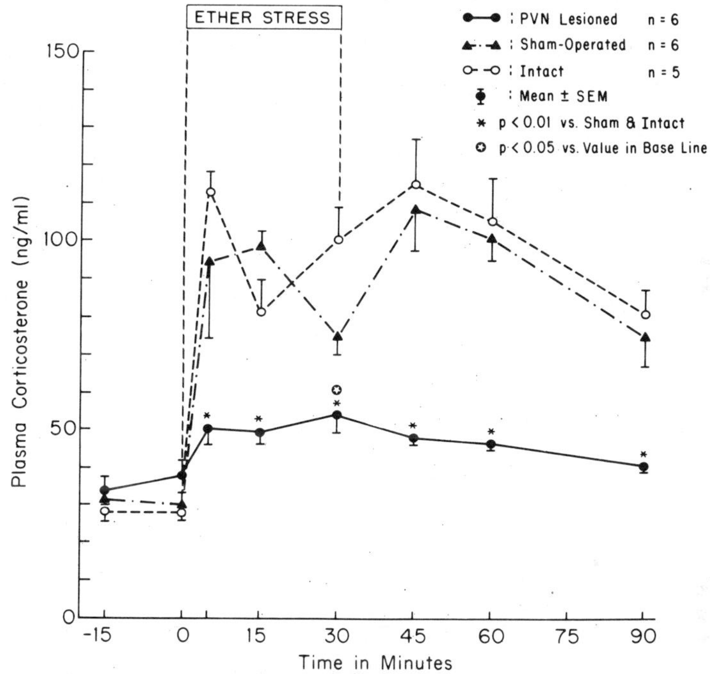

Fig. 8: Effect of ether-stress on serum corticosterone illustrates the stress response, and the almost complete blockade of the effect by lesions of the paraventricular nucleus, the site of origin of the CRF and vasopressinergic neurons projecting to the median eminence.

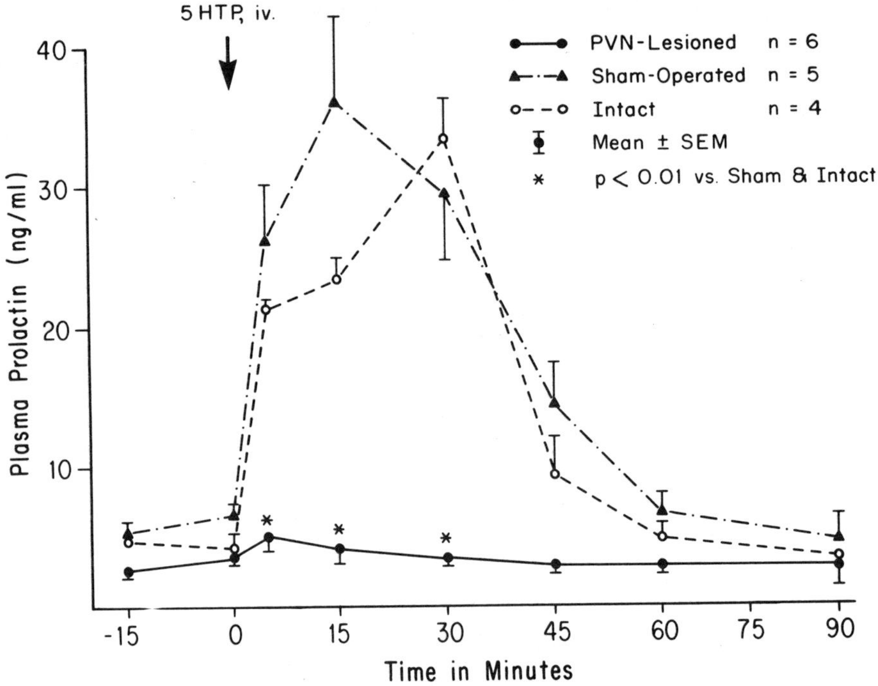

Fig. 9: Effect of PV lesions on 5HTP induced PRL release.

Although it seems most attractive to attribute the stress-increased release of PRL to VIP (and PHI), there are several unsettled problems with this hypothesis. Most important is the fact that the median eminence of the normal rat contains very few immunoreactive VIP-containing nerve endings (41) and little immunoreactive VIP, which is in contrast to the plentiful distribution in the median eminence of the established releasing factors such as TRH and GHRH. However, PHI is more readily demonstrated in median eminence and several workers have documented the presence of VIP in hypophysial-portal vessels (40,42-46) and a change in its concentration after the serotonin-precursor tryptophan administration (40) and prostaglandin D2 (45). In at least one study anti-VIP was only marginally effective as was anti-PHI in inhibiting PRL responses to ether stress; combined administration of both antisera markedly but not completely reduced PRL response (39). In the fowl, VIP and PHI are more plentiful in the median eminence and secretion of PRL is suppressed by anti-VIP (47). During pregnancy (in the rat) the P-V VIP system becomes much more prominent (48).

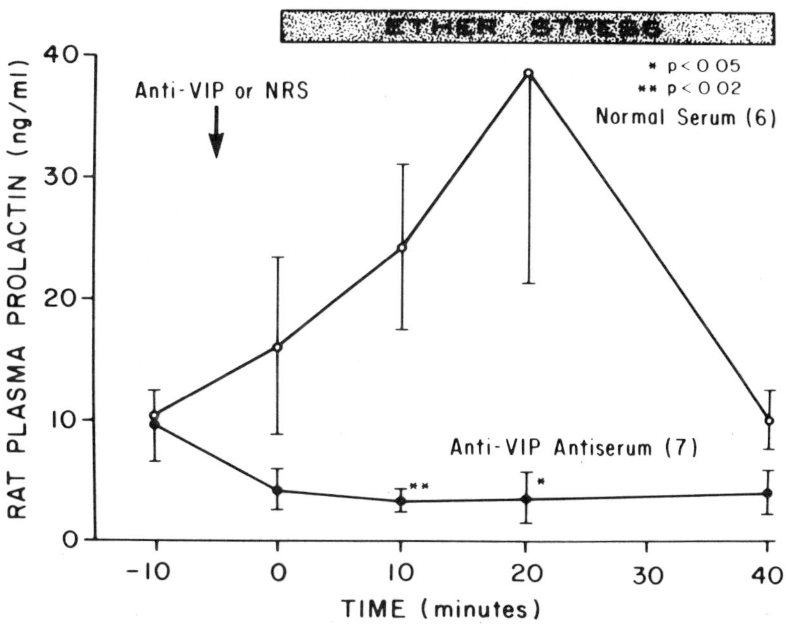

Fig. 10: Passive immunization with anti-VIP antibody blocked the acute response of PRL to ether-stress. This finding suggests that VIP is the principal involved in acute PRL release. (From Abe et al, 10, with permission).

An alternative (or complementary) hypothesis to explain the relationship between VIP and PRL secretion has been proposed by Hagen and colleagues (49). It previously had been shown that the prolactinotropes of the pituitary co-contain VIP (50); Hagen's group has shown that exposure of dispersed pituitary cells in culture to anti-VIP antisera reduces basal PRL release (49). They propose that VIP may act a paracrine regulator of PRL response, a model for regulation of pituitary secretion that has been emphasized by Dunef et al (51). Our group (Lam, Minamitani, Cacicedo) has similarly found that rat pituitary cells secrete VIP but our preliminary data suggest that it is not localized to prolactinotropes. It is possible, that VIP may be an important paracrine regulator of PRL secretion.

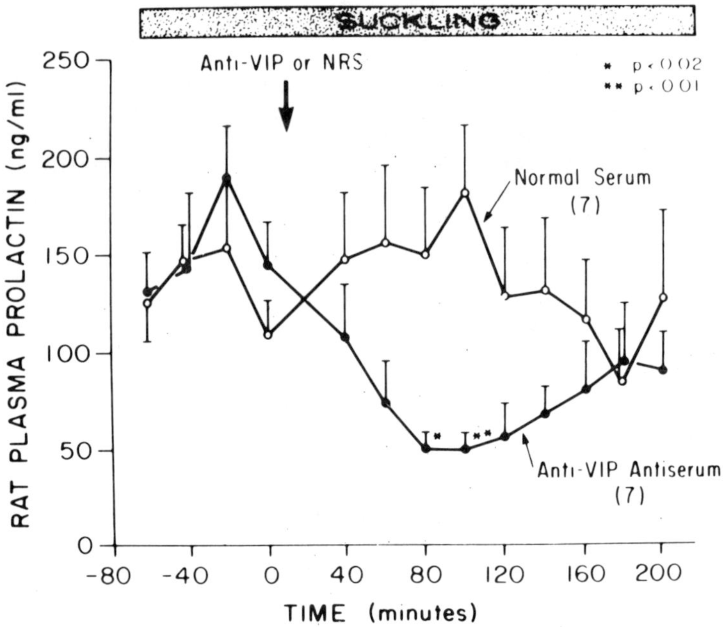

Fig. 11: Passive immunization with anti-VIP antibody partially lowers the serum level of PRL which administered during a suckling episode. This indicates that VIP is only partially responsible for maintenance of suckling-induce PRL levels. (From Abe et al, 10, with permission).

Stress and GH secretion

As in the case of PRL regulation, it was not until radioimmunoassays for GH became available that it was recognized that stress influenced GH secretion (52,53,54). In man, the predominant effect of virtually all stressful stimuli is the release of GH, whereas in the rodent, stress inhibits GH secretion. As has become clear with the discovery of somatostatin, GH secretion is subject to two interacting hypophysiotropic factors (Fig. 12), one excitatory, and the other inhibitory (55) which have been shown by direct measurements of portal vessel peptides (56) to exert reciprocal effects. One has to assume that in the rat and mouse, the inhibitory component predominates, while in the primate, the excitatory component prevails. In neither species of animal does this effect appear to be teleologically useful.

Within a few months of the development of the GH radioimmunoassay by Roth, Glick, Berson and Yalow (57) they showed that hypoglycemia induces GH secretion, and that the response depended upon an intact hypothalamic-pituitary connection (57,58). Subsequently, exercise, surgical procedures, trauma, and the anticipation of stress (Fig. 13a) were shown to increase GH release (52), and spontaneous fluctuations of GH secretion were found to occur, mainly at night during the first few hours of sleep (60,61). In rats, hypoglycemia, restraint, ether, and starvation were found to inhibit GH release (59), an effect that could be partially or completely reversed by passive immunization against somatostatin (62,63).

The monkey is an excellent model for the study of GH regulation in man. In this species, lesions of the anterior lip of the median eminence were shown to block reflex GH release following hypoglycemia (64) (Fig. 14). The location of these lesions subsequently has been shown to correspond to the distribution of the GRHergic pathways in the ventromedial nucleus (65,66)(Fig. 15). Frohman and colleagues (67), and Martin and colleagues (68) showed that electrical stimulation of the medial basal hypothalamus led to release of GH. A detailed analysis of the pathways involved was made by Martin et al who demonstrated the anatomical location within the limbic system that mediated the response, including major structures in the amygdala and hippocampus (67), important anatomical elements of the visceral brain, and presumably the pathway by which emotion influences Gh secretion.

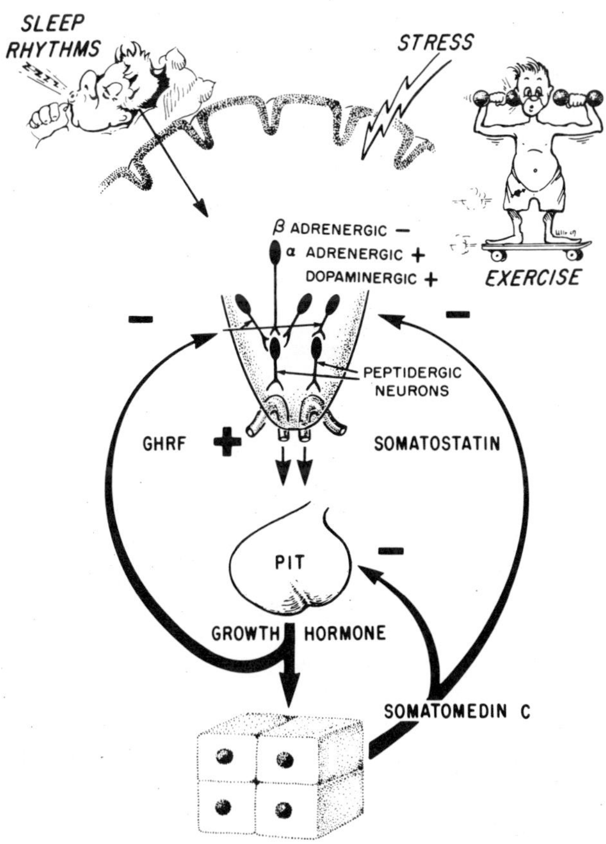

HYPOTHALAMIC - GROWTH HORMONE - SOMATOMEDIN AXIS

Fig. 12: Diagrammatic representation of factors regulating GH secretion. Most important is the interaction of Growth hormone releasing hormone (GHRH) with somatostatin in the determination of GH set point.

Fig. 12 legend continued... In addition, somatomedin C, a secretion of the liver under the influence of GH also acts as a negative feedback signal to the pituitary (it inhibits response to GHRH), and also stimulates the release of somatostatin. There also is evidence that GH acts on the hypothalamus to stimulate somatostatin release, and inhibits GHRH secretion. This system is further activated by way of hypothalamic biogenic amine, and cholinergic (not shown), and peptidergic (not shown) pathways. Through these connections, sleep, stress and excercise influence GH levels. (From Martin J.B. and Reichlin S. Clinical Neuroendocrinology 2nd ecition, FA Davis, Philadelphia, 1987.)

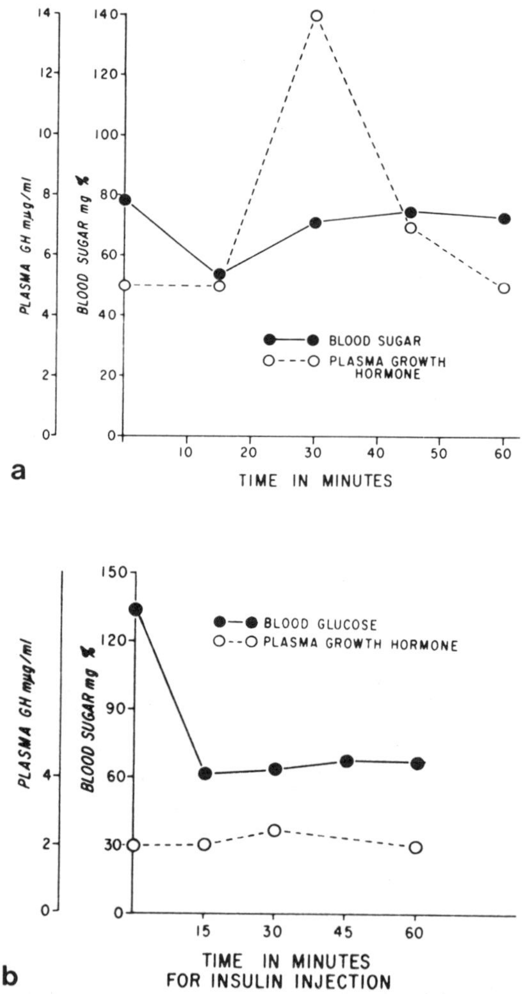

Fig. 13: A. GH response to insulin-induced hypoglycemia in a normal squirrel monkey B. Response is blocked in an animal with lesion involving the anterior lip of the median eminence. (From Abrams et al, 64, with permission).

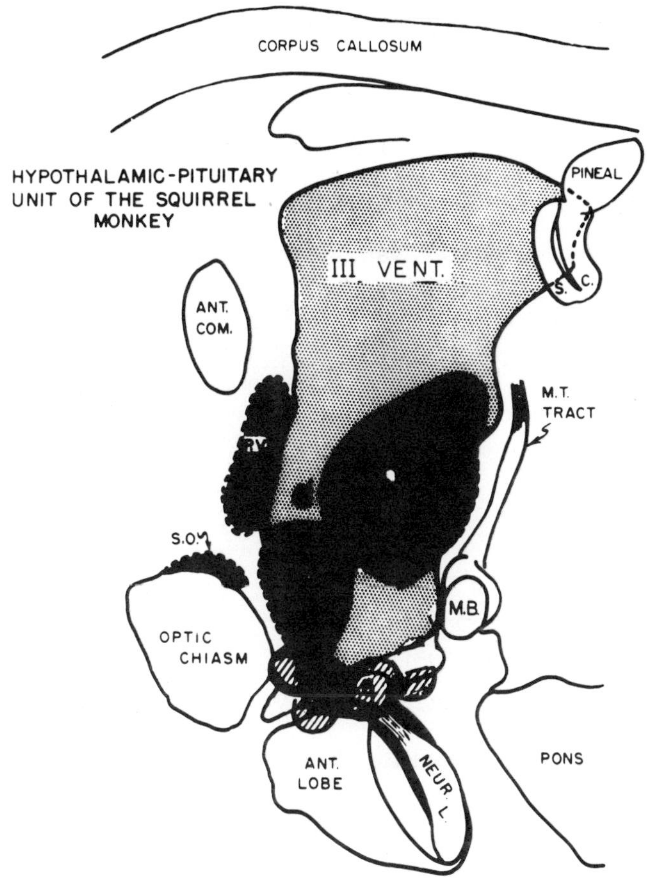

Fig. 14: Location of lesions in the squirrel monkey that block reflex GH release. (From Abrams et al, 64, with permission).

Of continuing interest has been the elucidation of the specific neuroactive materials that mediates these responses. At the level of the pituitary, (as noted above) growth hormone releasing factor (GRF) and somatostatin interact in regulation. The structures of the GRF peptides were elucidated by Ling et al (70), and Rivier et al (71), while the structure of Somatostatin was elucidated by Brazeau et al (72). The early history of knowledge of these substances has been reviewed extensively (74). Somatomedin C, a peptide secretion of the liver feeds back upon the pituitary to modify its responsiveness to GRH, thus serving as a feedback regulator (75,76).

Impinging upon the somatostatinergic and GRHergic neuronal systems are a variety of neuroregulators involving afferents from the midbrain noradrenergic and adrenergic cell bodies, serotonergic fibers from the raphe nuclei, and multisynaptic relays mediated by many different neurotransmitters and neuropeptides. If one looks only at the output side, i.e. the blood level of GH, which is a resultant of changes in both GRH and somatostatin, the following neurotransmitters appear to be involved in stimulation: dopamine (77) noradrenaline (a-adrenergic) (78), serotonin (79), adrenaline (80,81), histamine (81). A number of peptides can stimulate GH secretion, some effects exerted at the level of the

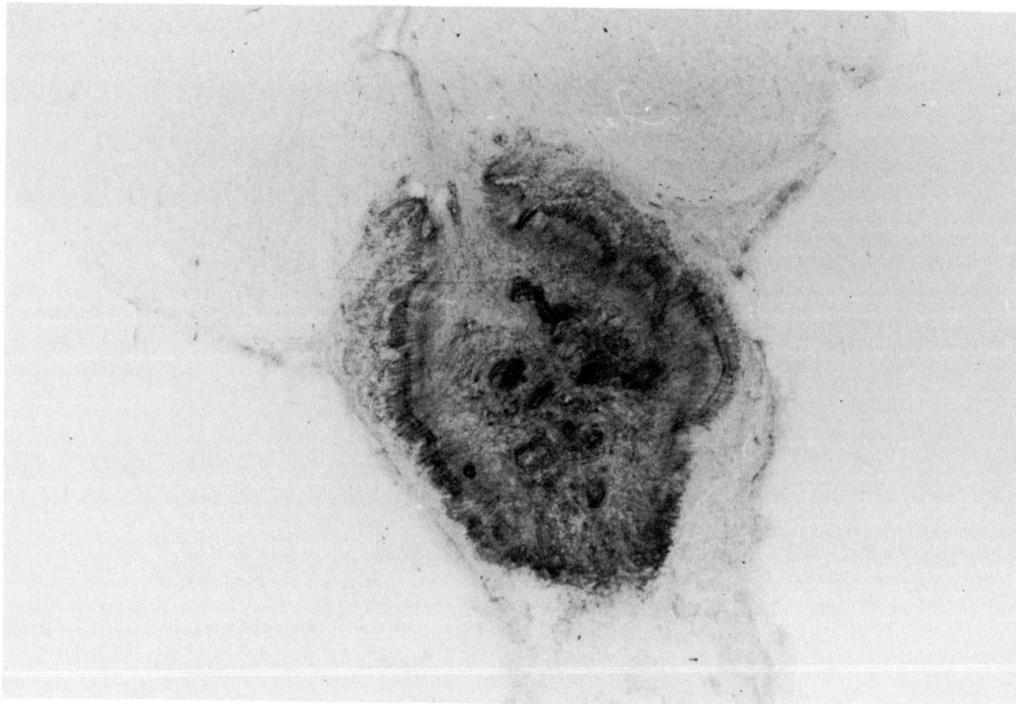

Fig. 15: Immunohistochemical demonstration of localization of GHRH in the stalk-median eminence region of a rhesus monkey. (Figure courtesy of Ronald Lechan).

pituitary, and others at the level of the hypothalamus, but their physiological role is uncertain. These include b-endorphin (82), other endogenous opioids (82,83) the enkephalins (83), TRH (84), VIP (85), motilin (86), galanin (87,88), a-MSH (88), vasopressin (90), neuropeptide Y (91), substance P (92), and neurotensin (92). Benzodiazepines can stimulates GH secretion through a direct action on pituitary BDZ receptors (93). GABA (94,95), CRH (95), CCK (97,98), and bombesin (99) can inhibit GH secretion at the level of the hypothalamus.

The role of acetylcholine as a mediator of GH regulation, particularly in man, now appears to be much more important than first thought. The earliest studies using cholinergic antagonists demonstrated that atropine did not block GH response to hypoglycemia (100); more recently it has been shown that all other secretory responses including other forms of stress, and the spontaneous nocturnal surges are blunted or abolished by muscarinic blockers (100,101). Even the response to GRH is blocked by atropine (102). The mechanism by which atropine brings this effect about is still unknown. The most reasonable hypothesis is that under normal conditions cholinergic pathways tonically inhibit somatostatin secretion, and that atropine and its analogs reverse this effect. Indeed, studies utilizing incubated fragments of hypothalamus have shown that ACH is inhibitory to somatostatin release (103), but other workers, including ourselves have found that ACH stimulates somatostatin secretion from dispersed hypothalamic cells (104-107).

Taking into account all of the findings to date, it seems reasonable to propose that in man, the acute stress response of GH release involves the direct effects on the pituitary of GRH, and possibly vasopressin and VIP, brought about at the central level by activation of these tuberinfundibular systems by a-noradrenergic, adrenergic, serotonergic, dopaminergic, b-endorphinergic, and enkephalinergic impulses. In addition, somatostatin is probably inhibited (reciprocally) by cholinergic, and possibly endorphin and enkephalin fibers. These neurotransmitters are probably involved in all stresses except for hypoglycemia which appears not to involve the somatostatin inhibitory component.

REFERENCES

1. Macleod RM 1976 Regulation of prolactin secretion. In: Martini L, Ganong WF, (eds) Frontiers in Neuroendocrinology. Raven Press, New York Vol 4, 169 194.

2. Martin JB, Reichlin S. Clinical Neuroendocrinology. Second Edition. FA Davis, Philadelphia pg 201-232, 1987.

3. McNeill AS, Robinson CAF Houston, M, Howie, PW: Release of oxytocin and prolactin in response to suckling. BMJ 286:257-259, 1983.

4. Hwang P, Guyda H, Friesen J. A radioimmunoassay for human prolactin. Proc Natl Acad Sci USA 68: 1902-1906, 1971.

5. Boyd, AE, Reichlin,S. Neural control of prolactin secretion in man. Psychoneuroendocrinology 3:113-130, 1978.

6. Neill JD, 1980 Neuroendocrine regulation of prolactin secretion. In: Martini L, Ganong WF (eds) Frontiers in Neuroendocrinology. Raven Press, New York Vol 6: 129-155.

7. Nesse, RM, Curtis, GC, Brown GM. et al. Anxiety induced by flooding therapy for phobias does not elicit prolactin secretory response. Psychosom Med 42:25-31, 1980.

8. Rose, RM. Overview of endocrinology of stress. I : Neuroendocrinology and psychiatric disorder Brown GM, Koslow SH, Riechlin S, (eds). Raven Press 1984, 95-122.

9. Ganong WF. Neurotransmitter mechanisms underlying stress responses. In: Neuroendocrinology and psychiatric disorder. Brown GM, Koslow SH, Reichlin S. (eds) Raven Press 1984: 133-144.

10. Abe H, Engler D, Molitch ME, Bollinger-Gruber J, Reichlin S. 1985. Vasoactive intestinal peptide is a physiological mediator of prolactin release in the rat. Endocrinology 116: 1383-1390.

11. Gibbs DM, Neill JD. 1978. Dopamine levels in hypophysial stalk blood in the rat are sufficient to inhibit prolactin secretion _in vivo_. Endocrinology 102: 1895-1900

12. Demarest KT, Moore KE, Riegle GD. 1985. Acute restraint stress decreases tuberoinfundibular dopaminergic neuronal activity: evidence for a different response in male versus female rats. Neuroendocrinology 41: 504-510.

13. Schally AV, Redding TW, Arimura A, Dupont A, Linthicum GL 1977, Isolation of Gamma-amino butyric acid from pig hypothalami and demonstration of its prolactin release-inhibiting (PIF) activity _in vivo_ and _in vitro_. Endocrinology 100: 681-691.

14. Nicolics K, Mason AJ, Szonyi E, Ramachandran J, Seeburg PH, 1985. A prolactin-inhibiting factor within the precursor for human gonadotropin-releasing hormone. Nature 316: 511 517.

15. Khorram O, McCann SM. On the presence of a nondopaminergic, peptidergic prolactin release-inhibiting factor in hypothalamic extracts of infantile rats. Neuroendocrinology 44: 65-69, 1986.

16. Valverde RC, Chieffo V, Reichlin S. Failure of reserpine to block ether-induced release of prolactin: Physiological evidence that stress induced prolactin release is not caused by acute inhibition of PIF secretion. Life Sci. 12: 327-335, 1973.

17. Krulich L. Central neurotransmitters and the secretion of prolactin, GH, LH, and TSH. Annu Rev Physiol. 41:603-615, 1979.

18. Valverde C, Chieffo V, Reichlin S. 1972: Prolactin releasing factor in porcine and rat hypothalamic tissue. Endocrinology 91: 982-993, 1972.

19. Sheward WJ, Fraser HM, Fink G. 1985. Effects of immunoneutralization of thyrotrophin-releasing hormone on the release of thyrotrophin and prolactin during suckling or in response to electrical stimulation of the hypothalamus in the anaesthetized rat. J.Endocrinol. 106: 113-119, 1985.

20. Fink G, Koch Y, Benaroya N, (1983). TRH in hypophysial portal blood: Characteristics of release and relationship to thyrotropin and prolactin secretion. In Thyrotropin-Releasing Hormone, pp 127-143 eds. E.C. Griffiths and G.W. Bennett. New York: Raven Press.

21. Rostene WH. (1981) Neurobiological and neuroendocrine functions of vasoactive-intestinal (VIP). Progress in Neurobiology 22: 103-129, 1984.

22. Frawley LS, and Neill JD. (1981). Stimulation of prolactin secretion in rhesus monkeys by vasoactive intestinal polypeptide. Neuroendocrinology 33:79-83.

23. Charpentier G, Bercovia JP, Jednak P, Talbot JN, Mutt V, Rosselin G. (1982). VIP test in prolactinomas and in normal subjects. Biomedicine 36:105-107.

24. Werner S, Hulting AL, Hokfelt T, Eneroth P, Tatemoto K, Mutt V, Marode L, Wunsch E. Effect of peptide PHI-27 on prolactin release <u>in vitro</u>. Neuroendocrinology 37, 476-478, 1983.

25. Kaji H, Chihara K, Abe H, Minamitani N, Kodama H, Kita T, Fujita T, Tatemoto K. Stimulatory effect of peptide histidine isoleucine amide 1-27 on prolactin release in the rat. Life Sci 35:641-647, 1984.

26. Lumpkin MD, Samson WK, McCann SM. 1983. Hypothalamic and pituitary sites of action of oxytocin to alter prolactin secretion in the rat. Endocrinology 112:1711-1712.

27. Itoh N, Obata K, Yanaihara N, Okamoto H. 1973. Human preprovasoactive intestinal polypeptide contains a novel PHI-27-like peptide, PHM-27. Nature 304:547-549.

28. Hokfelt T, Schultzberg M, Lundberg JM, Fuxe K, Mutt V, Fahrenkrug J, Said SI. 1982. Distribution of vasoactive intestinal polypeptide in the central and peripheral nervous system as revealed by immunocytochemistry. In: Vasoactive Intestinal Peptide, pp.65-86. Ed. S.I. Said. New York; Raven Press.

29. Minamitani N, Minamitani T, Lechan RM, Bollinger-Gruber J, Reichlin S. Paraventricular nucleus mediates prolactin secretory responses to restraint stress, ether stress, and 5-hydroxy-L-tryptophan injection in the rat. Endocrinology 120: 860-877, 1987.

30. Hokfelt Y, Fahrenkrug J, Tatemoto K, Mutt V, Werner S, Hulting AL, Terenius L, Chang KJ. 1983. The PHI(PHI 27)/ corticotropin-releasing factor/enkephalin immunoreactive hypothalamic neuron: possible morphological basis for integrated control of prolactin, corticotropin, and growth hormone secretion. Proc Natl Acad Sci 80:895-898.

31. Swanson LW, Sawchenko PE, Rivier J, Vale WW. 1983. Organization of ovine corticotropin-releasing factor immunoreactive cells and fibers in the rat brain: an immunohistochemical study. Neuroendocrinology 36:165-186.

32. Olschowka JA, O'Donohue TL, Mueller GP, Jacobowitz DM. 1982. Hypothalamic and extrahypothalamic distribution of CRF-like immunoreactive neurons in the rat brain. Neuroendocrinology 35:305-308.

33. Bruhn TO, Plotsky PM, Vale WW. Effect of paraventricular lesions on corticotropin-releasing factor (CRF)-like immunoreactivity in the stalk-median eminence: Studies on the adrenocorticotropin response to ether stress and exogenous CRF. Endocrinology, 1984, 114:57-62.

34. Swanson LW, Sawchenko PE. 1980. Paraventricular nucleus: a site for the integration of neuroendocrine and automatic mechanisms. Neuroendocrinology 31:410-417.

35.Swanson LW, Sawchenko PE. Hypothalamic intergration: organization of the paraventricular and supraoptic nuclei. Ann Rev Neurosci 1983. 6:269-324.

36. Minamitani N, Minamitani T, Lechan RM, Bollinger-Gruber J, Reichlin S. Role of the paraventricular nucleus in the pituitary-adrenal responses to restraint stress, ether stress, and serotonergic stimulation by 5-HTP in conscious male rats. In preparation.

37. Kaji H, Chihara K, Abe H, Kita T, Kashio Y, Okimura Y, Fujita T. 1985. Effect of passive immunization with antisera to vasoactive intestinal polypeptide and peptide histidine isoleucine amide on 5-hydroxy-L-trytophan-induced prolactin release in rats. Endocrinology 117:1914-1919.

38. Ohta H, Kato Y, Shimatsu A, Tojo K, Kabayama Y, Inoue T, Yanaihara N, Imura H. Inhibition by antiserum to vasoactive intestinal polypeptide (VIP) of prolactin secretion induced by serotonin in the rat. Eur J Pharmacol, 1985. 109:409-412.

39. Kaji H, Chihara K, Kita T, Kashio Y, Okimura Y, Fujita T. 1985. Administration of antisera to vasoactive intestinal polypeptide and peptide histidine isoleucine attenuates ether-induced prolactin secretion in rats. Neuroendocrinology 41:529-531.

40. Shimatsu A, Kato Y, Matsushita N, Katakami H, Yanaihara N, Imura H. Stimulation by serotonin of vasoactive intestinal polypeptide release into rat hypophysial portal blood. Endocrinology 111:338-340, 1982.

41. Loren I, Emson PC, Fahrenkrug J, Bjorklund A, Alumets J, Hakanson R, Sundler F. (1979). Distribution of vasoactive intestinal polypeptide in the rat and mouse brain. Neuroscience 4:1953-1976.

42. Said SI, Porter JC, (1978) vasoactive intestinal peptide release into hypophysial portal blood. Life Sciences 24:227-230.

43. Shimatsu A, Kato Y, Inoue T, Christofides ND, Bloom SR, Imura H. Peptide histidine isoleucine and vasoactive intestinal polypeptide-like immunoreactivity coexist in rat hypophysial portal blood. Neurosci. Let 43:259 262, 1983.

44. Shimatsu A, Kato Y, Matsushita N, Katakami H, Yanaihara N, Imura H. 1981. Immunoreactive vasoactive intestinal polypeptide in rat hypophyseal-portal blood. Endocrinology 108: 395-398.

45. Shimatsu A, Kato Y, Matsushita N, Ohta H, Kabayama Y, Yanaihara N, Imura H. Prostaglandin D2 stimulates vasoactive into intestinal polypeptide release into rat hypophysial portal blood. Peptides 1984 5:395-398.

46. Brar AK, Fink G, Maletti M, Rostene W. Vasoactive intestinal peptide in rat hypophysial portal blood: effects of electrical stimulation of various brain areas, the oestrus cycle and anaesthetics.J. Endocr. 1985, 106:275-280.

47. Macnamee MC, Sharp PJ, Lea RW, Sterling RJ, Harvey S. Evidence that vasoactive intestinal polypeptide is a physiological prolactin-releasing factor in the Bantam hen. Gen Comp Endocrinol 62:470-478, 1986.

48. Mezey E, Kiss JZ. 1985. Vasoactive intestinal polypeptide containing neurons in the paraventricular nucleus may participate in regulating prolactin secretion. Pro Natl Acad Sci, USA. 82:245-247.

49. Hagen TC, Arnaout MA, Scherzer WJ, Martinson DR, Garthwaite TL. Antisera to vasoactive intestinal polypeptide inhibit basal prolactin release from dispersed anterior pituitary cells. Neuroendocrinology 1986, 43:641-645.

50. Morel G, Besson J, Rosselin G, Dubois PM. Ultrastuctural evidence for endogenous vasoactive intestinal peptide-like immunoreactivity in the pituitary gland. Neuroendocrinology 1982, 34:85-89.

51. Denef C, Baes M, Schramme C. Paracrine interactions in the anterior pituitary: Role in the regulation of prolactin and growth hormone secretion. In: Frontiers in Neuroendocrinology Vol 9. Ed. E.D. Ganong, L. Martini (eds) Raven Press, New York, 1986. 115-148.

52. Reichlin S. Regulation of somatotrophic hormone secretion. In Handbook of Physiology Section 7: Endocrinology Volume IV, The Pituitary Gland and Its Neuroendocrine Control Part 2, Ed. E. Knobil, W.H. Sawyer. 1974, 405-447.

53. Brown GM, Seggie JA, Chambers JW, Ettigi PG. Psychoneuroendocrinology and growth hormone: A review. Psychoneuroendocrinology 3:131 153, 1978.

54. Quabbe H. Hypothalamic control of GH secretion: pathophysiology and clinical implications. Acta Neurochirurgica 75:60-71, 1985.

55. Tannenbaum GS. Physiological role of somatostatin in regulation of pulsatile growth hormone secretion. Adv. Exp Med Biol, 188:229-260, 1985.

56. Plotsky PM, Vale WW. Patterns of growth hormone-releasing factor and somatostatin secretion into the hypophysial-portal circulation of the rat. Science 230:461, 1985.

57. Roth J, Glick SM, Yalow RS, Berson SA. Hypoglycemia: a potent stimulus to secretion of growth hormone. Science 140:987-988, 1963.

58. Roth JS, Glick SM, Yalow RS, Berson SA. Secretion of human growth hormone: Physiological and experimental modifications. Metabolism 12:577-579, 1963.

59. Schalch DS. The influence of physical stress and exercise on growth hormone and insulin secretion in man. J. Lab. Clin. Med. 69:256 269, 1967.

60. Quabbe H.J, Helge H, Kubicki S. Nocturnal growth hormone secretion: correlation with sleeping EEG in adults and pattern in children and adolescents with non-pituitary dwarfism, overgrowth and with obesity. Acta Endocrinologica (Kbh) 67:767-783, 1971.

61. Takahashi Y, Kipnis DM, Daughday WH. Growth Hormone secretion during sleep. J. Clin Invest. 47:2079-2090, 1968.

62. Tannenbaum GS, Epelbaum J, Colle E, Brazeau P, Martin JB. Antiserum to somatostatin reverses starvation-induced inhibition of growth hormone but not insulin secretion. Endocrinology 102:1909-1914, 1978.

63. Terry LC, Willoughby JO, Brazeau P, Martin JB, Patel Y. Antiserum to somatostatin prevents stress-induced inhibition of growth hormone secretion in the rat. Science 192:565-567, 1976.

64. Abrams RL, Parker ML, Blanco S, Reichlin S, Daughaday WH. Hypothalamic regulation of growth hormone secretion. Endocrinology 78:605-613, 1966.

65. Bloch B, Brazeau P, Ling N, et al. Immunohistochemical detection of growth hormone-releasing factor in brain. Nature 301:607-608, 1983.

66. Lechan RM, Lin HD, Ling N et al. Distribution of immunoreactive growth hormone releasing factor (1-44)NH2 in the tuberoinfundibular system of the rhesus monkey. Brain Res 309:55-61, 1984.

67. Frohman LA, Bernardis LL, Kant K. Hypothalamic stimulation of growth hormone secretion. Science 162:580-582, 1968.

68. Martin JB. Plasma growth hormone (GH) response to hypothalamic and extrahypothalamic electrical stimulation. Endocrinology 91:107-115, 1972.

69. Martin JB. Hypothalamic regulation of growth hormone secretion. In: Secretory Tumors of the Pituitary Gland. Black, P McL., Zervas NT, Ridgway EC, Martin JB (eds). Raven Press New York, N.Y., 109-134, 1984.

70. Ling N, Esch F, Bohlen P, Brazaeu P, Wehrenberg WB, Guillemin R. Isolation, primary structure, and synthesis of human hypothalamic somatocrinin: growth hormone-releasing factor. Proc Natl Acad Sci (USA) 81:4302-4306, 1984.

71. Rivier J, Speiss J, Thorner M, Vale WW. Characterization of a growth hormone-releasing factor from a human pancreatic islet tumour that caused acromegaly. Nature 300:276-8, 1982.

72. Brazeau P, Vale W, Burgus R. et al. Hypothalamic polypeptide that inhibits the secretion of immunoreactive pituitary growth hormone. Science 179:77-79, 1973.

73. Thorner MO, Cronin MJ. Growth Hormone-releasing factor: Clinical and Basic studies. In: Neuroendocrine Perspectives, Muller EE, MacLeod RM, Frohman LA, (eds). Elsevier, Amsterdam, 95-144, 1985.

74. Frohman LA, Jansson J-O. Growth Hormone-releasing hormone. Endocrine Rev. 1, 223-253, 1986.

75. Wehrenberg WB, Ling N, Bohlen P, Esch F, Brazeau P, Guillemin R. Physiological roles of somatocrinin and somatostatin in the regulation of growth hormone secretion. Biohem Biophys Res Commun 109:562-567, 1982.

76. Ceda GP, Davis RG, Rosenfeld R.G, Hoffman AR. The growth hormone (GH)-releasing hormone (GHRH)-GH-Somatomedin axis: Evidence for rapid inhibition of GHRH-ELICITED GH release by insulin-like growth factors I and II. Endocrinology 120:1658-1662, 1987.

77. Boyd AE III, Lebovitz HE, Pfeiffer JB. Stimulation of human growth hormone secretion by L-DOPA, New Engl J Med 283:1425-1429, 1970.

78. Laron Z, Topper E, Gil-Ad I. Oral clonidine a simple, safe and effective test for growth hormone secretion. In: Evaluation of Growth Hormone Secretion. Pediat. Adolesc. endocr. 12:103-115, (Karger, Basel, 1983).

79. McCann SM, Krulich L, Ojeda SR et al. Neurotransmitters in the control of anterior pituitary function. In. Fuxe K, Hokfelt T, Luft R. Central regulation of the endocrine system. N.Y. Plenum Press 329-348: 1979.

80. Terry LC, Crowley WR, Johnson, MD. Regulation of episodic growth-hormone secretion by the central epinephrine system: studies in the chronically cannulated rat. J. Clin Invest 69:104-112, 1982.

81. Terry LC. Neuropharmacologic regulation of anterior pituitary hormone secretion in man. In: Hormone-secreting tumors, J.R. Givens, ed. Year Book Medical Publishers, Inc. Chicago, 27-44, 1982.

82. Grossman A, Clement-Jones V, Besser GM. Clinical implications of endogenous opioid peptides. In. Neuroendocrine Perspectives ed Muller EE, Macleod RM. Frohman LA. Elsevier, Amsterdam 243-294, 1985.

83. Chihara K, Arimura A, Coy DH, Schally AV. 1978. Studies on the interaction of endorphins, substance P, and endogenous somatostatin in growth hormone and prolactin release in rats. Endocrinology 102:281-290.

84. Bluet-Pajot M.T, Durand D, Drouva SV, Mounier F, Pressac M, Kordon C. Further evidence that thyrotropin-releasing hormone participates in the regulation of growth hormone secretion in the rat. Neuroendocrinology 44:70-75, 1986.

85. Chihara K, Iwasaki J, Minamitani N, Kaji H, Matsukura S, Tamaki N, Matsumoto S. Fujita T. Effect of vasoactive intestinal polypeptide on growth hormone secretion in perifused acromegalic pituitary adenoma tissues. J. Clin. Endocrinol Metab 54:773-779, 1982.

86. Samson WK, Lumpkin MD, Nilaver G, McCann SM. Molitin: A novel growth hormone releasing agent. Brain Res Bull 12:57-62, 1984.

87. Bauer FE, Ginsberg L, Venetikou M, MacKay DJ, Burrin JM, Bloom SR. Growth hormone release in man induced by galanin, a new hypothalamic peptide. 2(8500) 192-5, 1986.

88. Ottlecz A, Samson WK, McCann SM. Galanin: Evidence for a hypothalamic site of action to release growth hormone. Peptides 7, 51-53, 1986.

89. Schlegel W, Vecsey A, Eberle AE, Chretien M, Seidah NG, Zahnd GR. Peptides of the neurointermediary lobe of rat pituitary stimulate GH secretion in vitro. Mol Cell Endocrinol 1983, 47-55.

90. Malacara Jm, Valverde RC, Reichlin, S. Elevation of plasma radioimmunoassyable growth hormone in the rat induced by porcine hypothalamic extract. Endocrinology 91:1189-1198, 1972.

91. McDonald JK, Lumpkin MD, Samson WK, McCann SM. Neuropeptide Y affects secretion of luteinizing hormone and growth in ovariectomized rats. Proc Natl Acad Sci USA 82:561-564, 1985.

92. Vijayan E, and McCann SM, 1980. Effects of substance P and neurotensin on growth hormone and TSH release in vivo and in vitro. Life Sci. 26:321-327.

93. Laakmann G, Treusch J, Schmauss M, Schmitt E, Treusch U. Comparison of growth hormone stimulation induced by desimipramide, diazepam and metaclazepam in man. Psychoneuroendocrinology. 1982, 141-146.

94. Steardo L, Iovino M, Monteleone P, Agrusta M, Orio F. Pharmacological evidence for a dual gabaergic regulation of growth hormone release in humans. Life Sci 39:979-985, 1986.

95. Willoughby JO, Jervois PM, Menadue MF, Blessing WW. Activation of GABA receptors in the hypothalamus stimulates secretion of growth hormone and prolactin. Brain Res 374:119-125, 1986.

96. Rivier C, Vale W. Involvement of corticotropin-releasing factor and somatostatin in stress-induced inhibition of growth hormone secretion in the rat. Endocrinology, 2478-82, 1985.

97. Karashima T, Okajima T, Kato K, Ibayash H. Suppresive effects of cholecystokinin and bombesin on growth hormone and prolactin secretion in urethane-anesthetized rats. Endocrinol J, 31:539-547, 1984.

98. Blackard WG, Waddell CC, 1969. Cholinergic blockade of growth hormone responsiveness to insulin hypoglycemia. Proc. Soc. Exp. Biol. Med. 131:192-196.

99. Scarpignato C, Tirelli F, Pontiroli AE. Bombesin inhibits growth hormone response to insulin-induced hypoglycemia in humans. Brain Res. 371:187-189, April 16, 1986.

100. Casanueva FF, Villaneuva L, Diequez C, Catranes JA, Diaz Y, Szoke B, Scanlon MF, Schally AV, Fernandez-Cruz A. Atropine blockade of growth hormone (GH)-releasing hormone-induced GH secretion in man is not exerted at pituitary level. J Clin Endocrinol Metab, 1986, 186-91.

101. Casanueva FF, Villanueva L, Cabranes JA, Cabezas-Cerrato J, Fernandez-Cruz A. Cholinergic mediation of growth hormone secretion elicited by arginine, clonidine, and physical exercise in man. J Clin Endocrinol Metab, 1984, 526-30.

102. Casaneuva FF, Villaneuva L, Diaz Y, Devesa J, Fernandez-Cruz A, Schally A-V. Atropine selectively blocks GHRH-induced GH secretion without altering LH, FSH, TSH, PRL, & ACTH/Cortisol secretion elicited by their specific hypothalamic releasing factors. Clin Endocrinol, 1986, 319-323.

103. Richardson SB, Hollander CS, D'Eletto R, Greenleaf PW, Thaw C. 1980. Acetylcholine inhibits the release of somatostatin from rat hypothalamus _in vitro_. Endocrinology 107:122-129.

104. Robbins RJ, Sutton RE, Reichlin S. 1982. effects of neurotransmitters and cyclic AMP on somatostatin release from cultured cerebral cortical cells. Brain Res. 234:377-386.

105. Peterfreund RA, Vale WW. 1983. Muscarinic cholinergic stimulation of somatostatin secretion from long-term dispersed cell cultures of fetal rat hypothalamus: inhibition by gamma aminobutyric acid and serotonin. Endocrinology 112:526-534.

106. Peterfreund RA, Vale WW. Somatostatin secretion from the hypothalamus In: Advances Exp Med Biol 188: 183-200, 1985.

107. Robbins R. Somatostatin and the cerebral cortex. In Advanc Exp Biol 188:201-216, 1985.

STRESS AND REPRODUCTION: PHYSIOLOGIC AND PATHOPHYSIOLOGIC INTERACTIONS BETWEEN THE STRESS AND REPRODUCTIVE AXES

Douglas Rabin, Philip W. Gold, Andrew N. Margioris, and George P. Chrousos

From the Developmental Endocrinology Branch
NICHD, NIH and the Biological
Psychiatry Branch, NIMH, ADAMHA
Bethesda, MD

INTRODUCTION

Stress is a ubiquitous feature of life. Although its prevalence as a factor in health and disease is difficult to ascertain, it is arguably a significant risk to the well being of the overall population. Stress and the response to stressful events, on the other hand, have played an important part in human survival and evolution. From earliest times, people have dealt with harsh elements and adverse situations requiring both behavioral and physiological responses to protect themselves and their societies.

At a fundamental level, mammalian species have developed a quite stereotypic repertoir of behavioral and physiologic responses to attain adaptation during stress (1). It appears that the CRH and locus coeruleus-norepinephrine systems in the brain represent the principal biological effectors of the stress response or general adaptation syndrome, as first described by Selye over 40 years ago (2). Set into motion during states of threatened homeostasis, these effectors act directly within the CNS to facilitate adaptive neural pathways subserving arousal and attention and inhibiting non-adaptive ones subserving feeding, sexual behavior, and reproduction (3). In the periphery, the CRH and locus coeruleus-norepinephrine systems, respectively, via the catecholamines and the glucocorticoids, act to redirect blood flow to the CNS and to mobilize fuel for immediate action.

The stress response can affect reproductive function in multiple ways. Generally, stress leads to suppression of reproductive function, which is presumably an adaptive response to conserve energy during hardship. For example, certain physiological states or disease states - such as depression and anorexia nervosa - are associated with abnormal activation of the stress responsive systems and are frequently associated with reproductive hypofunction. However, there also are several reproductive periods or states which may directly affect the function of the reproductive system. For example, states or periods of reproductive importance - such as puberty, infertility, premenstrual tension syndrome, menopause, pregnancy, labor and delivery and the puerperium - may act as the stressor that activates the stress systems.

In this chapter wc review the many interactions between stress and reproductive function from both of these viewpoints. A brief review of reproductive function preceeds this.

A BRIEF OVERVIEW OF REPRODUCTIVE FUNCTION

Reproduction is the culmination of individual reproductive maturation and the sexual interaction between the male and female of a given species (4). After birth, there is first a period of active reproductive suppression (childhood). This is followed by puberty, during which the reproductive organs mature and become capable of fertilization and conception. Following the advent of male sperm production and female ovulatory menses, reproduction is possible and at the appropriate time, the association of a sperm with an ovum results in fertilization. This is followed by implantation of the blastocyst in the womb, gestation, and delivery. Complementing this successful reproductive event is the initiation of an appropriate bonding relationship between parent and child, which is important to infant development. Finally, in the female, development of menopause marks the end of reproductive capability.

At birth, gonadotropin levels are relatively high and remain so for approximately two years. They are associated with some degree of testicular and ovarian function. They then fall to basal levels, which are quite low until the onset of puberty. Puberty is marked by an initial elevation of leuteinizing hormone (LH), first noted at night, and then by increasing plasma concentrations of follicle stimulating hormone (FSH). Eventually, the coordinated pulsatile release of LH and FSH result in mature testicular and ovarian function. In the male, FSH regulates the development of the seminiferous tubules and the ability to produce sperm. LH stimulates the actual production of sperm and of testosterone, which together with 5α- dihydrotesterone, a peripheral conversion product, cause masculinization. In the female, FSH and LH secretion result in ovarian production of estradiol and progesterone. Estradiol stimulates feminization. After several years, regular ovulatory cycles occur as the ovary maintains a coordinated production of estradiol and progesterone, which will last until menopause.

In the male, the development of masculinization, libido and testicular function are markers of the normal progression through puberty and the establishment of normal reproductive ability. These features are associated with the maturation of the hypothalamic-pituitary-testicular axis. One can measure testosterone, sperm count, and sexual performance in the evaluation of fertility. However, it is the production of offspring which attests to full reproductive capability.

In the female, reproductive maturation is marked by the development of feminization, libido, and normal ovarian function. The hypothalamic-pituitary-ovarian axis however, demonstrates more complex changes than the male reproductive axis. Ovulation is the result of a well-defined interaction that requires the pulsatile release of gonadotropin releasing hormone (GnRH), which in turn causes the pulsatile release of LH and of FSH. FSH stimulates the ovary to develop a dominant follicle. At the periovulatory period, estrogen feedback at the pituitary changes from negative to positive and in conjunction with a rise in progesterone assists the midcycle surge of LH which results in ovulation. This signals the end of the follicular phase of the menstrual cycle. The luteal phase, which follows ovulation, is named for the corpus luteum. Production of progesterone and estrogen by the corpus luteum during this phase of the cycle prepares the uterine endometrium for implantation. If conception does not occur the endometrium is shed during menses and the cycle is renewed. If conception is achieved, the corpus luteum supports pregnancy for 8 to 10 weeks, at which time the placental production of human chorionic gonadotropin (HCG) takes over the maintenance of the corpus luteum. HCG levels decline by 90 days gestation and the continued placental production of estrogen, progesterone and other hormones and growth factors supports fetal nourishment and growth until delivery.

The placenta produces corticotropin releasing hormone (CRH), resulting in high circulating plasma CRH concentrations in the pregnant mother (5-7). This hormone may drive the maternal pituitary-adrenal system during pregnancy. In addition the placenta secretes POMC-derived peptides primarily α-MSH and ß-endorphin and to a lesser extent ACTH (8). In pregnant women, bound and free plasma cortisol are significantly increased over the levels in nonpregnant women (9-17). Maternal cortisol is poorly suppressed by exogenous dexamethasone throughout pregnancy (9, 18).

The onset of labor may be signaled by the maturation of the fetus, although this is as yet unclear (19). It appears that labor is associated with intrauterine production of prostaglandins, arising from fetal membrane phospholipids, stimulated by phospholipase A2 (4,20). In conjunction with fetal oxytocin, prostaglandins initiate the first stage of labor (21). During labor maternal plasma concentrations of CRH increase and after delivery they rapidly decline to undetectable levels (7,22). Delivery of the fetus initiates the postpartum reproductive obligations of the mother, primarily those of bonding with and nurturing the infant.

The hypothalamic-pituitary-ovarian axis is suppressed post partum and does not recover for 4 to 6 weeks (22,23). A normal concomitant of the first 2 weeks postpartum also appears to be the relative suppression of the hypothalamic-pituitary-adrenal axis (9).

Following the reproductive years, women enter a period of menopausal transition during which there is a decline in ovarian function and estradiol production (25). This period may last for 5 - 10 years and ends with menopause, which is associated with high gonadotropin levels, low estrogen levels, and the concomitant changes associated with hypoestrogenemia.

MECHANISMS OF STRESS-INDUCED SUPPRESSION OF REPRODUCTIVE FUNCTION

Activation of the hypothalamic-pituitary-adrenal (HPA) axis in stress can affect reproductive function at all three levels of the hypothalamic-pituitary-testicular/ovarian axes as well as at the end-organs for the sex steroids (Figure 1). Suppression of reproductive function by concomittant activation of the HPA-axis or by exogenous administration of CRH or glucocorticoids has been well documented (26-29). Examples include the suppression of hypothalamic GnRH by CRH and glucocorticoids and of pituitary gonadotropins by cortisol. Cortisol also has a direct suppressive effect on the testicular production of testosterone, although it does not appear to suppress ovarian function directly (30-34). In addition, when dexamethasone is administered to immature rats it can inhibit the uterotrophic effect of estradiol (35). Moreover, opioid peptides, administered exogenously or produced during stress or after administration of CRH, also suppress GnRH secretion. While this effect can be reversed by the prior administration of naloxone, intracerebroventricular CRH may act directly or through an opioid mechanism that cannot be antagonized by naloxone (36,37). Serotonin, a hypothalamic CRH stimulator can suppress the LH surge at proestrus in the rat (38). Alpha-MSH, GABA agonists, and neuropeptide Y have also been implicated in the suppression of reproductive function, by preventing the proestrus LH surge in rats (39-41). Prolactin which can be elevated by stress, may alter the midcycle LH surge in humans, causing infertility correctable by bromocriptine (42).

In the rat, stress during pregnancy has been shown to impede the normal masculinization of the male offspring of stressed mothers (43,44). This effect may be due to production of higher than normal levels of CRH and cortisol resulting in the suppression of gonadal testosterone production at a critical time period in the fetus. In addition, stress hyperresponsive (Maudsley) rats demonstrate smaller litter size and an increased likelihood of maternal cannibalization (45). The biochemical characterization of this phenomenon is incomplete. In sheep, cortisol may precipitate premature labor, although this does not appear to occur in primates (46,47).

Generalized effects of stress mediated via starvation, war, crowding and climatic change have also been demonstrated to suppress reproductive function in both human and animal species (48-51). This suppression may be due to activation of the HPA axis and to other factors not well characterized.

PHYSIOLOGIC AND PATHOPHYSIOLOGIC STATES WHICH SUPPRESS REPRODUCTIVE FUNCTION

As noted above, both physical and emotional stress are capable of altering reproductive function. Exercise is an example of a frequently self-imposed physical stress which causes well-defined changes in the HPA-axis which may suppress reproductive function. Athletic training is associated with progressive incremental work loads which have been shown to increase levels of ACTH, cortisol, and presumably CRH (52). Men and women athletes demonstrate reductions in testosterone and estrogen production, respectively (53,54). In exercising women, secondary amenorrhea may occur in up to 60% of some reported series (54). Excessive exercise may be accompanied by weight loss, which independently has been associated with menstrual abnormalities (see below) (55,56).

Famine and inadequate caloric intake have also been shown to cause changes in reproductive function (4,57). Secondary amenorrhea ranges from 40 to 100 percent in patients who are starving and/or in those on low protein, vegetarian diets (57). The precise mechanism(s) for this effect are not known. Malnutrition, however, is clearly associated with increased activity of the HPA axis, which may cause reproductive hypofunction (58-60).

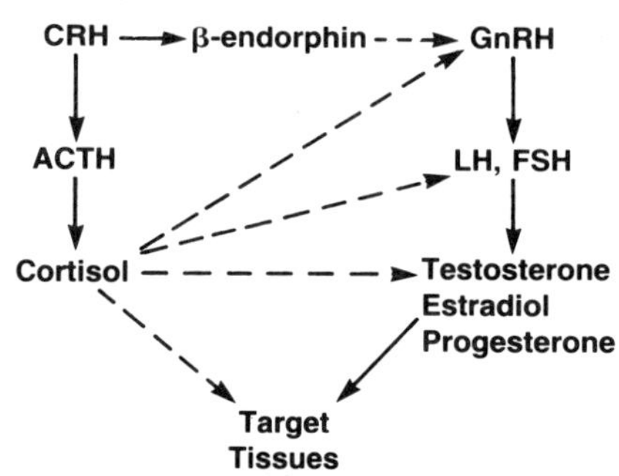

Figure 1. Interaction of the hypothalamic-pituitary-adrenal axis and reproductive axis. Solid lines represent stimulation; interrupted lines represent inhibition.

Systemic illness and hospitalization can increase cortisol production (4). Perhaps best described after surgery and recovery from anesthesia, cortisol and ACTH levels are significantly elevated and transient suppression of the hypothalamic-pituitary-testicular/ovarian axes occurs (61,62). Patients with chronic disease - such as cystic fibrosis, cancer, chronic renal failure, and inflammatory bowel disease - demonstrate stress elevations of ACTH and cortisol as well as concomittant suppression of gonadotropins. These patients may also be on systemic glucocorticoid therapy, which may also suppress reproductive function (30-32).

Stressful life situations which may be associated with elevations of ACTH and cortisol are associated with amenorrhea (63,64). Examples include the commencement of a new job, starting college, or being on death row awaiting execution (63-66). Female prisoners in the latter group have been reported to have a 100 percent prevalence of amenorrhea (49). Women inmates in a second World War concentration camp had a 54 percent prevalence of amenorrhea (66). "Life events" by their very nature are diverse and as stressors they can cause multiple biochemical responses, thus, the determination of a direct association between the effects of the HPA axis on the HPO axis is difficult. However, when measured in nonhuman primates or in human studies, stress-induced elevation of cortisol and secondary suppression of reproductive function are significantly correlated (67-69).

Cushing's syndrome is a pathologic condition associated with excessive cortisol secretion or exogenous glucocorticoid administration (70). Adrenal overproduction of cortisol, which may be the result of a pituitary or nonpituitary tumor secreting ACTH, the result of autonomous production of cortisol by adrenal tumors, or due to pharmacologic administration of glucocorticoid therapy can duplicate changes induced by stress (70). Diminished libido and suppression of reproductive function occur in both men and women with Cushing's syndrome (32). However, not all patients demonstrate the same degree of reproductive suppression.

Depression is a syndrome encompassing the feeling of being sad and blue and the inability to experience pleasure; it often is accompanied by crying and a sense of fear and anxiety. Depressed patients have strong biochemical evidence of a hyperactive HPA axis (71-74). They have a blunted ACTH response to exogenous CRH and they do not suppress their plasma cortisol levels after the administration of dexamethasone (72,73). Few studies have compared the prevalence of reproductive dysfunction in depression, in men or women, although anecdotal reports suggest a strong association between the two. Weight and appetite changes associated with depression, as well as depression's effects on other neurotransmitters, may also be important in suppressing reproductive function, although this is not well known (74).

Anorexia nervosa, an increasingly prevalent disorder in adolescent and young adult women and to some degree in adolescent and young adult men, is associated with a complete suppression of reproductive function. The etiology of this disorder is uncertain, but theories suggest that stress plays an important role. These patients are noted to have high ACTH and cortisol levels, not suppressible by dexamethasone. They also have a blunted ACTH response to ovine CRH (75). Two subsets may exist (76). One in whom amenorrhea occurs prior to weight loss and a second in whom weight loss occurs prior to amenorrhea. The distinction between these groups may be early as opposed to late elevation of CRH, however, there appears to be a continuum of disease marked by temporally increasing CRH concentrations and subsequent menstrual suppression. Weight recovery alone does not always correct the amenorrhea. Male runners, who demonstrate obsessive weight control, mild hypothalamic hypogonadism, change in libido, and similar psychological features with anorexia, may represent an analog of anorexia nervosa (76).

REPRODUCTIVE STATES ASSOCIATED WITH EMOTIONAL STRESS

Puberty is a period of striking physiological and psychological change. The term "adolescence" has been employed to express the psychosocial maturation process that goes along with puberty. The stress induced by the body's physical and physiological adjustment to adulthood is poorly understood and few studies have looked at the effects due to this stressor. (see Susman, et. al., this volume)

Another reproductive state associated with emotional stress is the syndrome of pre-menstrual tension (PMS). Symptoms of this disorder, which is distinguished from depression, include irritability, mood change, compromised daily performance, as well as weight, appetite, and perceptual changes during the immediate pre-menstrual period. Recovery occurs rapidly with the onset of menses (77-80). We postulate that this disorder is related to increased activation of the HPA axis and/or increased sensitivity to its function (83). In the luteal phase progesterone, a weak glucocorticoid antagonist, may stimulate the HPA-axis to increase hypothalamic CRH secretion. Women who develop pre-menstrual tension may be those who are more sensitive to this increased CRH. Their rapid recovery, with the onset of menses, may relate to the precipitous decline of progesterone and subsequently of CRH. No hormonal changes have been found which are consistently associated with this syndrome (84).

Pregnancy and the postpartum period also are associated with significant physiological and emotional stresses which may manifest in a group of depressive syndromes. Frequently, pregnancy related depression also results in disruption of maternal nurturing and support of the newborn (85-102). "Postpartum blues," which occurs in the first 2 weeks postpartum and may compromise maternal-infant bonding, may be related to a relative insufficiency of cortisol. The enormous psychological significance of the recently completed 9 month pregnancy, a time when cortisol and CRH are elevated, represents a response to both maternal expectations and the demands of motherhood, as well as to the remarkable maternal physiological changes which occur. This is especially true for primigravidas, in whom there is a higher incidence of postpartum depressive syndromes.

High levels of cortisol, which are pronounced in the third trimester and at delivery, sharply decline postpartum, while CBG declines more slowly (9,11). These lower levels of unbound cortisol (steroid withdrawal) may be the factor responsible for the development of the "postpartum blues" (9). As many as 80 percent of postpartum women develop this syndrome which is marked by tearfulness, fatigue, and feelings of not being able to accommodate the newborn infant(s)' needs (98).

With the recovery of the HPA-axis, which presumably occurs at about 3-6 weeks post delivery, frank "postpartum depression" may develop. This syndrome, occurring in up to 10 - 20 percent of pregnancies is marked by moderate to severe depression, a feeling of worthlessness, as well as the classic symptoms of depression, including appetite and sleeping changes. In addition, it can compromise the normal maternal-infant relationship (96). This syndrome may last for up to one year. The fact postpartum depression occurs when the HPA axis is recovering suggests that it may be associated with an abnormal adjustment to renewed CRH secretion. The finding that patients with PMS may have a higher incidence of postpartum depressive syndromes coincides with our hypothesis that an abnormal sensitivity to CRH in the premenstruum may be related to the symptoms of PMS (97).

The effects of postpartum depression not only extend to the mother, but also to the infant. Postpartum depression seriously compromises maternal function, affecting both nurturing and bonding and studies suggest that the ensuing development of infants whose mothers had a postpartum depressive episode is abnormal (83,84,94,96). Women who develop a first postpartum depressive syndrome are at higher risk for developing a second episode. If this syndrome is a continuum commencing with "postpartum blues", then this would suggest that the lower cortisol levels in the immediate postpartum period may result in a rebound elevation of CRH which could reset the HPA axis and cause greater CRH and cortisol production therafter. The more prolonged course of "postpartum depression" also might compromise future reproduction. If a woman has a negative psychological reaction to her own postpartum period, she may be discouraged from attempting a future pregnancy. Furthermore, no data exist on the possible disruption of the spousal relationship by pregnancy-related stress, although marital difficulties have been associated with postpartum depression (96).

Finally, menopause is defined as the cessation of menses for 1 year; the period of transition to menopause is associated with hot flashes, flushing, palpitations and thermal dysregulation in 40 - 70 per cent of patients. These symptoms usually remit even if estrogen replacement therapy is not provided. Reaching the end of one's reproductive life would seem to represent a significant stressor, however, the menopause is not associated with a greater prevalence of depression.

There are few studies which link the adverse symptoms of the menopause to stressful events. However the symptomatology and the signs of the menopause, including flushing and abnormal temperature regulation, as well as the secretion of LH in conjunction with these changes, argue for the possible secretion of CRH as the mediator of these symptoms. If this is the case then the menopause also might act as an endogenous stressor.

CONCLUSION

Stress itself and in response to reproductive events alters the normal reproductive function of men and women, and affects the lives of children and infants. Not all of these effects are well understood. The pervasive interaction between the hypothalamic-pituitary-adrenal and the hypothalamic-pituitary-ovarian and testicular axes, perhaps once a key to survival, represents a perturbation we are increasingly learning how to recognize, to study, and to treat.

REFERENCES

1. Munck A, Guyre PM, Holbrook NJ 1984 Physiological functions of glucocorticoids in stress and their relation to pharmacological actions. Endocr Rev 5:25.
2. Selye H 1936 The stress syndrome. Nature 138:32.
3. Munck A and Guyre PM: Glucocorticoid physiology pharmacology and stress, in Steroid Hormone Resistance, ed. by GP Chrousos, DL Loriaux and M Lipsett, Advances in Experimental Biology and Medicine Vol 196, 1986, Plenum Press, NYC, NY
4. Yen SSC and Jaffe RB: Reproductive Endocrinology, 1986, WB Saunders Co, Philadelphia PA
5. Sasaki A, Sato S, Murkami O, Go M et al 1987 Immunoreactive corticotropin-releasing hormone present in human plasma may be drived from both hypothalamic and extrahypothalamic sources. J Clin Endocrinol Metab 65:176.
6. Goland RS, Wardlaw SL, Stark RI, Brown LS et al 1987: High levels of corticotropin-releasing hormone immunoactivity in maternal and fetal plasma during pregnancy. J Clin Endocrinol Metab 63:1199.
7. Sasaki A, Liotta AS, Luckey MM, Margioris AN, Suda T, Kreiger DT 1984: Immunoreactive corticotropin-releasing factor is present in human maternal plasma during the third trimester of pregnancy, J Clin Endocrinol Metab 59:812
8. Margioris A, Grino M, Protos P, Gold PW, Chrousos G 1987 Corticotropin releasing hormone and oxytocin POMC peptide secretion from the human placenta in vitro. J Clin Endocrinol Metab (in press)
9. Jolivet A, Blanchier H, Gautry JP 1974 Blood cortisol variations during late pregnancy and labor. Am J Obstet Gynecol 119:775.
10. Willcox DL, Yovich JL, McColm SC, Schmitt LH 1985 Changes in total and free concentrations of steroid hormones in the plasma of women throughout pregnancy: effects of medroxyprogesterone acetate in the first trimester. J Endocrinol 107:293.
11. Carr BR, Parker Jr CR, Madden JD, Macdonald PC et al 1981 Maternal plasma adrenocorticotropin and cortisol relationships throughout human pregnancy. Am J Obstet Gynecol 139:416.
12. Peterson RE 1983 Corticosteroids and corticotropins. In: Fuchs F, Klopperz A (eds) Endocrinology of Pregnancy, Vol 3, Harper and Row, p 112.
13. Nolten WE, Lindheimer MD, Ruekert PA, Oparil S, Ehrlich EN 1980: Diurnal patterns and regulation of cortisol secretion in pregnancy, J Clin Endocrinol Metab 51:446.
14. Burke CW 1969 Biologically active cortisol in plasma of oestrogen-treated and normal subjects. Br Med J 2:798.
15. Plager JE, Schmidt KG, Staubitz WJ 1964 Increased unbound cortisol in the plasma of estrogen-treated subjects. J Clin Invest 43:1066.

16. Midgeon CJ, Kenny FM, Taylor FH Cortisol production Rate. VIII. Pregnancy. J Clin Endocrinol 28:661.
17. Pepe GJ, Townsley JD 1975 Cortisol metabolism in the baboon during pregnancy and at the postpartum period. Endocrinology 96:587.
18. Evans MI, Chrousos GP, Mann DW, Larsen JW et al 1985 Pharmacologic suppression of the fetal adrenal gland in utero. JAMA 253:1015.
19. Fuchs A-R, Fuchs F, Husslein P, Soloff MS et al 1982 Oxytocin receptors and human parturition: a dual role for oxytocin in the initiation of labor. Science 215:1396.
20. Casey ML, MacDonald PC 1984 Endocrinology of preterm birth. Clin Obstet Gynecol 27:562.
21. Casey ML, MacDonald PC, Mitchell MD 1985 Despite a massive increase in cortisol secretion in women during parturition, there is an equally massive increase in prostaglandin synthesis. A paradox? J Clin Invest 75:1852.
22. Morishita H, Higuchi K, Nakago K, Hashimoto T et al 1984 Hypothalamic function in women during the first month post-partum. Acta Endocrinol 105:145.
23 Iszizuka B, Quigley ME, Yen SSC 1984 Postpartum hypogonadotropinism: evidence for increased opioid inhibition. Clin Endocrinol 20:573.
24. Sherman BM, West JG, Korenman SG 1976 The menopausal transition: analysis of LH, FHS, estradiol, and progesterone concentrations during menstrual cycles of older women. J Clin Endocrinol Metab 42:629.
25. Rivier C, Rivier J, Vale W 1986 Stress-induced inhibition of reproductive functions: role of endogenous corticotropin-releasing factor. Science 231:607.
26. Rivier C, Vale W 1984 Influence of corticotropin-releasing factor on reproductive functions in the rat. Endocrinology 114:914.
27. Ogle TF 1977 Modification of serum luteinizing hormone and prolactin concentrations by corticotropin and adrenolectomy in ovariectomized rats. Endocrinology 101:494.
28. Baldwin DM, Sawyer CH 1974 Effects of dexamethasone on LH release and ovuulation in the cyclic rat. Endocrinology 94:1397.
29. Kakakura M, Takebe K, Nakagawa S 1975 Inhibition of luteinizing hormone secretion induced by synthetic LRH by long-term treatment with glucocorticoids in human subjects. J Clin Endocrinol Metab 40:774.
30. Bambino TH, Hsueh AJW 1981 Direct inhibitory effect of glucocorticoids upon testicular luteinizing hormone receptor and steroidogenesis in vivo and in vitro. Endocrinology 108:2142.
31. MacAdams MR, White RH, Chipps BE 1987 Reduction of serum testosterone levels during chronic glucocorticoid therapy. Ann Intern Med 104:648.
32. Cumming DC, Quigley ME, Yen SSC 1983 Acute suppression of circulating testosterone levels by cortisol in men. J Clin Endocrinol Metab 57:671.
33. Lescoat G, Lescoat D, Garner D 1984 Dynamic changes in plasma luteinizing hormone and testosterone after stress in the male rat. Influence of adrenalectomy. Can J Physiol Pharmacol 62:1231.
34. Rose RM, Bernstein IS, Gordon TP 1975 Consequences of social conflict on plasma testosterone levels in Rhesus monkeys. Psychosom Med 37:50.
35. Rabin D, McCLure E, Brandon D, Liapi C, Chrousos GP, Glucocorticoids inhibit estradiol-mediated uterine growth: Role of the uterine estradiol receptor. Abstract submitted to the Endocrine Society Meeting 1988.
36. Quigley ME, Yen SSC, 1980: The role of endogenous opiods in /lh secretion during the menstrual cycle J Clin Endocrinol Metab 51:179-181.
37. Rivier C, Vale W, 1984: Influence of corticotropin-releasing factor on reproductive functions in the rat, Endocrinol 114:914-921.
38. Koh T, Kinoshita F, Tsukada T, Tsujii S et al 1983 Serotonin mediated inhibition of pulsatile luteinizing hormone (LH) release during the immobilization stress in castrated rats. Proc 65th Ann Meet Endocrinol Soc 19:109.
39. Khorram O, deCastro JCB, McCann SM 1985 Stress-induced secretion of alpha-

melanocyte-stimulating hormone and its physiological role in modulating the secretion of prolactin and luteinizing hormone in the female rat. Endocrinology 117:2483.

40. Clark JT, Kalra PS, Kalra SP 1985 Neuropeptide Y stimulates feeding but inhibits sexual behavior in rats. Endocrinology 117:2435.

41. Adler BA, Crowley WR 1986 Evidence for gamm-aminobutyric acid modulation of ovarian hormonal effects on luteinizing hormone secretion and hypothalamic catecholamine activity in the female rat. Endocrinology 118:91.

42. Ben-David M, Schenker JG, 1983 Transient hyperprolactinemia: A correctable cause of idiopathic female infertility, J Clin Endocrinol Metab 57:442-4.

43. Pollard I, Dyer SL 1985 Effect of stress administered during pregnancy on the development of fetal testes and their subsequent function in the adult rat. J Endocrinol 107:241.

44. Pollard I, 1986: Prenatal stress effects over two generations in rats, J Endocrinol 109:239-244.

45. Broadhurst PL, 1960: Experiments in psychogenetics: applications of biomedical genetics to the inheritance of behavior. In: Eysenck HJ (ed) Experiments in personality, Vol. 1: Psychogenetics and Psychopharmacology, Routledge and Kegan Paul, London, pp 1-102.

46. Bassett JM, Thorburn GD 1969: Foetal plasma corticosteroids and the initiation of parturition in the sheep. J Endocrinol 44:285.

47. Liggins GC, Fairclough RJ, Grieves SA, Kendall JW et al 1973 The mechanism of parturition in the ewe. Recent Prog Horm Res 29:111.

48. Smith CA 1947 Effects of maternal undernutrition upon the newborn infant in Holland (1944-1945). J Pediatr 30:229

49. Drew FL 1961 The epidemiology of secondary amenorrhea. J Chron Dis 14:396

50. Christian J, Lloyd J, Davis D, 1964: The role of the endocrines in the self-regulation of mammalian population, Recent Prog Horm Res 21:501-578.

51. Harvey PW, Chevins PFD 1984 Crowding or ACTH treatment of pregnant mice affects adult copulatory behavior of male offspring. Horm Behav 18:101.

52. Luger A, Deuster P, Kyle SB, Gallucci WB, Montgomery LC, Gold PW, Loriaux DL, Chrousos GP, 1987: Acute hypothalamic-pituitary-adrenal responses to the stress of tradmill exercise: Physiologic adaptations to physical training, New Engl J Med 316:1309-1315.

53. Cummings DC, Rebar RW, 1983: Exercise and reproductive function in women, Am J Industr Med 4:113-125.

54. Bullen BA, Skrinar GS, Beitins IZ, Carr DB et al 1984 Endurance training effects on plasma hormonal responsiveness and sex hormone excretion. JAppl Physiol:Respirat Environ Exercise Physiol 56:1453.

55. Frisch RE, Wyshak G, Vincent L, 1980: Delayed menarche and amenorrhea in ballet dancers, New Engl J Med 303:17-19.

56. Warren MP 1980 The effects of exercise on pubertal progression and reproductive function in girls. J Clin Endocrinol Metab 51:1150.

57. Leonard PJ 1973 Malnutrition and the hypothalamic-pituitary-adrenal axis. In: Gardner LI and P Amacher(eds) Endocrine Aspects of Malnutrition, KROC Foundation, St. Ynez, California, pp 355-362.

58. Alleyne GAO, Young VH 1967: Adrenal function in children with severe protein-calorie malnutrition, Clin Sci 33:189-200.

59. Rao KSJ, Srikentia SG, Gopalan G 1968: Plasma cortisol levels in protein-calorie malnutrition, Arch Dis Child 43:365-367.

60. Smith SR, Bledsoe T, Chhetri MK 1975: Cortisol metabloism and the pituitary-adrenal axis in adults with protein-calorie malnuturition, J Clin Endocrinol Metab 40:43-52.

61. Udelsman R, Norton JA, Jelenich SE, Goldstein DS et al 1987 Responses of the hypothalamic-pituitary-adrenal and renin-angiotensin axes and the sympathetic system during controlled surgical and anesthetic stress. J Clin Endocrinol Metab 64:986.

62. Fuller GB, Hobson WC, Reyes FI, Winter JSD, Faiman C 1984 Influence of

restraint and ketamine anesthesia on adrenal steroids, progresterone, and gonadotropins in Rhesus monkeys (41825). Proc Soc Exp Biol Med 175:487.
63. Mason JM 1975 A historical view of the stress field. J Human Stress 1:6.
64. Bachman GA, Kemmann E, 1982 Prevalence of oligomenorrhea and amenorrhea in a college population, Am J Obstet Gynecol 144:98-102.
65. Pettersson F, Fries H, Nillius SJ, 1973: Epidemiology of secondary amenorrhea I. Incidence and prevalence rates, Am J Obstet Gynecol 117:80-86.
66. Bass F 1947 L'amenorrhee au camp de concentration de Terezin. Gynaecologia 123:211.
67. Adams MR, Kaplan JR, Koritnik DR 1985 Psychosocial influences on ovarian endocrine and ovulatory function in Macaca fascicularis. Physiol Behav 35:935.
68. Collu R, Gibb W, Ducharme JR 1984 Effects of stress on the gonadal function. J Endocrinol Invest 7:529.
69. Facchinetti F, Nappi G, Cicoli C, Micieli G et al 1986 Reduced testosterone levels in cluster headache: a stress-related phenomenon? Cephalagia 6:29.
70. Loriaux DL, Cutler GB Jr 1986 Diseases of the adrenal glands. In: Clinical Endocrinology, ed by Kohler PO; John Wiley & Sons, Inc., Vol 5, p 167-238.
71. Linkowski P, Mendlewicz J, Leclercq R, Brasseur M et al 1985 The 24-hour profile of adrenocorticotropin and cortisol in major depressive illness. J Clin Endocrinol Metab 61:429.
72. Gold PW, Loriaux DL, Roy A, Kling MA et al 1986 Responses to corticotropin releasing hormone in the hypercortisolism of depression and Cushing's Disease. N Engl J Med 314:1329.
73. Caroll BJ 1981 The dexamethasone suppression test in melancholia. Br J Psychiatry 140:292.
74. Rubin RT, Poland RE 1984 Pituitary-adrenocortical and pituitary-gonadal function in affective disorder. In: Brown GM et al (eds) Neuroendocrinology and Psychiatric Disorders, Raven Press, New York, p 151.
75. Gold PW, Gwirtsman H, Avgerinos PC, Nieman LK et al 1986 Abnormal Hypothal-amic-pituitary-adrenal function in anorexia nervosa. N Engl J Med 314:1335.
76. Kays WH, Gwirtsman HE, George DT, Ebert MH, Jimerson DC, Tomai TP, Chrousos GP, Gold PW, 1987: Elevated cerebrospinal fluid levels of immunoreactive corticotropin-releasing hormone in anorexia nervosa: Relation to state of nutrition, adrenal function, and intensity of depression, J Clin Endocrinol Metab 64:203-8.
77. Halbreich U, Endicott J 1981 Possible involvement of endorphin withdrawal or imbalance in specific premenstrual syndromes and postpartum depression. Med Hypotheses 7:1045.
78. Nadelson CC, Notman MT, Ellis EA 1983 Psychosomatic aspects of obstetrics and gynecology. Obstet Gynecol 24:871.
79. Watts JFF, Butt WR 1985 Hormonal studies in women with premenstrual tension. Br J Obstet Gynecol 92:247.
80. Clare AW 1985 Invited Review: Hormones, behaviour and the menstrual cycle. J Psychosom Rev 29:225.
81. Meltzer ES, Kumar R 1985 Puerperal mental illness, clinical features and classification: a study of 142 mother-and-baby admissions. Br J Psychiatry 147:647.
82. Wrate RM, Rooney AC, Thomas PF, Cox JL 1985 Postnatal depression and child development. Br J Psychiatry 146:622.
83. Naslund B, Persson-Blennow I, McNeil TF, Kaij L 1985 Offspring of women with nonorganic psychosis: mother-infant interaction at three and six weeks of age. Acta Psychiatr Scand 71:441.
84. Persson-Blennow I, Naslund B, McNeil TF, Kaij L 1986 Offspring of women with nonorganic psychosis: mother-infant interaction at one year of age. Acta Psychiatr Scand 73:207.
85. Gard PR, Handley SL, Parsons AD, Waldron G 1986 A multivariate

investigation of postpartum mood disturbance. Br J Psychiatry 148:567.
86. Saks BR, Frank JB, Lowe TL, Berman W et al 1985 Depressed mood during pregnancy and the puerperium: clinical recognition and implications for clinical practice. Am J Psychiatry 142:728.
87. Arnetz BB, Fjellner B 1986 Psychological predicators of neuroendocrine responses to mental stress. J Psychosom Res 30:297.
88. Wolkind S 1981 Depression in mothers of young children. Arch Dis Child 56:1.
89. Hopkins J, Marcus M, Campbell SB 1984 Postpartum Depression: A critical review. Psychol Bull 95:498.
90. Alder Em, Cox JL 1983 Breast feeding and post-natal depression. J Psychoso Res 27:139.
91. Handley SL, Dunn TL, Baker JM, Cocshott C et al 1977 Mood changes in puer-perium, and plasma tryptophan and cortisol concentrations. Br Med J 2:18.
92. Handley SL, Dunn TL, Waldron G, Baker JM 1980 Tryptophan, cortisol and puerperal mood. Br J Psychiatry 136:498.
93. Watson JP, Elliott SA, Rugg AJ, Brough DI 1984 Psychiatric disorder in pregnancy and the first postnatal year. Br J Psychiatry 144:453.
94. Davidson J, Robertson E 1985 A follow-up study of post partum illness, 1946-78. Acta Psychiatr Scand 71:451.
95. Schopf J, Bryois C, Jonquiere M, Scharfetter C 1985 A family hereditary study of post-partum "psychoses". Eur Arch Psychiatr Neurol Sci 235:164.
96. Kruckman L, Asmann-Finch C 1986 Postpartum Depression: A Research Guide and International Bibliography, Garland Publishing, Inc., New York & London.
97. Melges FT, 1968 Postpartum psychiatric syndromes, Psychosomatic Medicine, 1968, 30:95-108.
98. True-Soderstrom BA, Buckwalter KC, Kerfoot KM 1983: Postpartum Depression. Matern Child Nurs J 12:109-118.

HUMAN PLACENTA AND THE HYPOTHALAMIC-PITUITARY-ADRENAL AXIS

Andrew N. Margioris, Michel Grino, Douglas Rabin, and George P. Chrousos

Developmental Endocrinology Branch
National Institute of Child Health and Human Development
National Institutes of Health, Bethesda, MD 20892

INTRODUCTION

It has been suggested since the early 50's that the human placenta may influence the hypothalamic-pituitary-adrenal (HPA) axis during pregnancy by secreting an ACTH-like hormone (1-3). A number of recently published findings, regarding the expression of the proopiomelanocortin (POMC) and the corticotropin releasing hormone (CRH) genes in the human placenta, as well as the documentation of the *in vitro* synthesis and release of CRH, β -endorphin, α -MSH, and ACTH by perifused human placental fragments and human placental trophoblasts in culture, substantiate this hypothesis. This chapter reviews these new findings and speculates on their possible role in the regulation of the HPA axis during pregnancy.

MATERNAL PLASMA CORTISOL DURING PREGNANCY

Free and total maternal plasma cortisol rise progressively during pregnancy (4-7). The circardian periodicity of plasma cortisol persists but the evening levels do not decline as much as in non-pregnant women and in women on estrogen treatment (5,8,9). Dexamethasone administration (2 mg/ day x 3 or 7 days) during the third trimester of pregnancy does not suppress maternal plasma and urine free cortisol to the degree achieved in normal non-pregnant women (5,10).

Since estrogens and progesterone increase during pregnancy it would be proper to review first the effect of these steroids on the HPA axis of nonpregnant women.

Effect of estrogens on the HPA axis

Estrogens increase the plasma level of transcortin (11) which is the main cause of the increased levels of total plasma cortisol during pregnancy. Estrogens can also increase plasma cortisol by decreasing its metabolic clearance rate probably by direct action on liver enzymes involved in the inactivation of glucocorticoids (12); in fact 17-hydroxysteroids (17-OHS) can decrease by almost 50% following estrogen administration (13). In contrast,there is only a minimal decrease of 17-OHS during pregnancy in spite of the high levels of plasma estrogens (4,14,15). It has also been suggested that estrogens may be able to increase (by an unknown mechanism) plasma free cortisol, thus contributing to the decrease of plasma ACTH found in women on oral contraceptives (16).

Effect of progesterone on the HPA axis

Progesterone does not affect the concentration of plasma transcortin (17), but it exhibits some affinity for it thus competing with cortisol for its binding sites which means that high plasma progesterone levels can displace transcortin bound cortisol (18). In addition, progesterone has a direct effect on the pituitary and possibly on the hypothalamus as well. Progesterone added to primary anterior pituitary cultures decreases ACTH release (19). The net progesterone effect on the HPA axis is presumambly a decrease of plasma ACTH by at least two mechanisms: first by displacing cortisol from transcortin thus increasing plasma free cortisol which will feedback on the HPA axis, and second by directly inhibiting pituitary ACTH secretion. Indeed, the concentration of maternal plasma ACTH is slightly suppressed early in pregnancy, presumambly as a direct result of the combined effect of the increased levels of plasma estrogens and progesterone. It should be pointed out however that plasma ACTH increases as pregnancy progresses, in spite of the continually increasing estrogens and progesterone (6,16). The progressively rising levels of plasma ACTH during pregnancy, in spite of the increasing levels of plasma estrogens, progesterone and free cortisol, and the documended resistance of cortisol secretion to low doses of dexamethasone point to an additional factor influencing the HPA axis during pregnancy. Evidence is now accumulating that the human placenta may be the cause of this phenomena by secreting both CRH and POMC-derived peptides.

PLACENTAL AND PLASMA POMC PEPTIDES

Placental POMC-derived peptides

Immunoreactive (IR)-ACTH, IR- β -endorphin, and IR- α -MSH are present in human placenta extracts (10, 20-24). β -Endorphin has been shown to be synthesized *in vitro* by primary cultures of human placental trophoblastic cell (25). The reported placental concentration of each POMC-derived peptide varies with the extraction technique and antibody used. In general, it is several orders of magnitude less than that reported for the anterior pituitary. In particular, the reported concentration of placental IR-ACTH is around 1-4 ng/g of tissue wet weight (wt wt) (10,21,23) while the anterior pituitary concentration of ACTH is around 1 ug/ mg and its content 0.5-2 ug (26,27). Thus, the placental concentration of IR-ACTH is 1/1.000.000th that of pituitary but because of the size of human placenta (average weight is 500 g) its total IR-ACTH content is 1/250th to 1/500th that of anterior pituitary.

The reported concentration of placental IR- β -endorphin is 1-2ng/g wet weight, of IR- β -lipotropin 3-9 ng/g and of IR- α -MSH 0.2-0.6 ng/g (22,24).

Gel filtration of placental extracts shows that the IR-ACTH is composed of at least three different molecular weight (mol wt) components, the first eluting in the void volume, the second coeluting with synthetic ACTH, and the third coeluting with synthetic α -MSH; the placental IR- β - endorphin is composed of a void volume fraction, a second peak coeluting with β -lipotropin and a third one coeluting with synthetic human β -endorphin (23,28). Recently, a POMC-mRNA was identified in both human and rat placenta (29), though its size is different from that of rat anterior pituitary. The difference in size found between placental and pituitary POMC-mRNA is not due to difference in the length of the poly (A) tail. S_1 nuclease mapping analysis showed that these POMC-mRNAs have the same nucleotide seqn ence coding for ACTH, β -lipotropin and the untranslated 3' region; it seems that the difference is due to an altered splicing in the regipn coding for the N-terminal part of the precursor peptide.

Maternal plasma POMC-derived peptides during pregnancy

During the firs,few weeks of pregnancy the maternal plasma ACTH is significantly lower than that of nonpregnant women, but it rises progressively from 23±n4.6 pg/mL at 12 weeks to 59±16 pg/mL at 37 weeks (6,10). The diurnal variation of maternal plasma ACTH and free cortisol is similar with that retorted for nonpregnant women (5,6,8). Maternal plasma β -endorphin was reported to rise during pregnancy (30), though a subsequent study was unable to document such a rise (31). This discrepancy seems to be due to a shift of the β -lipotropin to β -endorphin ratio during pregnanry, with the level of β -endorphin progressively increasing while β -lipotropin is decreasing (7); thus, if a β -endorvhin antiserum with 100% crossreactivity towards β -lipotropin is used no difference in plasma IR- β -endorphin can be documented while a more specific antiserum (with lower or no β -lipotropin crossreactivity), will presumably show a clear β -endorphin risr. Finally, IR- α -MSH has been reported to be present in maternal plasma during the third trimester of pregnancy, while it is undelectable in the plasma of nonpregnant women (24).

Placental ACTH and opioid receptors

Human placenta has specific opioid binding sitH s (32,33). These binding sites have low affinity with morphine and enkephalins, but very strong affinity with the kappa receptor agonists.

5Fetal plasma POMC-derived peptides

Most of the values reported in the literature were measured in umbilical cord plasma. A significant aegative correlation has been reported between umbilical arterial pH, pO_2 and IR- β -endorphin (34).

Effect of glucocorticoids on plasma and placental POMC peptides

Dexamethasone administration (8 mg/day x 2 days) to pregnant women at term and to nonpregnant controls (21), showed that the concentration of ACTH in the plasma of the pregnant women failed to decrease to the same degree as in the control group (nonpregnant women). The placental content of IR-ACTH did not significantly change after dexamethasone administration (21).

IR-CRH IN PLASMA AND PLACENTA

During pregnancy, maternal plasma not only contains progressively rising levels of IR-ACTH, IR- β -endorphin, IR- α -MSH, bound and free cortiso but in addition also progressively increasing concentrations of IR-CRH as well.

Plasma IR-CRH in nonpregnant women

Low levels of IR-CRH are detectable in the plasma of nonpregnant women (35-37). Plasma IR-CRH shows no diurnal variation, and is not suppressible by low doses of dexamethasone (35-37). It has been postulated that most of this plasma IR-CRH is of extrahypothalamic origin (35,37).

Maternal plasma IR-CRH during pregnancy

Maternal plasma IR-CRH increases progressively with pregnancy to high levels (from few pg/mL to several ng/mL) (37-42). This plasma IR-CRH appears to be of placental origin, since its plasma levels correlate well with the concentration of placental IR-CRH and it rapidly decreases after delivery (35-42).

Gel chromatography of plasma from pregnant women at term, shows that this IR-CRH consists of at least two different molecular weight immunoreactive substances. The first has a mol wt of approximately 10 K daltons, while the second coelutes with synthetic human (h) CRH (37-39). Plasma IR-CRH is also bioactive, its bioactivity to immunoreactivity ratio is close to one (37,39).

Placental IR-CRH

Human placenta contains relatively high (10-20 ng/g wt wt) concentrations of IR-CRH (37,39). The gel chromatography profile of placental IR-CRH is similar with the profile of maternal plasma IR-CRH i.e. it contains high molecular weight IR-CRH in addition to the hCRH-sized IR-CRH. The later IR-CRH (the one which coelutes with the synthetic hCRH on gel chromatography), elutes in a single peak on subsequent PR-HPLC exhibiting

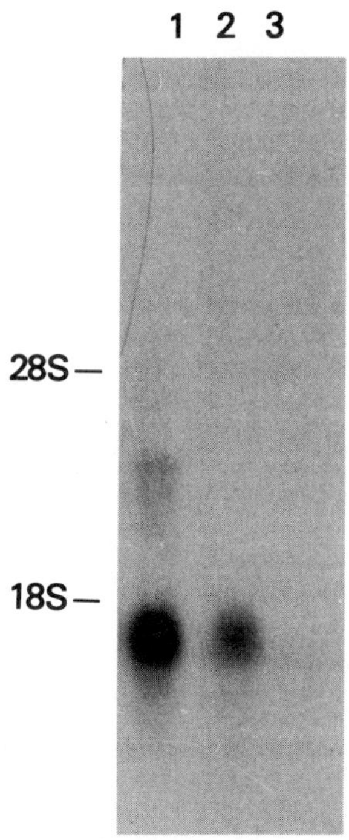

Fig. 1. Northern blot hybridization analysis of RNA isolated from human placenta (lane 1), rat hypothalamus (lane 2), and rat liver (lane 3). The amount applied was 5 ug of total RNA (lane 1), and 5 ug of poly (A) RNA (lanes 2,3). (From reference 43).

the retention time of synthetic hCRH (43). This further supports the postulate that this form of placental IR-CRH is similar to or identical with synthetic hCRH. Recently we have found on Northen Blot analysis of human term placental poly (A) RNA, a CRH-mRNA (43); this mRNA has the same size as rat hypothalamic CRH mRNA.

Human placenta secretes IR CRH *in vitro*. Gel filtration chromatography of placental perifusion effluent shows that only the synthetic hCRH-sized IR-CRH is secreted while the high mol wt IR-CRH is only present in the placental tissue extracts (44).

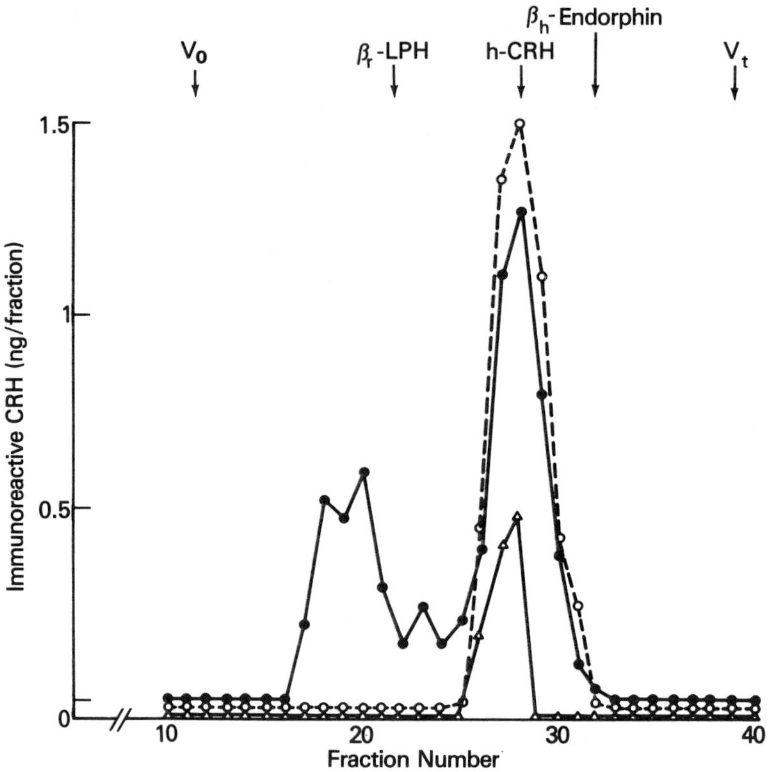

Fig. 2. Gel filtration Chromatography of IR-CRH from human placenta extracts (o---------o), and from pooled placental perifusion effluent chromatographed (o--------o), and subsequently rechromatographed (-------). The arrows indicate the void volume (V_o), and the elution position of rat- β -lipotropin, h- β -endorphin, and the total volume (V_t). (From reference 44).

A short exposure to a dlaolarizing dose of KCl at the end of each perifusion caused a several fold increase of the released IR-CRH. A four hour exposure of perifused plaf Rntal fragments to 10^{6} mol/L dexamethasone failed to decreased the basal and the KCl-induced secretion of placental IR-CRH (44).

Significance of placental and plasma CRH during pregnancy

Placental CRH may act on the maternal pituitary causing the increased levels of plasma ACTH and free cortisol during pregnancy, as well as the resistance to dexamethasone suppression observed in this condition. Two lines of evidence support such a role: a) the level of maternal plasma IR-CRH during the third trimester of pregnancy is above the minimal effective concentration of synthetic hCRH necessary to stimulate pituitary ACTH secretion in vivo (45), and b) maternal plasma IR-CRH has been shown to be also bioactive (37,39).

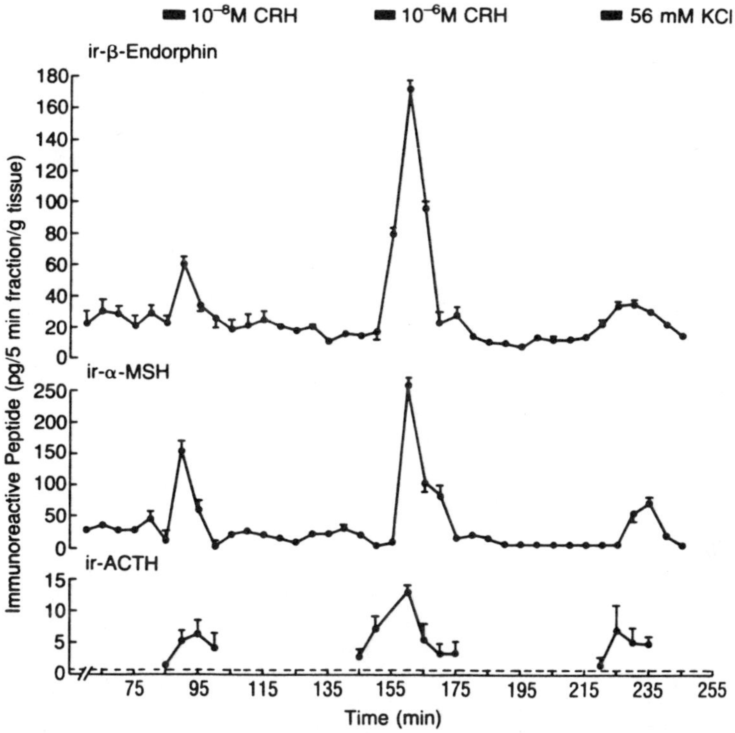

Fig. 3. Effect of synthetic hCRH and KCl on placental POMC-derived peptide secretion. Each point represents the main ($\pm$ SD) peptide concentration per effluent fraction of four chambers of two different placentae (From reference 44).

We have recently suggested an additional role for placental IR-CRH, that of a regulator of placental POMC-peptide secretion via a paracrine or autocrine action. Indeed, a brief exposure of the perifused placental fragments to CRH induced an acute increase of all placental POMC-derived peptides (44).

SUMMARY AND CONCLUSIONS

Maternal plasma bound and free cortisol rises during pregnancy, while maternal plasma IR-ACTH is initially low but soon rises inspite of the further increase of plasma free cortisol. This rise of plasma ACTH during pregnancy, can not be attributed to changes of plasma levels of estrogens or progesterone. It is possible that the human placenta is responsible for the rise in maternal plasma ACTH during pregnancy. There are two possible mechanisms by which this effect could occur: 1) by the placental secretion of CRH into the maternal circulation, which stimulates the maternal pituitary to secrete ACTH, and 2) the secretion of placental POMC-derived peptides. Recent data indicate that the human placenta is capable of both of these actions: A) The POMC and CRH genes are expressed in human placenta; B) the human term placenta is able to secrete both CRH and POMC-derived peptides in vitro; C) the CRH present in the plasma of pregnant women is bioactive and in sufficient levels to be effective on maternal pituitary; D) synthetic hCRH can stimulate the release of placental POMC peptides in vitro. We conclude that the human placenta may be a modulator of the HPA axis during pregnancy in a number of possible ways. Additional experimental work should clarify the intriguing interaction between the HPA axis and the human placenta during pregnancy, labor and delivery.

REFERENCES

1. Jailer JW, Knowlton S Adrenal-like activity in a pregnant Addisonian J Clin Invest 29:1430-1436, 1950.
2. Silman FG, Bergman FG, ACTH activity of placental extracts J Obstet Gynecol Br Empire 60:123-125, 1953.
3. Assali NS, Hamermesz J Adrenocorticorticotropic substances from human placenta Endocrinology 55:561-567, 1954.
4. Rosenthal HC, Slannwhite WR Jr, Sandberg AA Transcortin: A corticosteroid-binding protein of plasma.X. Cortisol and progesterone interplay and unbound levels of these steroids in pregnancy J Clin Endocrinol 29:352-367, 1969.
5. Nolten WE, Ruecker PA Elevated free cortisol index in pregnancy: Possible regulatory mechanisms Am J Obstet Gynecol 139:492-498, 1981.
6. Carr Br, Parker CR Jr., Madden JD, MacDonald PC, Porter JC Maternal plasma adrenocorticotropin and cortisol relationship throughout pregnancy Am J Obstet Gynecol 139: 416-422, 1981
7. Abou-Samra AB, Pugeat M, Dechaud H, Nachury L, Bauchareb B, Ferre-Montauge M, Tourniaire J Increased plasma concentration of N-terminal β -lipotropi and unbound cortisol durng pregnancy Clin Endocrinol 20:221-228, 1984.
8. Nolten WE, Lindhiemer MD, Rueckert PA, Oparel S, Ehrlich EN, Diurnal patterns and regulation of cortisol secretion in pregnancy J Clin Endocrinol Mateb 51:466-472, 1980.
9. Patric J, Challis J, Cambell K, Carmichael L, Natale R, Richardson R, Circadian rhythms in maternal plasma cortisol and estriol concentrations at 30 to 31, 34 to 35, and 38 to 39 weeks gestational age Am J Obstet Gynecol 136:325-333, 1980.
10. Rees LH, Buurke CW, Chard T, Evans SW, Letchwork AT Possible placental origin of ACTH in normal human pregnancy Nature 254:620-622, 1975.
11. Katz FH, Kappas A The effect of estradiol and estriol on plasma levels of cortisol and thyroid hormone-binding globulins and aldosterone and cortisol secretion rates in man J Clin Invest 46:1768-1777, 1967.
12. Kumagai A, Otomo M, Yano S, Takeuchi N, Nishino K, Ueda H, Ko S,

Kitamura M Inhibition of cortisol metabolism in the liver by other steroids Endocrinol Japon 6:86-92, 1959.

13. Robertson ME, Stiefel M, Laidlaw JW The influence of estrogens on the secretion, dispotion and biologic activity of cortisol J Clin Endocrinol 19:1381-1398, 1969.
14. Appleby JI, Norymberski JK The urinary excretion of 17-hydroxycorticosteroids in human pregnancy J Endocrinol 15:310-319, 1957.
15. Cohen M, Stiefel M, Reddy WJ, Laidlaw JW JC The secretion and disposition of cortisol during pregnancy J Clin Endocrinol 18:1076-1092, 1958.
16. Carr BR, Parker CR Jr., Madden JD, MacDonald PC, Porter JC Plasma levels of adrenocorticotropin and cortisol in women receiving oral contraceptive steroid treatment J Clin Endocrinol Metab 49:346-349, 1979.
17. Layne DS, Meyer CJ, Vaishwonar PS, Pincus G The secretion and metabolism of cortisol and aldosterone in normal and in steroid treated women J Clin Endocrinol 22:107-118, 1962.
18. Laatikainen T, Pelkonen J, Apter D, Ranta T Fetal and maternal serum levels of steroid sulfates, unconjugated steroids, and prolactin at term pregnancy and in early spontaneous labor J Clin Endocrinol Metabol 50:489-494, 1980.
19. Vale W, Rivier C, Yang L, MinicK S, Guillemin R Effects of purified hypothalamic corticotropin-releasing factor and other substances on the secretion of adrenocorticotropin and β -endorphin-like immunoreactivities in vitro Endocrinology 103:1910-1915, 1978.
20. Genazzani AR, Fraioli F, Hurlimann J, Fioretti P, Felder C Immunoreactive ACTH and cortisol plasma levels during pregnancy: detection and partial purification of corticotropin-like placental hormone: human chorionic corticotropin Clin Endocrinol 4:1-14, 1975.
21. Liotta AS, Osathandh R, Ryan KJ, Krieger DT Presence of corticotropin in human placenta: Demonstration of in vitro synthesis Endocrinology 101:1552-1558, 1977.
22. Nakai Y, Nakao K, Oki S, Imura H Presence of immunoreactive β -lipotropin and β -endorphin in human placenta Life Sci 23: 2013-2018, 1978.
23. Odagiri E, Sherrell BJ, Mount CD, Nicholson WE, Orth DN Human placental immunoreactive cortocotropin, lipotropin, and β -endorphin; evidence of a common precursor Proc. Natl Acad Sci USA 76:2027-2031, 1979.
24. Clark D, Thooly AJ, Shuster S, Bowers H Immunoreactive α -MSH in human plasma in pregnancy Nature 273:163-164, 1978.
25. Liotta AS, Houghten R, Krieger DT Identification of a β -endorphin -like peptide in cultured human placental cells Nature 295:593-595, 1982.
26. Krieger DT Placenta as a source of "brain" and "pituitary" hormones Biolog of reprod 26:55-71, 1982.
27. Krieger DT, Liotta AS, Brownstein MJ, Jimmerman EA ACTH, β -lipotropin and related peptides in brain, pituitary and blood Recent Prog Horm Res 36:272-344, 1980.
28. Liotta AS, Krieger DT In vitro biosynthesis and comparative posttranslational processing of immunoreactive precursor corticotropin / β -endorphin by human placental and pituitary cells Endocrinology 106:1504-1511, 1980.
29. Chen CLC, Chang CC, Krieger DT, Bardin CW Expression and regulation of proopiomelanocortin-like gene in the ovary and placenta: comparison with the testis Endocrinology 118:2382-2388, 1986.
30. Cenazzani AR, Facchinetti F, Parrini D β -lipotropin and β -endorphin plasma levels during pregnancy Clin Endocrinol 14:409-418, 1981.
31. Goland RS, Wardlow SL, Stark RI, Frantz AG Human plasma β -endorphin during pregnancy, labor and delivery J Clin Endocrinol Metab 52:74-78,1981.

32. Porthe G, Valette A, Moisand A, Tafani M, Cros J Localization of human placental opiate binding sites on the syncytial brush border membrane Life Sciences 31:2647-2654, 1982.
33. Valette A, Tafani M, Porthe G, Pontonnier G, Cros J Placental Kappa binding sites: interaction with Dynorphin and its possible implication in hcg secretion Life Sciences 33:523-526 (sup I), 1983.
34. Wardlaw SL, Stark RI, Baxi L, Frantz AG Plasma β -endorphin and β -lipotropin in human fetus at delivery: correlation with arterial pH and pO_2 J Clin Endocrinol 49:888-891, 1979.
35. Sasaki A, Sato S, Murakami O, Go M, Inone M, Shimizu Y, Hanew K, Andoh N, Sato I, Sasano N, Yoshinaga K Immulvreactive corticotropin-releasing hormone present in human plasma may be derived from both hypothalamic and extrahypothalamic sources eClin Endocrinol Metab 65:176-182, 1987.
36. Suda T, Tomori N, Yajima F, Sumitomo T, Nakagami Y, Ushiyanma T, Demura H, Shizume K Immunoreactive corticotropin-releasing factor in human plasma J Clin Invest 76:2026-2029, 1985.
37. Cunnah D, Jessop DW, Besser GM, Rees LH Measurement of circulating corticotropin-releasing factor in man J Endocrinol 113:123-131, 1987.
38. Sasaki A, Liotta AS, Luckey MM, Margioris AN, Suda T, Krieger DT Immunoreactive corticotropin-releasing factor is present in human maternal plasma during the third trimester of pregnancy J Clin Endocrinol Metab 59:812-814, 1984.
39. Sasaki A Shinkawa O, Margioris AN, Liotta AS, Sato S, Murakami O, Go M, Shimizu Y, Hanew K, Yoshinaga K Immunoreactive corticotropin-releasing hormone in human plasma during pregnancy, labor and delivery J Clin Endocrinol Metab 64:224-229, 1987.
40. Goland RS, Wardlaw SL, Stark RI, Brown LS Jr., Frantz AG High levels of coticotropin-releasing hormone immunoreactivity in maternal and fetal plasma during pregnancy J Clin Endocrinol Metab 63:1199-1203, 1986.
41. Linton EA, McLean C, N Kruseman AC, Tilders FJ, Van der Veen, Lowry PJ, Direct measurement of human plasma corticotropin-releasing hormone by "two-site" immunoradiometric assay J Clin Endocrinol Metab 64:1047-1053, 1987.
42. Cambell EA, Linton EA, Wolfe CDA, Scraggs PR, Jones MT, Lowry PJ, Plasma corticotropin-releasing hormone concentration during pregnancy and parturition J Clin Endocrinol Metab 64:1054-1059, 1987.
43. Grino M, Chrousos GP, Margioris AN The corticotropin releasing hormone gene is expressed in human placenta Biochem Bioph Res Commun 148:1208-1214, 1987.
44. Margioris AN, Grino M, Protos P, Gold PW, Chrousos GP Corticotropin releasing hormone and oxytocin stimulate the release of placental proopiomelanocorting peptides J Clin Endocrinol Metab (in press)
45. Schurmeyer TH, Avgerinos PC, Gold PW, Gallucci WT, Tomai TP, Cutler GB, Loriaux DL, Chrousos GP Human corticotropin-releasing factor in man: Pharmacokinetic properties and dose-response of plasma adrenocorticotropin and cortisol secretion J Clin Endocrinol Metab 59:1103-1107, 1984.

INDIVIDUAL DIFFERENCES AND THE STRESS RESPONSE: STUDIES OF A WILD PRIMATE

Robert M. Sapolsky

Department of Biological Sciences
Stanford University
Stanford, CA 94305

and

Institute of Primate Research
National Museums of Kenya
Karen, Nairobi, Kenya

INTRODUCTION

It has been half a century since Selye's first paper on stress and the General Adaptation Syndrome. As is apparent from this volume, stress physiology is now a rigorous and credible biomedical discipline. It is not clear, however, if those would have been the first attributes to come to mind at the mention of the subject during its earlier days. Much of this credibility has emerged with the uncovering of precise mechanisms to explain the relationships between environmental or emotional perterbations and physiological processes. Certainly, sensitive physicians long before Selye recognized that a patient's emotional state could influence disease outcome. Few, one suspects, imagined that emotion and disease would be shown to be linked by precise biochemical events.

Much of our progress has been to understand those links. By now, we have a reasonably solid knowledge of how, for example, glucocorticoids lyse lymphocytes or suppress inflammation. However, we have made nowhere near as much progress in uncovering mechanisms in another branch of stress physiology. Whether one is a physician, confronted with the mystery of how, against all logic, a patient fails to thrive, or a researcher confronted with the exasperating variability of study subjects, we only minimally understand individual differences. We all go through lives of stressors and challenges. Why do some succumb to stress-related diseases while others cope and adapt? Certainly, we differ as to the stressors to which we are exposed, but given the same stressors, we differ dramatically in our means of psychologically and physiologically coping. What mechanisms underlie these individual differences? How do they relate to behavior, psychological makeup or social status? In effect, what is the psychobiology of our differing vulnerabilities to stress?

These are difficult questions. In studying relationships between individual differences in behavior and in stress physiology in rodents, one might monitor various physiological events precisely. However, there are limits in extrapolating from the psychic life of a rat to our own. If captive primates are studied, one must be very attuned to their natural behavior and demographics (as well as be uncommonly well-funded) in order to provide housing that does not distort their behavior drastically. Otherwise, one far too readily winds up modeling the psychobiology of human imprisonment. Finally, in epidemiologic studies of humans, one may uncover correlates of rich and subtle psychologic processes. However, it can be quite difficult to also uncover the physiologic underpinnings of these correlates. For example, consider the range of speculations which seek to explain the reported connection (62) between a major depression and an increased risk of cancer decades later.

In studying individual differences in stress physiology, I have attempted a rather different approach to the issues. I have studied the relationship between the social rank of a male primate and the quality of his endocrine stress response. However, rather than utilize captive primates, I have studied a population of primates in their natural habitat, living freely in a national park in East Africa.

THE OLIVE BABOON AND THE STRATIFICATION OF HIS SOCIETY

My study subject has been the olive baboon (Papio anubis), a large Old World monkey widely distributed throughout Africa. The study site has been one of the last untouched expanses of woodland-grassland savannah in which vast numbers of mammals still live undisturbed, the Serengeti of Tanzania and Kenya. Within this protected region, baboons live in highly-organized social groups of 20-200 individuals. It is the richness of the Serengeti ecosystem that makes the study of these resident animals particularly appropriate. These baboons need only a small percentage of each day to meet their nutritional requirements. Furthermore, infant mortality is low. Finally, because of their size and social organization, the baboons are rarely threatened seriously by predators (46). Much like in our own society, few of these baboons experience major environmental stressors. Instead, there is the time and energy available to indulge in creating social stressors. In effect, within these troops, if you are a baboon and you are feeling miserable, it is most likely due to the machinations of another baboon.

The hub around which these machinations revolve is the dominance hierarchy. Resources and individuals in the world of the male baboon are limited. There are only finite amounts of food (especially desirable meat from prey species) and limited numbers of other baboons to groom you or mate with you. There is a limit to how many spots are shaded from the tropical sun or offer protection from a predator. Dominance hierarchies emerge from the competition for and unequal distribution of these resources. Obviously, each resource item can be contested with overt aggression, with the winner of the interaction claiming the desired item. However, such overt aggression is rare among primates. When there has been an aggressive interaction and a decisive victor has emerged, it is rare for the next confrontation between the pair to be settled in a similarly aggressive manner. Instead, a series of conventionalized gestures have evolved, allowing the previous winner to remind the loser of the prior outcome and to hint at potential violence with a display of his canines, or a stereotyped lunge. These interactions also allow the prior loser to re-express his subordinance with crouching or perhaps presenting his buttocks. These features--that all individuals are not equally capable of gaining access to contested resources, and that one can save much energy and minimize risk by recognizing these differences without having to test them each time--combine to form dominance hierarchies.

Among male baboons, these hierarchies are fairly linear, and the rewards of dominance are dramatic--among the twenty or so mature males in a troop, it is the top four or five who account for 95% of the meat consumption, and do 95% of the matings (28,46,53). Ranks shift with time, as the physical strengths of a male (as well as, most probably, his degree of motivation) change over his lifespan. Nevertheless, hierarchies will remain static for long periods, as the status quo is maintained by the conventionalized rituals of dominance and subordinance. At times, a subordinate forgoes the rituals and instead overtly challenges a dominant individual. And on occassion, the outcome differs from the previous time, and the rankings change.

As noted, attaining and maintaining high rank can revolve around explosive moments of aggression, and olive baboons have among the highest rates of escalated aggression observed in non-human primates. And, as noted, males more frequently rely upon the conventionalized gestures that reinforce the direction of dominance in a dyad. But subtler strategies are also available besides violence and the threat of violence. Males may incessantly harass a sexual consortship between a rival and an estrus female--never overtly challenging the male, but shadowing the individual for days. In some cases, this leads to confrontation between the pair of males, but with a comparable frequency, the consorting male may ultimately relinquish the consortship (53). The male may slowly develop non-sexual affiliations with females--what has been legitimately referred to as "friendships"--carrying their infants, sustaining long mutual grooming sessions. That female is then less likely to cooperate when in a consortship with another male, frequently manipulating the circumstances to transfer the consortship to the "friend" (64). Finally, males may not necessarily compete solitarily, but may form a cooperative coalition with another male which, when stable (a relatively rare event) is essentially unbeatable (44).

Thus, for the male baboon, social competition involves months of building social affiliations, days of inexorably harassing an opponent, or seconds of dangerous violence. The successful male seems to combine strength, good health, optimal amounts of aggression and bravery, intelligence, and skill at forming and abandoning alliances. What I set out to test was whether the physiology of a high-ranking male--especially the physiology intrinsic to the stress response--distinguishes him from his subordinate conspecifics. (For more detailed descriptions of baboon social life, consult 28,31,46,47,64 and 67).

To do so, one must first observe the behaviors of these animals, using the rigorous quantitative approaches pioneered by earlier field primatologists. These will tell you the rates and directions of interactions and reveal the dominance hierarchy at that particular time. It is then necessary to capture the individuals. To do so, I anesthetize them with a dart fired from a blowgun, under extremely controlled circumstances. Subjects must be darted at the same time of day, to avoid circadian fluctuations in hormones. To avoid skewing normative values, animals cannot be darted if they have recently been ill, injured, in a fight or have mated. For some of the metabolism studies described, they cannot yet have eaten that morning. The baboon must have his back turned when he is darted, so that he never sees the blowgun; this precludes the possibility of activating the stress response in anticipation of the darting. Finally, the first blood sample must be obtained within a few minutes of the onset of the stressful anesthetic effects, so that hormone concentrations in that first sample have not yet changed significantly from basal values (51).

This approach makes it possible to anesthetize animals under controlled and comparable conditions. Subjects are then given a variety of tests of endocrine function, plasma samples are collected, and animals are allowed to recover from the anesthesia before being returned to their troop. These studies have uncovered a number of dramatic rank-related differences in the quality of the stress response in these animals.

THE TESTICULAR AXIS

One of the hallmarks of the stress response is the disruption of sexual physiology and behavior in essentially every species examined. Among males, a consistent manifestations of this is a decline in testosterone concentrations. Not surprisingly, this pattern is observed among these baboons. In examining a number of different baboon troops over numerous years, basal testosterone concentrations are remarkably consistent. However, two cases of suppression have been noted, namely during a period of sustained social instability (which will be discussed below at length), and during the African drought of 1984. During each of these times, basal testosterone concentrations were approximately halved (53,56).

Therefore, both an ecologic and a social stressor can inhibit the testicular axis. In addition, the stressor of the darting process itself suppresses the axis (the stressor most probably not being the pain of the intramuscular injection, but rather the muscle spasms beginning just before unconsciousness [51]). In that circumstance, LH concentrations plummet within an hour, followed shortly afterward by declining testosterone concentrations (54).

It has been possible to gain a fairly detailed understanding of the neuroendocrine mechanisms that mediate the inhibition of the testicular axis in these baboons. The suppression at the level of LH secretion appears to be mediated by stress-induced opiate secretion. Blockade of opiate receptors with high doses of naloxone prevents the stress-induced decline in LH concentrations. A body of *in vitro* and *in vivo* studies indicate that, in fact, opiates are likely to bring this about by directly inhibiting hypothalamic LHRH secretion (11,48); this has not been possible to test directly in these subjects. The other major stress-responsive endocrine axes are unlikely to contribute to inhibiting LH secretion. Blockade of sympathetic catecholamine secretion following darting stress with the ganglionic blocker chlorisondamine fails to prevent the decline in LH concentrations. Similarly, inhibition of glucocorticoid secretion with the adrenocortical steroid synthesis inhibitor metyrapone also fails to prevent the decline (Sapolsky and Krey, in prep.). Furthermore, administration of exogenous glucocorticoids does not dampen pituitary sensitivity to LHRH (54). Finally, we have been unable to detect an increase in prolactin concentrations in response to darting, making it unlikely that this hormone contributes to the inhibition (Sapolsky and Petraglia, unpubl.) (despite its demonstrated potential for inhibiting the testicular axis during stress [49]).

The opiate effect at the level of the pituitary or higher is not the only point at which the testicular axis is disrupted in these animals during stress. As evidence, while naloxone administration potentiates LH secretion following darting stress, it is nonetheless incapable of preventing the decline in testosterone secretion. Thus, a stress-induced inhibition of testicular sensitivity to LH must also occur. This appears to be mediated by the elevated glucocorticoid concentrations that are found following darting. Exogenous administration of glucocorticoids causes a prompt decline in testosterone secretion. As already noted, this is not accomplished by inhibition of pituitary sensitivity to LHRH. Rather, the steroid inhibits testicular sensitivity to LH (54).

Thus, opiate action at the pituitary or higher, as well as glucocorticoid inhibition at the testes contribute to the inhibition of the testicular axis of these animals during stress. The potency of these regulatory effects varies among baboons according to their social rank; the result is a dramatic rank-related difference in their response to stress. Specifically, while all males ultimately have diminished testosterone concentrations 6-8 hours after darting, high-ranking males (by the criteria of success in approach-avoidance interactions [53]) are actually capable of increasing their testosterone concentrations during the first few post-stress hours. This is in sharp contrast to subordinates, in whom concentrations promptly decline (55). The neuroendocrine mechanisms underlying this rank-related pattern are coming to be understood. The individual differences do not occur at the pituitary level, as both dominant and subordinate males show an identical decline in LH concentrations during the immediate post-stress period (55). Instead, they appear to occur at the testicular level, involving the aforementioned inhibition by glucocorticoids of testicular responsiveness to LH. High-ranking males do not secrete less inhibitory glucocorticoids than do subordinates. Instead, the testes of dominant males are less sensitive to the suppressive actions of glucocorticoids (54). This would, perhaps, account for why the testosterone concentrations of these males might decline more slowly than in subordinates. A seemingly novel catecholaminergic mechanism appears to account for why androgen concentrations can actually RISE during that time. In high-ranking males, inhibition of sympathetic catecholamine secretion during stress with chlorisondamine prevents the typical rise in testosterone concentrations. In contrast, the drug has no effect on the testicular axis of subordinates (55).

Thus, relative buffering from the suppressive effects of glucocorticoids and a novel catecholaminergic potentiation of testosterone secretion combine to produce a unique gonadal response to stress observed only in dominant males. At present, it is not known if the protection from the glucocorticoid inhibition is due to a diminution of glucocorticoid receptor concentrations in the testes. It is also not clear whether catecholamines directly stimulate testosterone secretion from the testes or increase the flow of blood and LH through the testes by dilating the testicular vasculature (precedent for both mechanisms exist [22,29]); finally, it is not known whether the unique pattern of high-ranking males is due to greater sympathetic outflow during stress or to enhanced testicular sensitivity to catecholamines. What is important, for the purposes of this chapter, is that the manner in which the gonadal axis in these males initially responds to stress can be diametrically opposite, depending upon the social rank of the individual.

It is important to consider possible functional consequences of this rank-related pattern. A high- and low-ranking male tensely face each other, perhaps in contesting a sexual consortship. What are the consequences of testosterone concentrations rising steadily in the former, and declining in the latter such that after an hour, the dominant individual might have three times the circulating androgen concentrations of the subordinate? These differences are unlikely to effect reproductive physiology--they are simply too rapid and the end-organs show little evidence of being sensitive to three-fold differences in concentrations (13,15,37,49). More likely would be a relative potentiation of aggressiveness in the dominant male. This effect is likely to be limited, as limbic mediation of aggression appears to be only somewhat more sensitive to (relatively) small differences in androgen concentrations than is the reproductive axis (17). Most likely would be a rank-related difference in muscle metabolism, given the known time-course and sensitivity with which androgens influence muscle physiology (39).

It should be noted that while these dominant males are uniquely capable of elevating testosterone concentrations during stress, they do not have the highest basal concentrations (51,55). An early and often mis-interpreted study demonstrated that in a recently formed social group of rhesus monkeys, dominant males had the highest basal testosterone concentrations (50). However, it has become clear that that is an attribute of unstable dominance hierarchies--recently formed, recently reorganized, or atypically aggressive. As such captive groups are allowed sufficient time to form stable dyadic relations, the correlation between dominance and high testosterone concentrations disappears (19,26,34). As will be discussed below, among these wild baboons during periods of social instability and hierarchical reorganization, it is indeed high-ranking males who have the highest rates of aggression and the highest basal testosterone concentrations. However, during the stable periods discussed so far, neither is an attribute of dominance.

THE ADRENOCORTICAL AXIS

As is no doubt familiar to every reader, adrenocortical activation and enhanced glucocorticoid secretion are hallmarks of the response to virtually every stressor. Therefore, it is not surprising that the stressor of darting causes a prompt and significant rise in cortisol concentrations in every baboon examined over the years. A far more interesting case of elevated cortisol secretion is observed among these animals. Consistently, subordinate males have elevated basal cortisol concentrations, relative to dominant individuals (51,52). The magnitude of the hypersecretion in subordinates can be quite marked, with AM basal values even in excess of 30 μg/100 ml (in contrast to dominant individuals, who typically have concentrations ranging from 10-20 μg/100 ml). In addition to the basal hypersecretion of the steroid, subordinate males show a relatively sluggish adrenocortical response to the stressor of darting. Although males of both ranking cohorts ultimately attain the same plateau concentrations of cortisol, the rate of rise of cortisol concentrations during the early post-stress period among subordinates is less than that in dominant males (51).

For those familiar with baboon social behavior, it is not surprising that subordinate males should have basal concentrations of cortisol in the range typically seen in response to an overt stressor. As discussed above, baboon social interactions are frequently competitive, aggressive and unpredictable, and the subordinate male is subject to the highest rates of social stressors. When they obtain a highly desirable food item, such as a prey species, they are the least likely to be able to hold on to it. Furthermore, even when they are feeding upon something as quotidian as a tuber, they are frequently displaced from the item. Their sexual consortships and grooming bouts are the most likely to be disrupted. Finally, even if they choose not to involve themselves in overtly agonistic challenges to dominant males, they are subject to the highest rates of displaced aggression--being attacked as an "innocent bystander" when a nearby dominant male looks for an outlet for his aggression (53). Thus, their social lives are filled with lack of control or predictability over contingencies, hallmarks of a psychologically stressful milieu (69). In agreement with this interpretation, it has been reported frequently in other species that within a stable dominance hierarchy, subordinate individuals tend to have the highest circulating glucocorticoid concentrations and/or enlarged adrenals. This has been observed among primates (1,2,16,25,38,61,63,70), rodents (4,5,8,14,36,45,65), wolves (20) and spotted hyenas (Frank, Glickman and Sapolsky, in prep.).

The mechanisms underlying the hypercortisolism and sluggish adrenocortical responsiveness in the subordinate male are being unravelled. Dominant and subordinate individuals do not differ in their clearance rates of 3H-cortisol, indicating that they must differ in their cortisol secretory rates (52). The rank-related differences are not a peripheral phenomenon--males of differing ranking cohorts do not differ in their adrenal responsiveness to an ACTH challenge (52). Instead, the hypercortisolism of the subordinate males appears to arise because their pituitaries or brains are less sensitive to glucocorticoid feedback inhibition. As evidence, they show the classic marker of hypercortisolism in the human depressive--dexamethasone resistance (52). This includes the longest latencies after dexamethasone administration before showing a significant decline in cortisol concentrations, as well as both lesser absolute and relative declines in concentrations over the post-dexamethasone period. This basal hypercortisolism and blunted responsiveness to feedback regulation is, as mentioned, strikingly similar to that observed in the depressive. Similar stigmata are noted in both humans and rodents maintained under chronic stress (the rodent, of course, hypersecreting corticosterone, rather than cortisol) (6,10,33,58). Recent studies suggest that the adrenocortical hypersecretion and dexamethasone resistence of both the depressive human (22,30,43) and of the chronically-stressed rat (59) are due to CNS, rather than pituitary dysfunction. We have proposed a model whereby stress-induced down-regulation of hippocampal glucocorticoid receptors desensitizes the structure to circulating glucocorticoids, attenuates its normal role as an inhibitor of the adrenocortical axis and thus eventuates in hypersecretion and feedback insensitivity (60). In the case of the subordinate baboon, it is not yet clear whether the hypersecretion arises from desensitization of the pituitary or the CNS; however studies are underway modeled on those of Gold and colleagues (23; which implicated the CNS in depressive hypercortisolism) to clarify this point.

From the perspective of information theory, the adrenocortical axis of the subordinate male baboon is inefficient. The signal is amplified during basal periods and the speed with which the signal is further amplified during an acute stressor is attenuated. Thus, the signal to noise ratio is quite poor. From the perspective of everything known about pathophysiologic consequences of excessive exposure to elevated glucocorticoid concentrations (35), the adrenocortical axis of the subordinate male baboon is pathogenic. Are there any deleterious consequences to this hypersecretion?

There may be at least two, the first involving a consistent rank-difference in cholesterol transport. In both human and non-human primates, atherosclerosis and coronary heart disease are directly associated with serum low density lipoprotein (LDL) concentrations and inversely with concentrations of high density lipoproteins (HDL). Cholesterol association with LDL appears to promote atherosclerosis via the accumulation of LDL-cholesterol on arterial walls and the subsequent formation of foam cells and arterial lesions. In contrast, HDL appears to be anti-atherogenic by mediating "reverse cholesterol transport" from arterial walls to the liver for catabolism (9). In these baboons, HDL-cholesterol concentrations are consistently suppressed in subordinate males. This pattern has been observed in a number of troops over numerous years. In addition, concentrations of the major apolipoprotein associated with HDL (apoprotein A-I) are also suppressed. There is no rank-difference in concentrations of LDL-cholesterol or of its major apoprotein (57). In the primate, individual variation in these cholesterol measures are most readily explained by differences in gender, testosterone concentrations, diet, gene pool, degree of senescence and extent of exposure to stress (7,18,24,27,32,40). We have been able to eliminate all but the last factor as likely sources of variability. First, males are obviously all of the same gender and basal testosterone concentrations do not correlate with any of the cholesterol measures. Males of differing ranks have essentially the same diet in this population and spend equal time feeding. Meat comprises less than .5% of their diet (46); while it is true that it is dominant males who do the bulk of the meat eating, a significant increase in dietary cholesterol in the baboon eventuates in a different pattern of cholesterol measures than observed in dominant males (41). Because ranks change over time (discussed below), high-ranking males in one year's study may be subordinates at some later point. Therefore, the two ranking-cohorts cannot come from different gene pools. Finally, senescence cannot contribute to the suppressed HDL-cholesterol concentrations in subordinate males, as the trait is as pronounced among subordinates who are subadult as in those who are older adults. Instead, the increased rate of exposure to social stressors among subordinates seems to be the most likely cause of the rank-related pattern. Basal cortisol concentrations are inversely correlated with concentrations of

HDL-cholesterol and Apoprotein A-I (57). It is not yet clear whether there is causality in this correlation. However, it has been shown that chronic exposure to elevated glucocorticoid concentrations will raise the HDL-cholesterol:LDL-cholesterol ratio (3,66), as observed here in subordinate males.

Ongoing studies suggest a second important area in which the hypercortisolism of the low-ranking male may have functional consequences. As is well-known, stress can disrupt immune function, and a large proportion of this suppression is mediated by glucocorticoids. The steroids decrease circulating lymphocyte number; in some species, this is accomplished by sequestering of cells out of the circulation, whereas in "steroid-sensitive" species, lymphocytes are actually lysed. Furthermore, glucocorticoids inhibit the release of, and target tissue sensitivity to, Interferon and Interleukins 1 and 2, all intercellular mediators of immune activation. Finally, they stimulate certain angiogenic factors which promote tumor ingrowth (12,21,42). In agreement with these patterns, circulating lymphocyte numbers decline some 60% during the 8 hours following darting. This phenomenon appears to be glucocorticoid-mediated, as it can be prevented by metyrapone administration. That suggests that subordinate males, with their elevated basal cortisol concentrations, might have suppressed basal lymphocyte counts, and this has been observed recently in both troops in which it was examined (Sapolsky, in prep.). The magnitude of the suppression is approximately 40%, and lymphocyte counts are inversely correlated with basal cortisol concentrations. Current studies examine whether the immune systems of subordinate males are functionally impaired, and whether the elevated cortisol concentrations are indeed a causal mechanisms for the suppressed lymphocyte counts.

BEHAVIORAL/ENDOCRINE CORRELATES: IS THERE CAUSALITY IN THE RELATIONSHIP?

These studies have shown that testicular and adrenocortical physiology differ dramatically according to the social rank of the individual baboon. An emerging body of studies suggest that these differences may have implications for immune function and metabolism. Thus, there is a dramatic relationship between social rank and physiology in these animals. But is there indeed a relationship; do the unique endocrine traits of the dominant male really arise from his dominance? As noted, high-ranking males tend to be of prime age, in good health and well-fed. Do, for example, low basal cortisol concentrations result from those traits and are, in reality, unrelated to the fact that that well-fed male also happens to be socially dominant? Asked differently, will those well-fed, healthy, prime-aged males show these unique endocrine features regardless of their social milieu?

This is a difficult question to answer in the field, as one tries at all costs to avoid manipulating the study subjects. However, in 1981, a spontaneous period of social instability arose which cast a great deal of light on these issues. At the start of the season, the highest-ranking male was under considerable pressure from a cohort of subordinates, ranks 2-7. Shortly after that, those six formed a cooperative coalition, seriously injured the highest-ranking male, and he dropped out of all social interactions. The coalition promptly disintegrated, and three months of sustained social instability ensued. Throughout that time, those six males were clearly in the dominant cohort, supplanting individuals of ranks 7 and lower. However, rank relations among themselves shifted continuously and unpredictably. Rankings changed daily; the rates of dominance reversals rose dramatically. The rates of consortship harassments rose and the ability of a dominant male to maintain a sexual consortship declined. High-ranking males, unlike in all other years, had the highest rates of aggressive interactions. Feeding was suppressed, coalitions were formed frequently and disintegrated nearly as fast.

Throughout this period, all of the usual advantages of dominance in a stable hierarchy--high degrees of control and predictability over contingencies--were lost. If, despite all those changes in the psychological and social milieu in which they were dominant, they showed the same exact unique endocrine patterns observed in every other year, then endocrine function would probably not be very sensitive to the emotional state of the male.

Instead, an endocrine pattern was observed which was distinct from that of any other year. Among these combating dominant males, basal cortisol concentrations were significantly elevated, and cortisol secretory responsiveness to stress was inhibited. Furthermore, high-ranking males lost their ability to elevate testosterone concentrations during the early post-stress period (53).

These studies demonstrate that the unique physiological features of high-ranking males do not merely arise from the developmental or nutritional correlates of their dominance. Not only is their physiology sensitive to the actuality of their social dominance, but the context in which that dominance occurs can profoundly influence the physiology. In effect, there is no single physiologic picture of social dominance; dominance in a stable hierarchy is dramatically different from dominance in an unstable one, from both a behavioral and physiologic standpoint (26,34,49,50). The 1981 period of instability demonstrates that when the psychologic advantages of dominance are lost (without loss of dominance itself), the physiologic advantages are also lost.

CONCLUSIONS AND FUTURE DIRECTIONS

Thus, among these primates, who you are, what your place is in your society, and what sort of society it is appear to have everything to do with your physiology, both under basal and stressed circumstances. Furthermore, one may argue at this stage that these rank-related differences in physiology are of consequence, that the patterns observed in dominant males seem to be the most adaptive. These findings may not be surprising, given the fairly similar conclusions that have arisen from careful study with laboratory animals living in stable social groups. Nevertheless, they are at least comforting, as they confirm that the patterns uncovered with captive animals also exist within the relatively subtle range of variation found in groups of primates in their natural habitat, exposed to their normal arrays of stressors.

However, these studies should not merely exist to validate those of captive primates. Rather, they will soon be able to offer a richness unlikely to be available within the laboratory. These will arise in at least two areas:

--Anyone studying baboons is struck with the individuality of these animals, their personalities and the range of competitive strategies that they employ. Yet, in these studies to date, these individualities have been collapsed into the relatively stilted dichotomy of labeling a male either as in the dominant or subordinate cohort. As is apparent to any primatologist, there are tremendously different STYLES of each. For example, among dominant males, they may achieve and maintain dominance through the skillful development of cooperative coalitions, or they may compete individually. They may or may not have high rates of agonistic confrontations with opponents and displacement of aggression upon subordinates. Ultimately, they may relinquish their high rank voluntarily, or it may require an incapacitating injury from an opponent (See F. de Waal's "Chimpanzee Politics", 68, for a brilliant analysis of these factors in chimpanzee social behavior). As these free-living subjects are followed over the years, with new individuals transfering into study troops, there will soon be sufficient numbers of cases of each style to see if they have their own physiological correlates.

--Males differ in rank, and the style with which they attain and maintain their rank. In addition, dominance rankings change dramatically over the lifespan, and the longitudinal patterns differ among males. Some may rise abruptly and achieve high rank early in adulthood only to be badly injured and rapidly plummet in status. Others may spend years ensconsed just below the most competitive stratum of the dominant cohort. Others may never achieve anything resembling high rank. Can adolescent endocrine profiles predict these eventual adult behavioral profiles? Will, for example, males who were very high ranking and those who were always subordinate differ in the lengths and qualities of their old ages?

These are obviously not easy questions to answer. However, the sheer individuality of these animals, which makes such studies scientifically worth pursuing, also makes the process of the pursuit so gratifying.

ACKNOWLEDGMENTS

I am grateful for the financial support of the Harry Frank Guggenheim Foundation, and of the Life Sciences Research Foundation, of which I was a Mathers Fellow during parts of this study. I thank the Office of the President and the Ministry of Tourism and Wildlife, Republic of Kenya, for permission to conduct this research. Invaluable field assistance was provided by Richard Kimoso Kones, Josiah Masau, Francis Milili, Francis Onchiri, Hudson Oyaro, Dianne Rich and Reed Sutherland.

REFERENCES

1. M. Adams, J. Kaplan, T. Clarkson, and D. Koritnik, Ovariectomy, social status and atherosclerosis in cynomolgus monkeys, Atheroscler. 5:192 (1985).
2. M. Adams, J. Kaplan, and D. Koritnik, Psychosocial influences on ovarian endocrine and ovulatory function in Macaca fascicularis, Physiol Behav 35:935 (1984).
3. D. Adlersberg, L. Schaefer, and S. Drachman, Development of hypercholesterolemia during cortisone and ACTH therapy. J. Am. Med. Assoc. 144:909 (1950).
4. J. Archer, Effects of aggressive behavior on the adrenal cortex in laboratory mice, J. Mammal. 51:327 (1970).
5. S. Barnett, Competition among wild rats, Nature 175:126 (1955).
6. A. Baumgartner, K. Graf, and I. Kurten, The dexamethasone suppression test in depression, in schizophrenia and during experimental stress, Biol. Psychiat. 20:675 (1985).
7. D. Bradley, J. Wingerd, D. Petitti, R. Krause, and S. Ramcharan, High-density-lipoprotein cholesterol and oral contraceptin, N. Eng. J. Med. 299:17 (1978).
8. F. Bronson, and B. Eleftheriou, Chronic physiologic effects of fighting on mice, Gen. Comp. Endo. 4:9 (1964).
9. M. Brown, and J. Goldstein, Lipoprotein metabolism in the macrophage: implications for cholesterol deposition in atherosclerosis, Ann. Rev. Biochem. 52:223 (1983).
10. D. Ceuleman, H. Westenberg, and H. Van Praag, The effect of stress on the DST, Psychiat. Res. 14:189 (1984).
11. M. Ching, Morphine suppresses the proestrus surge of GnRH in pituitary portal plasma of rats, Endocrinology 112:2209 (1983).
12. T. Cupps, and A. Fauci, Corticosteroid-mediated immunoregulation in man, Immuno. Rev. 65:133 (1982).
13. J. Davidson, C. Camargo, and E. Smith, Effects of androgen on sexual behavior in hypogonadal man. J. Clin. Endo. Metab. 48:955 (1979).
14. D. Davis, and J. Christian, Relation of adrenal weight to social rank of mice, Proc. Soc. Exp. Biol. Med. 94:728 (1957).

15. D. Demassa, E. Smith, and B. Tennent, The relationship between circulating testosterone levels and sexual behavior, Horm. Behav. 8:275 (1977).
16. F. Dessi-Fulgheri, P. Messeri, and C. Prisca, A study of testosterone, estradiol, cortisol and prolactin in a socially intact group of Japanese macaques, Anthropol. Contemp. 4:123 (1981).
17. A. Dixson, Androgens and aggressive behavior in primates: A review, Agg. Behav. 6:37 (1979).
18. B. Flow, and G. Mott, Genetic mediation of cholesterol metabolism in the baboon (Papio cynocephalus), Atherosclerosis 41:403 (1982).
19. G. Eaton, and J. Resko, Plasma testosterone and male dominance in a Japanese macaque (Macaca fuscata) troop compared with repeated measures of testosterone in laboratory males, Horm. Behav. 5:251 (1974).
20. M. Fox, and R. Andrews, Physiologic and biochemical correlates of individual differences in behavior of wolf cubs, Behavior 46:129 (1973).
21. J. Folkman, R. Langer, C. Haudenschild, and S. Taylor, Angiogenesis inhibition and tumor regression caused by heparin or a heparin fragment in the presence of cortisone, Science 221:719 (1983).
22. A. Frankel, and E. Ryan, Testicular innervation is necessary for the response of plasma testosterone levels to acute stress, Biol. Repro. 24:491 (1981).
23. P. Gold, L. Loriaux, A. Roy, M. Kling, J. Calabrese, C. Kelolner, L. Nieman, R. Post, D. Pickar, W. Gallucci, P. Avgerinos, S. Paul, E. Oldfield, G. Cutler, and G. Chrousos, Responses to corticotropin-releasing hormone in the hypercorticsolism of depression and cushing's disease, N. England J. Med. 314:1329 (1986).
24. R. Goldberg, D. Rabin, A. Alexander, G. Doelle, G. Getz, Suppression of plasma testerone leads to an increase in serum total and high density lipoprotein cholesterol and apoprotein A-I and B, J. Clin. Endo. Metab. 60:203 (1985).
25. M. Golub, E. Sassenrath, and G. Goo, Plasma cortisol levels and dominance in peer groups of rhesus monkey weanlings, Horm. Behav. 12:50 (1982).
26. T. Gordon, R. Rose, and I. Bernstein, Seasonal rhythm in plasma testosterone levels in the rhesus monkey (Macaca mulatta): A three-year study, Horm. Behav. 7:229 (1976).
27. S. Grundy, Circulation 69:5 (1984).
28. G. Hausfater, Dominance and reproduction in baboons: A quantitative analysis, Contrib Primatol, Vol 7, Karger, Basel (1975).
29. N. Hodsin, The nerves of the testis, epididymis, and scrotum, in: "The Testis," D. Johnson, W. Gome, and N. Vandermark, eds., Academic Press, NY (1970).
30. F. Holsboer, U. von Bardeleben, A. Gerken, G. Staller, and O. Muller, Blunted corticotropin and normal cortisol response to human corticotropin-releasing factor in depression. N. Engl. J. Med. 311:1127 (1984).
31. A. Jolly, "The Evolution of Primate Behavior," 2nd edition, MacMillan, New York (1986).
32. W. Kannel, and T. Gordon, "The Framingham Study," NIH Publication 74-599 (1974).
33. E. Katz, H. Weiner, and T. Gallagher, Beta-Endocrphin immunoreactivity and acute behavioral distress in children with leukemia, J. Nerv. Ment. Dis. 170:72 (1982).
34. E. Keverne, R. Meller, and J. Eberhart, Dominance and subordination: Concepts or physiological states?, in: "Advanced Views in Primate Biology," V. Chiarelli, and J. Corruccini, eds., Springer-Verlag, NY (1982).
35. D. Krieger, "Cushing's Syndrome", Springer-Verlag, Berlin (1982).
36. C. Louch, and M. Higginbotham, The relation between social rank and plasma corticosterone levels in mice, Gen. Comp. Endo. 8:441 (1967).

37. P. Lundberg, and L. Wide, Sexual function in males with pituitary tumors, Fert. Steril. 28:175 (1978).
38. K. Manogue, D. Candland, and A. Leshner, Dominance status and adrenocortical reactivity to stress in squirrel monkeys (Saimiri scuireus), Primates 16:457 (1975).
39. S. Max, and J. Toop, Androgens enhance in vivo 2-deoxyglucose uptake by rat striated muscle, Endocrinology 113:119 (1983).
40. H. McGill, The relationship of dietary cholesterol to serum cholesterol concentrations and to atherosclerosis in man, Am. J. Clin. Nutr. 32:2664 (1979).
41. G. Mott, C. McMahan, J. Kelley, C. Farley, and H. McGill, Influence of infant and juvenile diets on serum cholesterol, lipoprotein cholesterol, and apolipoprotein concentrations in juvenile baboons (Papio sp.), Atherosclerosis, 45:191 (1982).
42. A. Munck, P. Guyre, and N. Holbrook, Physiological functions of glucocorticoids in stress and their relation to pharmacological actions, Endo. Rev. 5:25 (1984).
43. C. Nemeroff, E. Widerlov, and G. Bisette, Elevated concentrations of CSF corticotropin-releasing factor-like immunoreactivity in depressed patients. Science 226:1342 (1984).
44. C. Packer, Reciprocal altruism in Papio anubis, Nature 265:441 (1977).
45. N. Popova, and E. Naumenko, Dominance relation and the pituitary-adrenal system in rats, Anim. Behav. 20:108 (1972).
46. J. Popp, "Male Baboons and Evolutionary Principals." Ph.D. Dissertation, Harvard University, (1978).
47. T. Ransom, "Beach Troop of the Gombe." Bucknell Press, Lewisburg, (1981).
48. D. Rassmussen, J. Liu, P. Wolf, and S. Yen, Endogenous opioid regulation of gonadotropin-releasing hormone release from the human fetal hypothalamus in vitro, J. Clin. Endo. Metab. 57:881 (1983).
49. R. Rose, Psychoendocrinology, in: "Textbook of Endocrinology," J. Wilson, ed., 7th edition, Saunders, Philadelphia (1985).
50. R. Rose, J. Holaday, and I. Bernstein, Plasma testosterone, rank and aggressive behavior in male rhesus monkeys, Nature, 231:366 (1971).
51. R. Sapolsky, The endocrine stress-response and social status in the wild baboon, Hormones and Behav., 16:279 (1982).
52. R. Sapolsky, Individual differences in cortisol secretory patterns in the wild baboon: role of negative feedback sensitivity, Endocrinology 113:2263 (1983).
53. R. Sapolsky, Endocrine aspects of social instability in the olive baboon (Papio anubis), Am. J. Primatology, 5:365 (1983).
54. R. Sapolsky, Stress-induced suppression of testicular function in the wild baboon: Role of glucocorticoids, Endocrinology 116:2273 (1985).
55. R. Sapolsky, Stress-induced elevation of testosterone concentrations in high-ranking baboons: Role of catecholamines, Endocrinology, 118:1630 (1986).
56. R. Sapolsky, Endocrine and behavioral correlates of drought in the wild baboon, Am. J. Primatology, in press (1986).
57. R. Sapolsky, and G. Mott, Social subordinance in wild baboons is associated with suppressed high density lipoprotein-cholesterol concentrations: A possible role for chronic social stress, Endocrinology, submitted.
58. R. Sapolsky, L. Krey, and B. McEwen, Glucocorticoid-sensitive hippocampal neurons are involved in terminating the adrenocortical stress response. Proc. Nat. Acad. Sci. USA 81:6174 (1984).
59. R. Sapolsky, L. Krey, and B. McEwen, The neuroendocrinology of stress and aging: The glucocorticoid cascade hypothesis, Endocrine Rev 7:284 (1986).
60. R. Sapolsky, and B. McEwen, Why dexamethasone resistance? Two possible neuroendocrine mechanisms, in: "HPA Physiology and Pathophysiology," A. Schatzberg and C. Nemeroff, eds., Raven Press, NY (1987).

61. E. Sassenrath, Increased adrenal responsiveness related to social stress in rhesus monkeys, Horm. Behav. 1:283 (1970).
62. R. Shekelle, W. Raynor, and A. Ostfeld, Psychological depression and 17-year risk of death from cancer, Psychosom. Med. 43:117 (1981).
63. C. Shively, and J. Kaplan, Effects of social factors on adrenal weight and related physiology of Macaca fascicularis, Physiol. Behav. 33:777 (1984).
64. B. Smuts, "Sex and Friendship in Baboons," Aldine Press, NY (1986).
65. C. Southwick, and V. Bland, Effect of population density on adrenal glands and reproductive organs of CFW mice, Am. J. Physiol. 197:111 (1959).
66. M. Stern, O. Kolterman, J. Fries, H. McDevitt, G. Reaven, Adrenocortical steroid treatment of rheumatic diseases, Arch. Intern. Med. 132:97 (1973).
67. S. Strum, Agonistic dominance in male baboons: An alternative view, Int. J. Primatology 3:175 (1982).
68. F. de Waal, "Chimpanzee Politics," Harper and Row, New York (1982).
69. J. Weiss, Somatic effects of predictable and unpredictable shock, Psychosom. Med. 32:397 (1970).
70. U. Yodyingyuad, C. Riva, D. Abbott, J. Herbert and E. Keverne, Relationship between dominance hierarchy, cerebrospinal fluid levels of amine transmitter metabolites (5-hydroxyindole acetic acid and homovanillic acid) and plasma cortisol in monkeys, Neurosci. 16:851 (1985).

PSYCHOLOGICAL AND SOMATIC FACTORS IN WOMEN WITH PAIN DUE TO PELVIC CONGESTION

Richard Beard*, Philip Reginald* and Shirley Pearce**

*Department of Obstetrics and Gynaecology
St Mary's Hospital Medical School
London W2
**Department of Psychology
University College
London WC1

SUMMARY

Chronic pelvic pain in association with dilated pelvic veins and polycystic ovaries is a common complaint in women in the reproductive years, regardless of whether or not they have been pregnant. These women have a higher incidence of deaths and illnesses amongst family members and close relatives which makes them pay more attention to physical complaints such as pain which may make them more susceptible to environmental "stress" of daily life. A hypothesis is put forward which would link the effects of stress to the somatic changes observed in women with what is here described as the pelvic pain syndrome.

INTRODUCTION

Pelvic pain is the single most frequent complaint of women referred to hospital for a gynaecological opinion (Henker 1979, Morris and O'Neill 1958). Table 1 shows it was also found to be the commonest reason for a laparoscopy in a survey carried out by the Royal College of Obstetricians and Gynaecologists in 1978.

Table 1. Indications for diagnostic laparoscopy (Royal College of Obstetricians and Gynaecologists (London) Survey, 1978

Indication for Laparoscopy	n	%
Pain	10,825	51.7
Infertility	9,048	43.1
Infertility and pain	186	0.9
Menorrhagia	84	0.4
Ovarian enlargement	389	1.9
Lost IUCD	156	0.7
Other	283	1.3
Total	20,971	100

It is also well recognised that only about 30-50 per cent of women with chronic pelvic pain have a lesion visible on laparoscopy to which their pain can be ascribed (Gillibrand 1981). Clinical experience suggests that this figure may be considerably lower since the association of pain with "pinhead" deposits of endometriosis or from adhesions from a past attack of pelvic inflammatory disease, is not reliable evidence of causality.

There is a vast literature on the subject of chronic pelvic pain in women extending over at least the last 150 years. In 1831, Gooch made some perceptive observations on this condition noting that this pain occurred most commonly just prior to menstruation and was associated with an "exquisitely tender" and bulky uterus. He recognised that the condition was not a form of chronic pelvic inflammation but rather ascribed it to pelvic neuralgia. Since that time many causes have been proposed for this condition which we have called pelvic pain syndrome (PPS) the chief of which are summarised in Table II.

Undoubtedly, the major reason for the lack of progress in defining the cause of this condition over the last century has been the necessity of opening the abdomen of an otherwise healthy young woman. The relatively recent advent of laparoscopy, pelvic venography and more recently of pelvic ultrasound has afforded the possibility of more systematic and in depth investigation of the condition. As such, it has proved possible to exclude a number of the proposed causes in Table II.

There is no evidence that infection is a common cause. Studies by Missanelli (1982) and an unpublished study by ourselves have failed to isolate organisms from the pouch of Douglas. Laceration of the broad ligament presumably as a result of trauma as proposed by Allen and Masters (1955) is an uncommon cause, if it is a cause at all, because such lacerations are rarely found at laparotomy (Renaer et al. 1980) and the clinical syndrome of chronic pelvic pain occurs with equal frequency in women who have not been pregnant and those who have (Beard et al. 1984). Sclerocystic ovaries as described by Taylor (1949a) is a synonym for thickening of the capsule of the ovaries with multiple underlying cysts.

Table II. Summary of conditions proposed as causes for the pelvic pain syndrome (PPS)

Chronic parametritis. (Freund 1933, Schultze 1875)	Pelvic congestion. (Allen 1971, Taylor 1949a,b,c, Duncan and Taylor 1952)
Broad ligament laceration. (Allen and Masters 1955)	Pelvic sympathetic syndrome. (Theobald 1951)
Sclerocystic ovary. (Taylor 1949a, Netter 1953)	Dilated pelvic veins (varicosities). (Dudley 1888, Topolanski-Sierra 1958), Chidikel and Edlundh 1968, Kaupilla 1970)
Broad ligament neuritis. (Young 1938)	
Chronic pelvic pain without obvious pathology. (Renaer et al. 1980)	Dilated pelvic veins and congestion. (Beard et al. 1984)

Our ultrasound studies (Adams et al., Presentation at British Congress of Obstetrics and Gynaecology, Cardiff, April 1986) showed that 60 per cent of women with pelvic pain and dilated pelvic veins have polycystic ovaries, and that occasionally the cysts are as large as those shown by Taylor (1949a). Nevertheless, although the ovaries are the most tender organs in the pelvis of these women (Beard et al. 1986) they are not the only pelvic organs to be affected. It is our view that, while the ovary is a part of the pathophysiology of this condition, it is not the primary cause.

Dilated Pelvic Veins

Many authors have observed the presence of dilated pelvic veins in association with this condition. A century ago both Tait (1883) and Dudley (1888) made clinical and postmortem observations. With the advent of pelvic venography (Topolanski-Sierra 1958) the presence of these veins in women with pelvic pain was confirmed by numerous authors (Hughes and Curtis 1962, Kaupilla 1970, Beard et al. 1984). Some doubt exists as to whether the description "varicosity" is justified. Kaupilla (1970) described the histological changes as "fibrosis of the tunica intima, hypertrophy of the muscular layer of tunica media, and cellular proliferation of the capillary endothelium ..." eg changes commonly found in lower limb varicosities (Ochsner and Mahoner 1939). However, our preliminary studies have shown that in women with dilated veins on venography, who have been treated with medroxyprogesterone acetate for 4 to 6 months, repeat venography at the end of this period revealed that the pelvic veins had narrowed and that congestion was no longer present. For this reason, at present, it seems sensible to avoid the use of the term "varicosity".

Pelvic Congestion as a Cause of Pelvic Pain

Our studies (Beard et al. 1984) have shown that 84 per cent of women with a history of pelvic pain of more than 6 months duration with no obvious pathology on laparoscopy have dilated veins (most commonly the ovarian vein) and pelvic congestion. It seems likely that these veins are the cause of the pain because:

i) Standing exacerbates and lying alleviates the pain.
ii) Postcoital ache and congestive dysmenorrhoea are usually common associated symptoms.
iii) The administration of dihydroergotamine during an acute attack of pain narrows the pelvic veins, clears the congestion, and relieves the pain.

Factors Responsible for Dilated Pelvic Veins

Anatomical. The complex venous drainage system in the pelvis comprises many thin-walled unsupported veins which are devoid of valves and have weak attachments between the adventitia and supporting connective tissue (von Peham and Amreich 1934) - quite different from veins elsewhere in the body. Teleologically, the likely explanation for these singular characteristics is the need to accommodate the considerable increase in blood flow from the uterus that occurs in pregnancy. Hodgkinson (1953) reported that in pregnancy the ovarian vein dilated to more than three times its non-pregnant diameter. The ready ability of these veins to adapt to the demands of pregnancy may provide a clue to the abnormal dilation commonly found in pregnancy. The peculiar anatomical characteristics of the pelvic veins makes them particularly liable to dilate in the non-pregnant state, given the appropriate stimulus.

Hormonal. The most likely factor responsible for dilating pelvic veins is a hormone (Stearns and Sneeden 1965) - a vasoactive agent which originates from the ovary and probably from the placenta. This is supported by the observations of many authors that pelvic congestion is a condition of women confined to the reproductive years (Gordienne and Lhoist 1979). McCausland et al. (1961) considered that venous distension in pregnancy was caused by oestrogen and progesterone. On balance, it seems that, if it is one of these hormones, the most likely candidate is oestrogen. Oestrogen is known to be a potent vasoactive hormone, increasing blood flow by up to 40 per cent of baseline values for more than two hours (Meschia 1983). We have observed, from venographic studies, that pelvic vein distension can always be detected from as early as day 6 of the menstrual cycle on the side of the developing follicle, well before the secretion of progesterone by the corpus luteum.

If the hormone is oestrogen, the high concentration of this steroid in ovarian vein blood, which can sometimes be one hundred times that in peripheral blood (Baird and Fraser 1975) would explain the local effects on the pelvic veins and in particular on the veins in the hilum of the ovary, which are not apparent elsewhere in the body.

The next question is where is the oestrogen coming from in high concentrations and with such a sustained effect as to produce the chronically dilated veins and congestion we have demonstrated (Beard et al. 1984)? Clearly one can only speculate but it could be that the polycystic ovaries that have been found in 60 per cent of women with PPS are the source of the hormone. This does not, however, explain why the remaining 40 per cent of the women with apparently normal ovaries have dilated veins and only further studies can elucidate this matter.

Psychological Factors. The early failure to find an organic basis to pelvic pain has led many to believe it is a psychological disorder. The high prevalence of emotional disturbance, and in particular, sexual dysfunction, in women with PPS, is a frequent observation (Duncan and Taylor 1952, Benson et al. 1959, Beard et al. 1977). This led Cohen et al. (1953) to state "The predominance of gynaecologic and abdominal operations is probably related to the high frequency of abdominal pain and menstrual symptoms in hysteria". This suggests that the authors considered that the pain was more imagined than real or else that hysteria may heighten physiological discomfort in some women so that it is interpreted as unbearable pain. All pain conditions are an interaction between physical and psychological processes. It was Duncan and Taylor (1952) who postulated another possibility that the effect of prolonged emotional disturbance can result in somatic changes mediated through the autonomic nervous system. Support for such a concept is provided in daily life by the well known effect of emotion on the peripheral vasculature (blushing and blotching). At the same time, it is likely that the prolonged experience of pain is in itself stressful.

Our group has attempted to identify psychological differences between groups of pelvic pain patients by comparing those with recognisable pathology at laparoscopy and those with no pathology (pelvic congestion) (Pearce 1986). In a sample of 101 patients with chronic pelvic pain of more than 6 months duration, no differences were found between women with pathology and no-pathology on measures of personality using the Eysenck Personality Questionnaire (Eysenck and Eysenck 1964) and mood (profile of mood states, McNair et al., 1971). However, specific differences in attitudes to illness, sexual problems and exposure to death and illness do emerge from our comparison of pathology and no-pathology (pelvic congestion) groups. Since the groups are matched for duration of pain, it is unlikely that the differences that emerge are a consequence of pain and hence they may well play a causal role. We found that women with pelvic pain and no detectable pathology (PPS) report a greater number of deaths and illness among family members and close friends. Support for this observation is provided by the study of Duncan and Taylor (1952) who found that 29 of 36 women with pelvic pain syndrome had an unfavourable family environment including death of a parent before the age of 12, separation of the parents and mental or physical disability in one of them. It may be that the exposure to illness especially in childhood causes these women to pay greater attention to their physical state. This possibility is supported by the finding that women with PPS score higher on the "disease conviction" scale of the illness behaviour questionnaire (Pilowsky 1978). Hence, women in the no-pathology group may more readily interpret pelvic sensations associated with events such as ovulation and menstruation as painful. Pain arising from pelvic vein dilation would also be perceived as more painful.

The concept that women with pelvic pain are more likely to label pelvic sensation as painful clearly does not explain the development of the pathology we are currently observing, namely pelvic congestion, dilated veins, polycystic ovaries and, most recently, the finding of an enlarged uterus and thickened endometrium (Adams et al. unpublished). Nor does it satisfactorily explain the exquisite local tenderness of these organs particularly during an acute attack of pain. Psychological processes have, however, been implicated in the aetiology. The possibility that "stress" may produce long-standing somatic effects has to be considered. Duncan and Taylor (1952) showed that pelvic blood flow undergoes a prolonged and sustained increase when women with pelvic pain discussed anxiety provoking topics. They ascribed this change to the autonomic nervous system. Unfortunately this work needs to be repeated because it was uncontrolled. In considering the mechanisms involved, the indirect evidence we have put forward that it is an ovarian hormone, and probably oestrogen, which is responsible for the pelvic vascular changes, suggests the involvement of the hypothalamic-pituitary-ovarian pathway.

Treatment Methods

The influence of both psychological and physical factors on the genesis of pelvic congestion is illustrated by recent research on treatment methods.

Psychological Approaches. One of the earliest papers to discuss psychological treatment approaches to chronic pelvic pain is that of Beard et al. (1977). Although no formal outcome measures were taken, many of them apparently responded well to relaxation training provided by the psychiatrist.

We have recently completed a controlled trial of treatment using psychological methods in which stress analysis and pain analysis were compared with a minimal intervention control group. Women allocated to the "stress analysis" group received a form of cognitive and behavioural stress management and relaxation training. Discussion of the pain and its associated difficulties and events was discouraged and the focus was directed towards identifying current worries and concerns apart from the pain. Treatment started with a semi-structured interview to assess potential areas of stress such as financial, marital, housing, and the problem areas identified. The women were also asked to keep a daily record of the main concerns they had. The therapist's aim throughout was to identify cognitive strategies used in response to identified stressors, and to discuss alternative responses. In addition, the use of Jacobsonian relaxation strategies in stressful situations was encouraged. The "pain analysis" treatment involved close monitoring of her pain and associated antecedent and consequent events. Therapy is aimed, in collaboration with the patient, to identify patterns associated with pain episodes and alternative strategies for avoiding or reducing pain episodes were discussed with the patients. These strategies were determined on an individual basis and included both cognitive, behavioural and environmental manipulations. In addition, graded exercise programmes were instituted for each patient.

Spouses were encouraged to prompt and reinforce "well behaviours" and, where possible, were encouraged to become involved in the exercise programme. Patients allocated to the control group were given the same explanation as the treatment groups about the frequency of their disorder and the absence of a proven medical or surgical treatment approach.

A range of measures were used to assess outcome, including mood, pain intensity ratings, behavioural disruption and gynaecologists' "blind" ratings of the extent to which the patient was affected by the pain. On all measures of outcome both treatment groups performed significantly better than the minimal intervention control group at 6 months follow-up.

There was no significant difference in the improvement shown by all three groups at three months follow-up. This is an interesting observation since it demonstrates, with reference to the minimal intervention group, the power of the investigation and explanation procedures. The fact that by 6 months the two treatment groups were performing more effectively suggests that some active process of coping with stress and pain were assimilated and used to advantage in the follow-up period. Clearly, longer follow-up periods are required to assess the performance of these changes. However, at least at the 6 month period, it can be concluded with reasonable confidence that those women receiving psychological treatments were functioning at a more adaptive level.

Medical Treatment. The conclusion that an ovarian hormone(s) is responsible for the dilatation of the pelvic veins suggested to us that suppression of ovarian activity is likely to lead to pain relief. We therefore initiated a pilot study using Medroxyprogesterone acetate (Provera - Upjohn) 30 mg/day, to temporarily induce amenorrhoea.

Twenty-one women with pelvic pain and dilated veins, were treated for 6 months. The intensity of pain they experienced was assessed on a visual analogue scale before treatment and at 3 and 6 months after starting the treatment. Of the 21 women treated, 16 had stopped menstruating and all but one showed marked diminution in pain intensity. The remaining 5 women continued to have breakthrough bleeding and none of them showed any significant diminution in pain intensity. This suggests that if ovarian function is not suppressed sufficiently to cause amenorrhoea, dilation of the pelvic veins persists with the attendant pelvic pain. We have preliminary evidence that the pain returns some time after treatment is stopped. A controlled trial is now in progress to test the validity of these observations.

Hypothetical Considerations

The foregoing brief description of a condition that has defied elucidation for many years raises the possibility that it is a stress-related disorder. As has already been mentioned, there is evidence that women with pelvic congestion do have abnormal attitudes to illness, sexual problems and attitudes to death which may originate from unfavourable childhood experiences. The possibility exists that these women may be so sensitive to "stress" because of their conditioning that they fail to accommodate the resulting somatic disturbance in a normal manner. Undamped impulses which occur repeatedly acting through a number of pathways may eventually produce long-standing changes in receptor organs - in this case, the genital organs and vasculature. Clinical experience is that it is usually possible to relate the onset of pelvic pain with an acute crisis in the life of these women, such as the loss of a much-loved parent, an unsatisfactory marriage, or a financial crisis. Commonly these women also have some disturbance of their menstrual cycle, and the onset of sexual dysfunction. The finding in our study (Pearce 1986) that "stress-analysis" was the most successful form of psychotherapy in reducing the intensity of pain experienced by women with pelvic pain likely to be due to congestion is indirect evidence in support of this concept.

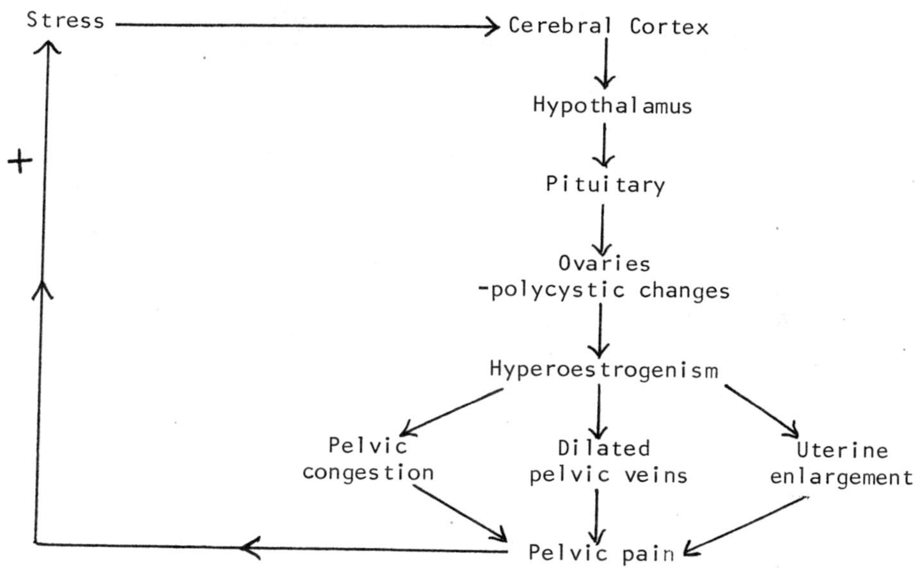

Figure 1. Hypothesis advanced to explain aetiology of PPS

Figure 1 is a schematic representation of our hypothesis. Stress is seen to be acting through the cerebral cortex on the hypothalamus. The hypothalamus in turn causes the pituitary via gonadotrophin releasing factors to secrete follicle stimulating hormone (and possibly luteinising hormone) irregularly. The result is that follicle development is interfered with by the inappropriate release of gonadotrophins. The ovaries are hyperstimulated with the development in some cases of polycystic ovaries. A state of hyperoestrogenism is created which may be due to excessive secretion or alternatively to unopposed action of oestrogen in the absence of progesterone. Pelvic congestion develops, one of the effects of which is to produce chronic pain which in turn is an additional form of stress, thereby creating a kind of vicious circle. Acute attacks of pain to which women with pelvic congestion are occasionally subject, cannot be explained by this mechanism but may be caused by effects mediated through the autonomic nervous system.

REFERENCES

Allen, W. M., 1971, Chronic pelvic congestion and pelvic pain, Am. J. Obstet. Gynecol., 109:198-202.

Allen, W. M., and Masters, W. H., 1955, Traumatic laceration of uterine support, Am. J. Obstet. Gynecol., 70:500-513.

Baird, D. T., and Fraser, I. S., 1975, Concentration of oestrone and oestradiol in follicular fluid and ovarian venous blood of women, Clinical Endocrinology, 4:259-266.

Beard, R. W., Belsey, E. M., Lieberman, M. B., and Wilkinson, J. C. M., 1977, Pelvic pain in women, Am. J. Obstet. Gynecol., 128:566-570.

Beard, R. W., Highman, J. H., Pearce S., and Reginald, P. W., 1984, Diagnosis of pelvic varicosities in women with chronic pelvic pain, Lancet ii:946-949.

Beard, R. W., Reginald, P. W., and Pearce, S., 1986, Pelvic pain in women - Clinical Algorithm, Br. med. J., 293:1160-1162.

Benson, R., Hanson, K., and Matarazzo, J., 1959, Atypical pelvic pain in women: gynecologic psychiatric considerations, Am. J. Obstet. Gynecol., 77:806-823.

Chidikel, N., and Edlundh, K. O., 1968, Transuterine pelvic venography with particular reference to pelvic varicosities, Acta Radiologica, 7:1-12.

Cohen, M. E., Robins, E., Purtell, S. S., and Altman, M. W., 1953, Excessive surgery in hysteria, J. A. M. A., 151:977.

Dudley, A., 1888, Varicocele in the female. What is its influence on the ovary?, N. Y., Med. J., Aug. 18:174-177.

Duncan, C. H., and Taylor, H. C., 1952, A psychosomatic study of pelvic congestion, Am. J. Obstet. Gynecol., 64:1-12.

Eysenck, H. J., and Eysenck, S. B. G., 1964, "Manual of the Eysenck Personality Inventory," University of London Press, London.

Freund, R., 1933, Die krankheiten des beckenbindegewebes, in: "Handbuch der Gynaekologie," Vol. 8, Part 1, pp. 446-717, J. Veit and W. Stoeckel, eds., Bergmann, Leipzig.

Gillibrand, P. N., 1981, The investigation of pelvic pain. Communication on the scientific meeting of chronic pelvic pain: a gynaecological headache, R. C. O. G., London.

Gooch, R., 1831, On some of the most important diseases peculiar to women. Republished by Sydenham Society, London, 1859.

Gordenne, W., and Lhoist, R., 1979, Value of uterine venography in the diagnosis of pelvic pain in women, J. Belge Radiol. - Belgisch Tijdschr. Radiol., 62:223-230.

Henker, F. O., 1979, Diagnosis and treatment of non-organic pelvic pain, South Med. J., 72:1132-1134.

Hodgkinson, C. P., 1953, Physiology of the ovarian veins during pregnancy, Obst. Gynec., 1:26-37.

Hughes, R. R., and Curtis, D. D., 1962, Uterine phlebography, Am. J. Obstet. Gynecol., 83:156-164.

Kaupilla, A., 1970, Uterine phlebography with venous compression. A clinical and roentological study, Acta Obst. Gynec. Scand., 49:33-34.

McCausland, A. M., Hyman, C., Winsor, T., and Trotter, A. D., Jr., 1961, Venous distensibility during pregnancy, Am. J. Obstet. Gynecol., 81: 472-478.

McNair, D. M., Lorr, M., and Droppleman, L. F., 1971, Manual for the Profile of Mood States. Educational and Industrial Testing Service, San Diego, California.

Meschia, G., 1983, Circulation to female reproductive organs, in: "Handbook of Physiology," Section 2, Vol. III, Part I, J. T. Shepherd, F. M. Abbond, and S. R. Geiger, eds., American Physiological Society, Bethesda.

Missanelli, J. S., 1982, Chronic pelvic pain: Intrapelvic fluid as a possible aetiological factor, Journal of American Osteopath Association, 7: 492-495.
Morris, N., and O'Neill, D., 1958, Outpatient Gynaecology, Br. med. J., 2: 1038.
Netter, A., 1953, l'Ovarite sclero-kystique, l'Expansion Scientifique Francaise, Paris.
Ochsner, A., and Mahoner, H. R., 1939, Varicose Veins, Mosby, St Louis.
Pearce, S., 1986, A Psychological Investigation of Chronic Pelvic Pain in Women, Ph.D. Thesis, University of London.
Pilowsky, I., 1978, Psychodynamic Aspects of the Pain Experience, in: "The Psychology of Pain," R. A. Sternbach, ed., Raven Press, New York.
Renaer, M., Nijs, P., Van Assche, A., and Vertommen, H., 1980, Personal observations and a review of the problem, Europ. J. Obst. Gynec. Reprod. Biol., 10:415-463.
R. C. O. G. Gynaecological Laparoscopy, 1978, Report of the working party of the confidential enquiry into gynaecological laparoscopy, G. Chamberlain, and I. C. Brown, eds., London.
Schultze, B. S., 1875, Arch. Gynak., 8, 134-180, quoted by Taylor (1949).
Stearns, H. C., and Sneeden, U. D., 1966, Observations on the clinical and pathological aspects of the pelvic congestion syndrome, Am. J. Obstet. Gynecol., 94:718-732.
Tait, L., 1883, The Pathology and Treatment of the Diseases of the Ovaries, William Wood and Company, New York.
Taylor, H. C., 1949a, Vascular congestion and hyperemia I. Psychologic basis and history of the concept, Am. J. Obstet. Gynecol., 57:211-230.
Taylor, H. C., 1949b, Vascular congestion and hyperemia II. The clinical aspects of the congestion - fibrosis syndrome, Am. J. Obstet. Gynecol., 57:637-653.
Taylor, H. C., 1949c, Vascular congestion and hyperemia III. Etiology and therapy, Am. J. Obstet. Gynecol., 57:654-668.
Taylor, H. C., 1954, Pelvic pain based on a vascular and autonomic nervous system disorder, Am. J. Obstet. Gynecol., 67:1177-1196.
Theobald, G. W., 1951, Pelvic sympathetic syndrome, J. Obstet. Gynec. Brit. Emp., 58:733-761.
Topolanski-Sierra, R., 1958, Pelvic phlebography, Am. J. Obstet. Gynecol., 76:44-52.
Von Peham, H., and Amreich, J., 1934, Operative Gynecology, Vol. 1, Lippincott Company, Philadelphia.
Young, J., 1938, Lower abdominal pains of cervical origin, their genesis and treatment, Br. med. J., 1:104-111.

EMOTIONAL STRESS AND PSYCHIATRIC ILLNESS

NEUROCHEMICAL AND ELECTROPHYSIOLOGICAL EVENTS UNDERLYING STRESS-INDUCED DEPRESSION IN AN ANIMAL MODEL

Jay M. Weiss and Peter E. Simson

Department of Psychiatry
Duke University Medical Center
Durham, NC 27710

AN ANIMAL MODEL OF DEPRESSION

Exposing rats to uncontrollable shock produces behavioral and vegetative changes which bear considerable similarity to what is seen in human depression. This particular procedure represents the most widely-studied animal model of psychopathology. As a result of these studies, similarity of this model to human depression has been shown with respect to etiology, symptomatology, and responsiveness to treatment. These similarities will now be reviewed briefly.

Etiology. One of the primary attributes of uncontrollable shock is that it is a highly stressful condition (e.g., 1,2,3). (Hence the model is referred to hereafter as "stress-induced depression.") Stressful conditions have been widely proposed to increase vulnerability to depression in humans (reviewed in 4; see also 5,6,7). Moreover, numerous studies indicate that the shock must be uncontrollable in order to produce symptoms of depression, since exposure to equal amounts of controllable shock will not produce depressive symptomatology (2,8-11). Depressed individuals often report that they feel helpless, hopeless, or unable to control events, and this feeling of uncontrollability has been hypothesized to be quite significant in bringing about depression (12,13).

Additional results have recently come to light which appear to give further support for modeling of depression by exposing subjects to stressful conditions that are uncontrollable. These findings derive from a naturally-occurring situation where humans are exposed to an uncontrollable stressor, which results in a high incidence of clinical depression. In a series of four papers appearing recently in the Journal *Pain*, depression was found to be highly prevalent in human patients suffering from chronic pain (14-17). In the basic cohort of 71 individuals studied, 57 (i.e., 80% of the patients suffered) were found to be depressed as defined by the Research Diagnostic Criteria (major depression, minor depression, or intermittent depression). The study found that only 10 of the patients evidenced depression prior to the onset of pain, so that depression developed *after* the onset of the pain in most cases. Anxious mood and symptoms of anxiety were often concommitants of depression, as has also been found in the stress-induced depression model (18,19; see also discussion at conclusion of this section).

Symptomatology. Based on 37 publications (2,8,9,19-52), animals exposed to uncontrollable shock show:

(a) weight loss and decreased intake of food and water,
(b) decreased ability to produce active behavior in numerous situations,
(c) decreased ability to compete with other animals and loss of normal aggressiveness,
(d) decreased grooming and play activity,
(e) decreased responding for appetitive rewards,
(f) decreased responding for rewarding brain stimulation,
(g) deficits in ability to make correct choices in an attentional situation,
(h) decreased sleep behavior marked especially by early morning waking.

The effects of uncontrollable shock on the list above correspond directly with 6 symptoms of depression in the DSM-III manual: Poor appetite and significant weight loss, Psychomotor alteration, Loss of energy or fatigue, Loss of interest in usual activities, Sleep changes, and Indecisiveness. The only DSM-III symptoms not represented in the above list are two that require verbal reports from patients (i.e., Feelings of worthlessness, Recurrent thoughts of death and suicide), these being obviously inapplicable to an animal model. The DSM-III requires the presence of 4 symptoms for a diagnosis of major depression.

Effects of Uncontrollable Shock		DSM-III Symptoms of Depression
Decreased food and water consumption weight loss	1.	1. Poor appetite and significant
Decreased body weight	2.	2. Psychomotor alteration
Decreased spontaneous motor activity	3.	
Decreased active behavior in situations requiring such responding	4.	3. Loss of energy or fatigue
Decreased competitive behavior	5.	4. Loss of interest in usual activities
Decreased grooming activities	6.	
Increased number of errors in a choice/discrimination situation	7.	5. Sleep changes
Decreased responding for "rewarding brain stimulation"	8.	6. Indecisiveness
Disturbance of sleep, particularly early morning awakening	9.	
		7. Feelings of worthlessness
		8. Recurrent thoughts of death and suicide

Responsiveness to Treatment. Studies have shown that depression of behavior produced by uncontrollable shock is eliminated by electroconvulsive shock, monoamine oxidase inhibitors, tricyclic antidepressants, and atypical antidepressants (53-57). Behavioral depression produced by uncontrollable shock is not reversed by a single application of drugs but requires several days drug administration (54,55,57), thus paralleling human depression which responds to chronic but not acute administration of antidepressants. The most comprehensive pharmacological assessment was carried out by Sherman, Sacquitne, & Petty (56), who found that behavioral depression induced by uncontrollable shock was reversed by a variety of antidepressants (imipramine, desipramine, amitriptyline, nortriptylene, doxepin, iprindole, mianserin, iproniazid, and nialamide [all tricyclics except the last two]), and not reversed by anxiolytics, phenothiazines, certain stimulants, and tranquilizers.

Further comments on the model. The foregoing briefly describes how the model corresponds to human depression with respect to etiology, symptomatology, and responsiveness to treatment. As stated earlier, this research represents the largest body of work to date relating human depression to an animal model. However, a few additional points need to be made concerning this model.

Different uncontrollable-shock procedures. First, the summary presented above is synthesized from studies that have used different uncontrollable-shock procedures. These procedures are essentially of two types -- procedures that employ (a) moderate intensity shocks of long duration or (b) high intensity shocks of brief duration. As we have pointed out previously (19,58,59), it is not yet clear that the two types of procedures produce all of the same effects. The proposed research utilizes high intensity, brief duration shocks. All of the statements made above regarding etiology, symptomatology, and responsiveness to treatment apply to this procedure except for the fact that the extensive battery of tricyclic antidepressant drugs and atypical antidepressants have yet to be tested on this procedure. However, the list of confirmed symptoms produced by this shock procedure is presently longer than the list of symptoms produced by the other shock procedure; the proposed procedure produces vegetative symptoms (body weight loss, decreased food and water intake, sleep reduction marked by early morning awakening) not yet reported for the other type.

Short-lasting effects. The second point to be made is that much of the depression-like phenomena produced by uncontrollable shock are short-lived, most symptoms dissipating in 48-72 hours with the shock procedure used. The short-lived nature of the depression-like features is a drawback to the stress-induced depression model, as depression in humans clearly is longer lasting.

Anxiety as an aspect of the model. Finally, it is important to note that the model described above apparently generates anxiety as well as depression. As has been described in more detail elsewhere (18,19), animals subjected to uncontrollable shock are tense and vocalize a good deal in conjunction with showing behavioral depression. Moreover, given the relationship between behavioral disturbance in this model and neural activity in the Locus Coeruleus (see below), it is also likely that anxiety would occur in the model in view of theoretical formulations linking Locus Coeruleus activity and anxiety (60). Consequently, this model is considered to produced depression with anxiety as an important associated feature.

NEUROCHEMICAL BASIS OF STRESS-INDUCED DEPRESSION

Given the parallels described above between stress-induced depression in rats and clinical depression in humans, the neurochemical basis of stress-induced depression in rats has been studied extensively in the hope of gaining some insight into the pathophysiology underlying clinical depression. Initial research (1968-1976) indicated that behavioral depression in this model was produced by alteration of norepinephrine (NE) concentration in the brain (e.g.,21,47,61-64). More recent research has indicated that large depletions of NE in the Locus Coeruleus (LC) are found when stress-induced depression of active behavior is observed (51,65,66,67), and this change appears to mediate this reduction of active behavior (51,67,68). To explain how large-magnitude depletion of NE in the LC might mediate behavioral depression, we have offered the following hypothesis: When stressful conditions produce NE depletion of large magnitude in the LC region, release of the amine in the LC is decreased. Since the primary target of NE released in the LC region is alpha-2 receptors, decreased release results in an understimulation of alpha-2 receptors in the LC region. It is this understimulation, or "functional blockade," of alpha-2 receptors in the LC that is thought to be responsible for stress-induced depression.

Several studies have now been carried out to test the foregoing hypothesis, and findings consistent with it have been obtained. First, behavioral depression similar to that seen after uncontrollable shock can be produced without use of shock by pharmacological blockade of alpha-2 receptors in the LC region (69). Second, stress-induced behavioral depression can be eliminated by microinfusion of clonidine into the LC, which presumably substitutes for the stimulation of alpha-2 receptors normally accomplished by NE (68). Third, infusion of pargyline into the LC after exposure to uncontrollable shock, which rapidly eliminates NE depletion in the LC region, will eliminate behavioral depression when NE concentration in the brain region returns to normal (67).

An important deduction that follows from the hypothesis described in the previous paragraph is that stress-induced behavioral depression arises from increased activity of LC neurons. That is, if alpha-2 receptors in the LC are inadequately stimulated, LC cells will show increased activity (i.e., depolarize more often) because stimulation of alpha-2 receptors inhibits LC activity. As a result, stress-induced behavioral depression is thought to derive from *increased* release of NE from the terminals on LC axons. Consequently, post-synaptic adrenergic receptors in brain regions to which the LC projects mediate behavioral depression by higher-than-normal activation. As one way to test this deduction, post-synaptic adrenergic agonists -- phenylephrine for stimulating alpha-1 receptors and isoproterenol for stimulating beta receptors -- were infused into the ventricular system of the brain. Both agonists were found to depress active behavior, though they appeared to be effective at different infusion sites in the ventricular system (69). It can be noted that tricyclic antidepressants and monoamine oxidase inhibiting drugs are hypothesized to accomplish their antidepressant effects by downregulating beta receptors (70-73); this is quite consistent with the pathophysiology of stress-induced depression revealed in the animal model.

ELECTROPHYSIOLOGICAL RECORDING OF LC ACTIVITY

As stated above, the hypothesis set forth to explain stress-induced depression states that the depression arises because alpha-2 inhibitory receptors in the LC are stimulated subnormally. Consequently, in stress-induced depression LC neurons should be disinhibited. Perhaps the most direct way to test this prediction is to record electrophysiological activity of LC neurons. Such studies have recently been carried out, and the results of these experiments will now be described.

Function of alpha-2 receptors in the LC. As a prelude to these studies, it was necessary to consider the function of the alpha-2 receptors in the LC. When stimulated, alpha-2 receptors potently inhibit firing of LC cells (74,75). Originally, it was thought that the inhibition produced by alpha-2 receptors was responsible for diminished neural activity that occurs immediately following a burst of firing of an LC cell (called "Post-Stimulus Inhibition," or "PSI") (76,77,78), but this function is no longer attributed to alpha-2 receptors by the investigators who first proposed it (76,79). Instead, these investigators suggest that alpha-2 receptors exert a tonic inhibitory influence on LC activity (79). (It should be mentioned that other investigators contend that alpha-2 receptors do exert some influence on PSI [80].) Searching for a different domain wherein alpha-2 receptors might influence LC activity, we have recently found that these receptors have a marked effect on responsivity of LC cells to excitatory stimuli (81).

The effect of alpha-2 receptors in regulating the responsivity of LC cells is illustrated by the example presented in Figure 1. Cells in the LC, which are excited by a wide variety of environmental inputs including novel and stressful stimuli, were excited in this case by compression of the contralateral hind paw (PC). The activity of LC cells in response to PC is greater following blockade of alpha-2 receptors by idazoxan. This effect was also obtained with microinfusion of idazoxan into the LC region, thus showing that the effect was due to blockade of alpha-2 receptors in that brain region and not elsewhere in the body. Similar results were also obtained with (a) alpha-2 blockade by yohimbine, which showed that the effect was not restricted to a particular alpha-2 blocking drug, and (b) excitation of LC activity by peripheral nicotine injection, which showed that the effect was not restricted to a particular method of exciting the LC. The finding that alpha-2 receptors regulate responsivity of LC cells is consistent with a tonic influence mediated through alpha-2 receptors.

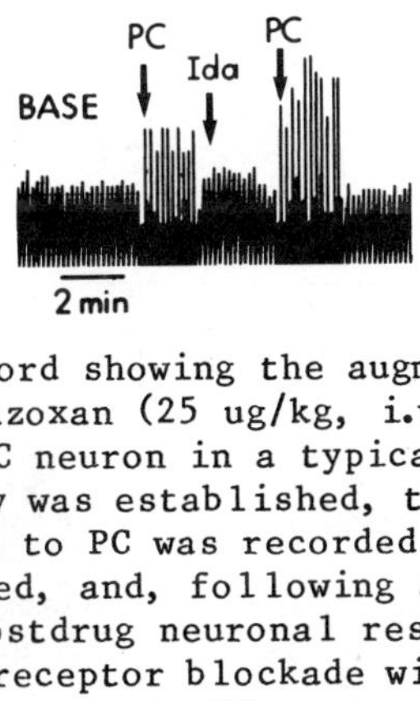

Figure 1. Strip-chart record showing the augmentation by the alpha-2 adrenergic antagonist idazoxan (25 ug/kg, i.v.) of the response to paw compression (PC) of an LC neuron in a typical subject. After a baseline rate of activity was established, the magnitude of the response of the LC neuron to PC was recorded on 10 trials. Idazoxan (Ida) was then administered, and, following a period for recording of baseline activity, the postdrug neuronal response to PC tested. Note that, following alpha-2 receptor blockade with 25 ug/kg idazoxan, responsiveness of the LC neuron to PC was augmented while baseline activity of the neuron was unchanged.

Figure 2 shows the relationship between dose of idazoxan and increase in responsivity of the LC to excitatory input as well as increase in baseline firing rate of the LC. It is clearly evident that elevated responsivity of LC neurons to excitatory input occurs at doses of an alpha-2 blocking agent far below the doses required to increase baseline firing rate of LC cells.

Summarizing the foregoing discussion of the function of alpha-2 receptors in the LC, (1) alpha-2 receptors strongly inhibit LC activity, and (2) while alpha-2-mediated inhibition may contribute to PSI, it appears to play an important role in determining (tonically?) the responsivity of LC cells to excitatory inputs.

LC activity in stress-induced depression. In the experiment described in the remainder of this paper, we measured the electrophysiological activity of LC cells to determine whether activity of these cells was altered in animals showing behavioral depression in the model described above. If decreased active behavior in this model results from decreased

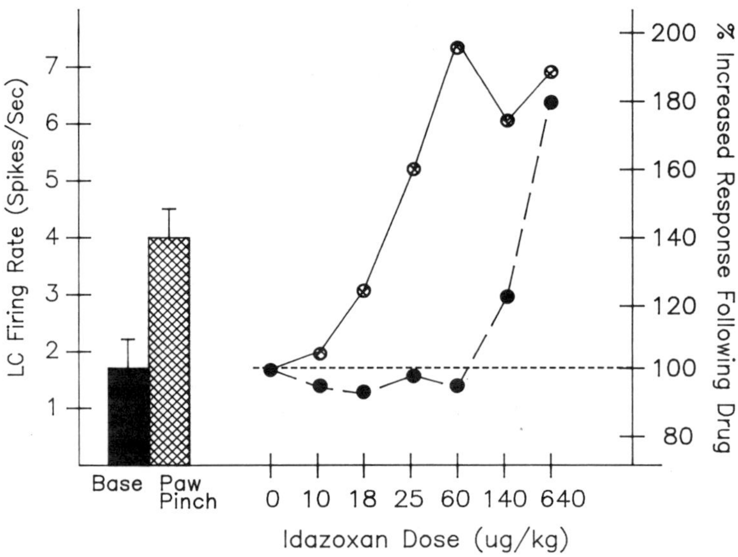

Figure 2. At left is shown the average (and standard error) LC neuron firing rate at baseline (Base) and in response to paw compression (Paw Pinch). At right is shown the baseline and evoked activity of the LC as a function of increasing dose of idazoxan. Although spontaneous activity (solid circles) of LC neurons did not increase until a dose of 140 ug/kg (i.v) of idazoxan was reached, the responsiveness of LC neurons to PC (open circles) increased following as little as 18 ug/kg idazoxan. Each dosage of idazoxan was administered to four animals utilizing the procedure outlined in Figure 1, and no animal received more than one dose.

stimulation, or "functional blockade," of alpha-2 receptors in the LC region, then LC neurons in such animals should be disinhibited. In view of recent studies showing the important influence of alpha-2 receptors on responsivity of LC cells, we wished to determine if this disinhibition would particularly appear as increased responsiveness of LC cells to excitatory input as has been found to occur with alpha-2 blockade.

The basic design of the study was as follows: One group of male Sprague-Dawley rats (Stress group) was subjected to an uncontrollable stressor (tail shock) on a schedule that has been found to induce behavioral depression (51,67,68), while another group was placed in the same apparatus without the stressor given (No-stress group). Following exposure to these conditions, animals underwent one of two procedures. In one procedure, activity of LC neurons was measured electrophysiologically two hours later. Ten animals (5 Stress and 5 No-stress) underwent this procedure. In the second procedure, LC activity was also measured two hours later but active behavior was quantified in a 15 minute swim test just prior to electrophysiological measurement. Eighteen animals (11 Stress and 7 No-stress) underwent this procedure. Electrophysiological recordings of LC neurons were obtained under chloral hydrate anesthesia using standard extracellular single unit recording techniques. Both the spontaneous activity of LC neurons and the magnitude of the response of LC neurons to an excitatory stimulus, compression of the contralateral hind paw (PC), were measured. A total of 103 LC cells were recorded from in the 28 animals.

Turning to the results, the usual depression of active behavior produced in this model was seen in the animals of the present study. In the activity test following exposure of animals to the Stress or No-stress condition (second procedure), activity was reduced in Stress animals. To quantify activity, an activity score (AS) is computed by subtracting seconds of floating from seconds of struggling during the swim test. Animals in the Stress group showed significantly less activity (AS = -306 + 67.0) than animals in the No-stress group (AS = -91.2 + 24.9; $p < .025$).

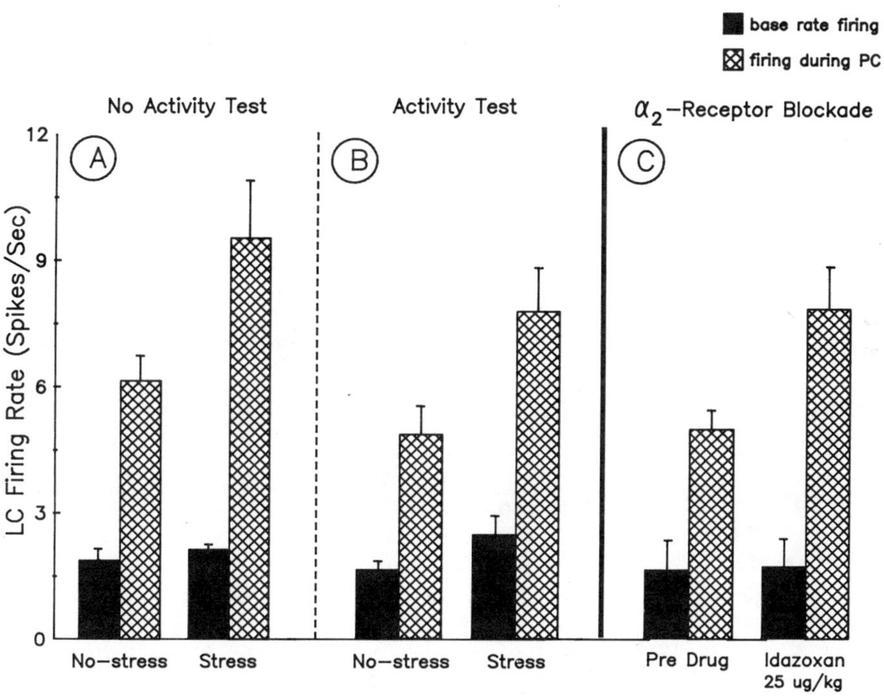

Figure 3. Effect of exposure to uncontrollable stress on the spontaneous and evoked activity of LC neurons in Stress and No-stress animals that were not given a swim test prior to electrophysiological measurement (Panel A) as well as those that were given the swim test (Panel B). The mean (and standard error) are shown. For the purpose of comparison, the effect of pharmacological blockade of alpha-2 receptors with idazoxan (25 ug/kg i.v.) on LC activity is shown in Panel C.

In Figure 3 (Panels A and B) is shown the spontaneous activity of LC neurons and responsivity of LC neurons to an excitatory stimulus (PC) in Stress and No-stress animals. As can be seen in this figure, regardless of whether or not animals underwent the activity test prior to electrophysiological testing, LC neurons of Stress animals showed a greater response to the excitatory stimulus of paw compression than did LC neurons of No-stress animals. In animals that underwent the first procedure --- that is, electrophysiological testing without a preceding swim test (Figure 3, Panel A) -- the responses of LC neurons to PC in animals in the Stress group were significantly larger ($p < .05$) than the responses of LC neurons to PC in animals in the No-stress group. Similarly, in animals that underwent the second procedure -- that is, swim testing prior to electrophysiological testing (Figure 3, Panel B) -- the responses of LC neurons to PC in animals in the Stress group were significantly larger ($p < .005$) than the responses of LC neurons to PC in animals in the No-stress group. Additionally, baseline LC activity appeared elevated in animals of the Stress group relative to baseline LC activity in the No-stress group, although this difference did not reach statistical significance.

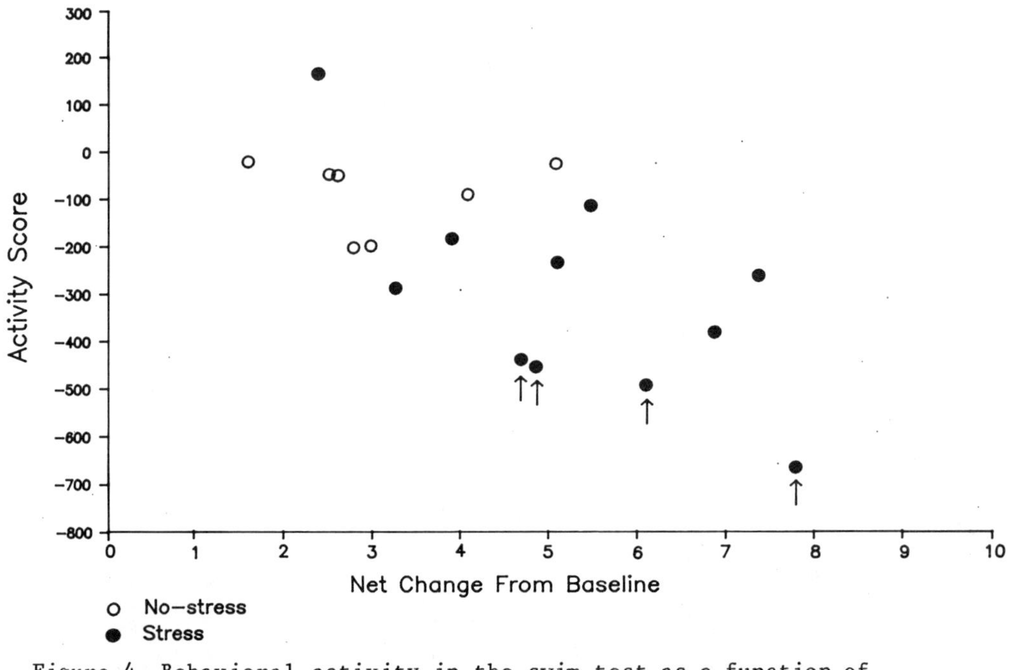

Figure 4. Behavioral activity in the swim test as a function of responsiveness of LC neurons to excitatory stimulation. Solid circles represent Stress animals (n=11), while open circles represent No-stress animals (n=7). The Activity Score was computed by subtracting total time (in seconds) engaged in floating behavior from total time engaged in struggling behavior. The abscissa (net change from baseline) represents the increase in LC firing (in spikes/sec) from baseline produced by paw compression. Note the high correlation ($r = -.70$) between the increase in LC activity that occurred in response to paw compression and the activity score in the swim test. Arrows indicate those animals whose LC neurons displayed aberations in spontaneous activity in addition to increased responsiveness to stimulation (see text).

Giving animals the activity test (i.e., swim test) prior to electrophysiological measurement made it possible to correlate behavioral activity with electrophysiological activity of the LC in these animals. Figure 4 shows, for each of the subjects that underwent this sequence of procedures, the relationship between the increase in LC activity that occurred in response to PC and its activity score in the swim test. The correlation between the measures was $r = -0.70$ ($p < .001$). Thus, the extent to which the LC cells showed an increased response to an excitatory stimulus during the single-unit recording procedure was highly correlated with the amount of behavioral depression that animals showed in the activity test just prior to electrophysiological measurement.

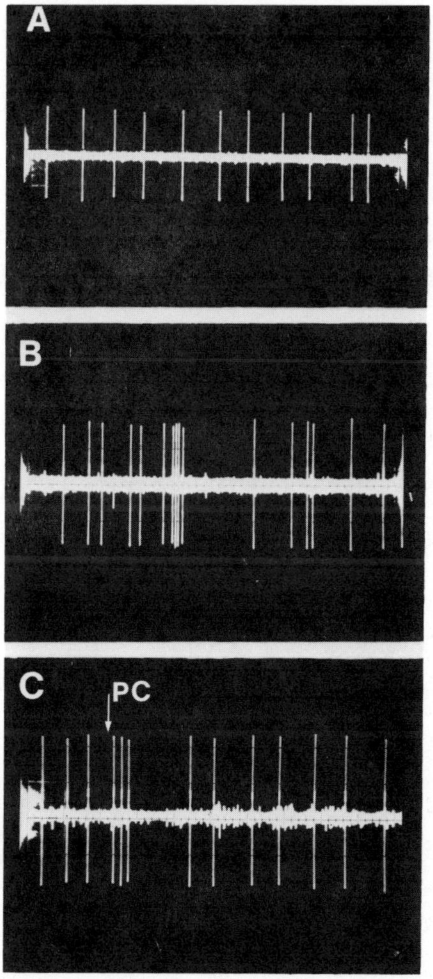

Figure 5. At top (A) is shown the spontaneous activity of a typical LC neuron in a No-stress animal. Note the slow, regular rate of firing. At middle (B) is shown the spontaneous activity of a LC neuron in one of four animals having the lowest Activity Score in the swim test, indicated by the arrows in Figure 4. Note the irregular firing rate, including the spontaneous burst of activity followed by a quiescent period. At bottom (C) is shown a typical LC neuron in a No-stress animal responding to the brief application of a noxious stimulus in the form of contralateral hind paw compression (PC).

Although baseline rate did not differ significantly in the Stress and No-stress groups, aberrations in the pattern of baseline firing were observed in some of the Stress animals. These proved to be the subjects that displayed the largest amount of behavioral depression, as reflected by the activity scores in the swim test (see Figure 4). In contrast to the steady LC firing rate typically observed in an anesthetized rat (Figure 5, top), cyclical changes in the baseline rate of firing for cells was seen. In addition, spontaneous activity in these animals was marked by bursts of firing of LC cells followed by periods of quiescence (Figure 5, middle). The spontaneous activity of LC cells in these four Stress animals bore remarkable resemblance to the evoked activity (i.e., response to PC) of a typical LC cell (Figure 5, bottom).

The elevated responsivity of LC cells that is seen in animals showing behavioral depression is similar to what is seen in normal animals when their alpha-2 receptors in the LC region are blocked pharmacologically (see Figure 3, Panel C). As mentioned previously, earlier work has suggested that behavioral depression in the model we are studying is mediated by reduced stimulation of alpha-2 receptors in the LC, so that the elevated responsivity seen in Stress animals is consistent with this. To further test for the possibility of "functional blockade" of LC alpha-2 receptors in

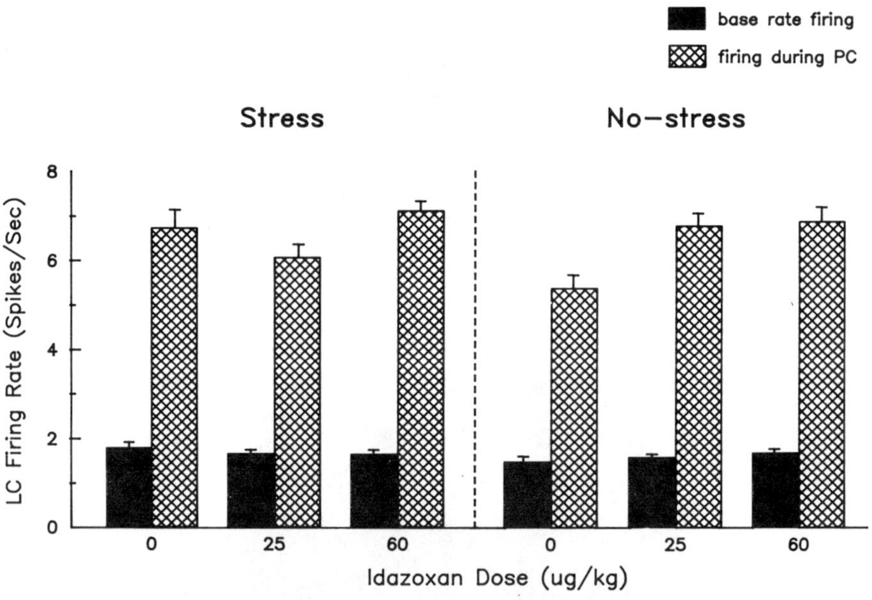

Figure 6. Effect of various doses of idazoxan on LC activity in Stress and No-stress animals. The mean (and standard error) are shown. In No-stress animals (right), idazoxan (25 and 60 ug/kg) significantly increased LC activity (relative to zero dose) that occurred in response to paw compression (PC), while these doses of the alpha-2 blocking agent had no such effect on LC activity in Stress animals (left). One interpretation of these results is that exposure to uncontrollable shock had already produced a functional blockade of alpha-2 receptors in the LC, making subsequent pharmacological alpha-2 receptor blockade inneffective.

this model of depression, we examined the consequences of pharmacological blockade of alpha-2 receptors on LC activity in Stress and No-stress animals. If alpha-2 receptors in Stress animals are functionally blocked as a result of exposure to the stressful condition, additional pharmacological blockade of alpha-2 receptors should, in these subjects, have less effect on responsivity of LC cells than would occur in No-stress animals. As can be seen in Figure 6, administration through a lateral tail vein of the alpha-2 blocker idazoxan was ineffective in augmenting the responsivity of LC cells in Stress animals whereas this drug produced the usual significant increase in responsivity of LC cells in No-stress animals. In that pharmacological blockade of alpha-2 receptors could not elevate responsivity of the LC in Stress animals, this provides further support for the notion that a functional blockade of alpha-2 receptors in the LC was present following exposure to an uncontrollable stressor.

SUMMARY AND CONCLUSION

This paper has described an animal (rodent) model of depression in which depression-like characteristics are produced by exposure of animals to stressful events that they cannot control. This model, called "stress-induced depression," appears to be mediated by stress-induced changes in brain norepinephrine (NE), and evidence now indicates that a large depletion of NE in the Locus Coeruleus region of the brain stem (LC) is critical for producing the behavioral disturbance seen in this model. To explain the functional significance of NE depletion in the LC, it has been suggested that this change results in decreased stimulation, (i.e., functional blockade) of alpha-2 receptors that normally inhibit firing of LC neurons. Consequently, LC neurons should be disinhibited in stress-induced depression. The final part of this paper describes findings from electrophysiological measurement of LC activity. Prior to measuring changes occurring in stress-induced depression, studies examined the basic role played by alpha-2 receptors in LC activity. These receptors were found to regulate the responsivity of LC neurons to excitatory stimuli, so that pharmacological blockade of these receptors increased the firing of LC neurons to excitatory input. When LC activity was then measured in stressed animals showing behavioral depression, responsivity of LC neurons to excitatory input was elevated in comparison to animals not exposed to stress. Also, administration of an alpha-2 blocking drug could not increase responsivity of LC neurons in stressed animals, which further indicated that alpha-2 receptors in the LC region were functionally blocked in these animals. Finally, the amount that LC neurons increased their firing to an excitatory stimulus (i.e., an index of alpha-2 receptor blockade) was highly correlated with the amount of behavioral depression seen in an activity test conducted just prior to electrophysiological measurement. These electrophysiological findings indicate that LC neurons are disinhibited in stress-induced depression, and that this disinhibition is particularly characterized by increased responsivity of LC neurons to excitatory input, which is indicative of alpha-2 receptor blockade. These findings further support the view that the LC is involved in stress-induced depression, and are consistent with a proposed mechanism that attributes behavioral disturbance in the model to disinhibition of LC neurons arising from sub-normal activation of inhibitory alpha-2 receptors.

ACKNOWLEDGEMENT. This research was supported in part by Public Health Service grant MH 40406. We thank Reckitt and Colman, Kingston-upon-Hull, England, for the generous donation of idazoxan hydrochloride used in these experiments.

REFERENCES

1. A. Tsuda, M. Tanaka, H. Hirai, and W.P. Pare, Effects of coping behavior on gastric lesions in rats as a function of predictability of shock, Japanese Psychological Research. 25:9-15 (1983).
2. J. Weiss, Effects of coping responses on stress, Journal of Comparative and Physiological Psychology. 65:251-260 (1968).
3. J. Weiss, Effects of coping behavior in different warning-signal conditions on stress pathology in rats, Journal of Comparative and Physiological Psychology. 77:1-13 (1971).
4. H. Anisman and R.M. Zacharko, Depression: The predisposing influence of stress, Behavioral Brain Science. 5:89-137 (1982).
5. E. Frank and B.D. Stewart, Treatment of depressed rape victims: An approach to stress-induced symptomatology, in: "Treatment of depression: Old controversies and new approaches," P.J. Clayton and J.E. Barrett, eds., Raven Press, New York, (1983).
6. M. Leff, J. Roatch, and W. Bunney, Environmental factors preceding the onset of severe depression, Psychiatry. 33:293-311 (1970).
7. C. Lloyd, Life events and depressive disorder reviewed, Archives of General Psychiatry. 37:541-648 (1980).
8. C. Corum and J. Thurmond, Effects of acute exposure to stress on subsequent aggression and locomotion performance, Psychosomatic Medicine. 39:436-443 (1977).
9. D. Redmond,Jr., J.W. Maas, H. Dekirmanjian, and R.E. Schlemmer, Changes in social behavior of monkeys after inescapable shock, Psychosomatic Medicine. 35:448-449 (1973).
10. M. Seligman and S.F. Maier, Failure to escape traumatic shock, Journal of Experimental Psychology. 74:1-9 (1967).
11. J. Weiss, P. Goodman, B.Losito, S. Corrigan, J.Charry, and W.H. Bailey, Behavioral depression produced by an uncontrollable stressor: relationship to norepinephrine, dopamine, and serotonin levels in various regions of rat brain, Brain Research Reviews. 3:167-205 (1981).
12. M. Seligman, Depression and learned helplessness, in: "The psychology of depression: Contemporary theory and research," R.J. Friedman and M.M. Katz, eds., V.H. Winston, Washington,D.C., (1974).
13. M. Seligman, "Helplessness: On depression, development, and death", W.H. Freeman, San Francisco (1975).
14. R. D. France, K. Krishnan, The dexamethasone suppression test as a biologic marker of depression in chronic pain, Pain. 21:49-55 (1985).
15. R. France, K. Krishnan, and M. Trainor, Chronic pain and depression. III. Family history study of depression and alcoholism in chronic low back pain patients, Pain. 24:185-190 (1986).
16. K. Krishnan, R.D. France, S. Pelton, U. McCann, J. Davidson, and B. Urban, Chronic pain and depression. I. Classification of depression in chronic low back pain patients, Pain. 22:279-287 (1985).
17. K. Krishnan, R. France, S. Pelton, U. McCann, J. Davidson, and B.J. Urban, Chronic pain and depression. II. Symptoms of anxiety in chronic low back patients and their relationship to subtypes of depression, Pain. 22: 289-294 (1985).
18. J. Weiss, W.H. Bailey, P.A. Goodman, L.J. Hoffman, M.J. Ambrose, S. Salman and J.M. Charry, A model for neurochemical study of depression, in: "Behavioral models and the analysis of drug action", M.Y. Spiegelstein and A. Levy, eds., Elsevier Scientific, Amsterdam (1982).
19. J. Weiss, P.G. Simson, M. Ambrose, A. Webster, L. Hoffman. Neurochemical basis of behavioral depression, in: "Advances in Behavioral Medicine" vol. 1, E. Katkin and S. Manuck, eds., JAI Press, Greenwich, Conn. (1985).
20. A. Altenor, E. Kay, and M. Richter, The generality of learned

helplessness in the rat, *Learning and Motivation*. 8:54-61 (1977).
21. H. Anisman, D. deCatanzaro, and G. Remington, Escape performance following exposure to inescapable shock: Deficits in motor response maintenance, *Journal of Experimental Psychology: Animal Behavior Processes*. 4:197-218 (1978).
22. H. Anisman, L. Sklar, Catecholamine depletion upon reexposure to stress: Mediation of the escape deficits produced by inescapable shock, *Journal of Comparative and Physiological Psychology*, 93:610-625 (1979).
23. H. Anisman, L. Sklar, Social housing conditions influence escape deficits produced by uncontrollable stress: Assessment of the contribution of norepinephrine, *Behavioral and Neural Biology*, 32:406-427 (1981).
24. J. Brady, D. Thornton, D. Fisher, Deleterious effects of anxiety elicited by conditioned pre-aversive stimuli in the rat, *Psychosomatic Medicine*, 24:590-595 (1962).
25. J. Freda and S. Klein, Generality of the failure-to-escape (helplessness) phenomenon in rats, *Animal Learning & Behavior*, 4:401-406 (1976).
26. H. Glazer and J. Weiss, Long-term interference effect: An alternative to "learned helplessness", *Journal of Experimental Psychology: Animal Behavior Processes*, 2:202-213 (1976).
27. H. Glazer and J. Weiss, Long-term and transitory interference effects, *Journal of Experimental Psychology: Animal Behavior Processes*, 2:191-201 (1976).
28. J. Irwin, A. Suissa, and H. Anisman, Differential effects of inescapable shock on escape performance and discrimination learning in a water escape task, *Journal of Experimental Psychology: Animal Behavior Processes*, 6:21-40 (1980).
29. R. Jackson, J. Alexander, and S. Maier, Learned helplessness, inactivity, and associative deficits: The effects of inescapable shock on response choice escape learning, *Journal of Experimental Psychology: Animal Behavior Processes*, 6:1-20 (1980).
30. R. Jackson, S. Maier, and P. Rapaport, Exposure to inescapable shock produces both activity and associative deficits in the rat, *Learning and Motivation*, 9:69-98 (1978).
31. J. Kelsey and M. Baker, Ventromedial septal lesions in rats reduce the effects of inescapable shock on escape performance and analgesia, *Behavioral Neuroscience*, 97:945-961 (1983).
32. J. Lawry, V. Lupo, J. Overmier, J. Kochevar, K. Hollis, and D. Anderson, Interference with avoidance behavior as a function of qualitative properties of inescapable shocks, *Animal Learning & Behavior*, 6:147-154 (1978).
33. T. Looney, and P. Cohen, Retardation of jump-up escape responding in rats pre-treated with different frequencies of noncontingent electric shock, *Journal of Comparative and Physiological Psychology*, 78:317 (1972).
34. S. Maier, C. Anderson, and D. Lieberman, Influence of control of shock on subsequent shock-elicited aggression, *Journal of Comparative and Physiological Psychology*, 81:94-100 (1972).
35. J. Overmier, Interference with avoidance behavior: Failure to avoid traumatic shock, *Journal of Experimental Psychology*, 78:340-343 (1968).
36. J. Overmier, and M. Seligman, Effects of inescapable shock on subsequent escape and avoidance learning, *Journal of Comparative and Physiological Psychology*, 63:23-33 (1967).
37. W. Pare, The effect of chronic environmental stress on stomach ulceration, adrenal function, and consummatory behavior in the rat, *Journal of Psychology*, 57:143-151 (1964).
38. W. Pare, Stress and consummatory behavior in the albino rat, *Psychological Reports*, 16:399-405 (1965).

39. J. Peters, and S. Finch, Short- and long-range effects on the rat of a fear-producing stimulus, Psychosomatic Medicine, 23:138-152 (1961).

40. S. Ritter, N. Pelzer, and R. Ritter, Absence of glucoprivic feeding after stress suggests impairment of noradrenergic neuron function, Brain Research, 149:399-411 (1978).

41. R. Rosellini, Inescapable shock interferes with the acquisition of an appetitive operant, Animal Learning & Behavior, 6:155-159 (1978).

42. M. Seligman, and G. Beagley, Learned helplessness in the rat, Journal of Comparative and Physiological Psychology, 88:534-541 (1975).

43. M. Seligman, and S. Maier, Failure to escape traumatic shock, Journal of Experimental Psychology, 74:1-9 (1967).

44. M. Seligman, R. Rosellini, and M. Kozak, Learned helplessness in the rat: Time course, immunization, and reversibility, Journal of Comparative and Physiological Psychology, 88:542-547 (1975).

45. A. Sherman, and F. Petty, Neurochemical basis of the action of antidepressants on learned helplessness, Behavioral and Neural Biology, 30:119-134 (1980).

46. E. Stone, Possible grooming deficit in stressed rats, Research Communications in Psychology, Psychiatry and Behavior, 3:109-115 (1978).

47. B. Sutton, G. Coover, and C. Lints, Motor debilitation, short- and long-term shuttlebox deficits, and brain monoamine changes following footshock pretreatment in rats, Physiological Psychology, 9:127-134 (1981).

48. J. Weiss, and H. Glazer, Effects of acute exposure to stressors on subsequent avoidance-escape behavior, Psychosomatic Medicine, 37:499-521 (1975).

49. J. Weiss, W. Bailey, L. Pohorecky, D. Korzeniowski, and G. Grillone, Stress-induced depression of motor activity correlates with regional changes in brain norepinephrine but not in dopamine, Neurochemical Research, 5:9-22 (1980).

50. J. Weiss, H. Glazer, L. Pohorecky, J. Brick, and N. Miller, Effects of chronic exposure to stressors on avoidance-escape behavior and on brain norepinephrine, Psychosomatic Medicine, 37:522-523 (1975).

51. J. Weiss, P. Goodman, B. Losito, S. Corrigan, J. Charry, and W. Bailey, Behavioral depression produced by an uncontrollable stressor: Relationship to norepinephrine, dopamine, and serotonin levels in various regions of rat brain, Brain Research Reviews, 3:167-205 (1981).

52. R. Zacharko, W. Bowers, L. Kokkinidis, and H. Anisman, Region-specific reductions of intracranial self-stimulation after uncontrollable stress: Possible effects on reward processes, Behavioral Brain Research, 9:129-141 (1983).

53. H. Glazer, J. Weiss, L. Pohorecky, and N. Miller, Monoamines as mediators of avoidance escape behavior, Psychosomatic Medicine, 37:535-543 (1975).

54. B. Leonard, Pharmacology of new antidepressants, Progress in Neuro-Psychopharmacology and Behavioral Psychiatry, 8:97-108 (1984).

55. F. Petty and A. Sherman, Reversal of learned helplessness by imipramine, Communications in Psychopharmacology, 3:371-373 (1979).

56. A. Sherman, J. Sacquitne, and F. Petty, Specificity of the learned helplessness model of depression, Pharmacology Biochemistry and Behavior, 16: 449-454 (1982).

57. J. Telner and R. Singhal, Effects of nortriptyline treatment on learned helplessness in the rat, Pharmacology Biochemistry and Behavior, 14:823-826 (1981).

58. H. Glazer and J. Weiss, Long-term and transitory interference effects, Journal of Experimental Psychology: Animal Behavior Processes, 2:191-201 (1976).

59. J. Weiss and P.G. Simson, Neurochemical basis of stress-induced depression, Psychopharmacology Bulletin, 21:447-457 (1985).

60. D. Redmond, Jr., Alterations in the function of the nucleus locus coeruleus: A possible model for studies of anxiety, in: "Animal models in psychiatry and neurology," I. Hanin and E. Usdin, eds., Pergamon Press, New York (1977).
61. H. Anisman, Time-dependent variations in aversively motivated behaviors: Non-associative effects of cholinergic and catecholaminergic activity, Psychological Review, 82:359-385 (1975).
62. H. Anisman, J. Irwin, and L.S. Sklar, Coping with stress, norepinephrine depletion and escape performance, Brain Research, 191:583-588 (1980).
63. J. Weiss, H.I. Glazer, and L.A. Pohorecky, Coping behavior and neurochemical changes: An alternative explanation for the original "learned helplessness" experiments, in: "Animal models in human psychobiology," G. Serban and A. King, eds., Plenum Press, New York and London (1976).
64. J. Weiss, E. Stone, and N. Harrell, Coping behavior and brain norepinephrine level in rats, Journal of Comparative and Physiological Psychology, 72:153-160 (1970).
65. H. Lehnert, D.K. Reinstein, B.W. Stowbridge, and R. Wurtman, Neurochemical and Behavioral Consequences of Acute, Uncontrollable Stress: Effects of Dietary Tyrosine, Brain Research, 303:215-223 (1984).
66. C. Hughes, T. Kent, J. Campbell, A. Oke, H. Croskell, and S. Preskorn, Cerebral blood flow and cerebrovascular permeability in an escapable shock (learned helplessness) animal model of depression, Pharmacology, Biochemistry & Behavior, 21:891-894 (1984).
67. P. Simson, J. Weiss, M. Ambrose, and A. Webster, Infusion of a monoamine oxidase inhibitor into the locus coeruleus can prevent stress-induced behavioral depression, Biological Psychiatry, 21: 724-734 (1986).
68. P. Simson, J. Weiss, L. Hoffman, and M. Ambrose, Reversal of Behavioral depression by infusion of an alpha-2 agonist into the locus coeruleus, Neuropharmacology, 25:385-389 (1986).
69. J. Weiss, P. Simson, L. Hoffman, M. Ambrose, S. Cooper, and A. Webster, Infusion of adrenergic receptor agonists and antagonists into the locus coeruleus and ventricular system of the brain: Effects on swim-motivated and spontaneous motor activity, Neuropharmacology, 25:367-389 (1986).
70. E. Stone, Subsensitivity to norepinephrine as a link between adaptation to stress and antidepressant therapy: An hypothesis, Research Communications in Psychology, Psychiatry and Behavior, 4:241-255 (1979).
71. F. Sulser, New perspectives on the mode of action of antidepressant drugs, Trends in Pharmacological Science, 1:92-94 (1979).
72. F. Sulser, Antidepressant drug research: Its impact on neurobiology and psychobiology, in: "Typical and atypical antidepressants: Molecular mechanisms," E. Costa and C. Racagni, eds., Raven Press, New York (1982).
73. J. Vetulani, R. Stawarz, J. Dingell, and F. Sulser, A possible mechanism of action of antidepressant treatments. Reduction in the sensitivity of the noradrenergic cyclic AMP generating system in the rat limbic forebrain, Naunyn-Schmiedeberg's Archives of Pharmacology, 293:109-114 (1976).
74. G. Aghajanian, J. Cedarbaum, and R. Wang, Evidence for norepinephrine-mediated inhibition of locus coeruleus neurons, Brain Research, 136:570-577 (1977).
75. J. Cedarbaum and G. Aghajanian, Noradrenergic neurons of the locus coeruleus: Inhibition by epinephrine and activation by the alpha-antagonist piperoxane, Brain Research, 112:413-419 (1976).
76. G. Aghajanian, C. Vandermaelen, and R. Andrade, Intracellular studies on the role of calcium in regulating activity and reactivity of

locus coeruleus neurons in vivo, Brain Research, 273:237-243 (1983).

77. J. Cedarbaum and G. Aghajanian, Catecholamine receptors on locus coeruleus neurons: pharmacological characterization, European Journal of Pharmacology, 44:375-385 (1977).

78. J. Cedarbaum and G. Aghajanian, Activation of locus coeruleus neurons by peripheral stimuli: Modulation by collateral inhibitory mechanism, Life Sciences, 23:1383-1392 (1978).

79. R. Andrade and G. Aghajanian, Locus coeruleus activity in vitro: Intrinsic regulation by a calcium-dependent potassium conductance but not alpha-2 adrenoceptors, Journal of Neuroscience, 4:161-170 (1984).

80. M. Ennis and G. Aston-Jones, Evidence for self- and neighbor-mediated inhibition of locus coeruleus neurons, Brain Research, 374: 299-305 (1986).

81. P. Simson and J. Weiss, Alpha-2 receptor blockade increases responsiveness of locus coeruleus neurons to excitatory stimulation, Journal of Neuroscience, 7(6):1732-1740 (1987).

IMPLICATIONS OF BEHAVIORAL SENSITIZATION AND KINDLING FOR STRESS-INDUCED BEHAVIORAL CHANGE

Robert M. Post, Susan R.B. Weiss, and Agu Pert

Biological Psychiatry Branch
National Institute of Mental Health
Bethesda, Maryland

INTRODUCTION

Psychomotor stimulant-induced behavioral sensitization and electrophysiological kindling are not animal models of stress with obvious face validity; however, we discuss them here as they may guide us to new principles concerning changes in long-term responsivity in behavior that may be relevant to some stress situations. Moreover, there is an extensive literature indicating that there are important interactions between psychomotor stimulant-induced behavioral sensitization and stress-induced behavioral alterations. Stimulants such as amphetamine and cocaine are thought to exert many of their actions by potentiating noradrenergic and dopaminergic neurotransmission. Stresses, in many instances, are also capable of acting on these catecholamine and related neurochemical substrates. Thus, it is not surprising that under some conditions, stresses may facilitate subsequent reactivity to psychomotor stimulants and, vice-versa, that psychomotor stimulant pretreatment may enhance stress responsivity. Seymour Antelman and colleagues have reviewed elsewhere (1,2) evidence for cross-sensitization between stress and stimulants. It is not the point of this paper to argue that this is a universal principle; we and others have observed in some instances that stresses may decrease subsequent stimulant-induced behavior. What we are suggesting is that, in many instances, stimulant-induced effects on short- and long-term alterations in behavior may be mediated by the same neurochemical systems that are relevant to the consideration of stress-induced changes in behavior.

Similarly, electrophysiological kindling is also not an ideal model for many types of stresses. In fact, several investigators have reported that stress-induced potentiation of catecholamine systems can actually inhibit kindled seizures rather than facilitate them. Nonetheless, processes revealed by kindling may, at times, be relevant to those affected by stresses, and a variety of stress-related neuropeptides may, themselves, be capable of inducing kindling or kindling-like changes in seizures and behavior. In addition, electrophysiological kindling itself appears to be associated with the release of a variety of neurotransmitters and peptides that are also affected by stress. These include not only opiates and vasopressin, but also, potentially, corticotropin-releasing factor.

Table 1

Characteristics of Behavioral Sensitization to Psychomotor Stimulants

1. Shorter latency to maximum response following repeated exposure.
2. Increased magnitude of behavioral response to same dose.
3. Selective increase in some behaviors, not others.
4. Alterations in responsivity are long-lasting (week to months).
5. Intermittent administration produces greater sensitization, compared with continuous administration.
6. Effects are dose-related.
7. Genetic component is evident in degree of sensitization.
8. Sensitization is not related to alterations in blood levels.
9. Sensitization shows environmental context dependency; i.e., is conditionable.
10. Cross-sensitization occurs among many stimulants and dopamine agonists.
11. Cross-sensitization occurs between stimulants and a variety of stresses (tail-pinch, shock, starvation).
12. Neurotransmitter and receptor changes postulated.
13. Hormones and peptides modulate and may produce behavioral sensitization.

It is also apparent that behavioral sensitization to psychomotor stimulants and electrophysiological kindling may involve different neurochemical and neuroanatomical pathways. We discuss these paradigms together in this context, since they illustrate the common principle that under certain circumstances increases in behavioral and physiological responsivity can be produced through repetition of the same stimulus over time.

Behavioral Sensitization to Cocaine

Various characteristics of behavioral sensitization to stimulants are listed in Table 1. Most of our studies have involved examining motor activity as an end-point; activity increases in magnitude and rapidity of onset following repeated doses of stimulants. The effect is dose-related and long-lasting (weeks to months depending on the number and magnitude of initial stimuli). There are marked differences in the degree of behavioral sensitization to cocaine among different genetic strains of mice (3), just as there are marked individual differences in the degree of stress responsiveness.

In one type of behavioral sensitization study, small doses of cocaine (10 mg/kg i.p.) are administered once daily to male Sprague-Dawley rats. As illustrated in fig. 1 (top panel), rats repeatedly administered this dose in the same test cage show an increased degree of locomotor activity. By day seven, animals are markedly more hyperactive to the same dose that originally produced only minimal degrees of locomotor activity, similar to that observed during normal exploratory activity in the pretreatment phase. When we tried to produce sensitization by injecting the animals in their home cage and then testing them in the activity meter, no sensitization effect was observed. In order to test this possible environmental context specificity more systematically, a group of animals was given saline in the test chamber and then injected with cocaine in their home cage. As illustrated in the lower panel of fig. 1, when these animals were injected with cocaine in the experimental test chamber (days 11, 13, 14), they showed lesser degrees of hyperactivity than animals that had experienced the cocaine hyperactivity repeatedly in the context of the test cage, despite identical handling, number of injections, and cage exposure of the two groups.

COCAINE PRETREATMENT IN DIFFERENT ENVIRONMENTAL CONTEXTS: COCAINE SENSITIZATION

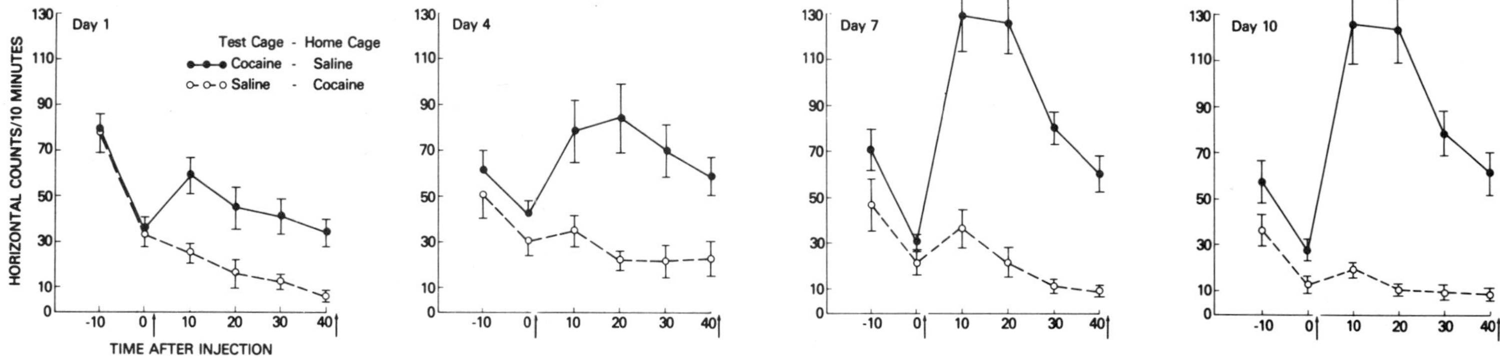

EFFECT OF ENVIRONMENTAL CONTEXT ON COCAINE SENSITIZATION

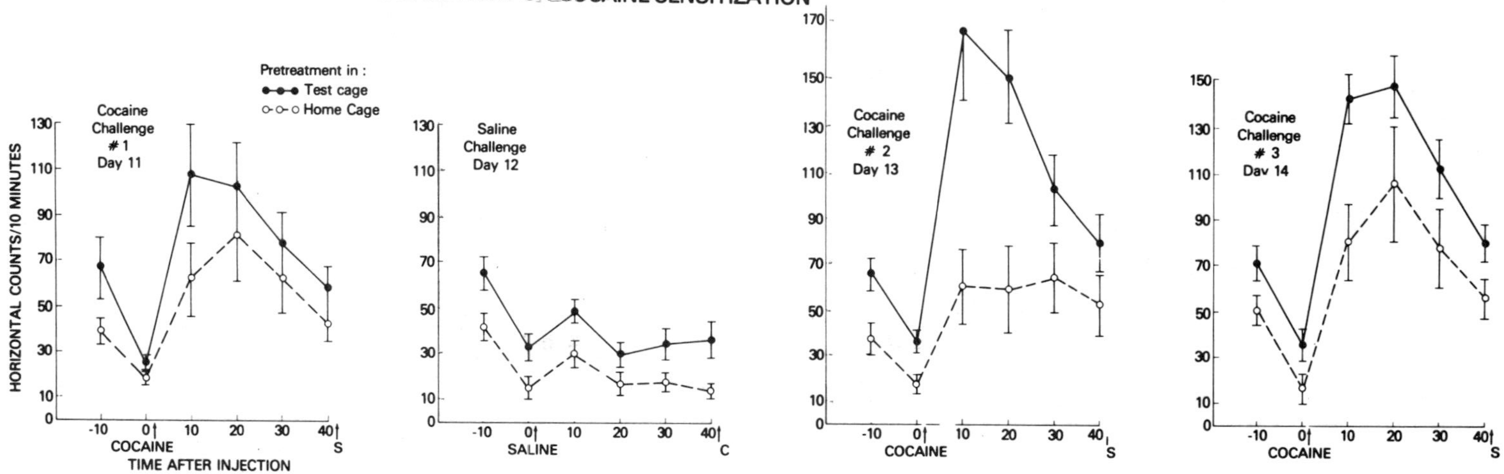

These data provide strong evidence for an environmental context component of cocaine-induced hyperactivity and indicate that cocaine-induced behavioral sensitization is, at least in part, related to conditioning. Activity observed after saline injection is also significantly elevated in the animals previously treated with cocaine in the test-cage context compared with animals receiving the same doses of cocaine in a different environment, as illustrated in the second panel on the lower part of the figure. Others have also noted a conditioned component of stimulant-induced behavioral sensitization to amphetamine (4,5) and to cocaine (6,7). Moreover, Hinson & Poulos (6) additionally demonstrated that repeated saline injections in the same context where animals had previously received cocaine were capable of desensitizing or deconditioning the cocaine sensitized animals, further suggesting that the increase in locomotor activity followed many of the rules of classical conditioning.

We have recently documented context-specific cocaine-induced behavioral sensitization following a large pretreatment dose on day 1 (cocaine, 40 mg/kg i.p.) and rechallenge with a small dose on day 2 (cocaine, 10 mg/kg i.p.) (8). As illustrated in fig. 2, animals pretreated with cocaine in the same plexiglas test cage show markedly increased locomotor activity to co-

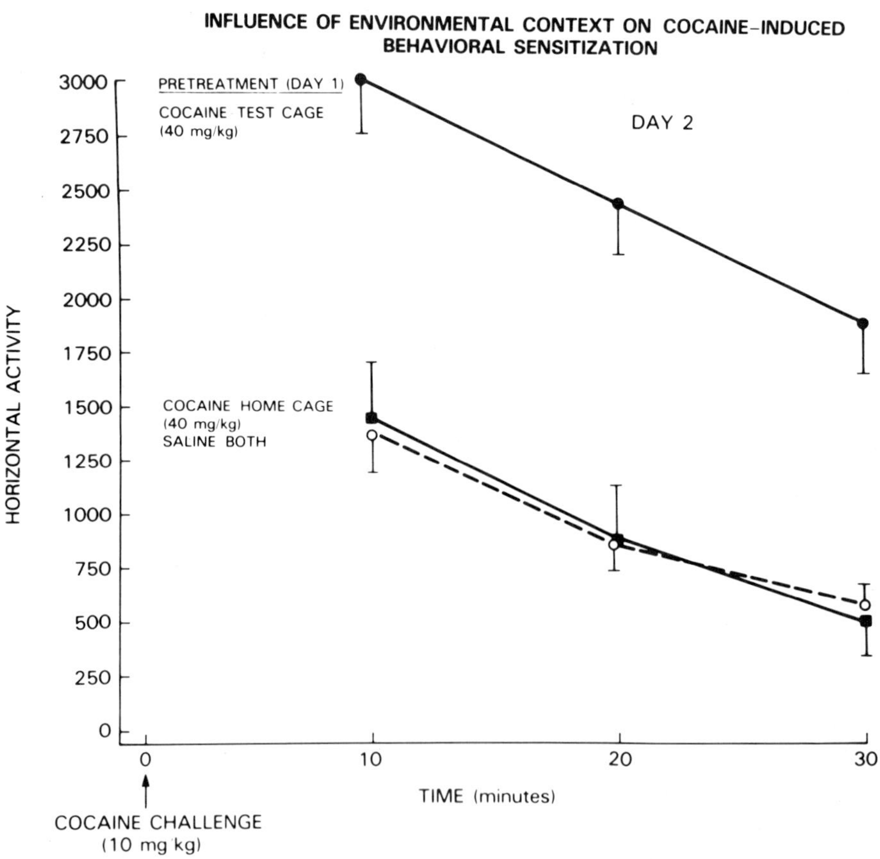

Fig. 2

caine on day 2 compared to saline-pretreated controls. We were surprised to find that animals pretreated with cocaine in a different test environment (home cage) showed no more locomotor activity than the saline controls. These data suggest that in this single, high-dose paradigm, the increase in cocaine-induced hyperactivity was entirely context-dependent and related to conditioning variables rather than to pretreatment with cocaine per se.

In order to test this effect more systematically, we varied the degree of similarity of the pretreatment and test environments. As illustrated in fig. 2a, the animals that received the pretreatment dose of cocaine in the same environment were more active on rechallenge than animals receiving the cocaine in increasingly dissimilar environments; i.e., in wire cages in a different room. These findings suggest that some of the principles of classical conditioning, including stimulus generalization, may also be applicable to cocaine-induced behavioral sensitization.

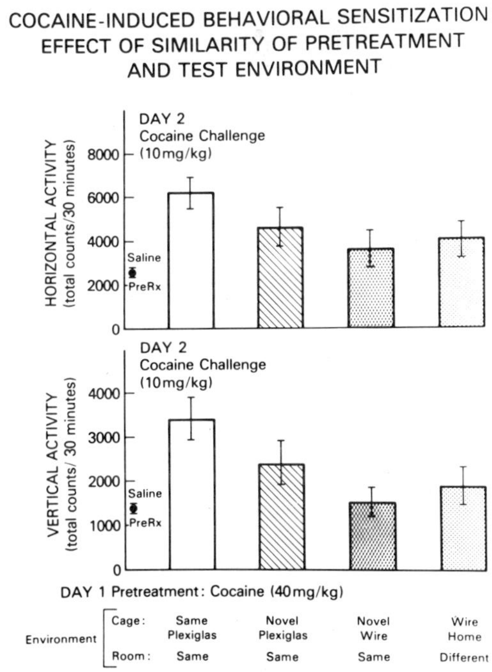

Fig. 2a

We also wanted to assess whether the experience of cocaine-induced behaviors in the test environment, rather than cocaine treatment per se, was necessary to behavioral sensitization. We, thus, pretreated animals with agents that would block cocaine-induced hyperactivity, through blockade of dopamine receptors (using the neuroleptic haloperidol) and non-specifically by increasing benzodiazepine and GABA-ergic tone with diazepam and muscimol, respectively. As illustrated in fig. 3, for haloperidol, we found that inhibition of cocaine-induced hyperactivity on day 1 did block subsequent behavioral sensitization to cocaine on day 2, again suggesting that the appearance of cocaine-induced behaviors, rather than cocaine injections per se, were important to the development of behavioral sensitization.

We have also replicated the initial findings of Beninger & Hahn (9), Beninger & Hertz (19), and Tadokoro & Kuribara (11), who reported that once stimulant-induced behavioral sensitization has been induced and fully developed, its expression is no longer blockable with neuroleptic pretreatment on subsequent test days. As illustrated in fig. 4, while haloperidol administered on day 1 is capable of blocking the development of cocaine-induced behavioral sensitization, when administered only prior to the test dose on day 2, it is not sufficient to do so. Thus, the neuroleptic is incapable of blocking the expression of cocaine-induced behavioral sensitization when administered in moderate (0.2 mg/kg) or in high sedating doses (0.5 mg/kg) prior to the cocaine dose on day 2.

These data would appear to be of considerable interest from several perspectives. They suggest that pharmacological intervention in stimulant-induced behavioral sensitization differs in different stages of its development and expression and imply that different neurochemical or neuroanatomical substrates are involved. To the extent that they provide a useful model for increasing behavioral pathology observed in response to repeated stress induction, they suggest that pharmacological intervention in stress-induced behavioral pathology may also differ as a function of development versus

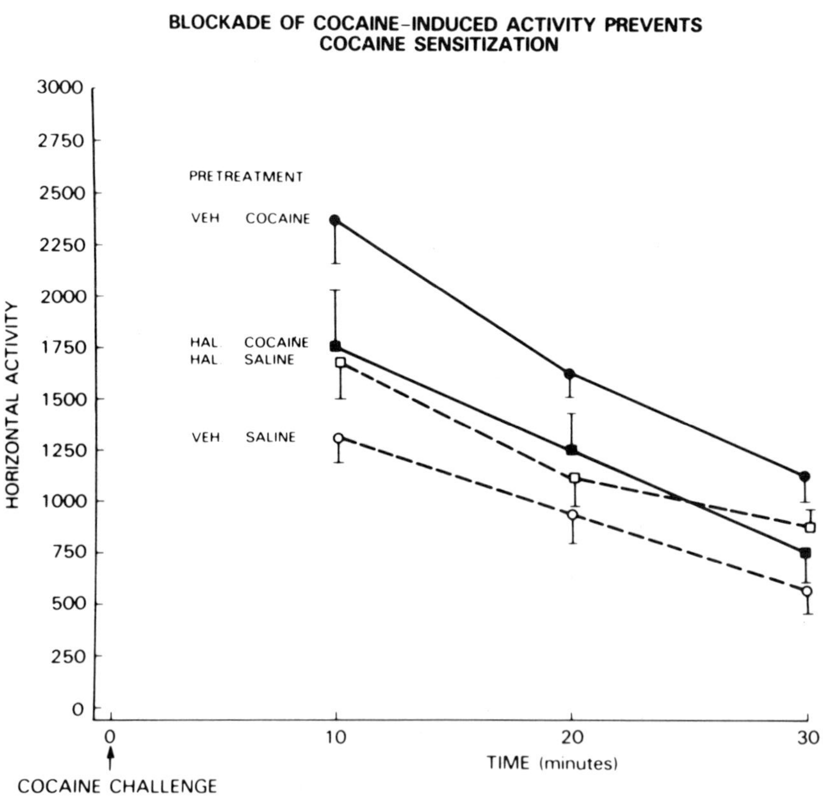

Fig. 3

expressive phases of these syndromes. Psychomotor stimulant-induced psychoses have long been considered a useful animal model for schizophrenic-like psychosis in man. A large literature documents many similarities in the presentation of chronic stimulant-induced psychosis and that of schizophrenia. To the extent that some components of the schizophrenic syndrome are adequately modeled by stimulant-induced processes and that behavioral sensitization is important to the increasing induction of behavioral pathology, our studies and those reviewed above would also suggest a possible mechanism for neuroleptic nonresponsiveness. That is, if some patients have experienced single or multiple episodes of untreated manic or schizophrenic excitement and psychosis, neuroleptic intervention in a subsequent episode may not effectively treat the component of the syndrome that is attributable to the expression of behavioral sensitization based on prior facilitation of circuits presumably involved in the augmented behavioral response. This would be analogous to findings observed in our sensitization experiment (figs. 3 and 4). We will return to this theme of potentially differential pharmacology, depending on stage of evolution of sensitized syndromes, in our discussion of electrophysiological kindling, where it is also clear that pharmacological interventions in the evolution of seizures differ as a function of stage of seizure development.

Gale et al (12) reported that high doses of neuroleptics sufficient to block the development of cocaine-induced behaviors, were not sufficient to block sensitization induced by massive doses of cocaine (160 mg/kg subcutaneous for 10 days). Thus, it would appear that while some doses of cocaine are associated with environmental context-dependent behavioral sensitization

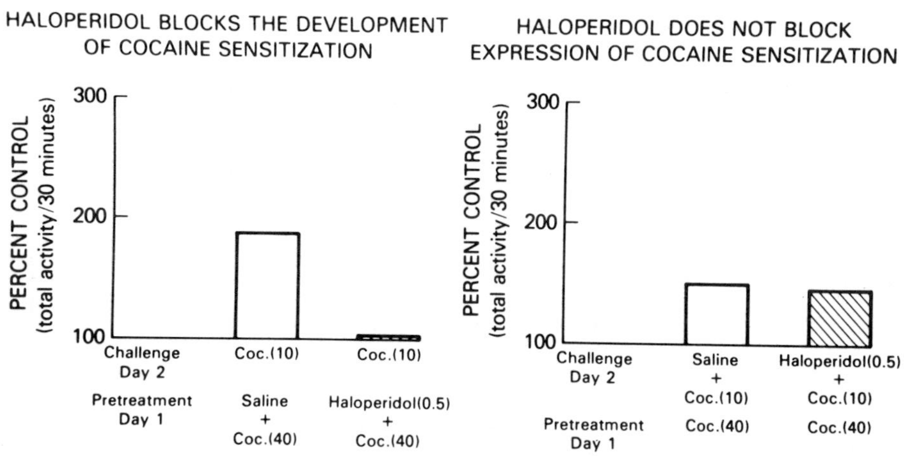

Fig. 4. In the left panel, the control group (open bar) was pretreated with two saline injections on day 1 and was challenged with cocaine (10 mg/kg) on day two; the second control group (shaded bar) received haloperidol and saline injections on day 1 and was challenged with cocaine (10 mg/kg) on day 2 (see fig. 3).

In the right panel the control group (open bar) was pretreated with saline on day 1 and then challenged with cocaine (10 mg/kg) on day 2; the second control group (shaded bar) was treated with saline on day 1 but received haloperidol and cocaine (10 mg/kg) on day 2.

and the development of the sensitization is neuroleptic-blockable, other very high dose paradigms are neuroleptic-insensitive. In addition, if high doses of cocaine are repeatedly administered, as discussed in detail in the next section, a pharmacological kindling effect can occur and, thus, seizures develop to a previously subconvulsant dose of cocaine. This phenomenon also occurs independent of environmental context and, apparently, independent of conditioning variables.

Thus, as summarized in fig. 5, there appears to be a gradient of effects of cocaine depending on the initial doses of drug, number of repetitions, and intervals between repetitions. In an oversimplified view, the data suggest that with increasing initial dose and increasing numbers of repetitions, the consequences to the organism are increasingly long-lasting, robust, and able to override initial conditioning variables. Thus, one or three doses of cocaine (10 mg/kg i.p.) are not associated with the induction of behavioral sensitization but 7-10 doses are. Similarly, a single dose of cocaine (40 mg/kg) induces behavioral sensitization that persists for a number of days, while three injections of this dose or ten injections of an even lower dose (10 mg/kg) may produce sensitization that lasts for weeks to months. The cocaine-induced sensitization to a single dose of 40 mg/kg is entirely context-dependent; however, with repetition of this dose for three injections, increasing amounts of stereotypic activity occur, such that context-dependent sensitization of stereotypy is observed, while the locomotor activity sensitization becomes context-independent. The stereotypic behavior produced also shows a component that is context-independent. Neuroleptics are also able to block the development of this sensitization to stereotypic behaviors and locomotor activity.

COCAINE		EFFECT									
Number of Injections	Dose mg/kg i.p.	Behavioral Sensitization	Duration	Activity Context Dep.	Indep.	Stereotypy Context Dep.	Indep.	Saline Conditioning	Sensitization Neuroleptic Independent	Seizure Kindling	Death
↑↑↑	COC_{65}									++	++
x 10 days	COC_{160} subcut. (K. Gale)	++			++		++		++		
↑↑↑↑↑↑↑↑↑↑	COC_{10}	++	months	++	0			++			
↑↑↑	COC_{40}	+		0	++	++	++	±			
↑	COC_{40}	++	days	++	0	0	0	0	0		
↑↑↑	COC_{20}	0	0	0	0	0	+				
↑↑↑	COC_{10}	0	0	0	0	0	0				
↑	COC_{10}	0	0								

Fig. 5

We dwell on the details of this schema (fig. 5) because we think it amply documents the evolving spectrum of different components of behavioral sensitization, suggesting different underlying mechanisms that are engaged as a function of dose and numbers of repetitions of the drug. To the extent that stress-responsive behavioral syndromes in animals and man also utilize mechanisms related to this stimulant-induced sensitization (1,2), similar variations in evolution of behavioral pathology may also occur as a function of magnitude of stress and frequency of repetition. As illustrated in the hypothetical schema in fig. 6, the duration and severity of effects of prior stress on behavior may vary in a fashion similar to that illustrated for cocaine-induced behavioral sensitization. Components of stress-induced behavioral syndromes may be susceptible to conditioning mechanisms but if the stress is severe enough in magnitude or in numbers of repetitions, more extensive neurologic changes may be induced; e.g., a dementia may occur which has been associated with enlarged ventricles (13,14).

Obviously, magnitude and frequency of repetition of the stress are only two of many variables that could explain much of the variance regarding quality, severity, and duration of behavioral symptoms following stresses. We illustrate in the starred column that the same set of stresses induced in different developmental periods may have differential developmental consequences. For example, Putnam et al (15) have postulated that repeated physical or sexual trauma during a critical period for personality development could lead to the development of multiple personality disorder while similar stresses occurring in an adult patient would not induce this syndrome.

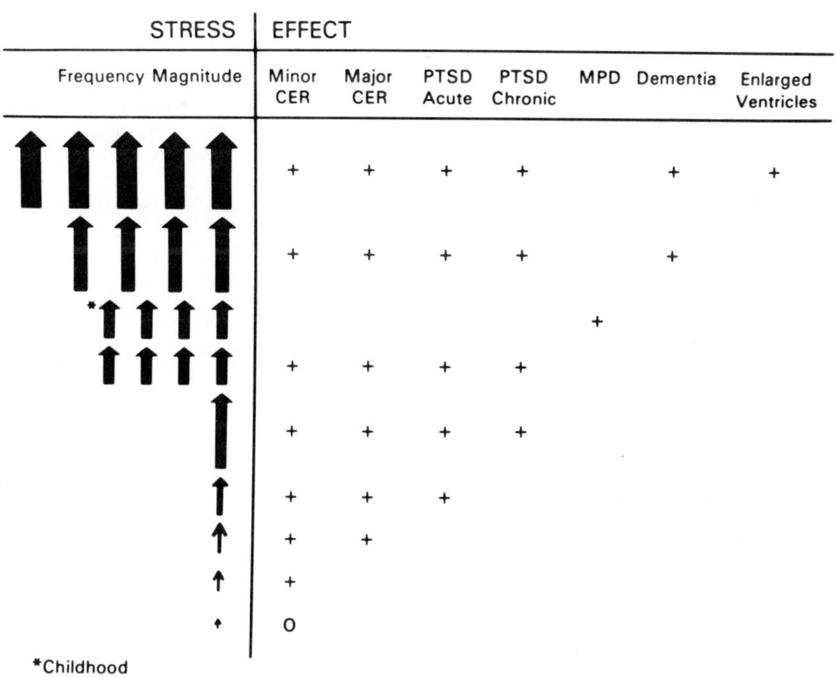

Fig. 6. Hypothetical schema of the relationship between repeated exposures to life stresses and development of behavioral pathology. CER = conditioned emotional response; PTSD = post-traumatic stress disorder; MPD = multiple personality disorder.

We present the hypothetical schema in fig. 6 for its potential heuristic value in comparison with a similar set of variables illustrated for cocaine-induced behavioral sensitization in fig. 5, not with the idea that there is ample or even preliminary documentation for each of the potential effects illustrated. We do feel, however, that the behavioral sensitization paradigm focuses on the long-lasting consequences of previous exposure to similar stimuli and the potential important role for environmental context and conditioning variables, as well as variables such as magnitude and frequency of repetition in the ultimate outcome for behavioral pathology.

Electrophysiological and Pharmacological Kindling

Major characteristics of electrophysiological kindling are summarized in Table 2. As originally described by Goddard and associates (16), kindling involves repeated stimulation of a given brain area (such as the amygdala) on an intermittent basis, and evokes an increasingly widespread electrophysiological effect and an increasingly robust behavioral effect until a major motor seizure occurs to a previously ineffective stimulation. This development of kindling can be produced in the amygdala in approximately two weeks following stimulation of the amygdala once a day for one second at several hundred μamps or just below the threshold for an afterdischarge. Once completed kindled seizures are stably achieved in this fashion, if they are repeated sufficiently (usually requiring 50-100 times in the rat), a last stage of kindling development may occur; i.e., that of spontaneity. In this phase, animals will show the occurrence of full-blown seizures without any exogenous electrophysiological stimulation; that is, they appear to show a true onset of spontaneous epilepsy.

We have observed a similar progression of events that we have labeled pharmacological kindling induced by the mixed local anesthetic/psychomotor stimulant cocaine or the pure local anesthetic lidocaine. We began using lidocaine in order to dissect out the properties of cocaine that may be attributable only to its local-anesthetic properties, since lidocaine is equipotent to cocaine as a local anesthetic but is not a psychomotor stimulant. Utilizing repeated administration of the same dose of lidocaine (65 mg/kg i.p.), we observed that an increasing number of animals demonstrated seizures to a previously subconvulsant dose. Moreover, the behavioral seizures and electrophysiological spiking recorded from the amygdala were similar in lidocaine-kindled animals to that observed with electrical kindling, and lidocaine-kindled animals showed faster onset of electrophysiological kindling than did saline-pretreated controls.

The lidocaine-induced behavioral seizures are associated with tonic-clonic movements of the head, trunk, and forepaws with rearing and falling. This convulsion is similar to that observed with electrophysiological kindling, although the lidocaine seizures persist intermittently for much longer periods of time. Glucose utilization is selectively increased during lidocaine-kindled seizures in many limbic areas of brain (amygdala, hippocampus) or their associated neural circuitry (claustrum, endopyriform nucleus, and perirhinal cortex) (17). Repeated administration of cocaine also leads to a pharmacological kindling process. However, in contrast to lidocaine-kindled seizures, which are readily tolerated by the rat, cocaine-induced seizures tend to be lethal. In a series of studies, we have observed most of the lidocaine-treated animals surviving weeks or months of seizures; in contrast, almost all of the cocaine-treated animals die after their first or second cocaine-induced seizure. Elsewhere, we have discussed in detail the possible importance of cocaine-induced behavioral sensitization and pharmacological kindling to the evolution of neuropsychiatric syndromes following repeated cocaine administration in man (18-21).

Table 2

ELECTRICAL KINDLING: MAJOR CHARACTERISTICS*

1. Repeated stimulations
   a. Initially subseizure threshold
   b. Intermittent
2. Local afterdischarges and seizure activity
   a. Increases in amplitude, frequency
   b. Increase in duration
   c. Increase in complexity of wave form
   d. Increase in anatomical spread
3. Replicable sequence of seizure stages
   Behavioral arrest, blinking and masticatory movements, head nodding, opsithotonis, contralateral then bilateral forelimb clonus, rearing and falling
4. Discharges kindle in quantum jumps
5. Limbic system kindled more readily than cortex
6. Full-blown kindled seizures recapitulate the history of convulsion development
7. Transfer effects to secondary sites; kindling facilitated in other sites even after primary site destroyed
8. Interference: secondary kindled site interferes with primary site kindling
9. No toxic or neuropathological changes evident: kindling is a transsynaptic process
10. Relatively permanent change in connectivity; a kindled animal will still seize after a one-year seizure-free interval
11. Interictal spikes and spontaneous epileptiform potentials develop
12. Seizure may develop spontaneously in chronically kindled animals
13. Efficacy of pharmacological interventions differs as a function of stage of kindling (i.e., development, completed, spontaneous)

* See Goddard et al (42) regarding 1,4-10; Wada & Sato (43) regarding 2-4,6; Wada et al (44) regarding 10-12; Racine (45) and Pinel & Rovner (46,47).

In addition to the alterations in mood, schizophrenia-like paranoid psychosis, and the development of seizures, we highlight the increasing reports of cocaine-induced panic attacks for this discussion. Some 50% of callers to the Cocaine Hotline reported experiencing cocaine-related panic attacks. In a series of patients being studied in an outpatient clinic, many have reported the onset of panic attacks after repeated use of cocaine (T. Uhde, B. Vittone and M. Geraci, unpublished observations, 1986). One of their patients in particular continued his cocaine use for approximately six months, with frequent induction of cocaine-induced panic attacks. After many repetitions, a panic attack occurred spontaneously without the use of cocaine, causing this patient to become concerned that the cocaine-related panic attacks had now evolved on their own and become manifest in an unpredictable form. Despite discontinuing cocaine use, the patient continued to experience spontaneous panic attacks that were associated with the rapid development of severe generalized anxiety and profound agoraphobia. This case would appear to offer an interesting potential illustration of how behavioral sensitization and kindling concepts could help explain the progressive evolution of panic symptomatology. Repeated use of drug which originally did not produce such effects could evolve into the production of spontaneous panic attacks, much like the phenomenon of spontaneity in kindling. Certainly, these and related cases of cocaine-induced panic attacks following repeated use (22,23) could serve as a template for further discussion of the sensitization, conditioning, and kindling variables that may be applicable to this apparent evolution in symptomatology. It is clear, in this patient, that cocaine-induced symptomatology did not

Table 3

SUBSTANCES THAT "KINDLE" SEIZURES

Local Anesthetics:
LIDOCAINE
Stimulants:
COCAINE
Acetylcholine Agonists (I.A.):
PHYSOSTIGMINE
GABA Antagonists (I.C.):
BICUCULLINE
Benzodiazepine Inverse Agonists:
FG-7142
"Peripheral-type" Benzodiazepine Ligands:
Ro 5-4864
Endogenous Opiates (I.A.):
beta-ENDORPHIN, ENKEPHALIN
Corticotropin-Releasing Hormone (I.C.V.)
CRH*

* CRH seizures occur after single, not repeated, drug administration. I.A. = intra-amygdala injection; I.C. = intra-cortical; I.C.V. = intracerebroventricular. Unabbreviated = intravenous or intraperitoneal injections

involve the development of full-blown seizures and, in this fashion, cannot be construed as a classic pharmacological kindling. Nonetheless, it would appear that many of the associated temporal aspects of the kindling phenomenon may be relevant. In addition, a classic pharmacological kindling process may appear in some patients where seizures occur as the full-blown kindled endpoint.

Although kindling and sensitization may involve different underlying mechanisms, temporal parameters, and behavioral endpoints, both may be relevant to the development of cocaine-induced neuropsychiatric dysfunction and may occur in parallel. However, we would continue to make the distinction between behavioral phenomena which might follow a sensitization-like time-course, and classical kindling phenomena which involve a seizure end-point, since many phenomena in neuropsychiatric illness involve processes that stop short of full-blown kindling. One should reserve the term kindling for phenomena that involve a seizure end-point, even though some of the principles elucidated by the kindling phenomenon may be relevant to behavioral pathology such as the development of spontaneity in cocaine-induced panic attacks.

These distinctions may also be important in relation to other substances which may be capable of inducing either behavioral sensitization or kindling-like phenomena. As listed in Table 3, a variety of exogenous and endogenous substances appear capable of inducing a kindling-like seizure process. Several of the substances on this list are of particular interest from a "stress" perspective. Peptides of both the endogenous opiate class (enkephalin and beta-endorphin) and corticotropin-releasing factor (CRF) have been implicated in aspects of behavioral sensitization or processes which lead to a kindling-like occurrence of major motor seizures. In some instances, repeated doses of morphine can lead to behavioral sensitization. Schuster et al (24) demonstrated in the mouse that, depending on the interval between injections, repeated morphine could produce either sensitization or tolerance to opiate-induced running behavior. With injections spaced at short intervals tolerance developed, while, with long intervals between injections, sensitization or increased locomotor running activity was observed. Sensitization phenomena have also been demonstrated with direct injection into brain. Kalivas et al (25) injected enkephalin directly into the ventral tegmental area and found that this produced increased motor activity upon repetition. Moreover, enkephalin-induced motor activity showed cross sensitization to stimulant-induced activation with amphetamine. Foot-shock stresses were also capable of activating opiate systems in this area of brain and of producing cross sensitization to amphetamine and enkephalin. Thus, it would appear that stresses may be capable of utilizing pathways similar to those involved in stimulant-induced behavioral sensitization (18,1,2,25,26).

In addition, it appears that, under other circumstances, opiates may themselves be capable of producing pharmacologically kindled seizures. Cain and collaborators (27) have documented that administration of enkephalinergic compounds directly into the amygdala were initially without effect but upon repetition, evoked full-blown major motor seizures that were remarkably like those achieved by electrical kindling of the amygdala. Moreover, they were naloxone-reversible. These findings have recently been replicated by Tanaka et al (28). Seeger, Pert, Ostrowski, Weiss and colleagues (29,30) have demonstrated that amygdala-kindled seizures appear to release opiate compounds, as documented by *in vivo* opiate receptor autoradiography with the ligand [^{3}H]diprenorphine. *In particular*, there is a marked decrease in binding in amygdala and nucleus accumbens following amygdala-kindled seizures compared to controls. This would indicate that the kindled seizure is releasing endogenous opiate substances which occupy opiate receptors making less available for binding of exogenously administered radioactive opiate ligands. It therefore appears that seizures may be capable of releasing endogenous opiate substances, and when opiate substances are administered intracerebrally, they are capable of engaging mechanisms related to behavioral sensitization and pharmacological kindling.

Since a variety of stresses appear to be capable of releasing endogenous opiate substances in animals and man, it is possible, given sufficient magnitude and frequency, that they might be capable of engaging mechanisms related to behavioral sensitization (25,26), if not pharmacological kindling (27,28). If one were to argue from the data on neuroleptic blockade of stimulant-induced behavioral sensitization, one would postulate that the development phases of behavioral sensitization to opiate-induced changes might be naloxone-reversible, for instance, but that the expression of sensitization may be resistant to such an intervention. This presents a directly testable experiment which could also have implications for the treatment of stress-related symptomatology that might initially involve opiate systems but become resistant to therapeutic intervention by naloxone once sensitization has occurred.

CRF administered intracerebroventricularly (i.c.v.) produces seizures that are electrophysiologically and behaviorally similar to those achieved by electrical kindling. The seizures occur some 4-8 hours after i.c.v. administration, leading Dr. Ehlers (31), who first described this phenomenon in 1983, to postulate that the seizures were "kindling themselves" over this interval. While this kindling mechanism remains to be directly tested and confirmed, the late developing seizure phenomenon and marked increases in aggressive behavior which also occur in this time frame are of considerable interest. Upon daily repetition of the i.c.v. CRF injections, tolerance appears to develop to the CRF effects (fig. 6a). In spite of this decrease in CRF-induced seizures and aggression following repetition of the i.c.v. injections, these CRF-treated animals are more sensitive to electrical kindled seizures compared with implanted, sterile water-injected controls. Thus, there appears to be cross sensitization between CRF-induced seizures and amygdala-kindled seizures (fig. 6b).

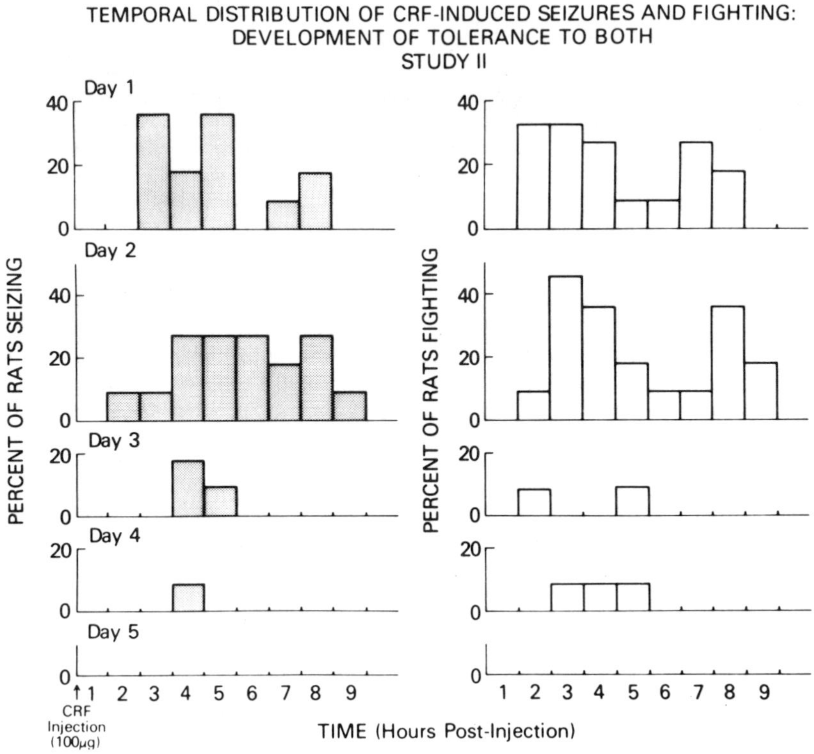
TEMPORAL DISTRIBUTION OF CRF-INDUCED SEIZURES AND FIGHTING:
DEVELOPMENT OF TOLERANCE TO BOTH
STUDY II
Day 1
Day 2
Day 3
Day 4
Day 5
PERCENT OF RATS SEIZING
PERCENT OF RATS FIGHTING
CRF
Injection
(100μg)
TIME (Hours Post-Injection)

Fig. 6a

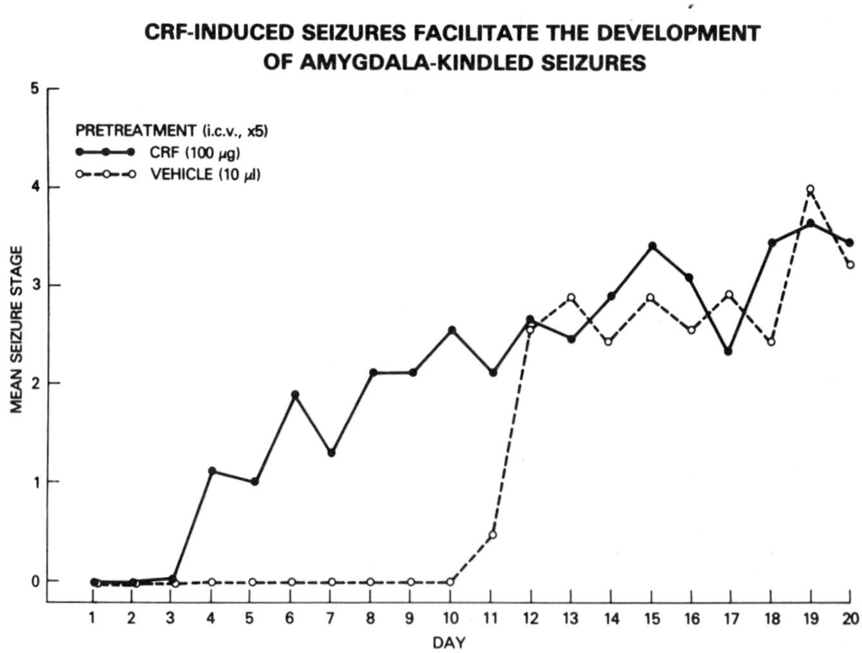
CRF-INDUCED SEIZURES FACILITATE THE DEVELOPMENT
OF AMYGDALA-KINDLED SEIZURES
PRETREATMENT (i.c.v., x5)
CRF (100 μg)
VEHICLE (10 μl)
MEAN SEIZURE STAGE
DAY

Fig. 6b

Earlier in this volume, Koob has described CRF-induced changes in locomotor activity and anxiety-like behavior. As illustrated in fig. 6c, we might conceptualize a continuum of CRF-induced changes depending on dose that progress from mild increases in arousal behavior and its associated decreased interest in reinforcing behaviors such as eating and sex, to more marked increases in locomotor activity, progressing to anxiety-like behavior and, with the largest doses, vicious, defensive-aggressive behavior, and, ultimately, limbic-type seizures. Again, this apparent continuum effect elucidated by increasing doses of i.c.v. CRF administration, is of potential clinical interest from several perspectives. Koob and others (32) have indicated that stress is capable of releasing CRF into the cerebroventricular system and, while it is far from clear that any stress could achieve levels of CRF required to induce the late-onset of seizures, it is possible that graded intensities of stress are capable of releasing graded amounts of CRF into selected brain regions and into ventricular CSF. Thus, some of the behavioral concomitants of this putative stress-induced CRF release could be affected by some of the variables that affect behavioral sensitization; for example, magnitude and number of repetitions. It is also of interest in this case that repeated CRF administration, at least at 24-hour intervals, is capable of producing tolerance to some end-points (e.g., CRF seizures and aggression) while sensitizing subsequent responsivity to other behaviors (such as facilitating subsequent electrically kindled seizures of the amygdala).

Gold and associates reviewed evidence that CRF might be hypersecreted in some neuropsychiatric conditions such as endogenous depression or primary affective illness (this volume). There do appear to be interesting parallels

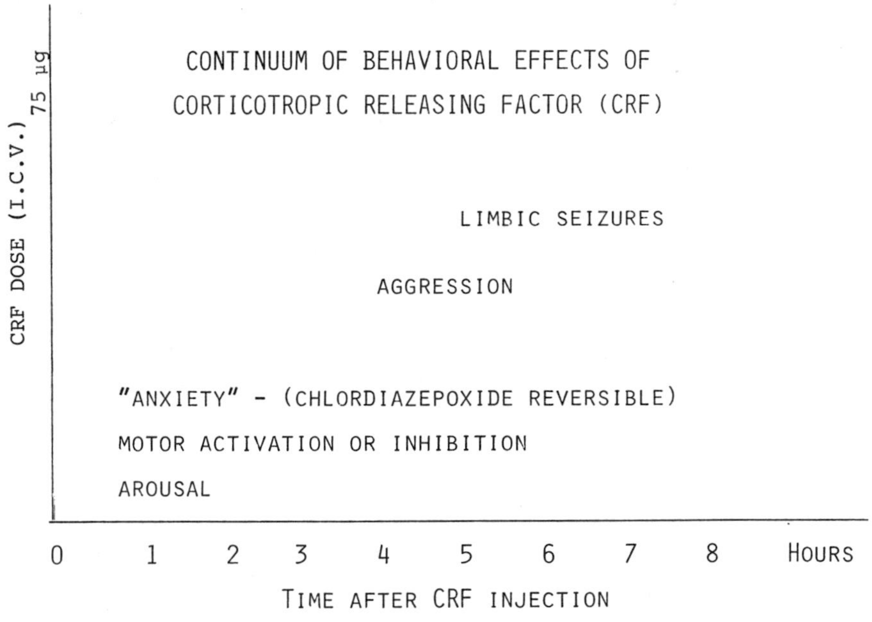

Fig. 6c

in the symptomatology of endogenous depression and changes that can be induced by i.c.v. CRF, including decreased appetite, sleep, changes in psychomotor activity, and social interaction and, in high doses, aggression and anxiety-like behavior as well as cortisol hypersecretion. Moreover, Kalin and colleagues (33) have demonstrated that CRF administered i.c.v. in unrestrained primates produced a posture typical of that assumed by monkeys in the despair phase of separation of infant or juvenile monkeys from their maternal or peer groups, respectively. The current analysis would suggest the utility of extending these formulations with a focus on the potential long-lasting consequences of exposure to high doses of CRF endogenously or exogenously achieved, and the possible differential development of both tolerance and sensitization phenomena, depending on the end-point studied and the frequency and intensity characteristics of the CRF release. The long lag between i.c.v. CRF administration and maximum effects on aggressive behavior and seizures should also provide an interesting model for considering long lags between the onset of stressors and maximum behavioral or physiological effects. Elucidation of the mechanism of this lag should also be of considerable interest for assessing possible pathophysiological processes that are occurring over this interval.

Thus, it is clear that under at least some pathological circumstances, endogenous peptide compounds of the opiate and CRF class may be capable of exerting profound effects on behavior and convulsive thresholds following single or repeated administrations. Elucidating the range of compounds which may show this phenomenon would be of considerable clinical and theoretical interest. Parenthetically, a kindling-like time course of seizures does not appear to occur following i.c.v. somatostatin or vasopressin (34).

One does, however, see the development of kindling-like seizures to repeated administration of ligands active at the central-type (C-type) and "peripheral-type" (P-type) benzodiazepine receptors. Repeated administration of the C-type benzodiazepine ligand, the beta-carboline FG 7142 (35), as well as the P-type ligand, Ro5-4864 (36), both appear capable of inducing seizures in an increasingly large percentage of animals to a dose which was previously subconvulsive. The P-type ligand may be proconvulsant by virtue of its additional effects at the TBPS site on the chloride channel, however (36-37). Since endogenous ligands have recently been postulated and preliminarily described for both receptor sites (38,39), it is possible that under some conditions, stresses may be capable of activating these systems with effects on endogenous P-type and C-type benzodiazepine receptors. This is particularly interesting from the perspective that the putative ligand for the endogenous C-type benzodiazepine receptor described by Guidotti and Costa et al (38) appears to have the characteristics of a pro-anxiety or an inverse agonist compound potentially, with properties similar to those of FG 7142. These data would, again, raise the possibility that given sufficient type and magnitude of stressful stimuli, sufficient amounts of these types of compounds could be released that would engage mechanisms relevant to behavioral sensitization or pharmacological kindling. Obviously, this is a highly speculative approach to systems which have only preliminarily been described, but given the fact that pharmacological doses of these ligands can produce a kindling-like time course, we again mention these hypotheses for their potential heuristic use.

Pharmacological Intervention as a Function of Stage of Kindling

As illustrated in fig. 7, there is increasing data to suggest that different stages in the kindling process are differentially susceptible to intervention by anticonvulsants. We recognize three global stages in the evolution of kindling phenomena. The first stage is development, where one progresses from subconvulsant stimuli to the initial development of full-blown seizures. The second stage is that of completed kindling, where the stimulus more or less reliably evokes full-blown seizures upon each repetition. The final, or third stage, is that of spontaneity where exogenous electrophysiological or pharmacological stimuli are no longer required in order to produce seizures; these may occur upon handling or spontaneously. As illustrated in fig. 7, Pinel (40) has observed that diazepam is highly effective in the development of completed stages of kindling but is without effect in blocking spontaneous amygdala-kindled seizures. Conversely, phenytoin is largely ineffective in the early stages of kindling, but does block seizures of the spontaneous variety. We have found that the anticonvulsant carbamazepine is highly effective in blocking the completed kindled seizure, but is without effect on the developmental stages of amygdala kindling, regardless of dose or route of administration in rats (41). This is an interesting paradox in that, with pharmacological kindling induced by lidocaine or cocaine, the opposite is observed. In this instance, carbamazepine is highly effective in blocking the developmental phase of lidocaine- and cocaine-kindled seizures, but is not effective when acutely administered in blocking seizures induced either by cocaine or lidocaine, and, in some instances, may increase lethality.

Thus, there appears to be a clearcut dissociation in the efficacy of pharmacological interventions as a function of the stage of kindling, and this may be quite different, even with the same anticonvulsant, among different types of kindled seizure evolution. The clinical implications of this model would obviously have the most direct relevance for different stages in the evolution of seizure disorders. For example, they would suggest that, to the extent that kindled seizures model the late onset of seizures in post-traumatic epilepsy, the conventional drugs employed in the attempted

PHARMACOLOGY OF KINDLING AS A FUNCTION OF STIMULUS TYPE AND STAGE OF DEVELOPMENT

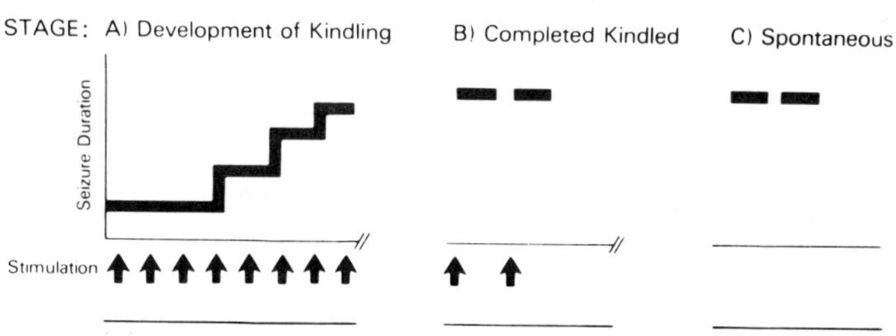

Drug	Effectiveness (Rat) A) Development of Kindling	B) Completed Kindled	C) Spontaneous
Amygdala Kindling (Electrical)			
Diazepam	+++	++	0
Phenytoin	0	+	+++
Carbamazepine	0	+++	?
Lidocaine Kindling (Pharmacological)			
Carbamazepine	+++	0	?
Diazepam	+++	+++	?

Fig. 7

Table 4
Parallels in the Phenomenology of Affective Illness to Kindling and Sensitization

Characteristic of Affective Illness	Characteristic of Kindling	Characteristic of Sensitization
1. Genetic component	+	+
2. Early experience predisposes	+	+
3. Mild alterations emerge as full blown episodes (threshold effects)	+	+
4. Episodes reach plateau and are similar in content and behavior over repeated recurrences	+	+
5. Repeated episodes of mania or depression may lead to emergence of opposite phase (conditioned compensatory reactions)	±	+
6. Onset of maximum disturbance occurs earlier in episode with repetitions	+	+
7. Vulnerability to recurrences and faster cycling	+	+
8. Early episodes may be precipitated, later appear "spontaneously"	+	
9. Lithium carbonate effective prophylaxis	-	±
10. Carbamazepine effective prophylaxis	±	-

+ = Observed; ± = Equivocal; - = Not observed.

SCHEMATIC ILLUSTRATION OF THE LONGITUDINAL AND PROGRESSIVE DEVELOPMENT OF RECURRENT AFFECTIVE ILLNESS

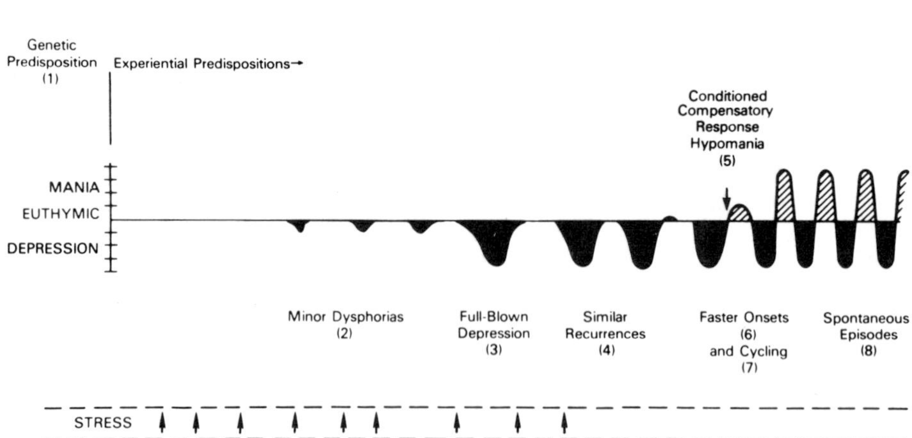

Fig. 8

prophylaxis of this disorder (usually phenytoin without success) should be reconsidered. It has also been observed in the treatment of parkinsonism that pharmacological effectiveness varies as a function of stage of illness. Therefore, we would wonder whether this principle of differential pharmacological effectiveness might not also apply to the treatment of different stages in the evolution, development, and late course of other neuropsychiatric syndromes, such as some stress-related syndromes and manic-depressive illness.

Elsewhere (20), we have described the possible relevance of sensitization and kindling phenomena for conceptualizing phenomena occurring in the course of manic-depressive illness. As illustrated in the schema in fig. 8, we have wondered whether the progressive evolution of various aspects of manic-depressive illness that have been noted since the early observations of Kraepelin, might not conform to sensitization and kindling-like phenomena. Some of the possible analogies are further summarized in Table 4. These include the observation that, with repetition, episodes may occur with greater severity and rapidity of onset and that the well intervals may decrease between successive episodes. Moreover, initial data are consistent with the notion that psychosocial stresses and precipitants may be more relevant to episodes early in the course of affective illness but, with sufficient repetition, the illness may proceed with its own autonomous course independent of such stresses, perhaps akin to the phenomena of spontaneity and electrical and pharmacological kindling. Clearly, many aspects of the schema are speculative and require further clinical investigation and direct testing. Nonetheless, discrete aspects of the phenomenology of manic-depressive illness may be reconsidered in light of sensitization and kindling phenomena, and, in particular, from the question of whether the psychopharmacology of

PSYCHOPHARMACOTHERAPY OF AFFECTIVE ILLNESS AS A FUNCTION OF TYPE AND STAGE OF DEVELOPMENT

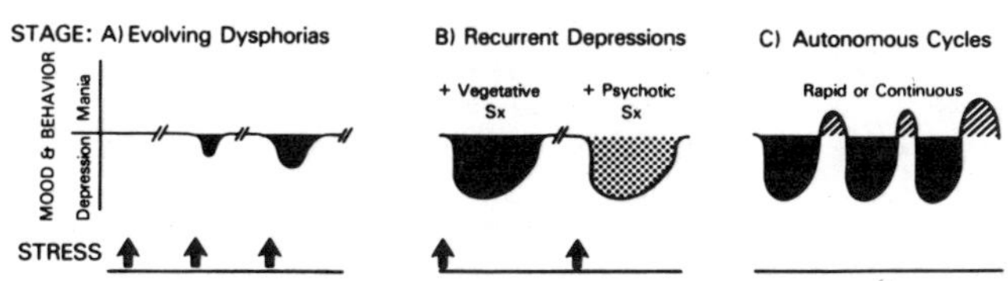

TREATMENT	A) Evolving Dysphorias	B) Recurrent Depressions + Vegetative Sx	B) Recurrent Depressions + Psychotic Sx	C) Autonomous Cycles
Psychotherapy	+++	++		(+)
Benzodiazepines	++	(++)	(+)	
Tricyclics	±	+++	+	−−
MAOI's	±	+++	(+)	−−
ECT		+++	+++	(+)
Tricyclics plus Neuroleptics			+++	
Lithium	±	+	+	(++)
Carbamazepine		(+)	(+)	+++
Lithium plus Carbamazepine				+++

Fig. 9

manic-depressive illness may not also differ as a function of stage in the course of illness in a fashion analogous to that clearly documented for the different stages in the evolution of electrophysiological kindling.

This hypothetical schema is illustrated in fig. 9 which suggests a series of studies to directly test this proposition. One might consider the possibility that early in the evolution of manic-depressive illness, psychotherapy and benzodiazepines might be effective in ameliorating mild stress-induced dysphorias and mood changes. However, there is substantial evidence that severe endogenous depressions may require more major pharmacological interventions, such as tricyclic antidepressants, monoamine oxidase inhibitors, or electroconvulsive therapy. The tricyclics and the monoamine oxidase inhibitors are, additionally, useful in the prophylaxis of unipolar depression. However, there is increasing evidence that these drugs may increase the likelihood of a manic switch and/or be associated with the induction of rapid cycling in recurrent bipolar affective illness. Lithium carbonate is useful in the prophylactic treatment of recurrent unipolar and bipolar affective illness, although there is increasing evidence that a small subgroup of patients, particularly with rapid cycling disorder, are less responsive to lithium carbonate. Initial data suggest that anticonvulsants, such as carbamazepine and, potentially, valproic acid, when used alone or in combination with lithium, may be useful in this latter stage of lithium-refractory rapid cycling manic-depressive illness. It remains to be directly tested whether these agents are also equally effective in earlier phases of manic-depressive illness.

Nonetheless, the schema presents a series of issues for further direct testing, such as whether lithium and carbamazepine are differentially effective in different stages in the evolution of manic-depressive illness and rapid cycling. It would appear that a variety of other possible clinical variables could also be important in the differential pharmacology of lithium versus carbamazepine, including response as a function of severity of illness, type of phenomenology, degree of rapid cycling (independent of course of illness), or even genetic subtype based on family history. We offer the course of illness possibility only as an area for further consideration and study. One might predict, for example, that episodes early in the course of illness would be more associated with psychosocial stresses than those later in the course of illness. Use of pharmaco-prophylaxis in the long-term treatment of manic-depressive illness would have an added rationale in order to prevent the sensitization phenomena from occurring, as well as to prevent episode recurrence. That is, to the extent that this model were relevant, one would predict that adequate pharmaco-prophylaxis with lithium or carbamazepine, when discontinued, should lead to patients' having a less rapid cycling course of illness compared with equivalently ill patients who were only treated for acute episodes once they became manifest.

This perspective would have the same implications for the consequences of stressful life experiences in the development of pathological syndromes. Sensitization and kindling perspectives would suggest the importance of early intervention and prophylaxis in this process in order to prevent some of the late-appearing and more pathological sequelae from occurring. In addition, sensitization and kindling phenomena may be relevant to stress-induced changes in behavioral pathology, from the standpoint of elucidating principles such as the role of context, conditioning, and time-course variables in the development and late appearance of behavioral dysfunction.

REFERENCES

1. S. M. Antelman, A. J. Eichler, C. A. Black, and D. Kocan, Interchangeability of stress and amphetamine in sensitization. Science 207: 329-331 (1980).

2. S. M. Antelman, and L. A. Chiodo, Amphetamine as a stressor, in: "Stimulants: Neurochemical, Behavioral and Clinical Perspectives," I. Creese, ed., Raven Press, New York (1983).
3. L. Shuster, G. Yu, and A. Bates, Sensitization to cocaine stimulation in mice. Psychopharmacology 52:185-190 (1977).
4. H. A. Tilson, and R. H. Rech, Conditioned drug effects and absence of tolerance to d-amphetamine induced motor activity. Pharmacol. Biochem. Behav. 1:149-153 (1973).
5. S. R. Schiff, Conditioned dopaminergic activity. Biol. Psychiatry 17:135-154 (1982).
6. R. E. Hinson, and C. X. Poulos, Sensitization to the behavioral effects of cocaine: modulation by pavlovian conditioning. Pharmacol. Biochem. Behav. 15:559-562 (1981).
7. R. M. Post, A. Lockfeld, K. M. Squillace, and N. R. Contel, Drug-environment interaction: context dependency of cocaine-induced behavioral sensitization. Life Sci. 28:755-760 (1981).
8. S. R. B. Weiss, D. Murman, R. M. Post, and A. Pert, Conditioning in cocaine-induced behavioral sensitization, in: "Abstracts, 16th Annual Meeting, Society for Neuroscience", Washington, D.C., Abstract #249.13, p. 914 (1986).
9. R. J. Beninger and B. L. Hahn, Pimozide blocks establishment but not expression of amphetamine-produced environment-specific conditioning. Science 220:1304-1306 (1983).
10. R. J. Beninger and R. S. Herz, Pimozide blocks establishment but not expression of cocaine-produced environment-specific conditioning. Life Sci. 38: 1425-1431 (1986).
11. S. Tadokoro and H. Kuribara: Reverse tolerance to the ambulation-increasing effect of methamphetamine in mice as an animal model of amphetamine-psychosis. Psychopharmacol. Bull. 22: 757-762 (1986).
12. K. Gale, Catecholamine-independent behavioral and neurochemical effects of cocaine in rats, in: "National Institute on Drug Abuse Research Monograph Series 54: Mechanisms of Tolerance and Dependence", C. Wm. Sharp, ed., U.S. Government Printing Office, Washington, D.C. (1984).
13. T. S. Jensen, J. K. Genefke, N. Hyldebrandt, H. Pedersen, H. D. Petersen, and B. Weile, Cerebral atrophy in young torture victims. N. Engl. J. Med. 307: 1341 (1982).
14. C. H. Kellner, P. P. Roy-Byrne, D. R. Rubinow, P. W. Gold, and R. M. Post, Cerebral atrophy in torture victims [Letter to the Editor]. N. Engl. J. Med. 308: 903-904 (1983).
15. F. W. Putnam, J. J. Guroff, E. K. Silberman, L. Barban, and R. M. Post, The clinical phenomenology of multiple personality disorder: review of 100 recent cases. J. Clin. Psychiatry 47: 285-293 (1986).
16. G. V. Goddard, D. C. McIntyre, and C. K. Leech, A permanent change in brain function resulting from daily electrical stimulation. Exp. Neurol. 25: 295-330 (1969).
17. R. M. Post, C. Kennedy, M. Shinohara, K. Squillace, M. Miyaoka, S. Suda, D. H. Ingvar, and L. Sokoloff, Metabolic and behavioral consequences of lidocaine-kindled seizures. Brain Res. 324: 295-303 (1984).
18. R. M. Post, Cocaine psychoses: a continuum model. Am. J. Psychiatry 132: 225-231 (1975).
19. R. M. Post, and N. R. Contel, Human and animal studies of cocaine: implications for development of behavioral pathology, in: "Stimulants: Neurochemical, Behavioral, and Clinical Perspective", I. Creese, ed., Raven Press, New York (1983).
20. R. M. Post, D. R. Rubinow, and J. C. Ballenger, Conditioning, sensitization, and kindling: implications for the course of affective illness, in: "Neurobiology of Mood Disorders", R. M. Post and J. C. Ballenger, eds., Williams & Wilkins, Baltimore (1984).

21. R. M. Post, S. R. B. Weiss, A. Pert, and T. W. Uhde, Chronic cocaine administration: sensitization and kindling effects, in: "Cocaine: Clinical and Biobehavioral Aspects", S. Fisher, A. Raskin, and E. H. Uhlenhuth, eds., Oxford University Press, New York (1987).
22. A. M. Washton, and M. S. Gold, Chronic cocaine abuse: evidence for adverse effects on health and functioning. Psychiatr. Ann. 14:-733-743 (1984).
23. A. M. Washton, and A. Tatarsky, Adverse effects of cocaine abuse, in: "Problems of Drug Dependence, 1983, NIDA Research Monograph No. 49", L. S. Harris, ed., U.S. Government Printing Office, Washington, D.C. (1984).
24. L. Shuster, G. W. Webster, and G. Yu, Perinatal narcotic addiction in mice: sensitization to morphine stimulation. Addict. Dis. 2: 277-292 (1975).
25. P. W. Kalivas, R. Richardson-Carlson, and G. Van Orden, Cross-sensitization between foot shock stress and enkephalin-induced motor activity. Biol. Psychiatry 21: 939-950 (1986).
26. P. W. Kalivas, Behavioral and neurochemical sensitization to endogenous enkephalin and environmental stress: role of the mesolimbic dopamine system, in Abstracts, Am. College of Neuropsychopharmacology, 25th Annual Meeting, Dec. 8-12, 1986, Washington, D.C., p. 63.
27. D. P. Cain, and M. E. Corcoran, Intracerebral beta-endorphin, met-enkephalin and morphine: kindling of seizure and handling-induced potentiation of epileptiform effects. Life Sci., 34: 2535-2543 (1984).
28. T. Tanaka, Y. Hamasaki, H. Takeshita, R. Kawahara, and H. Hazama, Chemical kindling with met-enkephalin and changes of brain opioid delta-receptor sensitivity after electrical kindling. Epilepsia 27: 596 (1986).
29. T. F. Seeger, G. A. Sforzo, C. B. Pert, and A. Pert, In vivo autoradiography: visualization of stress-induced changes in opiate receptor occupancy in the rat brain. Brain Res., 9: 303-311 (1984).
30. S. R. B. Weiss, T. F. Seeger, N. Ostrowski, R. M. Post, and A. Pert, Alterations in opiate receptor occupancy following amygdaloid kindling, in Abstracts, Society for Neuroscience, 15th Annual Meeting, Oct. 20-25, 1985, Dallas, Texas, p. 1068, Abs. #310.1.
31. C. L. Ehlers, S. J. Henriksen, M. Wang, J. Rivier, W. Vale, and F. E. Bloom, Corticotropin releasing factor produces increases in brain excitability and convulsive seizures in rats. Brain Res., 278: 332-336 (1983).
32. G. F. Koob, and K. Thatcher-Britton, The role of corticotropin releasing factor in the behavior response to stress, in Abstracts, American College of Neuropsychopharmacology (ACNP), Annual Meeting, Dec. 9-13, 1985, Maui, p 80.
33. N. H. Kalin, Behavioral effects of ovine corticotropin-releasing factor administered to rhesus monkeys. Fed. Proc., 44: 249-253 (1985).
34. D. P. Cain, J. Plant, S. Rouleau, and M. E. Corcoran, Failure to kindle seizures after repeated intracerebral administration of arginine vasopressin. Life Sci., 17: 985-989 (1986).
35. H. J. Little, D. J. Nutt, and S. C. Taylor, The effects of drugs acting at the $GABA_A$-receptor/ionophore after chemical kindling with the benzodiazepine receptor ligand FG 7142. Br. J. Pharmacol. 88: 507-514 (1986).
36. S. R. B. Weiss, R. M. Post, P. J. Marangos, and J. Patel, "Peripheral-type" benzodiazepines: behavioral effects and relationship to the anticonvulsant effects of carbamazepine, in: "Kindling III", J. Wada, ed., Raven Press, New York (1986).
37. B. A. Weissman, J. Cott, J. A. Jackson, G. T. Bolger, K. H. Weber, W. D. Horst, S. M. Paul, and P. Skolnick, "Peripheral-type" binding sites for benzodiazepines in brain: relationship to the convulsant actions of Ro 5-4864. J. Neurochem. 44: 1494-1499 (1985).

38. A. Guidotti, P. Ferrero, M. Fujimoto, R. M. Santi, and E. Costa, Studies on endogenous ligands (endacoids) for the benzodiazepine/-beta carboline binding sites. Adv. Biochem. Psychopharmacol. 41: 137-148 (1986).
39. R. C. Drugan, A. S. Basile, J. N. Crawley, S. M. Paul, and P. Skolnick, Inescapable tailshock stress decreases "peripheral-type" benzodiazepine binding in rat kidney, in Abstracts, Society for Neuroscience, 15th Annual Meeting, Oct. 20-25, 1985, Dallas. Abstract #80.14.
40. J. P. Pinel, Effects of diazepam and diphenylhydantoin on elicited and spontaneous seizures in kindled rats: a double dissociation. Pharmacol. Biochem. Behav., 18: 61-63 (1983).
41. S. R. B. Weiss, and R. M. Post, Carbamazepine and carbamazepine-10,11-epoxide inhibit amygdala kindled seizures in the rat but do not block their development. Clin. Neuropharmacol., in press.
42. G. V. Goddard, D. C. McIntyre, and C. K. Leech, A permanent change in brain function resulting from daily electrical stimulation. Exp. Neurol. 25: 295 (1969).
43. J. A. Wada, and M. Sato, Generalized convulsive seizures induced by daily electrical stimulation of the amygdala in cats. Neurology 24: 565-574 (1974).
44. J. A. Wada, M. Sato, and M. E. Corcoran, Persistent seizure susceptibility and recurrent spontaneous seizures in kindled cats. Epilepsia 15: 465-478 (1974).
45. R. Racine, Kindling: the first decade. Neurosurgery 3: 234-252 (1978).
46. J. P. J. Pinel, and L. I. Rovner, Electrode placement and kindling-induced experimental epilepsy. Exp. Neurol. 58: 335-346 (1978).
47. J. P. J. Pinel, and L. I. Rovner, Experimental epileptogenesis: kindling-induced epilepsy in rats. Exp. Neurol. 58: 190-202 (1978).

EFFECTS OF PSYCHOLOGICAL STRESS ON PITUITARY HORMONES IN MAN[a]

J.L. Meyerhoff, M.A. Oleshansky and E.H. Mougey

Neurochemistry & Neuroendocrinology Branch
Department of Medical Neurosciences
Division of Neuropsychiatry
Walter Reed Army Institute of Research
Washington, DC 20307-5100

INTRODUCTION

Psychological stressors such as stressful social interactions have been shown to elevate heart rate as well as plasma levels of epinephrine (EPI), norepinephrine (NE), and cortisol (CS) in humans (1-3). Exercise also produces increases in these hormones (4-6) as well as in plasma levels of the peptide hormones derived from proopiomelanocortin (POMC): adrenocorticotropin (ACTH), β-endorphin (β-EP) and β-lipotrophic hormone (β-LPH) (7-15). To determine whether a stressful social interaction would also increase plasma levels of these POMC-derived peptides, we examined the effect of an oral exam on the responses of these hormones as well as CS in a group of healthy young men. We also measured plasma levels of prolactin (PRL), a pituitary hormone not derived from POMC. Our experimental design controlled for any effects due to physical exertion.

BACKGROUND

Occupational, clinical and laboratory paradigms as well as sporting events have been used in studies of physiological responses to environmental stressors. Public speaking in an occupational environment (physicians presenting at a medical conference) was associated with increases in heart rate equal to those seen after moderate exercise (1). In more recent studies using similar paradigms, public speaking elicited increases in plasma levels of EPI and NE (2), and dexamethasone pre-treatment failed to suppress blood CS levels (16). Also, examinations have been reported to increase urinary excretion of NE, EPI and CS in students (17).

Endocrine changes have been reported in individuals anticipating various types of challenge. Resting heart rate and plasma CS levels were elevated in medical students during final exam week (3). In hospital settings, elevated plasma levels of CS and PRL were reported in patients

[a] The views of the authors do not purport to reflect the position of the Department of the Army or the Department of Defense (para 4-3, AR 360-5).

awaiting imminent surgical procedures (18,19). Elevated plasma CS was also found in subjects about to commence vigorous exercise (5), and increases in ACTH and β-EP (20) were seen in trained long distance runners 20 min before a race.

Physical exertion (bicycle ergometers, treadmill running and competitive running) produced elevations in plasma ACTH, β-EP, β-LPH and CS, as well as NE, EPI and PRL (2,5-13,15,21-25). Plasma CS and PRL were also reported to be markedly elevated after parachutists' training jumps - events combining physical activity and psychological arousal (26,27).

In addition to the "naturalistic" studies discussed above, responses to stressful stimuli have been studied in laboratory settings. Early studies in monkeys reported stress-induced increases in adrenal corticoid secretion (28). In studies using human volunteers, a mental arithmetic task was reported to produce increases in heart rate and blood pressure (29) as well as increases in plasma levels of NE, EPI and CS (30). The Stroop color-word conflict task elicited increases in blood pressure, NE and EPI in normal young men (31). In studies of patients with phobic symptoms, exposure to phobic stimuli increased anxiety, heart rate, blood pressure, plasma levels of EPI, NE, CS, growth hormone and insulin (32) and urinary excretion of CS (33).

THE ORAL EXAM AS AN EXPERIMENTAL PARADIGM

The oral exam is a rite of passage in many professions and occupations, including the military. The salient features of oral exams include: (a) public speaking, (b) formal presentation before a number of higher-ranking members of a profession, and (c) submitting to evaluation. In studying stress in an occupational environment, it is necessary to satisfy several conditions: the subjects must be protected, yet the paradigm should be stressful; the situation should be occupationally relevant, yet permit rigorous experimental control. The oral exam paradigm we describe below meets these criteria.

As part of their training and preparation for promotion boards, soldiers at the Walter Reed Army Institute of Research (WRAIR) are encouraged to appear before "Soldier of the Month" (SOM) boards - contests held monthly to select and honor an outstanding soldier. These boards are very similar to the actual promotion boards: rigorous, competitive, structured interviews conducted by a panel comprised of senior noncommissioned officers. A soldier must win promotion to staff sergeant in order to be eligible to remain in the Army for a full 20 year career with retirement benefits. Promotion depends on many factors (e.g. supervisor's evaluations, time-in-grade, physical fitness, etc.) but promotions to sergeant and staff sergeant also require successful personal appearances before promotion boards (promotion between other grades is determined solely by review of records). The SOM contest is patterned after the promotion board. The contest winner is selected on the basis of knowledge of job-related subject matter, personal appearance and quality of oral presentation. We have characterized this competitive oral exam as a model for studying endocrine responses to stressful social interactions in an occupational setting.

The winner of the SOM contest is awarded a $50.00 savings bond, a weekend pass, a certificate of achievement and an opportunity to compete at higher levels of competition in the future. In addition the winner is exempt from ancillary duties for one month. Also a press release is sent to the winner's hometown newspaper and his/her photograph is displayed in the lobby of the Institute. Candidates who do not win are subjected to no penalties and may enter subsequent competitions as often as they wish. Entering the SOM contest is not a requirement for promotion.

Subjects in the study were candidates in the monthly competitions who volunteered to serve as subjects after being fully informed about the experimental procedures. Being a subject was in no way a requirement or a factor in the outcome of the SOM contest. None of the subjects was a supervisee of any of the authors. This study was approved by the Human Subjects Research Review Board of the Office of the Surgeon General of the U.S. Army.

There are usually 5 candidates (corporals or sergeants) who are interviewed individually during each monthly contest. Eleven of the contestants volunteered to be studied during this competition and gave written informed consent for us to monitor their heart rates and draw blood samples during their exam. Each candidate was examined individually by the panel for approximately 30 min. The hierarchy of rank, enlisted career progression ladder and the numbers of enlisted personnel in each rank at the WRAIR are displayed in Figure 1. A social diagram of the exam situation is presented in Figure 2; the seating arrangement is shown, as are the respective ranks of the soldier and the examiners. The examiners were seated at a long table. The soldier knocked on the door of the exam room, entered, saluted and was directed to sit in a chair across the room from the panel. To minimize any effects of diurnal variation, the subjects were

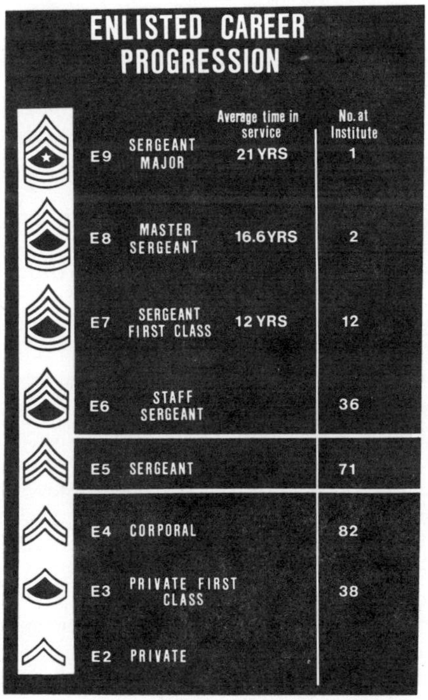

Figure 1. Enlisted career ladder, with numbers of soldiers at each step at the Walter Reed Army Institute of Research. The designations E2 through E9 refer to pay grades. The horizontal lines between corporal and sergeant and between sergeant and staff sergeant indicate steps in the career progression which require successful personal appearances before promotion boards. The soldier of the month exam is patterned after the promotion board.

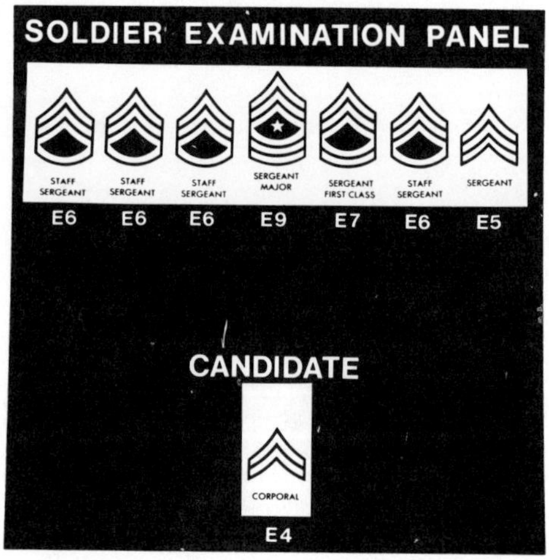

Figure 2. Social diagram of the oral exam situation, giving seating diagram with ranks.

always scheduled to appear before the panel at either 9:00 or 9:30 A.M. We used a nearby room as a waiting room to provide them a constant environment for 30 min before and after the exam.

The typical subject was male, 25 years old (range: 19-32), had 14 years of education (range: 11-17), was physically very fit, was a corporal or a sergeant in the U.S. Army with 31 months of active duty experience (range: 5-91), and was assigned as a research technician at the WRAIR.

We measured physiological (heart rate) and biochemical responses before, during and after the exam. Psychological responses were estimated via self-report questionnaires administered before and after the exam. Two hours before the exam, electrodes were placed on the subject's chest and connected via EKG leads to a Vitalog PMS-8 monitor (Vitalog Inc., Mountainside, CA) worn on a belt for continuous monitoring of heart rate. A sterile intravenous needle was then inserted into the anterior cubital vein of the left arm and blood was collected through non-thrombogenic tubing using a 3-speed blood withdrawal pump (Cormed Inc., Medina, NY, model #6-53R) also worn on the belt. A cylindrically-shaped plastic arm board was worn on the left arm to prevent flexion. Blood was withdrawn continuously at 0.1 ml/min throughout the procedure, except during discreet sampling intervals when pump speed was increased to 5 ml/min. All apparatus was concealed under the uniform jacket. Blood was sampled at 7 time points on the day of the exam (Table 1). Each blood sample consisted of two sequential 10 ml samples collected in chilled tubes containing EDTA and kept on ice until centrifuged at 4°C. The plasma samples from these two tubes were mixed and aliquoted into tubes containing 50 ul (22 T.I.U./ml) of aprotinin (a protease inhibitor) for β-LPH, β-EP and ACTH assays. These tubes, as well as a tube of untreated plasma for CS and PRL assays were frozen at -70°C until assayed. The values reported represent an integrated four minute blood collection beginning at the scheduled sampling point.

An extraction procedure was used to separate and purify β-EP and β-LPH, which were then quantified via RIA, using an endorphin kit supplied by

Table 1. Blood Sampling Schedule

SAMPLE	DESCRIPTION
I.V. + 22 MIN:	after i.v. inserted and 22 min sitting in medical room.
REST 7 MIN:	after walking 50 yards in hall, then sitting 7 min in waiting room.
REST 22 MIN:	after sitting 22 min in the waiting room, before walking to the exam room.
EXAM 7 MIN:	7 minutes after being seated before the panel of examiners.
EXAM 22 MIN:	22 minutes after being seated before the panel.
POST 7 MIN:	7 minutes after returning to the waiting room and being seated.
POST 22 MIN:	22 minutes after sitting in the waiting room.

Du Pont NEN Research Products. Plasma CS was measured by radioimmunoassay, using an antibody produced in our laboratory in rabbits against prednisolone-3CMO:BSA (34). Immuno Nuclear Corporation kits were used to assay human plasma for ACTH and PRL content.

Data were analyzed statistically by analysis of variance (ANOVA) for experiments with repeated measures. When the overall F-score was significant, the ANOVA was followed by a post-hoc test (Tukey's) for multiple comparisons to determine which time points showed significant differences.

EFFECT OF WALKING FOLLOWED BY SITTING 5 MINUTES

The logistics of the exam situation required that the subjects walk approximately 20 yards from the waiting room to the exam room. Because standing for 5 min has been reported to increase plasma levels of NE (4), and since we planned to take a blood sample 7 min after the subject was seated in the exam, the need to control for the effects of postural change was apparent. We therefore collected samples at 3 time points during the hour before the exam to determine if walking 50 yards would affect plasma hormone levels. Table 2 confirms the absence of any effect of postural change on the variables we measured. The first blood sample was taken after subjects were seated 22 minutes (I.V. + 22 MIN). They then walked 50 yards, were seated in the waiting room and a second blood sample was taken after they had been resting for 7 minutes (REST 7 MIN). This sample revealed no apparent effect of the preceding physical activity on any of the physiological or biochemical variables measured. These data permitted us to use the third sample (REST 22 MIN) as the point of comparison for the oral exam effects.

STRESSFUL INTERVIEW INCREASES HEART RATE AND PLASMA LEVELS OF PROLACTIN, CORTISOL AND POMC-DERIVED PEPTIDES IN MAN

As shown in Figure 3, the mean heart rate was 81 beats per min (bpm) at rest before the exam. This is higher than normally expected in a group of young, physically fit males. This higher-than-normal heart rate may reflect apprehension about the upcoming exam. In preliminary reports, we have noted increases in self-reported anxiety scores (Spielberger State Anxiety Scale) immediately before the exam compared to one week prior to the exam (35,36).

Table 2. Effect of Walking Followed by Sitting 5 Min

	"I.V. 22 MIN"	WALK 50 YARDS "REST 7 MIN"	"REST 22 MIN"
HR (bpm)	82 ± 3	80 ± 3	81 ± 2
ACTH (pg/ml)	21.0 ± 5.3	19.8 ± 4.8	20.6 ± 5.5
CS (μg/dl)	14.8 ± 0.8	14.0 ± 0.8	14.6 ± 1.1
β-EP (pg/ml)	6.0 ± 0.4	7.1 ± 1.0	8.7 ± 1.8
β-LPH (pg/ml)	120.4 ± 7.3	122.6 ± 9.4	131.8 ± 10.7
PRL (ng/ml)	6.2 ± 0.9	6.3 ± 1.0	6.8 ± 1.1

Data expressed as mean ± SEM, N = 11. No significant changes were seen. HR (bpm) = heart rate in beats per minute. ACTH = adrenocorticotropic hormone. CS = cortisol. β-EP = beta-endorphin. β-LPH = beta-lipotrophic hormone. PRL = prolactin.

Seven minutes after beginning the exam, the average heart rate increased 22 bpm to 103 bpm. Although this represents an increase of 27%, the increase seems even more significant when one considers that 22 bpm represents 3.7 times the standard deviation (5.97) of the average heart rate for the group before the exam.

This stressful interview also elicited rapid increases in plasma levels of the pituitary hormones ACTH, β-EP, β-LPH and PRL of 59%, 79%, 42% and 46%, respectively (Figures 4-7). Compared with values observed during the "rest" period just before the exam, all hormones except cortisol were significantly elevated 7 min after the subjects were seated in the examination situation and all but β-EP remained elevated after 22 minutes of the examination. It is interesting to note that the effects of this stressful social interaction elicited percent increases in plasma ACTH levels equal to those reported during moderate exercise (11). Plasma CS was significantly elevated (31%) after 22 minutes in the exam (Figure 8). This

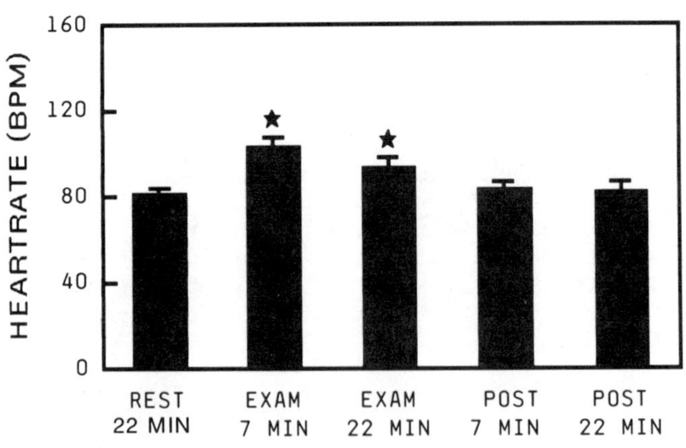

Figure 3. Heart rate response to stress of competitive oral exam. Data expressed as mean ± SEM, N = 11. Data were analyzed by ANOVA for experiments with repeated measures, followed by Tukey's test for multiple comparisons. * = significantly elevated ($p<0.05$) compared to all values not marked with an asterisk. BPM = beats per minute.

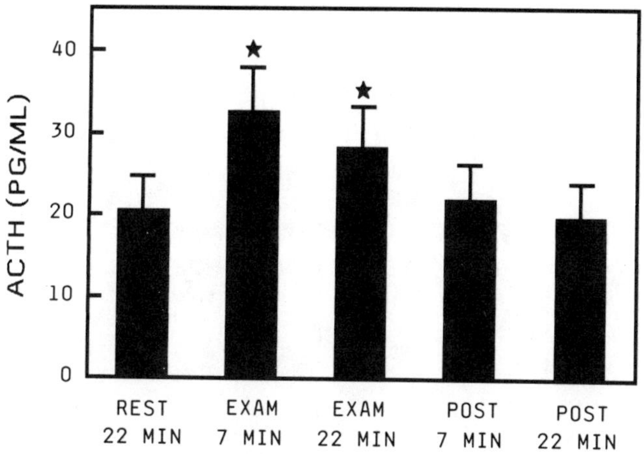

Figure 4. Response of plasma ACTH to stress of competitive oral exam. Data expressed as mean ± SEM, N = 11. Data were analyzed by ANOVA for experiments with repeated measures, followed by Tukey's test for multiple comparisons. * = significantly elevated ($p<0.05$) compared to all values not marked with an asterisk. ACTH = adrenocorticotrophic hormone.

followed by 15 min the maximal responses of the POMC-derived pituitary hormones. A similar difference in response times of ACTH and CS has been reported in studies of physical exertion (10). Seven minutes after leaving the examination room and sitting in the waiting room, heart rate, as well as levels of all POMC-derived hormones had declined to levels that were no longer significantly different from those seen before the exam. Plasma CS did not return to pre-examination levels until 22 minutes after the exam.

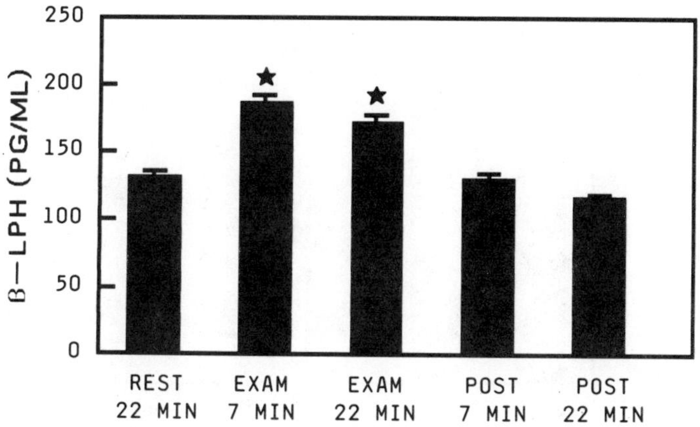

Figure 5. Response of plasma beta-lipotrophic hormone to stress of competitive oral exam. Data expressed as mean ± SEM, N = 11. Data were analyzed by ANOVA for experiments with repeated measures, followed by Tukey's test for multiple comparisons. * = significantly elevated ($p<0.05$) compared to all values not marked with an asterisk. β-LPH = beta-lipotrophic hormone.

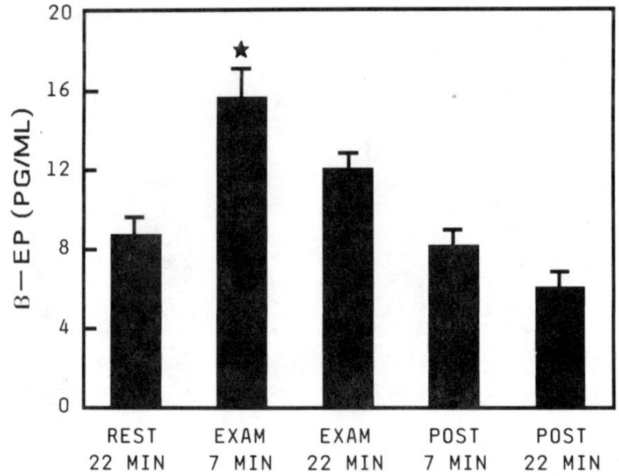

Figure 6. Response of plasma beta-endorphin to stress of competitive oral exam. Data expressed as mean ± SEM, N = 11. Data were analyzed by ANOVA for experiments with repeated measures, followed by Tukey's test for multiple comparisons. * = significantly elevated ($p<0.05$) compared to all values not marked with an asterisk. β-EP = beta-endorphin.

DISCUSSION

β-EP, ACTH and β-LPH are derived from a common precursor (37,38) and co-localized in the same secretory granules in anterior pituitary corticotrophs (39,40). β-LPH is a 91 amino acid peptide (41) which contains β-EP as the 61-91 segment (42). Plasma levels of the POMC-derived peptides might be affected by rate of enzymatic cleavage, as well as by the metabolic clearance rate which has been reported to be greater for β-LPH than for ACTH (43). In the present study "resting" plasma molar levels of β-EP, ACTH and

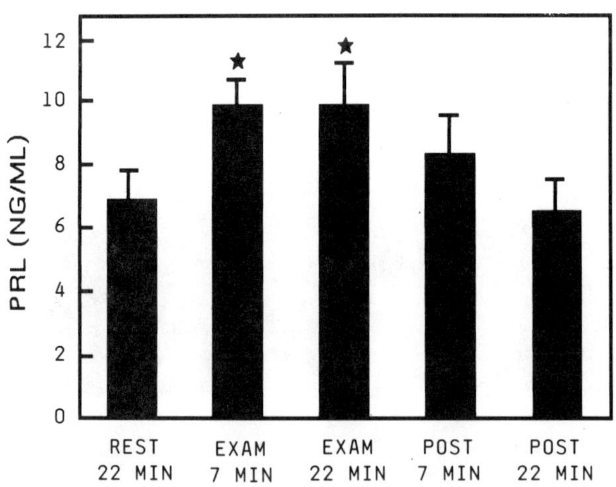

Figure 7. Response of plasma prolactin to stress of competitive oral exam. Data expressed as mean ± SEM, N = 11. Data were analyzed by ANOVA for experiments with repeated measures, followed by Tukey's test for multiple comparisons. * = significantly elevated ($p<0.05$) compared to all values not marked with an asterisk. PRL = prolactin.

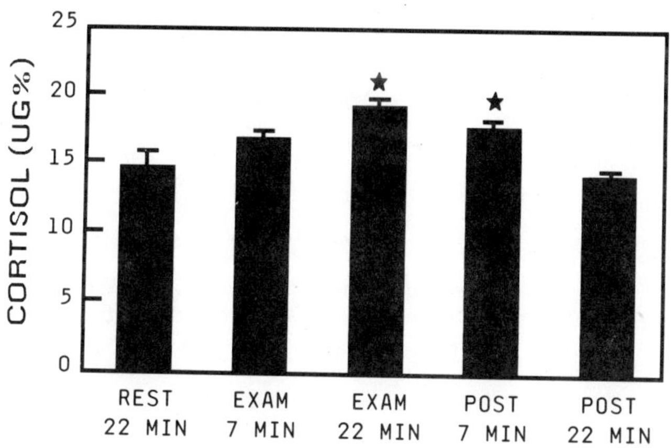

Figure 8. Response of plasma cortisol to stress of competitive oral exam. Data expressed as mean ± SEM, N = 11. Data were analyzed by ANOVA for experiments with repeated measures, followed by Tukey's test for multiple comparisons. * = significantly elevated ($p<0.05$) compared to all values not marked with an asterisk.

β-LPH were 2.5, 4.5 and 13.9 femtomoles per ml respectively. The molar ratio of β-EP/β-LPH was 0.18. Thus, plasma concentrations of β-LPH appear to be much higher than those of β-EP in man, where the intermediate lobe is very small (44). This relationship is reversed in the rat (45,46) where the intermediate lobe is well-developed and contains predominantly β-EP (47-50).

Hormones released during stress have multiple effects in the periphery including mobilization of energy reserves (51) via gluconeogenesis and lipolysis (52,53), as well as effects on the cardiovascular system (54) and on electrolyte balance (55). Given the present finding that PRL as well as β-EP is elevated in man during stressful situations, it is noteworthy that β-EP may facilitate PRL release (56). While the function of PRL release in response to stress is not understood, one possible physiological role for PRL might be stimulation of the immune system (57). In this capacity, PRL might serve as a counter-regulatory hormone to adrenocortical hormones, which can suppress immune function (58).

In addition to their effects on various peripheral targets, hormones are also reported to affect the brain. Adrenal cortical steroid binding in brain and pituitary has been extensively reported (59). Small amounts of peptides may enter the CNS from the periphery and ACTH and melanocyte stimulating hormone have been reported to exert effects on behavior (60). We have previously shown that a conditioned response can elevate plasma levels of β-EP in the rat (61). A stressful paradigm was reported to cause a naloxone-reversible analgesia in the rat - an effect which was dependent on an intact pituitary (62,63). Moreover, β-EP injected intravenously was shown to have analgesic potency (64). The possibility that β-EP of pituitary origin might reach the brain is supported by reports of retrograde blood flow in the pituitary stalk (65). Activation of endogenous opioids by stressful interactions in man might be inferred from the report that stress-induced analgesia in humans was reversed by naloxone administration (66). Just as opiates selectively dampen the affective component of pain (8,67), the endorphins might selectively modify the affective responses to other stressful situations (67). While not directly addressing the behavioral effects of pituitary peptides, the competitive oral examination model we have described in this paper would appear to be suitable for studying the effects of stressors on performance.

We have characterized a robust model for examining the effect of stressful social interactions on the responses of the endocrine and cardiovascular systems (68). The increase we found in plasma levels of PRL are particularly interesting, as we have not seen such increases during milder challenges such as mental arithmetic (Krantz, Oleshansky, Mougey & Meyerhoff, unpublished data). Thus, this model might also have utility for examining the effects of stressors on the immune system. In addition, our data suggest that hormonal responses in exercise studies might be influenced by the subjects' perceptions of the experimental situation.

CONCLUSIONS

A stressful social interaction can elicit rapid increases in heart rate and plasma levels of pituitary hormones. In the present study, a competitive oral examination produced increases in plasma levels of ACTH, β-EP, β-LPH and PRL of 59%, 79%, 42% and 46%, respectively. Increases in plasma cortisol were also seen. Further study is required to examine the relationship between these physiological responses and performance under stressful conditions.

ACKNOWLEDGEMENTS

The authors gratefully acknowledge the advice and consultation of Sergeant Major Bert J. Mueck, First Sergeant Lawrence D. Olmsted, Lieutenant Colonel Steven R. Hursh, Lieutenant Colonel Frederick J. Manning and Sergeant First Class Ed Marshall. We are also grateful for the interest and support of the soldiers and noncommissioned officers at the Walter Reed Army Institute of Research. We wish to thank the following individuals for excellent technical support: Lori Wittig, Rod Perez, Vinit Jain, Sergeant Thomasina Maxwell-Irving, Mr. Clint Wormley and Mr. Willie Gamble. We thank Dr. H. Ron Smith for consultation and administrative assistance, John R. Burge and Dr. Sue Hedges for excellent statistical consultation and Drs. Robert Smallridge, Henry Fein, Jean Kant and John Holaday for helpful comments on the manuscript.

REFERENCES

1. A. J. Moss and B. Wynar, Tachycardia in house officers presenting cases at grand rounds, Ann. Internal Med. 72(2):255-256 (1970).
2. J. E. Dimsdale, and J. Moss, Short-term catecholamine response to psychological stress, Psychosom. Med. 42(5):493-497 (1980).
3. W. R. Lovallo, G. A. Pincomb, G. A. Edwards, D. J. Brackett, and M. F. Wilson, Work pressure and the type A behavior pattern exam stress in male medical students, Psychosom. Med. 48(1/2):125-133 (1986).
4. C. R. Lake, M. G. Ziegler, and I. J. Kopin, Use of plasma norepinephrine for evaluation of sympathetic neuronal function in man, Life Sciences 18:1315-1326 (1976).
5. L. H. Hartley, J. W. Mason, R. P. Hogan, L. G. Jones, T. A. Kotchen, E. H. Mougey, F. E. Wherry, L. L. Pennington, and P. T. Ricketts, Multiple hormonal responses to graded exercise in relation to physical training, J. Appl. Physiol. 33(5):602-606 (1972).
6. A. Grossman, P. Bouloux, P. Price, P. L. Drury , K. S. L. Lam, T. Turner, J. Thomas, G. M. Besser, and J. Sutton, The role of opioid peptides in the hormonal responses to acute exercise in man, Clinical Science 67:483-491 (1984).

7. F. Fraioli, C. Moretti, D. Paolucci, E. Alicicco, F. Crescenzi, and G. Fortunio, Physical exercise stimulates marked concomitant release of β-endorphin and adrenocorticotropic hormone (ACTH) in peripheral blood in man, Experientia 36:987-989 (1980).
8. M. N. Janal, E. W. D. Colt, W. C. Clark, and M. Glusman, Pain sensitivity, mood and plasma endocrine levels in man following long-distance running: Effects of naloxone, Pain 19:13-25 (1984).
9. S. R. Gambert, T. L. Garthwaite, C. H. Pontzer, E. E. Cook, F. E. Tristani, E. H. Duthie, D. R. Martinson, T. C. Hagen, and D. J. McCarty, Running elevates plasma β-endorphin immunoreactivity and ACTH in untrained human subjects, Proc. Soc. Exper. Biol. & Med. 168:1-4 (1981).
10. M. J. Buono, J. E. Yeager, and J. A. Hodgdon, Plasma adrenocorticotropin and cortisol responses to brief high-intensity exercise in humans, J. Appl. Physiol. 61(4):1337-1339 (1986).
11. A. Luger, P. A. Deuster, S. B. Kyle, W. T. Gallucci, L. C. Montgomery, P. W. Gold, D. L. Loriaux, and G. P. Chrousos, Acute hypothalamic-pituitary-adrenal responses to the stress of treadmill exercise: Physiologic adaptations to physical training, New England J. Med. (in press, 1987).
12. M. Oleshansky, J. Zoltick, R. Herman, E. Mougey, and J. Meyerhoff, Neuroendocrine responses to maximal treadmill exercise, 1Psychiatry Research 16(4), Special Suppl.:72 (1986).
13. E. W. D. Colt, S. L. Wardlaw, and A. G. Frantz, The effect of running on plasma β-endorphin, Life Sciences 28:1637-1640 (1981).
14. D. L. Elliot, L. Goldberg, W. Watts, and E. Orwoll, Resistance exercise and plasma beta-endorphin/beta-lipotropin immunoreactivity, Life Sciences 34:515-518 (1983).
15. W. M. Bortz, P. Angwin, I. N. Mefford, M. R. Boarder, N. Noyce, and J. D. Barchas, Catecholamines, dopamine, and endorphin levels during extreme exercise, New England J. Med. 30(8):466-467 (1981).
16. A. Baumgartner, K.-J. Graf, and I. Kurten, The dexamethasone suppression test in depression, in schizophrenia, and during experimental stress, Biolog. Psychiatry 20:675-679 (1985).
17. M. Frankenhaeuser, M. R. Von Wright, A. Collins, J. Von Wright, G. Sedvall, and C.-G. Swahn, Sex differences in psychoneuroendocrine reactions to examination stress, Psychosom. Med. 40(4):334-343 (1978).
18. C. A. Czeisler, M. C. Moore Ede, Q. R. Regestein, E. S. Kisch, V. S. Fang, and E. N. Ehrlich, Episodic 24-hour cortisol secretory patterns in patients awaiting elective cardiac surgery, J. Clin. Endocrinol. & Metabolism 42(2):273-283 (1976).
19. B. Corenblum and P. J. Taylor, Mechanisms of control of prolactin release in response to apprehension stress and anesthesia-surgery stress, Fertility and Sterility 36(6):712-715 (1981).
20. C. M. Oltras, F. Mora, and F. Vives, Beta-endorphin and ACTH in plasma: Effects of physical and psychological stress, Life Sciences 40:1683-1686 (1987).
21. R. C. Smallridge, N. E. Whorton, K. D. Burman, and E. W. Ferguson, Effects of exercise and physical fitness on the pituitary-thyroid axis and on prolactin secretion in male runners, Metabolism 34(10):949-954 (1985).
22. J. R. Sowers, R. P. Raj, J. M. Hershman, H. E. Carlson, and R. W. McCallum, The effect of stressful diagnostic studies and surgery on anterior pituitary hormone release in man, Acta Endocrinologica 86:25-3 (1977).
23. D. B. Carr, B. A. Bullen, and G. S. Skrinar, Physical conditioning facilitates the exercise-induced secretion of beta-endorphin and beta-lipotropin in women, New England J. Med. 305:560-563 (1981).
24. M. J. Gawel, D. M. Park, J. Alaghband-Zadeh, and F. C. Rose, Exercise and hormonal secretion, Postgrad. Med. J. 55:373-376 (1979).
25. K. L. De Meirleir, L. Baeyens, M. L'Hermite-Baleriaux, M. L'Hermite,

and W. Hollmann, Exercise-induced prolactin release is related to anaerobiosis, J. Clin. Endocrinol. & Metabolism 60:1250-1252 (1985).

26. S. Levine, Cortisol changes following repeated experiences with parachute training, in: "Psychobiology of Stress: A Study of Coping Men," H. Ursin, E. Baade, and S. Levine, eds., Academic Press, New York (1978).

27. G. L. Noel, R. C. Dimond, J. M. Earll, and A. G. Frantz, Prolactin, thyrotropin, and growth hormone release during stress associated with parachute jumping, Aviation, Space & Environ. Med. 47(5):543-547 (1976).

28. J. W. Mason and J. V. Brady, Plasma 17-hydroxycorticosteroid changes related to reserpine effects on emotional behavior, Science 124:983 (1956).

29. J. D. Lane, A. D. White, and R. B. Williams, Jr., Cardiovascular effects of mental arithmetic in type A and type B females, Psychophysiology 21(1):39-46 (1984).

30. R. B. Williams, Jr., J. D. Lane, C. M. Kuhn, W. Melosh, A. D. White, and S. M. Schanberg, Type A behavior and elevated physiological and neuroendocrine responses to cognitive tasks, Science 218:483-485 (1982).

31. J.-H. Atterhog, K. Eliasson, and P. Hjemdahl, Sympathoadrenal and cardiovascular responses to mental stress, isometric handgrip, and cold pressor test in asymptomatic young men with primary T wave abnormalities in the electrocardiogram, British Heart J. 46:311-319 (1981).

32. R. M. Nesse, G. C. Curtis, B. A. Thyer, D. S. McCann, M. J. Huber-Smith, and R. F. Knopf, Endocrine and cardiovascular responses during phobic anxiety, Psychosom. Med. 47(1):320-332 (1985).

33. M. Fredrikson, O. Sundin, and M. Frankenhaeuser, Cortisol excretion during the defense reaction in humans, Psychosom. Med. 47(4):313-319 (1985).

34. E. H. Mougey, A radioimmunoassay for tetrahydrocortisol, Analytical Biochem. 91:566-582 (1978).

35. J. L. Meyerhoff, M. A. Oleshansky, E. H. Mougey, H. R. Smith, S. R. Hursh, L. K. Wittig, H. E. Wood, C. R. Glass, and D. B. Arnkoff, Increased heart rate and plasma level of beta-lipotrophin during a competitive oral examination, Neurosci. Abstr. 12:1480 (1986).

36. H. E. Wood, C. R. Glass, D. B. Arnkoff, H. R. Smith, J. L. Meyerhoff, and M. A. Oleshansky, An investigation of evaluation anxiety in an employment setting, Association for Advancement of Behavior Therapy, 20th Annual Convention, Chicago, IL, 14-16 Nov., 1986.

37. R. E. Mains, B. A. Eipper, and M. Ling, Common precursors to corticotropins and endorphins, Proc. natn. Acad. Sci. USA 74:3014-3018 (1977).

38. J. L. Roberts and G. Herbert, Characterization of a common precursor to corticotropin and β-lipotropin: Cell-free synthesis of the precursor and identification of corticotropin peptides in the molecule, Proc. natn. Acad. Sci. USA 74:4826-4830 (1977).

39. G. Pelletier, R. Leclerc, F. Labrie, J. Cote, M. Chretien, and M. Lis, Immunohistochemical localization of β-lipotropic hormone in the pituitary gland, Endocrinology 100:770-776 (1977).

40. E. Weber, K. H. Voigt, and R. Martin, Concomitant storage of ACTH- and endorphin-like immunoreactivity in secretory granules of anterior pituitary corticotrophs, Brain Res. 157:385-390 (1978).

41. C. H. Li, L. Barnafi, M. Chretien, and D. Chung, Identification and amino acid sequence of beta-LPH from sheep pituitary glands, Nature 208:1093-1094 (1965).

42. L. H. Lazarus, H. Ling, and R. Guillemin, Beta-lipotropin as a prohormone for the morphinomimetic peptides, endorphins and enkephalins, Proc. natn. Acad. Sci. USA 73:2156-2159 (1976).

43. A. S. Liotta, C. H. Li, G. C. Schussler, and D. T. Krieger, Comparative

metabolic clearance rate, volume of distribution and plasma half-life of human β-lipotropin and ACTH, Life Sciences 23:2323-2330 (1978).
44. W. Bloom and D. W. Fawcett, "A Textbook of Histology," W. B. Saunders Company, tenth edition (1975).
45. E. B. De Souza and G. R. Van Loon, Differential plasma β-endorphin, β-lipotropin, and adrenocorticotropin responses to stress in rats, Endocrinology 116(4):1577-1586 (1985).
46. J. L. Meyerhoff, E. H. Mougey, and G. J. Kant, Paraventricular lesions abolish the stress-induced rise in pituitary cyclic AMP and attenuate the increases in plasma levels of POMC peptides and prolactin, Neuroendo. 46:222-230 (1987).
47. F. E. Bloom, E. Battenberg, J. Rossier, N. Ling, J. Leppaluoto, T. M. Vargo, and R. Guillemin, Endorphins are located in the intermediate and anterior lobes of the pituitary gland, not in the neurohypophysis, Life Sciences 20:43-48 (1977).
48. B. A. Eipper and R. E. Mains, Structure and biosynthesis of proadrenocorticotropin/endorphin and related peptides, Endocrinol. Rev. 1:1-27 (1980).
49. R. Przewlocki, V. Hollt, K. H. Voight, and A. Herz, Modulation of in vitro release of β-endorphin from the separate lobes of the rat pituitary, Life Sciences 24:1601-1608 (1979).
50. I. Vermes, G. H. Mulder, P. G. Smelik, and F. J. H. Tilders, Differential control of β-endorphin/β-lipotropin secretion from anterior and intermediate lobes of the rat pituitary gland in vitro, Life Sciences 27:1761-1768 (1980).
51. J. W. Mason, Over-all hormonal balance as a key to endocrine organization, Psychosom. Med. 30:791-808 (1968).
52. G. Shanker and R. K. Sharma, β-Endorphin stimulates corticosterone synthesis, Biochem. Biophys. Res. Commun. 86:1-5 (1979).
53. W. O. Richter, R. J. Naude, W. Oelofsen, and P. Schwandt, In vitro lipolytic activity of β-endorphin and its partial sequences, Endocrinology 120(4):1472-1476 (1987).
54. J. W. Holaday, Cardiovascular effects of endogenous opiate systems, Ann. Rev. Pharmacol. Toxicol. 23:541-594 (1983).
55. H. Matsuoka, P. J. Mulrow, and C. H. Li, β-lipotropin: A new aldosterone-stimulating factor, Science 209:307-308 (1980).
56. K. H. Voight, D. Frank, E. Duker, R. Martin, and W. Wuttke, Dopamine-inhibited release of prolactin and intermediate lobe-POMC-peptides: Different modulation by opioids, Life Sciences 33:507-510 (1983).
57. B. L. Spangelo, N. R. Hall, and A. L. Goldstein, Evidence that prolactin is an immunomodulatory hormone. in: "Prolactin: Basic and Clinical Correlates," R. M. MacLeod, M. O. Thorner, and U. Scapagnini, eds., Liviana Press, Padova (1985).
58. K. Hirschhorn, F. Bach, R. Kolodny, I. Firschein, and N. Hashem, Immune response and mitosis of human peripheral blood lymphocytes in vitro, Science 142:1185-1187 (1963).
59. B. S. McEwen, G. Wallach, and C. Magnus, Corticosterone binding to hippocampus: Immediate and delayed influences of the absence of adrenal secretion, Brain Research 70:321-334 (1974).
60. A. J. Kastin, J. E. Zadina, W. A. Banks, and M. V. Graf, Misleading concepts in the field of brain peptides, Peptides 5(Suppl. 1):249-253 (1984).
61. J. L. Meyerhoff, B. N. Bunnell, and E. H. Mougey, Plasma β-endorphin in rats is increased by psychological stress, Neurosci. Abstr. 7:166 (1981).
62. S. Amir and Z. Amit, Endogenous opioid ligands may mediate stress-induced changes in the affective properties of pain related behavior in rats, Life Sci. 23:1143-1152 (1978).
63. S. Amir and Z. Amit, The pituitary gland mediates acute and chronic pain responsiveness in stressed and non-stressed rats, Life Sciences 24:439-448 (1979).

64. L. F. Tseng, H. H. Loh, and C. H. Li, β-Endorphin: Cross tolerance to and cross physical dependence on morphine, Proc. natl Acad. Sci. 73(11):4187-4189 (1976).
65. R. M. Bergland and R. E. Page, Can the pituitary secrete directly to the brain? (affirmative anatomical evidence), Endocrinology 102:1325-1338 (1978).
66. J. C. Willer, H. Dehen, and J. Cambier, Stress-induced analgesia in humans: Endogenous opioids and naloxone-reversible depression of pain reflexes, Science 212:689-690 (1981).
67. S. Amir, Z. Brown, and Z. Amit, The role of endorphins in stress: Evidence and speculations, Neuroscience and Biobehavioral Reviews 4:77-86 (1980).
68. J. L. Meyerhoff, M.A. Oleshansky, and E.H. Mougey, Psychological stress increases plasma levels of prolactin, cortisol and POMC-derived peptides in man, (submitted for publication).

PRECLINICAL AND CLINICAL EVIDENCE FOR THE INVOLVEMENT OF CORTICOTROPIN-RELEASING FACTOR IN THE PATHOGENESIS OF DEPRESSION

Mark A. Smith[1] and Charles B. Nemeroff[1,2]

Departments of Psychiatry[1] and Pharmacology[2]
Duke University Medical Center
Durham, North Carolina 27710

Long before its isolation and characterization in 1981 by Vale and colleagues (1), a factor such as corticotropin-releasing factor (CRF) was postulated to play a pathophysiological role in patients suffering from depression. Increased plasma cortisol concentrations have been observed in many patients with endogenous depression (2). Although the hypercortisolemia could theoretically result from abnormalities in the periphery-adrenals or pituitary--it was postulated that an unknown factor in the central nervous system (CNS) was responsible for the hyperactivity of the pituitary-adrenal axis (3). With the discovery of CRF, it is now possible to directly test the hypothesis that this peptide might be dysregulated in various psychiatric disorders.

What is the major rationale for study of the involvement of CRF in psychiatric disorders? There are primarily two major reasons. First, the neurotransmitters which are thought to be important in both depression and anxiety disorders are apparently the same ones that regulate CRF secretion from the hypothalamus. Second, CRF is the "master hormone" of stress, and the cephalic representative of the hypothalamic-pituitary adrenal (HPA) axis, and stress is believed to exacerbate or precipitate many psychiatric disorders, including depression. The preclinical studies elaborated below provide a further impetus for exploring the role of CRF in psychiatric disorders.

IN VITRO RELEASE OF CRF

The heterogeneous distribution of CRF and CRF receptors provides some evidence that this peptide functions as a neurotransmitter in the mammalian central nervous system. However, before CRF can be considered as a neurotransmitter, it must be demonstrated that physiological depolarizing agents stimulate CRF release. Using minced rat hypothalami and a sensitive and specific radioimmunoassay for CRF (4), we reported that high concentrations of potassium increase CRF release about 10 fold (5). The K^+ stimulated release is clearly dependent on extracellular calcium because EGTA (a calcium chelator) completely inhibited potassium-stimulated CRF secretion. Release of CRF was observed not only from hypothalamus, but also from other brain regions including amygdala, midbrain and, to a lesser degree, striatum.

Using minced brain slices, CRF release was stimulated by high $[K^+]$ or scorpion venom (which opens sodium channels to depolarize cells) but was unaffected by neurotransmitters such as acetylcholine (ACh) or serotonin (5HT). Therefore, as previously suggested (6), we postulated the necessity for the CRF neuron to be intact in order to respond to neurotransmitters. Thus we examined the effects of various neurotransmitters and drugs *in vitro* with the use of whole hypothalamic blocks dissected from the rat using rostral-caudal boundaries of the optic chiasm and mammillary bodies, respectively, and dorsally to a point just above the third ventricle. Two such blocks were placed in a test tube containing 0.6 ml of a modified Kreb's-Henseleit buffer which was gassed with 95% O_2/5% CO_2 at 37°, pH 7.4. The buffer was removed and replaced with fresh buffer every 10 minutes. After 50 minutes basal release was stabilized. The whole hypothalamus responded to 40 mM [K^+] by releasing CRF 1100% above the basal rate (Figure 1). In preliminary experiments, serotonin and bethanechol, a muscarnic cholinergic agonist, stimulated CRF release approximately 5 fold. These results are concordant with the hypothesis that the projections of CRF neurons from the paraventricular nucleus to the median eminence must be intact to respond to serotonin and acetylcholine agonists. Inhibiting effects on CRF release may be difficult to demonstrate because basal CRF secretion rates are near the limits of detection of the assay. Dysregulation of 5HT, ACh and norepinephrine (NE)-containing neurons have all been implicated in the pathophysiology of depression and may underlie the postulated abnormal CRF secretion seen in this disorder.

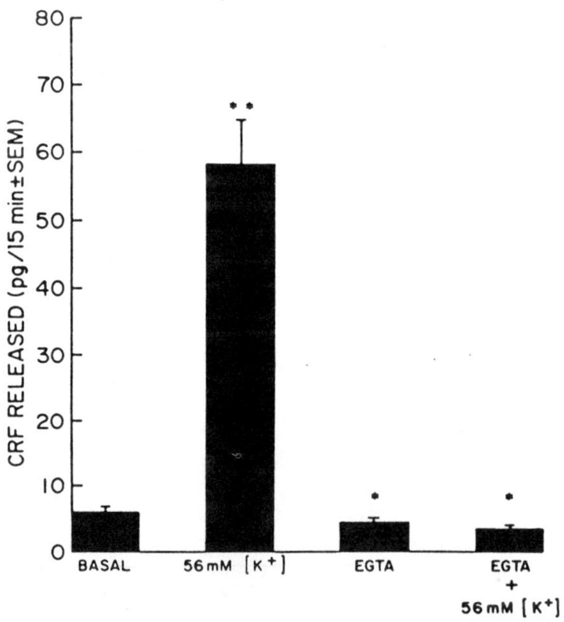

Fig. 1. Calcium-dependent release of CRF-LI concentration from the rat hypothalamus *in vitro*. CRF release into the medium was measured by RIA during a 15-min incubation at 37°C and is expressed as the average ± SEM of triplicate determinations from four separate experiments. CRF is expressed as picograms released from a single hypothalamus per tube during a 15-min incubation (average weight of hypothalamus, 50 ± 3 mg). *, $p < 0.05$; **, $p < 0.01$ (vs. basal release). (From Smith et al., 1986, with permission.)

CRF AND STRESS

With the discovery of each new hypothalamic releasing or inhibiting factor it becomes increasingly apparent that these peptides are widely distributed throughout the CNS and have many more actions than simply regulating anterior pituitary hormone secretion. Hence these substances have been referred to as "regulatory peptides" because they appear to influence many extrahypothalamic brain areas and consequently behavior. CRF is apparently no exception. As noted above, it is found in high concentrations in many brain areas including the limbic system, and these brain regions are thought to be important in affective disorders. CRF is also present in brain stem nuclei, such as the locus coeruleus (LC); the LC noradrenergic fibers project throughout the brain and have been implicated in affective disorders (7). These same areas are thought to mediate, at least in part, the behavioral, cardiovascular and cognitive responses to stress. This raises the possibility that CRF might mediate many components of the stress response within the CNS, as well as activation of the HPA axis.

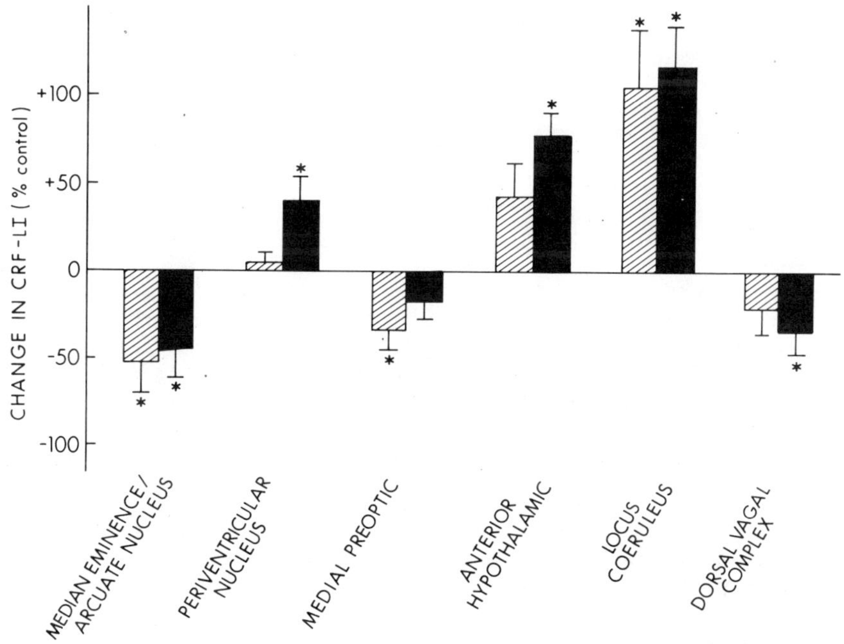

Fig. 2. Percent of stress-induced change in CRF-LI concentration from baseline. Hatched bars represent acute stress. Solid bars represent chronic stress. *Indicates significantly different from controls at $p < 0.05$. (From Chappell et al., 1986, with permission.)

In order to determine if stress alters CRF concentrations in the brain, we exposed rats to either 3 hours of immobilization at 4°C (acute stress) or 14 days of chronic unpredictable stressors designed to prevent adaptation (chronic stress) or no stress (control group) (8). Using the identical CRF RIA described above in the release studies, we measured the concentration of CRF in 36 brain regions microdissected by the method of Palkovits; only 6 brain areas showed significant differences in CRF concentration after stress (Figure 2). The concentration of CRF was reduced approximately 50% in the arcuate-median eminence after both acute and chronic stress suggesting that the peptide was released into the hypothalamo-hypophysial portal system. Chronic stress produced a decrease in CRF concentration in the dorsal vagal complex, suggesting that CRF may play a role in regulating autonomic nervous system responses to chronic stress.

The concentration of CRF in the LC was found to be markedly elevated after both acute and chronic stress. The LC contains both CRF cell bodies and terminals. The mechanism responsible for an increase in CRF content remains obscure. Measuring CRF mRNA changes may provide important information on alterations in CRF biosynthesis during stress. The rapid response to acute stress would suggest a post-translational effect, perhaps increased liberation of CRF from its prohormone. Although these studies cannot determine the precise role of CRF in the stress response, they do identify extrahypothalamic areas which may be critical in mediating the stress response. Electrophysiological evidence has also appeared which identifies the LC as an area where CRF may play a neuromodulatory role (9).

CLINICAL STUDIES

It is now well-established that many patients with endogenous depression exhibit hyperactivity of the HPA axis (2, 3). Increased urinary free cortisol excretion and cortisol nonsuppression after dexamethasone administration could be due to adrenal hypersensitivity to ACTH, pituitary insensitivity to the negative feedback by steroids, or abnormal hypersecretion of hypothalamic CRF. Gold and his colleagues have conducted a series of studies in which ACTH and cortisol responses to an intravenously administered CRF bolus was measured in patients with depression (10). The blunted ACTH response to CRF observed in depressed patients could be secondary to chronic hypersecretion of hypothalamic CRF either by leading to elevated plasma cortisol which suppresses pituitary ACTH secretion, or to down-regulation of pituitary CRF receptors. Thus the abnormality in the HPA axis of depressed patients may be suprapituitary - in the central nervous system.

In order to directly test the hypothesis that CRF is hypersecreted in patients with depression, we used our CRF RIA to measure the peptide in CSF of drug-free neuropsychiatric patients (11). The concentration of CSF CRF in our first study was significantly elevated in patients with major depression, compared to normal controls or to patients with schizophrenia or senile dementia (Figure 3). These results support the hypothesis that CRF is hypersecreted in patients with major depression. We have now confirmed these findings in a second study in which a large number of patients were included (12). Other studies indicate that CSF CRF is elevated not only in depressed patients but also in those with anorexia nervosa (13, 14). However, in general, CSF CRF levels are normal in schizophrenic patients, and they exhibit no abnormalities of the HPA axis as reflected in normal basal plasma cortisol concentrations and a normal plasma ACTH response to exogenous CRF (15, and Smith, Chappell, Krishnan and Nemeroff, unpublished observations).

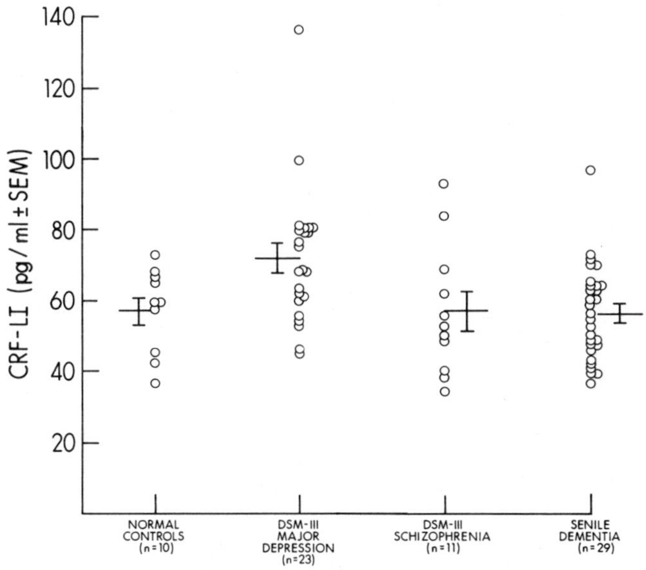

Fig. 3. Concentration of CRF-LI in CSF of normal controls (5 males and 5 females), DSM-III major depression (13 males and 10 females), DSM-III schizophrenia (8 females and 3 females), and senile dementia (DSM-III primary degenerative dementia or multi-infarct dementia, or both, 20 females and 9 males). Because the data distribution in the depressed patients was slightly skewed, the data were analyzed by both parametric (ANOVA) and Student-NewmanKeuls test) and nonparametric (Mann-Whitney U test). (From Nemeroff et al. [11], with permission)

The central nervous system origin of CRF in CSF is unclear; it may derive from extrahypothalamic brain regions (16). Pharmacological evidence suggests that CRF in CSF can be transported into the hypothalamo-hypophyseal portal circulation (17). It is likely that CRF is hypersecreted into this portal system during depression as it is during chronic stress. Indeed plasma CRF has now been reported to be elevated in patients with major depression (18).

If there is chronic hypersecretion of CRF in depression then it might be expected that this would result in decreases in brain CRF receptors. In a preliminary study we measured the density (Bmax) and affinity (Kd) of CRF binding sites in the frontal cortex of suicide victims and appropriate controls, using ^{125}I-Tyr^{0}-ovine CRF as the radioligand, in collaboration with Stanley and Andorn (19). Scatchard analysis revealed a significant decrease in CRF receptor density in the suicide group with no change in the apparent affinity. No significant correlations were found between the Bmax and sex, age or post-mortem delay. These findings suggest that CRF receptors are down-regulated in the frontal cortex of suicide victims, many of whom were depressed. The down-regulation of receptors is mot likely due to chronic hypersecretion of CRF.

DISCUSSION

What is the relationship between the neurotransmitters previously implicated in the pathophysiology of depression (NE, 5HT and ACh) and the postulated hypersecretion of CRF? The ability of acetylcholinesterase inhibitors such as physostigmine to precipitate depression and stimulate the HPA axis appears consistent (17), and thus increased synaptic availability of ACh might produce hypersecretion of hypothalamic CRF in depression. However the stimulatory effect of serotonin on CRF is at first glance inconsistent with the postulated overall decrease in 5HT turnover in depressed patients, unless one either postulates that the decreased 5HT turnover reflects exhaustion of 5HT reserves following chronic hyperactivity of 5HT neurons or differences in acute versus chronic depression. Perhaps the key then is norepinephrine (NE). CRF secretion is believed to be regulated by NE and epinephrine (E) in a complex manner with perhaps different adrenergic receptors mediating both stimulation and inhibition (18). Conversely, CRF appears to activate noradrenergic neurons in the LC and perhaps elsewhere in the brain stem (19).

What is the relationship between stress, CRF and depression? Studies with rats suggest that chronic stress can produce temporary alterations in the HPA axis similar to those seen in patients with major depression. We found that rats exposed to various stressors at unpredictable intervals over a 2 week period had elevated plasma corticosterone levels, increased adrenal weights, normal plasma ACTH, and decreased pituitary ACTH stores. In addition these rats had marked depletions of CRF in the arcuate-median eminence region. We concluded that chronic stress resulted in hypersecretion of CRF which was associated with the observed reduction of CRF concentration in the median eminence. This in turn chronically stimulated ACTH secretion which presumeably produced adrenal hypertrophy and hypercortisolemia. The pituitary appeared to remain normally responsive either due to negative feedback from corticosterone or down-regulation of pituitary CRF receptors. Gold and his colleagues have drawn similar conclusions concerning the HPA abnormalities in depressed patients: increased CSF CRF, hyperresponsive adrenals and a pituitary which responded appropriately to the negative feedback of high plasma cortisol concentrations (20).

Thus it seems that chronic stress can induce alterations in HPA axis homeostasis which mimic those seen in major depression. However, why would these HPA axis abnormalities persist in depression for weeks to months in the absence of an identifiable stressor? One could postulate that a subjective state such as guilt continues to serve as an internalized cognitive "chronic stressor". Another possibility is that a biochemical (genetic) predisposition renders certain individuals vulnerable to depression, or prevents a return to homeostasis, after stress. What that neurochemical abnormality might be remains to be determined, but it is likely that it affects CRF secretion at least indirectly. For example, if the LC was hyperactive in individuals vulnerable to depression and the increased NE turnover stimulated CRF secretion via putative post-synaptic adrenergic receptors in the hypothalamus (18, 21), then CRF might stimulate the LC further leading to a positive feedback loop and persistent hypercortisolemia.

An alternative hypothesis to explain HPA dysfunction in depression might involve glucocorticoid receptors. Occupation of glucocorticoid receptors in the hippocampus apparently causes inhibition of hypothalamic CRF activity. It is certainly possible that in depression glucocorticoid receptors are modulated differently in various parts of the brain with down-regulation in the hippocampus (resulting in a release from inhibition of CRF secretion) and no change in glucocorticoid receptor number in the pituitary (maintaining the normal negative feedback of adrenal steroids). This is precisely what happens in rats (22). Chronic stress in rats leads to a down-regulation of corticosterone receptors in the hippocampus which results in prolonged hypercortisolemia for several hours longer than normal after a subsequent acute stressor (23). This effect of chronic stress to prevent the termination of the adrenocortical stress response reverses to normal after about 1 week following the termination of the chronic stress. However in aged rats, the hippocampal neurons containing the corticosterone receptors are fewer in number and therefore the prolonged corticosterone response to a stressor is permanent. Could hippocampal neurons be killed by chronic stress either early in development which might predispose someone to an exaggerated CRF response to stress later in life, which may explain why depression often occurs after age 40? Answers to these and other questions will be important in understanding why CRF may be hypersecreted in depression.

Regardless of what may be causing CRF hypersecretion in depression, CRF itself may be relevant to the symptomatology of depression. When administered intracerebroventricularly to monkeys or rats, CRF increases emotionality and sympathetic nervous system activity and reduces sexual behavior and appetite (23-26). Many of these effects can be blocked by the CRF antagonist α-helical CRF_{9-41} (27). Development of a more potent CRF antagonist or an analog that could cross the blood-brain barrier might be of great importance in the development of a novel treatment for depression.

ACKNOWLEDGEMENTS

We are grateful to Shelia Walker for preparation of this manuscript. This research was supported by NIMH MH-42088, MH-40159, MH-40524 and MH-39415. CBN is the recipient of a Nanaline H. Duke Scholars Award from Duke University Medical Center.

REFERENCES

1. W. Vale, J. Speiss, C. Rivier, and J. Rivier, Characterization of a 41-residue ovine hypothalamic peptide that stimulates secretion of corticotropin and β-endorphin, Science 213:1394-1397 (1981).
2. E. J. Sachar, L. Hellman, D. K. Fukushima, and T. F. Gallagher, Cortisol production in depressive illness: a clinical and biochemical classification, Arch. Gen. Psychiatry 23:289-298 (1970).
3. B. J. Carroll, G. C. Curtis, and J. Mendels, Neuroendocrine regulation in depression, I. Limbic system-adrenal cortical dysfunction, Arch. Gen. Psychiatry 33:1039-1044 (1976).
4. W. Vale, J. Vaughan, G. Yamamoto, T. Bruhn, C. Douglas, D. Dalton, C. Rivier, and J. Rivier, Assay of corticotropin releasing factor, Meth. Enzymol. 103:565-577 (1983).
5. M. A. Smith, G. Bissette, T. A. Slotkin, D. L. Knight, and C. B. Nemeroff, Release of corticotropin-releasing factor from rat brain regions in vitro, Endocrinology 118:1997-2001 (1986).
6. M. T. Jones, E. W. Hillhouse, and J. Burden, Effect of various putative neurotransmitters on the secretion of corticotropin-releasing hormone from the rat hypothalamus in vitro - a model of the neurotransmitter involved, J. Endocrinol. 69:1-10 (1976).
7. L. W. Swanson, P. E. Sawchenko, J. Rivier, and W. W. Vale, Organization of ovine corticotropin-releasing factor immunoreactive cells and fibers in the rat brain: An immunohistochemical study, Neuroendocrinology 36:165-186 (1983).

8. P. B. Chappell, M. A. Smith, C. D. Kilts, G. Bissette, J. Ritchie, C. Anderson, and C. B. Nemeroff, Alterations in corticotropin-releasing factor-like immunoreactivity in discrete brain regions after acute and chronic stress, J. Neuroscience 6:2908-2914 (1986).
9. R. J. Valentino, S. L. Foote, and G. Aston-Jones, Corticotropin-releasing factor activates noradrenergic neurons of the locus coeruleus, Brain Res. 270:363-367 (1983).
10. P. W. Gold, G. Chrousos, C. Kellner, R. Post, A. Roy, P. Augerinos, H. Schulte, E. Oldfield, and D. L. Loriaux, Psychiatric implications of basic and clinical studies with corticotropin-releasing factor, Amer. J. Psychiat. 141:619-627 (1984).
11. C. B. Nemeroff, E. Widerlov, G. Bissette, H. Walleus, I. Karlsson, K. Eklund, C. D. Kilts, P. T. Loosen, and W. Vale, Elevated concentrations of CSF corticotropin-releasing factor-like immunoreactivity in depressed patients, Science 226:1342-1344 (1984).
12. C. M. Banki, G. Bissette, M. Arato, L. O'Connor, and C. B. Nemeroff, Cerebrospinal fluid corticotropin-releasing factor-like immunoreactivity in depression and schizophrenia, Amer. J. Psychiat. (in press).
13. M. Hotta, T. Shibasaki, A. Masuda, T. Imaki, H. Demura, N. Ling, and K. Shizume, The responses of plasma adrenocorticotropin and cortisol to corticotropin-releasing hormone (CRH) and cerebrospinal fluid immunoreactive CRH in anorexia nervosa patients. J. Clin. Endocrinol. Metab. 62:319-324 (1986).
14. W. H. Kaye, H. E. Gwirtsman, D. T. George, M. H. Ebert, D. C. Jimerson, T. P. Topai, G. P. Chrousos, P. W. Gold, Elevated cerebrospinal fluid levels of immunoreactive corticotropin-releasing hormone in anorexia nervosa: relation to state of nutrition, adrenal function, and intensity of depression. J. Clin. Endocrinol. Metab. 64: 203-208.
15. A. Roy, D. Pickar, A. Doran, O. Wolkowitz, W. Galluci, G. Chrousos, and P. Gold, The corticotropin-releasing hormone stimulation test in chronic schizophrenia, Amer. J. Psychiatry 143:1393-1397 (1986).
16. N. H. Kalin and W. T. McKinney, Diurnal rhythm of CRH concentration in primate CSF, Proc. Ann. Meeting of the Amer. Coll. Neuropsychopharm. 25:170 (1986).
17. Zs. Liposits, C. Phelix, and W. K. Paull, Spreading and neuronal uptake of intracerebroventricularly injected corticotropin-releasing factor (CRF) in the rat brain, as revealed by immunohistochemistry, Soc. Neuroscience Abst. 12:292 (1986).
18. E. Widerlov, R. Ekman, and C. Wahlestedt, Elevated corticotropin-releasing factor-like immunoreactivity in plasma from major depressives, Proc. Coll. Internat. Neuro-Psychopharmacologicum 25:P-207 (1986).
19. M. Owens and C. B. Nemeroff, The neurobiology of corticotropin-releasing factor: implications for affective disorders. in: Hypothalamic-Pituitary Adrenal Axis Physiology and Pathophysiology (A. F. Schatzberg and C. B. Nemeroff, Eds.), Raven Press, New York (in press).
20. D. S. Janowsky, S. C. Risch, D. Parker, L. Huey, and L. Judd, Increased vulnerability to cholinergic stimulation in affective disorder patients, Psychopharmacol. Bull. 16(4):29-31 (1980).
21. L. Rees, P. W. P. Butler, G. Gosling, and G. M. Besser, Adrenergic blockade and the corticosteroid and growth hormone responses to methylamphetamine, Nature 228:565-566 (1970).

22. R. J. Valentino, S. L. Foote, and G. Aston-Jones, Corticotropin-releasing factor activates noradrenergic neurons of the locus coeruleus, Brain Res. 270:363-367 (1983).
23. P. W. Gold, D. L. Loriaux, A. Roy, M. A. Kling, J. R. Calabrese, C. H. Kellner, L. K. Nieman, R. M. Post, D. Pickar, W. Gallucci, P. Augerinos, S. Paul, E. H. Oldfield, G. B. Cutler, and G. P. Chrousos, Responses to corticotropin-releasing hormone in the hypercortisolism of depression and Cushing's disease, N. Engl. J. Med. 314:1329-1335 (1986).
24. G. A. Smythe, J. E. Bradshaw, and R. F. Vining, Hypothalamic monoamine control of stress-induced adrenocorticotropin release in the rat, Endocrinology 113:1062-1071 (1983).
25. R. M. Sapolsky, L. C. Krey, and B. S. McEwen, Stress down-regulates corticosterone receptors in a site-specific manner in the brain, Endocrinology 114:287-292 (1984).
26. R. M. Sapolsky, L. C. Krey, and B. S. McEwen, Glucocorticoid-sensitive hippocampal neurons are involved in terminating the adrenocortical stress response, Proc. Natl. Acad. Sci. 81:6174-6177 (1984).
27. M. R. Brown, and L. A. Fisher, Corticotropin-releasing factor: effects on the autonomic nervous system and visceral systems, Fed. Proc. 44:243-248 (1985).
28. N. H. Kalin, Behavioral effects of ovine corticotropin-releasing factor administered to rhesus monkeys, Fed. Proc. 44:249-253 (1985).
29. D. J. S. Sirinathsinghji, L. H. Rees, J. Rivier, and W. Vale, Corticotropin-releasing factor is a potent inhibitor of sexual receptivity in the female rat, Nature 305:232-235 (1983).
30. J. E. Morley and A. S. Levine, Corticotropin-releasing factor, grooming and ingestive behavior, Life Sci. 31:1459-1464 (1982).
31. J. Rivier, C. Rivier, and W. Vale, Synthetic competitive antagonists of corticotropin-releasing factor: effect on ACTH secretion in the rat, Science 224:889-891 (1984).

THE PSYCHOLOGY OF STRESS: RUN SILENT, RUN DEEP

Donald S. Jewell
Clinical Psychologist, Washington, D.C.

Maureen Mylander
Writer-Editor, National Institutes of Health

Many experts define stress as the body's response to any demand (stressor) placed upon it. Still, many individuals are unsure whether stress means stressful events themselves or reactions to stressful events. It seems as if stress, in addition to being itself and the result of itself, is also the cause of itself.

Early investigations of stress as a dimension of human behavior were concerned mainly with "distress." Holmes and Rahe ranked 43 life events according to how stressful they are for most people.[1] These and other investigators subsequently discovered that while some people responded poorly to stress--becoming depressed or developing mild to serious psychosomatic illnesses--others who experienced "eustress" rose to new heights of performance. What are the implications of this paradox for those who are trying to explore and understand the stress phenomenon? Is the answer that some individuals are stress-free and others are not? Hardly. A quarterback may rise to the occasion in the last minute of a game and throw a perfect touchdown pass before 50,000 cheering fans. The same quarterback may not be able to rise to the occasion when faced with the demands of his spouse or girlfriend. Stress, therefore, is the response of the subject, not the causal attribute of any particular stimuli.

HIDDEN STRESSORS

Many environmental events produce stress reactions, but to widely varying degrees. One person may be devastated by the loss of a job, while another may be glad to be rid of it. The former may have been conditioned, by depression-minded parents, to believe that his occupation is the most important thing in his life and that he would be lucky to find somebody who would hire him. The latter may have been taught as a child that he is a worthy and qualified individual who always can get another job.

Stress, in this dimension, is the total physiological and psychological response of the organism to a stimulus that the organism interprets as a threat to its existence, whether the stimulus actually is a threat or not. Whether the result is "distress" or "eustress" depends on the learning and experience of the subject. It is our hypothesis, moreover, that successful management of stress and prevention of diseases that may result from mismanaged stress depend upon recognizing all its sources, including hidden stressors that lurk in the unconscious mind.

Hidden stressors are far more damaging than those at the surface. Most stress-inducing of all are thoughts and feelings--sometimes glimpsed in dreams--that lie hidden in the unconscious mind and that trace back to childhood desires to be loved and taken care of. The list includes overwhelming fear, anger, disgust, hatred, and love toward family members. These forbidden feelings are like the unseen mass of iceberg that floats below the water line: dangerous and uncontrollable. Thus, the jobless individual with low self-esteem inculcated from childhood will retreat into depression and doubt, and lower the likelihood that anybody would hire him. The person who learned during childhood that he is worthy of being hired will undertake a job search with anxious optimism. He will experience some stress, and his bodily systems will respond for a time with physiological reactions necessary to combat an acute stress reaction. But he probably will not develop a serious disease of stress unless the economic difficulties in his environment become chronic. He will cope, become employed, and go on with his life. The other individual, who believes that everything his parents have taught him is true, will experience the physiological reactions necessary to combat the immediate crisis of job loss, but they will not abate. He will then enter a state of chronic stress. Even if he eventually finds a job, his poor opinion of himself in terms of his early upbringing will quietly eat at him until his body surrenders. Unless he can revise this self-image, he may develop a disease consistent with his family history and genetic predisposition.

These two examples imply a continuum from environmentally-produced stress with minimal unconscious stimulation to unconsciously-produced stress with some external stimulation. To deal effectively with stress, a person must be aware not only of his environment, but also of his inner feelings and their meaning. The individual who loses his job and becomes clinically depressed is accustomed to feeling unworthy, cast out, quietly enraged, and helpless. If that person can remove or reduce the burden of these unconscious, childhood stressors--by understanding their emotional roots, unlearning old patterns of behavior, and replacing them with new and effective ones--he can markedly reduce, if not eliminate, the constant maladaptations to stressors that cause a chronic physiological response leading to illness. By contrast, the unconscious forces of childhood may be almost meaningless in situations where environmental forces are so overwhelming that they alone can produce chronic disease: witness the significantly higher rates of heart disease, cancer, and high blood pressure in black males.

Stimuli that produce stress responses appear to have certain similarities, not in terms of their sources, but in terms of the emotions or patterns of emotion these stimuli elicit. Whether environmental or internal, stimuli that cause stress tend to share these characteristics:

o They are vague. A black man may be aware that he cannot get a job, but he does not know why or what to do about it. A person responding to a stress producer originating in childhood usually is oblivious of this fact.

o They make the individual feel enraged.

o They make the individual feel hopeless and helpless.

It is this cycle of vagueness, rage, hopelessness, and helplessness, repeated over and over, that causes hormonal and other physiological reactions leading to the severe diseases of stress. Nor is this cycle easy to break. People from Westernized societies are taught from the cradle never to experience such emotions, especially as adults. Vagueness is forbidden since adults in general and men in particular are supposed to be "clear-headed" and sure of themselves. Rage implies loss of control, the epitome of "being a baby," and helplessness is seen as pouting and feeling sorry for one's self.

Thus, all stress--unlike fear, which is clear and identifiable--is characterized by vagueness, rage, hopelessness and helplessness, whether the stress is related to one's professional or personal life. A breeding ground for diseases of stress can be found in individuals in whom neither of these areas is receiving nourishment. Turning to one's work can help overcome the breakup of a marriage or close relationship. Isolating one's self in one's work cannot. The support of a loving spouse can help an unemployed person, but cannot put bread on the table or maintain that person's self-esteem over the long run.

JOB STRESS

One stress producer that seldom appears on rating scales involves day-to-day experience at work. Few other activities, including sleep, consume as much time: at least forty hours a week. If these forty-plus hours are spent in a job that is boring, unrewarding, machine-paced, or marred by nonsupportive supervisors and coworkers, the end result is a new kind of health hazard that only recently has been recognized as such.

Job stressors can harm the health of office and factory workers alike. This phenomenon was described more than fifty years ago when Hersey noted that even blue-collar workers felt emotional pressures from "the nature of the job, the amount accomplished, plant conditions, treatment by the foremen, (and) relations with one's fellow workers."[2]

Boredom, a common job stressor, afflicts people who have little control over the pace and style of their work. For example, operators of video display terminals are subjected to some of the most relentless physical and psychological pressures in the workplace. Their jobs are paced by machines that impose inflexibility, lack of privacy, and production quotas. Workers sit all day at a keyboard and process checks, insurance claims, address changes, and customer orders under the watchful eyes of supervisors who can verify, on their video display terminals, exactly how much work is done. It is the twentieth century counterpart of the nineteenth century sweatshop.

Burnout is a growing problem among people in the "helping professions" who toil, often under impossible workloads, for long periods without visible results. Physicians, nurses, social workers, police officers, teachers, criminal lawyers, and other human-service workers are especially vulnerable to burnout. So, too, are people responsible for the lives of others--such as air traffic controllers. They often lack either the time or the inclination to take vacations, days off, or memberships in health clubs. They often feel guilty or inadequate when they cannot take on even more work. Burnout occurs in three phases. First comes emotional exhaustion and a feeling of being drained, followed by a cynical period of asking "Why knock myself out?" Finally, the burned-out person becomes depressed and loses vitality and optimism not only about the job, but about other facets of life. Many people who experience job burnout are naive about organizations. When their high expectations about their employers are not met, they become cynical. If they remain on the job, they usually adopt a more practical view of the complexity and chaos of organizational life.

Another job stressor is the nightmare of working for a "crazy boss" who belittles employees (often in front of coworkers), pits one against another, keeps subordinates uninformed and off-balance, bullies, and exploits. Such a boss--who often is character-disordered--inflicts insidious physical and emotional damage. People who derive much of their personal identities from their jobs are especially vulnerable to abuse by a character-disordered boss. The fact that a person once depended on his parents for survival and now depends on his job--personified by his boss--for economic survival is no small factor in the powerlessness he may feel when dealing with such a "superior."

Powerlessness is felt by the head of a public agency who must please pressure groups and legislators and obey so many regulations that he has little control over the organization he supposedly heads. Powerlessness is felt by the mid-level manager who must meet someone else's deadlines and carry out policies he opposes. It is felt even by lower-level workers, such as police officers, who constantly collide with cumbersome administrative and legal systems. Working for a large organization magnifies feelings of powerlessness. More than 30 years ago Whyte described an individual who becomes part of a collective for the purposes of work.[3] According to Whyte, individuality is lost not only in the corporate and government workplaces, but in academia, the world of science, and other institutions upon which people depend for survival. Often without realizing it, people expect such institutions to take care of them. Considering the company retirement benefits, health care coverage, life insurance, savings and investment plans, recreation and vacation packages, and discount buying services, institutions indeed appear to do so. But the individual pays a high psychological price for these false securities.

Job stressors are an increasing problem even for successful people who are troubled about what it takes to cope with their jobs. Many individuals trade career success for close personal relationships. Others stifle their own needs in order to develop the character traits encouraged by their employers. Among this group are many "working wounded," who compromise their personal ethics to be successful.

Countless numbers of working people spend time and energy on jobs that do not help them attain their real goals in life. They arrive home physically and mentally exhausted after another day of joyless striving. Victims of the "Sisyphus complex," they resemble the mythical character who offended the gods and was condemned eternally to push a heavy boulder up a hill, only to have it roll down again just as it reached the top. The modern Sisyphus toils for a boss who does not utilize his skills or recognize his accomplishments. His work seems meaningless and his health, as a result, is at risk.

THE OVERWORK SYNDROME

Winston Churchill once said, "Those whose work and pleasures are one are fortune's favorite children." Even when a person is satisfied with his job, too much "job" can be harmful. The overwork syndrome afflicts people who have a compulsive need to work. Workaholics seem to lack an inner governor that tells them when they are exhausted, mentally and physically, and when they need rest and recreation. As teenagers, they worked three jobs at once, studied until dawn, and spent Saturday puttering with their automobiles. As full-fledged adult workaholics, they end up in doctors' offices because they tire easily; cannot sleep, concentrate, or make love; feel irritable, depressed and anxious; and suffer from stomach upsets and assorted aches and pains. Their complaint is not that they are working too hard, but that they cannot work harder. Sometimes the "overwork syndrome" causes symptoms that resemble other illnesses: fever of unknown origin, thyroid deficiency, chest pain, and "accidents" caused by lack of concentration. In severe cases, workaholics suffer from memory lapses, confusion, and suicidal thoughts. Overworkers are candidates for more serious medical problems, including alcoholism and drug dependency.[4]

The typical workaholic is a competitive, hurried, perfectionistic male. Overwork and preoccupation with work are common in male physicians, who, according to one survey, average about six waking hours a week with their spouses. This time usually is devoted to chores and "crisis conversations."[5] Doctors typically time vacations to coincide with medical meetings.

Workaholics tend to cope with fatigue and other difficulties by forcing themselves to work even longer and harder to catch up. They forego all forms of physical exercise and recreation, which makes them even less efficient and able to concentrate. Casualties of Parkinson's Law ("Work expands so as to fill the time available for its completion"), they need to stop and ask:

1. Am I overworking? A person whose family and friends complain that they seldom see him, who avoids social gatherings, who feels he is wasting time going to movies or dinner parties, or who consistently takes on more work than he comfortably can handle should answer "yes."

2. Why do I overwork? Some reasons include working to prove self-worth, to keep from feeling depressed, to avoid one's spouse and children, to avoid feeing guilty, to avoid delegating chores, or to avoid saying "no" to the demands of others.

3. Do I want to change? Change will not be easy, especially for those who overwork because of guilt or underlying personal problems such as avoiding a bad marriage or identifying with a parent who overworked or made unrealistic demands.

Workaholism often stems from poor work habits. Time-management experts suggest that a workaholic reassess his value system and work smarter--not harder--on the things he considers most important. One study of fifteen successful and healthy professionals who worked at least sixty hours a week showed that all were able to recognize and respond promptly to signs of fatigue. All were able to delay thinking about problems until they actually could deal with them. They exercised regularly, avoided drug and alcohol abuse, had stable family lives, maintained friendships, and enjoyed vacations.[6]

OTHER STRESSORS

When a stress-inducing event, large or small, is unexpected--a man loses his job or becomes a widower in his thirties instead of in his sixties or seventies--or when such events are cumulative and long-lasting, the consequences may be far more severe. The likelihood of becoming ill following exposure to a single stressor is relatively low. But the cumulative effects of several stressors may be beyond the body's restorative powers.[7] What counts is how much a person must adapt, and for how long.

Accumulated daily hassles also are harmful, sometimes as much as job problems, marital discord, and other major life events.[8] High on the "hassle scale" are misplaced car keys, too much to do, yard work, home maintenance, concern about weight, administrative chores, worries about investments and rising prices, and traffic jams. As the saying goes, "Rule no. 1 is, don't sweat the small stuff. Rule no. 2 is, it's all small stuff." The small stuff can erode emotional and physical health over the long run.

Chronic, long-term hassles are even more erosive. Consider the frustration of parents whose 27-year-old son does not leave home, the despair of a man who feels trapped in a bad marriage, or the repression of an executive who can shout at his wife after arguing with the boss, but not vice versa.[9]

DISTRESS SIGNALS

Several years ago, a psychologist meeting with nine corporate personnel directors asked each to describe a particularly stressful event in his life. After giving the executives a few minutes to ponder what intimacies they were willing to reveal, he announced that he actually had wanted them to describe how they had felt at the moment of being asked, since his real intent was to put them under stress. One by one, the personnel directors reported "butterflies," "nervousness," "light perspiration," "internal racing," "heart pounding," and "pressure on the back of the neck."[10]

Predictable body responses occur in reaction to a stressful experience. Whether the triggering event or stressor is "good" or "bad," the body undergoes a three-stage reaction: alarm, resistance, and exhaustion. During the alarm stage, when the threat is immediate (an oncoming car or trip to the podium), the body has a unique ability to release hormones that prepare it almost instantaneously for "fight or flight." Stage two is resistance: the body adapts to the stressful situation and is able to stop the effects of the stressor by releasing hormones. Stage three, exhaustion, occurs if the stressor is excessive or prolonged and if the body cannot repair the damage caused by both the stressor and the continued release of hormones. Prolonged adverse reactions to stressors produce unhealthy states that Selye calls "diseases of adaptation."[11]

The following are primitive, unconscious reactions to stressor are necessary for survival. They also are warning signals that sh heeded:

- o Pounding heart, shortness of breath, sweating, dry mouth
- o Frequent indigestion, diarrhea or urination
- o Frequent headaches, backaches, muscle spasms or fatigue
- o Susceptibility to colds and viruses
- o Accident-proneness
- o Chronically hostile or angry feelings
- o Tension (a grinding, steady reaction to a specific stressor) or anxiety (apprehension unrelated to any specific stressful event)
- o The feeling, "That's more than I can take!"

The absence of symptoms does not mean the absence of a stress reaction. Indeed, physical symptoms often are the way that a stress reaction unconsciously expresses itself, especially when symptoms occur suddenly with no apparent physical cause. One way to tune in to these forces is by "inner listening" to one's body. A "pounding" headache is a warning, perhaps, that a person is taking a pounding at work. Nausea and indigestion might mask distress over "not being able to stomach" the way somebody close is behaving. A person's back may hurt because his life hurts. Symptoms of disease often are the body's way of stating that its needs are not being met.

STRESS DISORDERS

Volumes have been written about the connection between stress reactions and disease. Boris Pasternak articulated the stress/illness connection in Doctor Zhivago:

> Your health is bound to be affected if, day after day, you say the opposite of what you feel, if you grovel before what you dislike, and rejoice at what brings you nothing but misfortune.

Or, as Woody Allen quipped in the film Manhattan, "I never get angry; I just grow a tumor."

But what about scientic proof of the mind-body link? It seems reasonable that a headache or upset stomach after an argument with one's spouse, or before a sexual encounter with a new partner, might be a physiological manifestation of stress. But what about more serious illnesses--such as heart disease or cancer--that take years to develop and cannot clearly be linked to adverse reactions to stressful events? Some people think poor health causes stress and not vice versa. While most people in stress-inducing situations probably do not get sick, evidence is growing that many disease states--including some of the top killing diseases in the U.S.--may be caused by chronic stress.

Scientists have long observed the link between stress and illness. During the Civil War, heart palpitations were so commonplace that they became known as "soldier's heart." During World War I, the phenomenon was called shellshock; in World War II, it was battle fatigue. By then Selye had suggested that chronic stress reactions cause long-term chemical changes in the body leading to high blood pressure, atherosclerosis, depressed immune function, and many other disorders.

Recent interest in the stress/illness connection began, in part, with crises faced by the hostages in Iran, people living near the radiation accident at Three Mile Island, Vietnam War veterans, earthquake victims, and people threatened by chemical waste hazards at Love Canal. The few available studies of disasters suggest that from 50 to 90 percent of the survivors who are physically unharmed in the disaster itself suffer some kind of acute emotional, physical, or psychosomatic consequence, and many suffer long-term changes in health.[12]

The evidence that such emotionally charged events damage health is so compelling that the National Academy of Sciences (NAS) has recommended that scientists move beyond questions of whether stress reactions can affect health and explore how they have such consequences.[13] Exposure to a stressful experience does not mean a person will get sick. For that to happen, several conditions must be present, according to the NAS report. First, there must be a stressor: death of a loved one, loss of a job, marital difficulties, financial crisis, or exposure to dangerous or boring work. Second, a physical reaction must take place. It may be elevated blood pressure, increased stomach acidity, chemical changes in the body, or weakened immunity. Third, there must be an absence of mediators such as supportive family relationships that alleviate the impact of the stressor.[14] Finally, the health consequence occurs, often after a lengthy period. Thus it is not a stressful event, but how a person experiences it, that eventually can make him sick or well.

Various investigators have linked adverse stress reactions to hypertension, peptic ulcers, coronary artery disease, sudden cardiac death, accidents, hyperthyroidism, cancer, liver dysfunction, bronchial asthma, minor infections, and colds.[15] For example:

o The physical consequences of job stressors have been extensively documented in air traffic controllers in the nation's busiest airports. These individuals were significantly more prone to high blood pressure and somewhat more prone to peptic ulcers and diabetes than individuals in less stressful jobs.[16]

o A study by Haynes and Feinleib showed that women clerical and sales workers who had nonsupportive bosses, trouble expressing anger, and lack of job mobility were twice as likely to develop coronary heart disease as nonclerical female workers and housewives.[17] Overall, the women in the study reported more symptoms of stress than did the men.

o Jacobs and Ostfeld's study of bereavement in 1977 showed excess mortality in the newly widowed, especially in younger persons and men. The elevated mortality continued about two years, but was highest for men during the first six months and highest for women during the second year of widowhood.[18]

o According to a more recent report on bereavement, men under age seventy-five who lose a mate or family member incur increased risk of death from accidents, cardiovascular disease, and some infectious diseases, especially during the first year.[19] Suicide rates rise during the first year, particularly among older widowers and single men who lose their mothers. Widowers who do not remarry incur an above-average death rate for six years. Health problems tend to be worse among survivors who already are in poor physical or mental health, who abuse alcohol or drugs, and who lack a social network.

Human loneliness is among the most important causes of premature death in America, according to Lynch who notes that divorced, single, and widowed individuals are up to ten times more likely than married people to die prematurely from heart disease, cancer, stroke, accidents, cirrhosis of the liver, suicide, pneumonia, diabetes--in short, just about every leading cause of death.[20] Divorced people, Lynch reports, are particularly vulnerable to all kinds of illness. While their grief is similar to that of the bereaved, the divorced incur the added burden of anger toward a spouse still living. "Quite literally," Lynch writes, "we must either learn to live together or increase our chances of prematurely dying alone."[21] Many men, in particular, rely on wives or girlfriends for social relationships and intimacy. When death or divorce occurs, they have no confidants. Men who are workaholics or who have highly successful careers often disconnect from their personal lives and are unable to sustain emotional relationships.

People who are socially isolated and whose human attachments are weak are more prone to illness and early death, according to the NAS report on Stress and Human Health.[22] Other studies show that people who die suddenly tend to relinquish control over their environment, feel helpless and hopeless before their deaths, and overwhelmed by the "giving-up/given-up" complex. There also is evidence that people who live in prolonged stressful situations and are powerless to resolve unsatisfactory work or family situations are at increased risk of heart failure. Even healthy individuals, trapped in situations from which there seems to be no escape, have been known to develop slowing of the heartbeat, heart arrhythmias, and loss of consciousness. "Learned helplessness" may weaken the body's ability to fend off disease, while ability to make and execute personal choices is critical to health and well-being.[23]

PEOPLE AT RISK

Many individuals, raised to hide their feelings, become victims of the "Pagliacci Syndrome." Like the operatic hero Canio in I Pagliacci (The Clowns), they cover despair and sadness with the characteristic light manner and smile of a clown. This cultural conditioning to ignore such feelings as grief, despair, anger, anxiety, jealousy, boredom, and uncertainty can leave them chronically out of synchrony with what they really feel behind their masks and what they express outwardly. Those at risk for stress disorders and depression include the bereaved, the divorced, the lonely, and people under job stress. Heavy workloads, too much responsibility, conflicts with coworkers, and mismatches between skills and job requirements raise the risk of illness. An accumulation of multiple stressors, such as getting married, moving, buying a house, and changing jobs can increase the risk of stress-induced illness.[24] The severest life strains supposedly befall women, the young, blacks, and people with low incomes. But results of a survey of mental disorders conducted by the National Institute of Mental Health in October 1984 showed that men and women suffer equally from serious emotional disorders, including anxiety and phobias, alcohol/drug abuse, and depression.

DEPRESSION

Winston Churchill called depression his "black dog." F. Scott Fitzgerald described it as the "dark night of the soul." Abraham Lincoln wrote in 1841: "If what I feel were equally distributed to the whole human family, there would not be one cheerful face on earth." Clinical depression, which probably causes more misery than any other single physical or psychological ailment, is among the most treatable--yet least treated--of all human afflictions. Depression, contrary to myth, is an illness, not a personal weakness. It affects at least one in four Americans at some point in their lives.[25] Yet most people fail to recognize depression in themselves or in others.

Like the common cold, periodic depression, sadness, "emptiness," or "the blues" is a normal part of the human condition. But when depression interferes with daily activities and lasts longer than two weeks, it is a more serious matter. While nobody is immune from clinical depression, some people are more vulnerable than others, especially those who react adversely to stressful events. The depression-prone include adolescents entering adulthood, the physically ill, people with drinking problems or family members with histories of depression, and people undergoing widowhood, separation, divorce, changes in job or residence, and other significant changes or losses. Even normally happy events such as marriage, childbirth, or success can trigger depression. Yet some people become depressed without a precipitating event. The peak years for depression are the twenties to forties overall, and the mid-fifties and sixties for men.[26]

Severe depression affects the entire body. Depressed people tend to feel hopeless, worthless, guilty, pessimistic, helpless, and withdrawn from others. They often become anxious and irritable and have trouble concentrating, reading, writing, conversing, and getting started in the morning. They may lose energy, ability to feel pleasure, and interest in their usual activities and in sex. In some people, periods of depression alternate with periods of mania, during which they may become frenzied rather than passive. They may work fourteen to eighteen hours a day, eat incessantly, and/or shift sexual activities into high gear. Depression also can cause insomnia, hypersomnia (too much sleep), fatigue, appetite loss, stomach upset, constipation, heightened sensitivity to emotional and physical pain, and thoughts--sometimes translated into action--of suicide. Depressed people are hard to live with. Having so little joy in their own lives, they bring little to others. As the poet John Donne observed, "No man is an iland, entire of itselfe; every man is a peece of the continent, a part of the main...." What the depressed need is reassurance that loved ones will stand by them and that the pain is a normal and--in all likelihood--temporary interlude.

Even seemingly peripheral undertakings are important in combatting depression and stress. Exercising, enjoying friends, and having fun, even while asserting one's right to feel badly, can produce harmony between a person under stress and his world. Serious loss, for example, calls for serious grieving. After the death of a loved one it takes at least a year to return to normal functioning, and the sorrow never completely goes away; it just softens with the passage of time. To not weep and grieve is to confine the forces of physiological stress within one's body. Free expression of these important emotions, on the other hand, creates a harmony of awareness that is conducive both to happiness and health.

When the blues last longer than two or three weeks, are severe, and cannot be related to an obvious event or loss, the depressed person should seek professional help from a physician, psychologist, health clinic, or local mental health association. Some people feel as if they were "born depressed," but fail to recognize that something is amiss because they have felt terrible all their lives. Serious depression of this kind will not disappear of its own accord. Yet only about half of the ten to fourteen million individuals who have a diagnosable depression seek treatment. The cost of this neglect is high: illness, disability, lowered productivity, absenteeism, and human misery beyond measure. Research scientists now theorize that many cases of depression are chronic and that people who avoid treatment and wait for their depression to "lift" may be inviting it to recur in the future.[27]

Tricyclic antidepressants have saved as many as 70 to 85 percent of depressed individuals from the depths of despair. Generally, moods begin to lift within two to three weeks after initial dosage.[28] Recent studies show that psychotherapy is just as effective as standard drug treatment in relieving acute depression, and have demonstrated the effectiveness of two forms of psychotherapy: 1) interpersonal therapy that focuses on problems with relationships, and 2) cognitive/behavioral therapy that helps patients change negative ways they have been thinking about themselves and their futures, and reinforces positive thinking and behavior they find more satisfying.

PUTTING PRESSURE IN ITS PLACE

People who handle pressure well understand what is happening to them and know how to minimize its worst effects. Ellis suggests a practical way of looking at stress: the A-B-C theory of emotion.[29] Individuals, he says, are the creators both of stressors and of their consequent reactions in the form of stress. For an emotional reaction such as tension or anxiety to occur, there must be the following:

A. An Activating Event or stress producer, such as a job problem.
B. A Belief System (which often is unconscious) about that event.
C. Consequent Feelings or actions.

Ellis' theory is that A does not cause C. It is B--or what a person consciously or unconsciously believes about A--that causes C. For example, a man learns that he has been passed over for promotion. This is the Activating Event. Next his unconscious Belief System about the meaning of

A comes into play. First he feels frightened, then angry. Then he starts worrying. He might even worry about the fact that he is worrying, further increasing his emotional load. But there are options. He can decide that being fired means he is inadequate and the victim of office politics. Or he can decide that being fired means that: 1) his performance does need improvement; 2) he needs new job skills; or 3) he might be happier in another job, with another boss, or even in another line of work.

A person's belief about A determines his Consequent Feelings. He can become depressed and dead-ended as he decides to hold onto his job while others pass him by. Or he can feel sad and disappointed about being passed over, and even anxious about a job or career change. But in the second case his distress will be temporary, will not endanger his health, and will motivate him to change.

There are other ways to minimize stress besides altering perceptions about what is stressful. One is to adopt certain attitudes and habits common to people who handle stress well. These individuals seem to have four major sources of strength: fitness, friends, fun, and flexibility.

Fitness

If a person is physically fit, he probably will feel mentally fit and able to handle stressful situations well. Fitness entails:

o Exercising regularly to increase flexibility, stamina, and strength and to let off steam. Physical exertion raises the body's level of beta endorphins that reduce anxiety and help make a person feel "mellow."

o Controlling weight and eating well-balanced meals.

o Getting enough sleep: seven to eight hours at least four times a week. To sleep less is to run on four cylinders instead of six or eight.

Friends

Recent research suggests that people with many social ties--whether through marriage, close friends and relatives, church membership, or other group activities--are better able to keep stress within tolerable limits. This means:

o Keeping in touch with others, not only through family and intimate relationships, but through self-help and other groups, and non-sexual friendships.

o Speaking openly about grief, anger, disappointment, and other important feelings, and crying. As Charles Dickens wrote in Oliver Twist, "It opens the lungs, washes the countenance, exercises the eyes, and softens down the temper."

o Giving and receiving affection regularly. Many experts have observed that touching can relieve pain and stress and promote more rapid healing. An arm around the shoulders, a pat, a handclasp, a hug can create a sense of well-being.[30] But many people, especially men, suppress their desire to be held.

o Learning to listen. Lynch reports that most people--especially those with high blood pressure--experience a significant rise in blood pressure when they speak, followed by a rapid drop in blood pressure and heart rate when they listen attentively to others.[31]

Fun

Enjoyment is an often-neglected way to break the negative power of a distressful situation by:

o Balancing work with recreation, scheduling breaks during the day, and spending lunchtimes or evenings doing something enjoyable.

o Taking a "mental health" day off at least once a month and escaping to a pleasant environment: in the woods, along a beach, beside a lake or river, or at home. There is evidence that even thinking about such peaceful surroundings can alleviate stress reactions.

o Learning to laugh. Laughter is a uniquely human behavior. Even a smile can have therapeutic value. In the words of King Solomon, "A merry heart doeth good like a medicine."

Flexibility

Openness to change can reduce stress because flexibility imparts a basic sense of control over one's life and a capacity for:

o Deciding what one really wants out of life, setting attainable goals, and undertaking one thing at a time according to a checklist with priorities.

o Giving in once in awhile and not always insisting on being right.

o Developing options. The person who feels "trapped" experiences more stress than the person who has a repertoire of healthy escapes and alternatives.

o Cultivating emotional resilience. Research shows that moods of people under intense distress one day tend to bounce back to normal the next. Scarlett O'Hara was right" Tomorrow is another day."

Some other antidotes to stress include:

o Concentrating on managing one's own life rather than wasting energy trying to change or "fix" family and friends.

o Not trying to alleviate stress with alcohol, drugs, smoking, overeating, overworking, fast driving, or fighting. These "solutions" keep a person from functioning independently and lead to even greater problems. Liquor and drugs reduce the perception of stress, but not stress itself.

o Learning stress reduction and relaxation techniques such as meditation, deep breathing, stretching and relaxing muscles, and brief naps. Relaxation is more than the absence of stress; it is a positive and satisfying feeling of peace of mind and simply of <u>being</u>.[32]

Hundreds--even thousands--of stress management techniques exist. They all work in the sense that by trying any one of them, a person moves from passivity to action against stress, from immobility to creativity, from helplessness to control, from despair to optimism, from ignoring to meeting personal needs. A person refuses, in other words, to give up.

Author Norman Cousins provides the classic example of how refusal to give up and positive thinking can affect health. Told that he had a one-in-500 chance of recovering from ankylosing spondylitis, he decided to beat the odds. He reasoned that if adverse stress reactions could cause illness, then positive emotions such as love, hope, faith, confidence, laughter, joy, and the will to live might promote health. He started the process of self-healing by checking out of the hospital and into a nearby hotel where his physicians could visit him. He then rented films of

"Candid Camera" and old Abbott and Costello movies. He discovered that twenty minutes of laughter afforded him a few hours of pain-free sleep. He then expanded his self-healing program with other measures that, during the next year, restored his health. "Since I didn't accept the verdict," Cousins later wrote, "I wasn't trapped in the cycle of fear, depression, and panic that frequently accompanies incurable illness.[33]

PSYCHOTHERAPY

Some people suffer anxiety, depression, and other emotional disorders because they have complex psychological problems. The danger signs include:

- Undue, prolonged anxiety
- Severe, long-lasting depression
- Abrupt changes in mood and behavior
- Persistent tension-caused physical symptoms
- A feeling of being trapped
- Suicidal thoughts
- Misuse of alcohol or drugs.

When psychological pain becomes great enough--often after a job crisis, divorce, or loss of a loved one--millions of men and women have sought professional help from psychiatrists, psychologists, social workers, psychiatric nurses, or other qualified counselors. The seekers are not insane, irrational, or even unable to function. They _are_ depressed, anxious, and unwilling to continue leading lives of quiet (or in some cases noisy) desperation.

Depressed patients progress according to how much effort and money they are willing to spend, how successfully they select a therapist who is interested, yet objective, and how well they develop a trusting relationship with the therapist. This relationship is the basic tool of psychotherapy.

There are two approaches to psychotherapy: 1) as short-term intervention for an immediate crisis such as death or divorce, or 2) as long-term therapy to explore underlying emotions, increase self-knowledge, modify behavior, build self-esteem, learn to manage stressful situations, and acquire a whole new set of tools for handling life's problems, large and small.

As psychotherapy begins, it often is difficult to determine whether the problem is short-term stemming from an immediate crisis, or if the immediate crisis is the end product of a lifetime of insidious maladaptive behavior. This is an important determination that may take several months to make; this calls for great skill on the part of the therapist and hard work for the client. If the immediate crisis is deemed the main problem, then supportive treatment, understanding, and the learning of new skills that aid behavioral change probably will suffice. But if the current crisis reflects a lifelong maladaptive pattern, the problem becomes one of greater intensity and severity. Severe feelings of hopelessness may recur in new situations, and the client may regress to old ways of behaving because the unconscious forces of childhood have not been sufficiently resolved.

Psychotherapists might use behavioral, cognitive, or short-term psychotherapy; psychoanalysis; group, marital, and family therapy; biofeedback; hypnosis; relaxation training; behavior modification; or sex counseling. Some problems, such as exploring the past, are better solved in individual therapy sessions. Difficulties getting along with other people are better handled in a group. Some therapists combine individual and group therapy.

An individual therapy session usually lasts fifty minutes and is scheduled weekly or more often if needed. Group therapy can last from one to several hours and usually includes eight or more people. Therapy may continue for months or, when problems are severe, for years. One study of three thousand individuals involved in various types of therapy showed that twenty-six to fifty-two sessions of psychotherapy--assuming the client actively cooperates in treatment--are sufficient for most emotional ailments. Other evidence shows that twenty cognitive therapy sessions over ten to twelve weeks will suffice for treating depression.[34]

Because psychotherapy is so expensive and personal a process, great care should be taken in choosing a therapist with whom the client feels comfortable. This entails:

o "Shopping around" for a qualified professional and getting referrals from respected friends, private physicians, county medical or psychological societies, local medical schools, health clinics, hospital psychiatric departments, family service/social agencies, community mental health clinics, or local chapters of a mental health association.

o Discussing fees, hours, and schedules. The client should state what he expects to gain from therapy, ask what method the therapist will use, and explore the therapist's background and values.

If, after a few sessions, the patient feels uncomfortable with the therapist, it is perfectly appropriate to look elsewhere. But three or four changes of therapists after treatment begins may mean the problem lies with the client, not with the therapist. Over the long run, one should expect a sense of progress despite the emotional discomforts of therapy.

Entering and staying in psychotherapy is a major life decision that deserves careful consideration and commitment. Perhaps the best analogy to what therapy offers in return appears in the Wizard of Oz, in which the scarecrow, cowardly lion, and tin woodsman finally realize that they already possess what they have traveled so far to find: brains, courage, and a heart. People who gain the most from psychotherapy:

o Recognize how they cause or contribute to their problems.

o Can define a single problem instead of saying, "Everything always goes wrong."

o Feel more comfortable with individuals who are struggling than with those who think they've "arrived."

o Are motivated to work toward personal change.

Psychotherapy helps the individual adapt to stress in relative harmony with the environment. A successful outcome must meet the demands of the individual and fit appropriately into the demands of the environment. Many people who undergo psychotherapy also experience improvements both in their mental and physical health. According to a study of health insurance claims by Federal employees, sponsored by the National Institute of Mental Health and reported in 1986, the use of mental health coverage may be an important way to reduce general health care costs. Successful management of stress, by whatever means, can produce "emotional hardiness," according to the National Academy of Sciences report on Stress and Human Health.[35] The emotionally hardy have a sense of purpose, feel in control of themselves, are open to change, and see stress-producing events as challenges rather than as threats. By mastering such challenges, they build skills for handling stress the next time it comes around. The way a person manages stress is an essential survival skill that can affect health and longevity, over the long term, as much as the way that person eats, exercises, sleeps, loves, and enjoys his working day.

REFERENCES

1. Thomas H. Holmes and Richard H. Rahe, "The Social Readjustment Rating Scale," J Psychosom Rsch, II:213-218 (1967).
2. Cited in Joel Makower, "Office Hazards: How Your Job Can Make You Sick," Washington, D.C., Tilden Press (1981).
3. William H. Whyte, Jr., "The Organization Man," N.Y.: Doubleday (1956).
4. John M. Rhoads, "Overwork," JAMA, 237:2615-2618 (1977).
5. "Physician Qualities Help at Work, Hinder Home Life," Family Practice News, (November 15-30, 1984), p. 6.
6. John Ingram Walker, "Stress of Practice: It Can Be Reduced," Dental Economics (June 1981).
7. National Academy of Sciences, "Stress and Human Health," New York, Springer (1982).
8. Richard S. Lazarus, "Little Hassles Can Be Hazardous to Health," Psychology Today (July 1981), pp. 58-62.
9. Nancy K. Schlossberg, "Counseling Adults in Transition: Linking Practice with Theory," N.Y., Springer (1984), p. 46.
10. Wall Street Journal, April 2, 1979, p. 1.
11. Hans Selye, "The Stress of Life," New York, McGraw-Hill (1956), pp. 66-67.
12. National Academy of Sciences, "Stress and Human Health," New York, Springer (1982), p. 26.
13. National Academy of Sciences, Institute of Medicine, "It's Clear that Stress Can Lead to Illness, but It's not Clear How," News Report (July 1982), p. 4.
14. Loc. cit., p. 5.
15. National Academy of Sciences, "Stress and Human Health," New York: Springer (1982), pp. 26, 318.
16. Sidney Cobb and Robert M. Rose, "Hypertension, Peptic Ulcer, and Diabetes in Air Traffic Controllers," JAMA, 224:489-492 (1973).
17. Suzanne G. Haynes and Manning Feinleib, "Women, Work, and Coronary Heart Disease: Prospective Findings From the Framingham Heart Study," Am J Pub Health 70:133-141 (1980).
18. Selby Jacobs and Adrian Ostfeld, "An Epidemiological Review of the Mortality of Bereavement," Psychosom Med, 39:344-349 (1977).
19. National Academy of Sciences, Institute of Medicine, "Bereavement: Reactions, Consequences, and Care," Washington, D.C., National Academy Press (1984).

20. James J. Lynch, "The Broken Heart: The Medical Consequences of Loneliness," N.Y., Basic Books (1977).
21. Op. Cit., p. 14.
22. National Academy of Sciences, "Stress and Human Health," N.Y., Springer (1982), p. xviii.
23. Judith Rodin, "Aging and Health: Effects of the Sense of Control," Science, 233:1271-1276 (September 19, 1986).
24. National Academy of Sciences, op. cit., pp. xviii, 26.
25. National Institute of Mental Health, "Depression: What We Know," DHHS Publication No. ADM 84-1318.
26. Op. cit., p. 20.
27. Op. cit., p. 51.
28. Op. cit., p. 52.
29. Cited in Cary L. Cooper and Roy Payne, eds., "Stress At Work," N.Y., John Wiley & Sons (1978), pp. 210-216.
30. Desmond Morris, "Intimate Behaviour," N.Y.: Random House (1971), p. 148.
31. James J. Lynch, "Language of the Heart," N.Y., Basic Books (1986), p. 5.
32. Herbert Benson, "The Relaxation Response," N.Y., Morrow (1976).
33. Norman Cousins, "Anatomy of An Illness As Perceived by the Patient," Boston, G.K. Hall (1979).
34. National Institute of Mental Health, "Depression: What We Know," DHHS Publication No. ADM 84-1318, p. 54.
35. National Academy of Sciences, "Stress and Human Health," N.Y., Springer (1982), pp. 164-165.

THE CLINICAL IMPLICATIONS OF CORTICOTROPIN-RELEASING HORMONE

Philip W. Gold[1], Mitchel A. Kling[1], Harvey J. Whitfield[1], Douglas Rabin[2], Andrew Margioris[2], Konstantine Kalogeras[1], Mark Demitrack[1], D. Lynn Loriaux[2], and George P. Chrousos[2]

From the Biological Psychiatry Branch, National Institute of Mental Health[1], and the Developmental Endocrinology Branch, National Institute of Child Health and Human Development[2]

INTRODUCTION

We now appreciate that the brain is the most prolific of all endocrine organs producing scores of neurohormones within and beyond the boundaries of the endocrine hypothalamus. The idea that the brain functions as a gland, however, is not new. Indeed, the evolution of thought leading to the identification of corticotropin releasing hormone (CRH) began around 400 B.C. (1,2). At this time, Hippocrates, in his work entitled De Glandulis, states explicitly, "The flesh of the glands is different from the rest of the body, being spongy and full of veins; they are found in the moist part of the body where they receive humidity ... and the brain is a gland as well as the mammae."

Following this farsighted postulate of Hippocrates, elucidation of the brain's endocrine functions proceeded at an extraordinarily slow pace. For instance, five hundred years passed before one of the principal target tissues of the neurohormones was inadvertently identified by Galen, which he mistook for a mucous (pituita) secretion of the brain (3). Sixteen hundred more years passed until the hypothalamic-pituitary system of portal vessels was described (4) and another two hundred more years elapsed until its functional significance was understood. It was then, in 1948, that Harris advanced the hypothesis that the hypothalamus produced humoral factors which traversed the hypophyseal portal system to regulate the secretion of anterior pituitary hormones into the systemic circulation (5).

By 1955, stimulated by Harris's postulate, the major groundwork for the identification of CRH was laid when Saffran and Schally showed that hypothalamic fragments possessed remarkable corticotropin (ACTH) releasing properties when incubated with pituicytes *in vitro* (6). Within 15 years, Vale et al. announced the sequence of a 41 amino acid peptide isolated from ovine hypothalami, which showed greater *in vivo* and *in vitro* corticotropin releasing potency than any previously identified synthetic or endogenous peptide (7). Shortly thereafter, Schally et al. described the partial composition of porcine CRH (8), and Rivier et al. the full sequence of rat CRH (rCRH) (9). Finally, Numa's group sequenced the genes of both ovine CRH (oCRH) and human CRH (hCRH) and deduced the amino acid sequence of the corresponding peptides (10,11). Surprisingly, both the rCRH and the hCRH appeared to be chemically identical. Moreover, oCRH and hCRH are also structurally similar, each containing 41 amino acids and showing 83% homology.

The sequencing and subsequent synthesis of CRH has greatly enhanced the capacity for clinical neuroendocrinologists to explore the hypothalamic-pituitary components of Cushing's disease and adrenal insufficiency. CRH is also of importance to clinical psychoneuroendocrinologists since its discovery occurred in the context of two other discoveries which have heightened their interest in the hypothalamic-pituitary-adrenal (HPA) axis and its regulation. First, many patients with major depression and anorexia nervosa (12,13) manifest a hyperactive pituitary-adrenal axis. In fact, the hypercortisolism seen in depressive illness and anorexia nervosa can be so severe that it can parallel that seen in Cushing's disease; in this regard, major depression has been referred to colloquially by some endocrinologists as a pseudo-Cushing's state. Second, it has been shown that ACTH is secreted synchronously with beta-endorphin, one of the principal endogenous opioid peptides (17), and that both hormones are contained within the sequence of a common precursor molecule, proopiomelanocortin (POMC) (18). CRH is the principal central signal for the cleavage of pituitary POMC into biologically active peptides.

CRH is also of interest to psychiatrists and neurobiologists for reasons other than its putative role in regulating the pituitary-adrenal axis. Hence, CRH is synthesized not only by the hypothalamus for transport by hypophyseal portal blood, but like other hypothalamic hormones, CRH is widely distributed and/or synthesized beyond the boundaries of the hypothalamus to play an important role in coordinating complex behavioral and/or physiological processes of survival value. Neuroanatomically, it has been shown that there are extensive hypothalamic aggregations of CRH cell bodies and terminal fields in the limbic system, cortex, and in close association with the central autonomic system and the locus ceruleus (19,20,21). This distribution of CRH within and beyond the hypothalamus provides an anatomical context for the observation that CRH can simultaneously activate and coordinate metabolic (22), circulatory (22), and behavioral responses that are adaptive in stressful situations (23-25). Hence, in the rat, intracerebroventricular (ICV) administration of CRH leads not only to activation of the HPA axis but also to activation of the sympathetic nervous system (22) and to several behavioral changes characteristic of the stress response, including decreased feeding (3) and sexual behavior (24), assumption of a freeze posture in a foreign environment (25) and increased exploration in familiar surroundings (25). In addition, in collaboration with Weiss and Post, our group has shown that CRH given ICV to the rat causes a marked increase in hostility and induces limbic seizures which show cross-sensitization with electrically kindled seizures (26).

Given CRH's significant role in HPA regulation and its intriguing effects on CNS function, we embarked on a series of clinical studies with CRH both in normal volunteers and in psychiatric patients with major psychiatric disorders whose illnesses are at times characterized by hypercortisolism (e.g., primary affective disorder, anorexia nervosa, panic anxiety disor der, and schizophrenia). In addition, we studied patients with Cushing's disease because the confluence of hypercortisolism and dysphoric mood in this disorder make it difficult or impossible to distinguish from major depression. In volunteers we hoped to examine the physiological relevance of CRH to pituitary-adrenal function in man as well as to explore the differential biological effects and pharmacokinetics of oCRH and hCRH under varying conditions. In our patient populations, we asked the following questions: (1) Can CRH help determine whether the hypercortisolism in depression reflects an alteration in the set-point for feed back inhibition of cortisol on ACTH secretion at the pituitary locus, versus the possibility of an alteration in the secretion of endogenous CRH; (2) Can CRH help in the differential diagnosis of the various hypercortisolemic psychiatric syndromes; (3) Can CRH help determine whether depression and Cushing's disease lie on a common pathophysiological continuum or represent distinct abnormalities of the HPA axis; (4) Is CRH of possible relevance to the overall symptom complex of major psychiatric illnesses such as depression and anorexia nervosa?

Relevance of CRH to Pituitary-Adrenal Regulation in Man

Even before the sequencing of CRH, this peptide was postulated to play a major role in the regulation of pituitary-adrenal function. According to this formulation, CRH is secreted in seven to ten discrete episodes per day (during unstressed conditions), which in turn produce a similar number of ACTH and cortisol secretory episodes. The majority of these pulses occur in the early morning hours. Since cortisol has a longer half-life than ACTH (and presumably CRH), the frequent cortisol pulses occurring in the early morning produce a situation in which a new cortisol secretory episode may be initiated before the last episode is terminated, producing a rise in cortisol with each successive ACTH pulse. Hence, with an increasing frequency of CRH and ACTH pulses, there emerges an increase in the level or amplitude of plasma cortisol secretion which we recognize as the early morning plasma cortisol surge. This phenomenon of the early morning rise in the amplitude of cortisol secretion is an example in which frequency modulation (i.e., timing of CRH, ACTH and cortisol pulses) is translated into amplitude modulation (i.e., amplitude or cortisol secretion).

Over the years (and particularly since the sequencing of CRH) this classic formulation has been modified by additional information. For instance, it has been known for some time that vasopressin is capable of stimulating the pituitary corticotroph cell, and it has even been postulated to be the major hypothalamic CRH peptide. It is now known that arginine vasopressin (AVP) itself is not as potent as hCRH 1-41 in stimulating ACTH secretion, but that AVP markedly potentiates CRH's corticotropin releasing properties (27). Moreover, other substances such as the catecholamines also potentiate CRH-induced ACTH secretion (28). In light of these data and because of a paucity of clinical studies with hCRH, one of the first studies our group undertook with this peptide was to explore its possible relevance to the physiologic regulation of the pituitary-adrenal axis in man.

To address this issue, we asked the following question: Could hCRH, given in pulses that simulate the expected pulsatile pattern restore the function of the pituitary-adrenal axis shown in patients with CRH deficiency [i.e., secondary adrenal insufficiency due to corticotroph-sparing (superpituitary) lesions] (30)? The patients included in this study had secondary adrenal insufficiency determined by a standard 48-hour ACTH stimulation test. Thus, during a standard three-hour evening oCRH test (*vide infra*), the plasma cortisol response of these patients was diminished while the plasma ACTH response was normal or exaggerated and showed a delayed pattern, suggestive of a hypothalamic lesion. Human CRH was given as a 1 ug/kg bolus eight times during the 24-hour period. The timing of each pulse of hCRH was chosen to correspond to the expected times of ACTH pulsation under naturalistic conditions. Hence, the majority of the pulses were given in the early morning hours to correspond with the a.m. cortisol surge. To a considerable degree, such a paradigm of hCRH pulsatile administration reproduced the normal amplitude and circadian variation of cortisol secretion in patients with hypothalamic-CRH deficiency . Moreover, the administration of eight pulses of synthetic hCRH, as described above, normalized 24-hour urinary 17-hydroxysteroid and urinary free cortisol secretion in these patients (30). Parenthetically, we have previously noted that a continuous infusion of oCRH in normal volunteers for 24 hours produces a pattern of cortisol secretion which includes preservation of a circadian rhythm though the amplitude is blunted compared to the naturalistic rhythm of that induced by the hCRH pulses (31). Thus, although basal circadian cortisol secretion may be dependent on endogenous CRH secretion, this circadian pattern of pituitary-adrenal function may also involve a component of a circadian rhythm in the responsiveness to CRH itself (31).

Development of a Clinically Applicable CRH Stimulation Paradigm

To ascertain the clinical applications of CRH, we initiated a series of studies in volunteers to assess the following questions: (1) Which peptide (oCRH or hCRH) might be best to use in acute challenges of the pituitary-adrenal axis; (2) What dose should be administered in these studies and for how long should hormonal responses be sampled; (3) What time of day would be best suited for the performance of dynamic stimulation of the human pituitary adrenal axis? To assess these questions, we conducted pharmacokinetic and dose-response studies with both oCRH and hCRH. Some of these studies were performed at different times of day to correspond to periods when the hypothalamic-pituitary-adrenal axis is normally quiescent or most active.

The first dose response study with oCRH in primates was performed by Schulte et al. in cynomolgus macaques (35). Corresponding studies in man yielding similar results were performed by Grossman et al. (36) and Orth et al. (37). These studies show that the lowest maximal stimulatory dose for cortisol secretion was 1 ug/kg; moreover, this dose produced clearcut plasma cortisol and ACTH secretion in all volunteers and experimental animals without detectable adverse effects. Of particular interest was the fact that the ACTH and cortisol responses to oCRH were prolonged, remaining clearly elevated at the end of the three-hour sampling period (35).

In our similar dose-response studies with hCRH in nonhuman primates (38) and man (39), Schuermeyer et al. noted a dose-dependent increase of plasma ACTH and cortisol con centrations with graded doses of hCRH (figs. 6,7). Peak plasma ACTH and cortisol responses to hCRH were significantly lower than those achieved by oCRH (38,39). More over, the ACTH and cortisol responses to hCRH were of much shorter duration than those with oCRH. Accordingly, comparisons for the integrated secretory responses of both ACTH and cortisol following hCRH administration indicate that oCRH is at least five times more potent than hCRH (38,39). This difference is mainly due to the longer-lasting effect of oCRH upon ACTH and cortisol secretion.

These longer-lasting effects of oCRH on ACTH and cortisol secretion can be presumably accounted for on the basis of the different pharmacokinetic properties of oCRH (40) and hCRH (38,39). Hence, in a study directly comparing the metabolic clearance rate of these two peptides in human volunteers, Schuermeyer et al. noted that hCRH is cleared from plasma much more rapidly than oCRH (38,39). On the basis of the relatively slower clear ance and more prolonged biological effects of oCRH, we elected to use this peptide for characterizing the functional integrity of the pituitary corticotroph in our clinical populations. We reasoned that the extra information provided by a pulse of oCRH might provide the kind of additional information that could be helpful in exploring the pathophysiology of hypothalamic-pituitary-adrenal activity in different patient subgroups. Moreover, we felt that this additional information could also be helpful in determining subtle differences in responses between patient subgroups and hence more helpful in establishing differential diagnoses between clinical entities. On the other hand, as illustrated by the studies in patients with hypothalamic CRH deficiency (30), the much more rapidly cleared hCRH seems far more suitable than oCRH for studies of pulsatile ACTH secretion.

An additional factor of relevance to the establishment of a clinically applicable CRH stimulation test is the determination of an optimal time of day for administration of the peptide to patient populations. To explore this question, we administered a 1 ug/kg bolus of oCRH at 9 a.m., near the time of day when the axis is most active, and at 8 p.m., when the axis is normally dormant. We found that owing to the lower baseline cortisol levels seen in the evening, the net integrated cortisol responses to oCRH are greater at this time (41). Hence,we decided our CRH stimulation test would consist of a 1 ug/kg bolus of oCRH given at the 8 p.m. time.

Clinical Studies with oCRH in Patients with Major Psychiatric Disorders

The first major finding utilizing CRH in psychiatry was made by our group when we noted that twelve of our drug-free depressed patients showed a significantly blunted ACTH response to oCRH (42). This finding was subsequently reproduced by Holsboer et al. in twelve subjects (44). We have continued to see this pattern in a larger series of depressed patients, and most recently have replicated our finding in a large series of thirty-two patients (43).

The finding of an attenuated response to oCRH in depression suggested that the pituitary corticotroph cell in depressed patients was appropriately restrained by the negative feedback effects of elevated cortisol levels (42,43,45). In light of the apparently normal corticotroph cell function in depressed patients, we first advanced the hypothesis that hypercortisolism in depression represents a defect at or above the hypothalamus which results in the hypersecret ion of endogenous CRH. To test this hypothesis, we attempted to replicate in normal controls a situation in which the pituitary corticotroph cell is exposed to excessive corticotropin releasing hormone. In order to accomplish this, we administered a continuous infusion of oCRH for 24 hours and evaluated the ACTH and corti-sol responses (31). Of interest is the fact that the circadian rhythm of cortisol was preserved despite the continuous administration of oCRH, suggesting that the pituitary corticotroph cell shows a diurnal sensitivity to exogenous CRH. This is of interest in light of the fact that the circadian rhythm of cortisol is generally preserved in depression (46). Of additional interest is the fact that the mean amplitude of cortisol secretion during continuous CRH infusion is elevated about 40% to 50%, and that the urinary free cortisol secretion during CRH infusion averaged 150 to 200 ug/day (31). Hence, the amplitude of plasma cortisol during the 24-hour period and the magnitude of urinary free cortisol hypersecretion is very similar during condi-tions of continuous administration of oCRH to controls and in the endogenously depressed state. We concluded, therefore, that a continuous CRH infusion to normal volunteers repro- duces the pattern and magnitude of hypercortisolism typically associated with depression. Additional data compatible with the idea that CRH is hypersecreted in depressed patients derives from the data of Nemeroff et al. who showed that the level of CSF CRH is elevated in depression (47). Although we could not demonstrate a significant elevation of CSF CRH in depressed patients, these subjects did manifest a significant positive correlation between post-dexamethasone cortisol levels and the CSF level of CRH (48). Moreover, we noted that CSF CRH is significantly higher in depressed patients who are dexamethasone non-suppressors than in dexamethasone suppressors.

Inspection of our ACTH and cortisol responses to CRH in depression revealed other salient features of HPA dysfunction in depressed patients. For instance, we noted that depressed patients showed a robust total and free cortisol response despite the very small ACTH released during CRH stimulation (42, 43,45). In fact, the free cortisol response to CRH was even greater in depressed patients than in controls (43). We surmise from these data that the adrenal cortex in depression has grown hyperresponsive to ACTH (42,43,45), compatible with the well-described phenomenon of progressive functional and anatomical hypertrophy of the adrenal cortex seen during either experimentally induced stress (49) or during the course of chronic and repeated hyperstimulation of the adrenal cortex by ACTH in man (50). This suggestion of adrenal hyperresponsiveness to ACTH in depression is compatible with the data of Amsterdam et al. which showed that chronically depressed patients manifest greater cortisol responses to a bolus of exogenous ACTH than normal subjects (51).

Although our depressed patients were hypercortisolemic, it is noteworthy that basal ACTH levels remained in the normal range (42,43,45). This "normal" plasma ACTH level in depression most likely reflects a normal corticotroph cell caught in the balance between forces (i.e., negative feedback exerted by a hyperactive adrenal cortex from below and a predominating excess of CRH drive from above). Hence, the corticotroph cell, though restrained by the negative feedback to secrete at a rate that produces ACTH levels in the normal range, is nevertheless sufficiently driven by CRH to promote excessive cortisol secretion by hyperplastic adrenals. Presumably, depressed patients would have shown elevated levels of ACTH in the beginning of their depressive illness (43).

We have also explored the pathophysiology of hypercortisolism in other major psychiatric disorders, including anorexia nervosa. We note that underweight patients with anorexia nervosa show hypercortisolism which is even more severe than that seen in depressed patients. However, like the depressed patients, these underweight anorexics manifest a markedly attenuated ACTH response to exogenous CRH (52). This finding strongly suggests that, like the depressed patients, the pituitary corticotroph cell in anorexia nervosa is appropriately restrained by the negative feedback effects of hyper- cortisolism, and also suggests that anorexics, like depressed patients, show a defect in the secretion of CRH. In support of this hypothesis is our finding that underweight patients with anorexia nervosa show significantly higher CSF CRH levels than controls When these underweight anorexic subjects were restudied after their weight had stabilized at 100% of normal body weight, their basal hypercortisolism had resolved, suggesting normalization of the central defect which resulted in the hypersecretion of endogenous CRH (36). Moreover, the level of CRH in the CSF also returned to normal at this time. However, despite normalization of the apparent central defect in CRH secretion which had produced their hypercortisolism, these normal weight anorexics contin-ued to show a markedly attenuated ACTH response to CRH (52). Although we cannot definitively account for this finding, it may represent the persistence of a functionally hypertrophied adre-nal cortex into this phase of short-term recovery. On the other hand, normal weight bulimic subjects studied for 10 days after voluntary absence from binging and vomiting, and a group of anorexia nervosa subjects who had maintained normal body weight for at least six months, showed normal basal ACTH and cortisol values in their responses to exogenous CRH (52).From these studies, we conclude that the basic pathophysiology of hypercortisolism anorexia nervosa is similar to that seen in the depressed phase of primary affective disorder and that subtle defects in hypothalamic-pituitary-adrenal function persist in patients with anorexia nervosa despite the return to eucortisolism after the short-term correc tion of the weight loss.

The apparently similar pathophysiology of hypercortisolism in depression and anorexia nervosa supports the notion that these two disorders lie on a pathophysiologic continuum. Clinical data also support this idea. For instance, both patients with depression and anorexia nervosa manifest not only hypercortisolism but also hypothalamic hypogonadism and anorexia. Moreover, many anorexia nervosa patients are depressed and show strong family histories for depression (53). Parenthetically, we observed a significant positive correlation between depression ratings and CSF CRH levels in anorexic patients, suggesting that CRH may be an important link between primary affective disorder and anorexia nervosa. We have also noted attenuated ACTH responses to CRH in hypercortisolemic patients with anxiety disorder (54), suggesting that these patientsshow a similar pathophysiology of HPA functioning to that seen in subjects with depression and anorexia nervosa.

In contrast to the three hypercortisolemic psychiatric subgroups who showed the blunted ACTH responses to CRH, a group of severely psychotic schizophrenic patients showed normal basal cortisol values and normal ACTH responses to exogenous CRH (55). These data are compatible with the previously published data of Nemeroff et al. showing normal CSF CRH in schizophrenia (47). Moreover, these data are compatible with previous studies which show that a smaller percentage of patients with schizophrenia than those with affective disturbance fail to suppress their cortisol levels after dexamethasone. Parenthetically, when eight of these drug-free schizophrenic patients were studied after treatment with fluphenazine, the ACTH and cortisol responses to CRH were similar to those seen during the drug-free state. In light of this eucortisolism and the normal ACTH responses to CRH in these psychotic schizophrenic patients, it is intriguing to speculate that perhaps they showed a defect in the translation of their enormous distress into activation of the CRH neuron. Additional work on exploring the perturbability of the CRH neuron in schizophrenic subjects following experimentally induced stress (i.e., hypoglycemic or exercise) would seem warranted to further test this hypothesis.

Putative Role of CRH and Related Components of the Stress Response in the Pathophysiology of Major Depression

The CRH and locus ceruleus norepinephrine systems represent the principal biological effectors of the generalized stress response, or general adaptational syndrome, as first described by Selye over forty years ago. Set into motion during states of threatened homeo--stasis, these effects act directly within the CNS to facilitate adaptive neural pathways subserving attention, arousal, and aggression, while inhibiting non-adaptive ones subserving vegetative functions such as feeding, sexual behavior, and reproduction. In the periphery, the CRH and locus ceruleus norepinephrine systems, acting through the catecholamines and glucocorticoids, act to re-direct blood flow to the CNS and to mobilize fuel for immediate action.

A potential role for CRH in the pathophysiology of affective illness is suggested by the data, noted above, that the icv administration of CRH produces many changes classically associated with depression (e.g., hypercortisolism, anorexia, decreased libido, hypothalamic hypogonadism, sympathetic activation, and changes in motor behavior). Experimental data that the icv administration of CRH in moderate to large doses produces effects which can be construed as anxiogenic (e.g., hyperresponsiveness to acoustic startle, decreased exploration, freezing behavior) (56,57) suggests a possible role for CRH in behavioral components of depression (especially melancholic depression, a severe form of depression associated with weight loss, dysphoric hyperarousal, and intense anxiety about self and the future).

Recent data also suggest that CRH may be relevant to the characteristically periodic, exacerbating course of major affective disorder. Sapolsky et al. have shown that experimentally induced hypercortisolism in the rat can damage or destroy glucocorticoid feedback upon the CRH neuron (58); in this regard, not only do older rats show significantly higher corticosterone levels that younger ones, but older, depressed patients show a higher incidence of the more severe melancholic form of depression as well as significantly higher cortisol levels than younger individuals with the illness (59). A further potential role for CRH in the periodic, exacerbating course of affective illness is suggested by data from our group that the icv administration of CRH produces limbic seizures which cross-sensitize with electrically kindled seizures (26). As a corollary, in clinical studies, our group has shown that procaine, which produces kindled seizures in experimental animals, can produce marked mood changes in patients with major depression and controls, in association with a dose-dependent activation of the pituitary-adrenal axis (60). Our *in vitro* studies show that procaine is a potent stimulus to hypothalamic CRH release, while carbamazepine, a limbic anticonvulsant effective in the treatment of major affective disorder, inhibits procaine-induced *in vitro* CRH release (Kling, MA et al., unpublished observations).

Putative Involvement of the Locus-Ceruleus-Norepinephrine (LC-NE) System in Major Depression and Its Potential Relationship to the CRH System

Like the CRH system, the LC-NE is a prime candidate to be involved in a syndrome associated with evidence of pathological arousal such as melancholic depression. Hence, electrical stimulation of the LC in unanesthetized primates produces intense anxiety, hypervigilence, and inhibition of exploratory behavior (61). Spontaneous locus ceruleus firing increases during threatening situations and diminishes during sleep, grooming, and feeding (62).

The weight of available evidence supports the idea that the LC-NE system is activated in melancholic depression. Hence, studies consistently show increased plasma NE and increased CSF and plasma MHPG in major depression (63-65). As a corollary, successful responses to antidepressant medication are consistently associated with decreases in CSF and plasma indices of noradrenergic hyperfunction (67). These clinical studies are compatible with pre-clinical data that MAO inhibitors and tricyclic antidepressants decrease the LC firing rate and reduce the level of NE metabolites in brain (67).

Clinical data from our group suggest linkage between the CRH and LC-NE systems in major depression, as evidenced by data that CSF CRH correlates positively with indices of LC-NE activation, including CSF NE and MHPG (68). Preclinical data from our laboratory and elsewhere support such linkage between these systems. Hence, while we have shown that NE significantly increases CRH release from an *in vitro* hypothalamic organ culture system (68), Valentino et al. have shown that CRH markedly increases the LC firing rate (69). Fur ther evidence to support a possible mutually reinforcing feedback loop between the CRH and LC-NE systems includes data that the icv administration of a CRH antagonist blocks the stimulatory effects of nitroprusside upon LC firing (70), while propranolol inhibits CRH-induced increases in locomotion (71).

The mechanism underlying the concomitant activation of the CRH and LC-NE systems in melancholic depression have not been definitively elucidated. However, several lines of recent evidence suggest that the most important role of glucocorticoid secretion during stress is to restrain or counter-regulate the effectors of the stress response in order to prevent their prolonged or excessive activation (72). Thus, glucocorticoids clearly restrain the CRH system, and may restrain the LC-NE system as well, analogous to their restraint of the sympathetic nervous system (73).

In a recent synthesis of data relevant to the mechanisms of melancholic depression, we suggested that melancholic depression may be associated with a failure of glucocorticoids to effectively counter-regulate the CRH and/or LC-NE system during stressful situations (74,75). Theoretically, many defects could cause such dysregulation, including an abnormality on the short arm of chromosome 11 (76) resulting in deficient glucocorticoid enhancement of tyrosine hydroxylase activity during stress (an effect which would be expected to decrease the LC firing rate by increasing the availability of NE in the region of the LC to activate inhibitory alpha 2 autoreceptors) (77). If this were the case, one would predict from early life onward that stress responses would be more frequent, intense, and prolonged, leading to potential damage to glucocorticoid receptor-containing cells mediating negative feedback upon the CRH neuron and sensitization of critical limbic sites. A reduction in the threshold for stress-mediated activation of hypothalamic-pituitary-adrenal and noradrenergic systems, as shown in a subset of children who hyperrespond to novel stimuli, could also theoretically predispose to major depressive illness.

It should be noted that major depression is a heterogeneous syndrome with many potential clinical presentations. In addition to the relatively consistent association of symptoms which constitute melancholic depression, another common cluster of symptoms compromises the syndrome of 'atypical' depression. In contrast to melancholic depression, which is associated with signs of arousal, such as insomnia and anorexia, atypical depression seems the antithesis, with hypersomnia and hyperphagia being prominent physiological concomitants. In this regard, our data in patients with Cushing's disease, whose depressions most frequently fall into the atypical category, reveal a pattern of significant decreases in both CSF

CRH (78) and the NE metabolite MHPG (M.A. Kling et al., unpublished observations). These data suggest that atypical depression, in contrast to melancholic depression, represents a possible pathological inactivation of the two major components of the stress response and mediators of arousal. Parenthetically, despite the apparent clinically and pathophysiologically antithetical aspects of melancholic and atypical depression, these syndromes may not be entirely unrelated. Hence, during the longitudinal course of recurrent major depression, some patients manifest melancholic depression during one phase and an atypical depression during another.

The CRH Stimulation Test: Implications for the Diagnosis and Pathophysiology of Hypercortisolism in Depresssion and Cushing's Disease

The hypercortisolism of depression can be of sufficient magnitude that it has been termed a pseudo-Cushing's state. Conversely, patients with Cushing's disease often show signs of clinical depression. Although there has been controversy over the years concerning the etiology of the hypercortisolism associated with affective illness and Cushing's disease, the overlap in the clinical and biochemical manifestations of these illnesses has prompted some to suggest that they share common pathophysiological features. Of clinical significance is the fact that patients with primary depression who may be hirsute or obese and who manifest high plasma and urinary free cortisol levels can be impossible to distinguish from patients with mild or early Cushing's disease. Indeed, depression can often be the first manifestation of Cushing's disease preceding the physical stigmata such as buffalo hump or purple striae by months or even years.

Data from our group and others show that despite profound basal hyercortisolism, patients with Cushing's disease show a marked hyperresponsiveness of the pituitary corticotroph cell to exogenous CRH (43,74-82). Thus, in contrast to patients with depression, who show a pituitary corticotroph cell normally responsive to the negative feedback effects of glucocorticoids, patients with Cushing's disease manifest a pituitary corticotroph cell which is grossly unresponsive to cortisol negative feedback effects. Our data also suggest that the differences in pituitary corticotroph cell function between depressed and Cushing's disease patients seem accompanied by differences in hypothalamic CRH neuron function. Specifically, we have shown that many of our patients with Cushing's disease whom we studied one week after selective transsphenoidal adenomectomy (at a time when basal ACTH and cortisol were uniformly undetectable) showed normal or nearly normal plasma ACTH responses to exogenous CRH. We surmise that the adrenal insufficiency in each of these postoperative patients reflects hypofunction of corticotropin releasing factor neurons which had been physiologically suppressed by exposure to the nega-tive feedback of their long-standing hypercortisolism. This formulation is supported by our recent finding that compared to depressed patients and controls, CSF CRH is significantly lower in Cushing's disease patients (78).

The differential pathophysiology of hypercortisolism which we propose for Cushing's disease and depression is manifested by the fact that responses to CRH in these disorders are in the opposite direction (e.g., an exaggerated ACTH response in Cushing's disease and a blunted one in depression. In all other diagnostic tests which have been utilized to differenti ate depression from Cushing's disease, such as the dexamethasone suppression test and serial urinary free cortisol determinations, responses and/or levels in depression and Cushing's disease were in the same direction. Thus, the CRH stimulation test may is in a unique position to assist in the differential diagnosis between depression and early Cushing's disease. Indeed, our data do show that the CRH stimulation test may be helpful in the differential diagnosis of these disorders, with approximately 25% of depressed patients showing ACTH responses to CRH in the Cushing's disease range and vice versa (43).

Conclusion

In summary, we have reviewed the physiological role of CRH in pituitary-adrenal function and advanced data regarding its putative role in the hypercortisolism of major depression and anorexia nervosa. We have also cited data regarding the central effects of CRH suggest ing a possible role for this peptide in the overall symptom complexes of major depression and anorexia nervosa. In addition, we have cited our data and data of others showing potentially impor-tant interactions betweeen the two major systems involved in responsiveness to stress ful situations (e.g., the CRH and LC-NE systems) and have hypothesized an important role for these interactions in the clinical and biochemical manifestations of melancholic and atypical depression. Finally, we have advanced data showing that the pathophysiologic locus of hypercortisolism in major depression and Cushing's disease is distinct, and have presented data illustrating the utility of the CRH stimulation test in their differential diagnosis.

REFERENCES

1. Zingerus T. Commentarii Hippocratis, Basileae: Episcoporium Opera; 1669; pp. 324-85.

2. Iason AH. The Thyroid Gland in Medical History. New York, Froben Press, 1946.

3. Medvei VC. A History of Endocrinology. Boston, MTP Press, 1984.

4. Lieutaud J. Essais Anatomiques, Contenant L'Histoire Exact de Toutes les Parties qui Composent le Corps de L'Homme, avec la Manier de Dissequer, Paris, Huart; 1742.

5. Harris GW. Neural control of the pituitary gland. Physiol. Rev. 1948; 28:134-79.

6. Saffran M, Schally AV, Bentey BG. Stimulation of the release of corticotropin from the adenohypophysis by a neurohypophyseal factor. Endocrinology 1955; 57:439-44.

7. Vale W, Spiess J, Rivier C, Rivier J. Characterization of a 41-residue ovine hypothalamic peptide that stimulates secretion of corticotropin and beta-endorphin. Science 1981; 213:1394-97.

8. Schally AV, Chang RC, Arimura A, Redding TW, Fishback JB, Vigh S. High molecular weight peptide with corticotropin-releasing factor activity from porcine hypothalami. Proc Natl Acad Sci USA 1981; 78:5197-201.

9. Rivier J, Spiess J, Vale W. Characterization of rat hypothalamic corticotropin releasing factor. Proc Natl Acad Sci USA 1983; 80:4851-55.

10. Furatani Y, Morimoto Y, Shubahara S, Noda M, Takahashi H, Hirose T, Asai M, Inayama S, Hayashida H, Miyata T, Numa S. Cloning and sequence analysis of cDNA for ovine corticotropin-releasing factor precursor. Nature 1983; 301:537-40.

11. Shibahara S, Morimoto Y, Furatani Y, Notake M, Takahashi H, Shimizu S, Horikawa S, Numa S. Isolation and sequence analysis of the human corticotropin-releasing factor precursor gene. The EMBO J 1983; 2:775-79.

12. Sachar EJ, Hellman L, Fukushima DK, Gallagher TF. Cortisol production in depressive illness: a clinical and biochemical classification. Arch Gen Psychiatry 1970; 23:289-98.

13. Carroll BJ, Curtis GC, Mendels J. Neuroendocrine regulation in depression. I. Limbic system - adrenocortical dysfunction. Arch Gen Psychiatry 1976; 33:1039-44.

14. Gerner GH, Gwirtsman HE. Abnormalities of dexamethasone suppression test and urinary MHPG on anorexia nervosa. Am J Psychiat 181; 138:650-53.

15. Stokes PE. Adrenocortical activation in alcoholics during chronic drinking. Ann NY Acad Sci 1973; 215:77-81.

16. Insel TR, Kalin WH, Guttmacher LB, Cohen RM, Murphy DL. The dexamethasone suppression test in obsessive-compulsive disorder. Psychiatry Res 1982; 6:153-60.

17. Guillemin R, Vargo T, Rossier J, Minick S, Ling N, Rivier C, Vale W, Bloom F. Beta-endorphin and adrenocorticotropin are secreted concomitantly by the pituitary gland. Science 1977; 197:1367-68.

18. Mains R, Eipper E, Ling N. Common precursor to corticotropin and endorphins. Proc Natl Acad Sci USA 1977; 74:3014-18.

19. Bloom FE, Battenberg ELF, Rivier J, Vale W. Corticotropin releasing factor (CRF): immunoreactive neurons and fiber in rat hypothalamus. Regul Pept 1;1982; 4:43-48.

20. Olschowka JA, O'Donohue TL, Mueller GP. The distribution of corticotropin releasing factor-like immunoreactive neurons in . Neuropharmacology 1982;35:305-308.

21. DeSouza EB, Perrin MH, Insel TR, Rivier J, Vale WW, Kuhar MJ. Corticotropin-releasing factor receptors in rat forebrain: autoradiographic identification. Science 1984:224;2449-51.

22. Brown MR, Fisher LA, Spiess J, et al. Corticotropin-releasing factor: actions on the sympathetic nervous system and metabolism. Endocrinology 1982; 111:928-31.

23. Britton DR, Koob GF, Rivier J, Vale WW. Intraventricular corticotropin releasing factor enhances behavioral effects of novelty. Life Sci 1982; 31:363-67.

24. Sirinathsinghji DJS, Rees LH, Rivier J, Vale WW. Corticotropin-releasing factor is a potent inhibitor of sexual receptivity in the female rat. Nature 1983; 305:232-35.

25. Sutton RE, Koob GF, LeMoul M et al. Corticotropin releasing factor produces behavioral activation in rats. Nature 1982; 297:331-33.

26. Weiss SRB, Post RM, Gold PW et al. Corticotropin releasing factor-induced seizures and behavior: interaction with amygdala kindling. Brain Res 1986;372:345-51.

27. Rivier C, Vale W. Modulation of stress-induced ACTH release by corticotropin-releasing factor, catecholamines and vasopressin. Nature (Lond.) 1983; 305:325-27.

28. Vale W, Vaughan J, Smith M, Yamamoto G, Rivier J, Rivier C. Effects of synthetic ovine corticotropin-releasing factor, glucocorticoids, catecholamines, neurohypophyseal, peptides, and other substances on cultured corticotropic cells. Endocrinology 1983; 113:1121-29.

29. Avgerinos PC, Schuermeyer TH, Gold PW, Nieman L, Udelsman R, Loriaux DL, Chrousos GP. Synthetic human CRH as a probe of pulsatile ACTH and cortisol secretion. In: Episodic Hormone Secretion (W. Crowley and J.F. Hofler, Eds.), New York, Churchill-Livingstone 1987, pp. 389-401.

30. Avgerinos PC, Schuermeyer TH, Gold, PW, Tomai TP, Loriaux DL, Sherins RJ, Cutler GP Jr, Chrousos GP. Pulsatile administration of human corticotropin releasing factor (hCRF) in patients with secondary adrenal insufficiency in restoration of normal cortisol secretory pattern. J Clin Endocrinol Metab 1986; 62:816-821.

31. Schulte HM, Chrousos GP,Gold PW, Booth JP, Oldfield TH, Cutler GB Jr and Loriaux DL. Continuous infusion of CRF in normal volunteers. Physiological and pathophysiological implications. J Clin Invest 1985;75: 1781-85.

32. Rivier C, Vale W. Modulation of stress-induced ACTH release by corticotropin releasing factor, catecholamines and vasopressin. Nature 1983; 305:325-27.

33. Plotsky PM, Rubin TO, Vale W. Hypophysiotropic regulation of adrenocorticotropin secretion in response to insulin-induced hypoglycemia. Endocrinology 1985; 117:323-29.

34. Axelrod J and Reisine TD. Stress hormones: Their interaction and regulation. Science 1984; 224:452-59.

35. Schulte HM, Chrousos GP,Oldfield EH, Gold PW, Cutler GB Jr, Loriaux DL. The effects of corticotropin releasing factor on the anterior pituitary function of stalk-sectioned cynomolgus macaques: dose response of cortisol secretion. J Clin Endocrinol Metab 1982; 55:810-12.

36. Grossman A, Perry L, Schally AV, Rees LH,Tomlin D, Coy D, Comary-Schull AM, Besser GM. The new hypothalamic hormone corticotropin releasing factor specifically stimulates the release of adrenocorticotropic hormone and cortisol in man. Lancet 1982; 8278:921-22.

37. Orth DN, Jackson RV, DeCherney GS, Debold CR, Alexander AN, Island DP,Rivier J, Rivier C, Spiess J, Vale W. Effect of synthetic ovine corticotropin-releasing factor: dose response of plasma adrenocorticotropin and cortisol. J Clin Invest 1983; 71:587-95.

38. Schuermeyer TH, Gold PW, Gallucci WT, Tomai TP, Cutler GB Jr, Loriaux DL, Chrousos GP. Effect and pharmacokinetic properties of the rat/human corticotropin releasing factor (r/h CRH) in rhesus monkeys. Endocrinology 1985; 117:300-306.

39. Schuermeyer TH, Avgerinos PC, Gold PW, Tomai TP, Gallucci WT, Cutler GB Jr, Loriaux DL, Chrousos GP. Human corticotropin releasing factor: dose response and time course of ACTH and cortisol secretion in man. J Clin Endocrinol Metab 1984; `59:1103-1108.

40. Schulte HM, Chrousos GP, Gold PW, Oldfield EH, Phillips JM, Munson PJ,Cutler GB Jr, Loriaux DL. Metabolic clearance rate and plasma half-life of radioiodinated ovine corticotropin releasing factor in a primate. J Clin Endocrinol Metab 1982; 55:1023-27.

41. Schulte HM, Chrousos GP, Oldfield EH, Gold PW, Cutler GB Jr, Loriaux DL. Corticotropin releasing factor administration in normal men: pituitary and adrenal responses in the morning and evening. Horm Metab Res 1985; 21:64-74.

42. Gold PW, Chrousos GP, Kellner CH, Post RM, Roy A, Avgerinos PC, Schulte HM, Oldfield EH, Loriaux DL. Psychiatric implications of basic and clinical studies with corticotropin releasing factor. Am J Psychiatry 1984; 141:619-27.

43. Gold PW, Loriaux DL, Roy A, Kellner CH, Post RM, Kling MA, Calabrese JR, Oldfield EH, Pickar D, Avgerinos PC, Paul SM, Cutler GB Jr, Chrousos PG. Physiology of hypercortisolism in depression and Cushing's disease. N Engl J Med 1986; 314:1329-35.

44. Holsboer F, Genken H, Stalla GK, Muller GH 1984 (Letter). Blunted ACTH responses to human CRH in depression. N Engl J Med 1984; 311:1127.

45. Gold PW and Chrousos GP. Clinical studies with corticotropin releasing factor: Implications for the diagnosis and pathophysiology of depression, Cushing's disease, and adrenal insufficiency. Psychoneuroendocrinology 1985; 10:401-19.

46. Halbreich V, Asnis GM, Slindledecker R, Zumoff B, Nathan RS. Cortisol secretion in endogenous depression II. Time-related functions. Arch Gen Psychiat 1985; 42:909-15.

47. Nemeroff CB, Widerlov E, Bissette G, Wallens H, Karllson I, Eklund K, Kilts G, Loosen P. Elevated immunorreactive corticotropin releasing hormone in depressed patients. Science 1984; 224:1342-44.

48. Roy A, Pickar D, Chrousos GP, Doran A, Paul SM, Loriaux DL, Gold PW. Post-dexamethasone cortisol levels in depression correlate positively with immunoreactive CSF CRH. Am J Psychiat. 1987; 114:641-45.

49. Selye H. Stress syndrome. A syndrome produced by diverse noxious agents. Nature (London) 1936; 138-32.

50. Renold AE, Jenkins D, Thorn GW. The use of intravenous ACTH. A study in quantitative adrenocortical stimulation. J Clin Endocrinol Metab 1952; 12:763-97.

51. Amsterdam JD, Winokur A, Abelman E, Lucki I, Rickels K. Cosyntropin (ACTH 1-24) test in depressed patients and healthy subjects. Am J Psychiat 1938; 140:907-909.

52. Gold PW, Gwirtsman H, Avgerinos PC, Nieman L, Jimerson D, Kaye W, Loriaux DL, Chrousos GP. Abnormal hypothalamic-pituitary-adrenal function in anorexia nervosa: Pathophysiological mechanisms in underweight and weight-recovered patients. N Engl J Med 1986; 314:1335-42.

53. Cantwell DP, Stuzenberger S, Burnough J et al. Anorexia nervosa. An affective disorder? Arch Gen Psychiat 1977; 34:1087-94.

54. Roy-Byrne PP, Uhde TW, Post RM, Gallucci WT, Chrousos GP, Gold PW. Blunted ACTH responses to ovine CRH in panic-anxiety disorder. Am J Psychiat 1986; 143:396-99.

55. Roy A, Pickar D, Doran A, Paul SM, Chrousos GP, Gold PW. Normal ACTH and cortisol responses to CRH in schizophrenia. Am J Psychiat 1986; 143:1393-97.

56. Koob GF. Behavioral actions of corticotropin releasing hormone in the central nervous system. In: Abstracts of the UCLA Symposia on Moelcular and Cellular Biology, Key stone, Colorado. J Cell Biochem, Suppl 12D, 1988;299.

57. Swerdlow NR, Geyer MA, Vail WW, Koob GF. CRF potentiates acoustic startle in rats: Blockade by chlordiazepoxide. Psychopharm 1986; 88:142-52.

58. Sapolsky RM, Krey LC, McEwen BS. Stress down-regulates corticosterone receptors in a site-specific manner in the brain. Endocrinology 1984; 114:287-92.

59. Halbreich U, Asnis GM, Zumoff B, Nathan RS. The effect of age and sex on cortisol secretion in depressives and normals. Psychiat Res 1984; 13:221-29.

60. Kling MA, Kellner CH, Post RM, Gold PW. Neuroendocrine effects of limbic activation by electrical, spontaneous, and pharmacological modes: Relevance to the pathophysiology of affective dysregulation in psychiatric disorders. Prog. Neuropsychopharmacol Biol Psychiatry 1987; 11:459-81.

61. Redmond DE. Current concepts II - New evidence for a locus coeruleus-norepinephrine connection with anxiety. Life Sci 1979; 25:2149-62.

62. Foote SL, Ashton-Jones FSL and Bloom FE. Impulse activity of locus coeruleus neurosis in awake rats and monkeys as a function of sensory stimulation and arousal. Proc Natl Acad Sci USA 1980; 77:3033-37.

63. Christiansten NJ, Vestergaard P, Sorenson T, Raphaelson OJ. Cerebrospinal fluid adrenaline and noradrenaline in depressed patients. Acta Psychiatrica Scand 1980; 61:178-82.

64. Wyatt RJ, Portnoy B, Kupfer DJ, Snyder F, Engelman K. Resting plasma catecholamine concentrations in patients with depression and anixiety. Arch Gen Psychiatry 19071; 24:65-70.

65. Esler M, Turbott J, Schwarz R, Leonard P, et al. The peripheral kinetics of norepinephrine in depressive illness. Arch Gen Psychiatry 1982; 39:295-300.

66. Linnoila M, Karoum F, Calil HM et al. Alteration of norepinephrine metabolism with desimpramine and zimelidine in depressed patients. Arch Gen Psychiatry 1982:39:1025-28.

67. Calogero AE, Gallucci WT, Chrousos GP, Gold PW. Effects of catecholamines upon rat hypothalamic corticotropin releasing hormone secretion *in vitro* : Clinical implications. J Clin Invest 1988, in press.

68. Calogero AE, Gallucci WT, Bernardini R, Saotis C, Gold PW, Chrousos GP. Effects of cholinergic agonists and antagonists on rat corticotropin releasing hormonal section *in vitro*.. Neuroendocrinology 1987; 47: 303-308.

69. Valentino R, Foote SL, Ashton-Jones G. Corticotropin-releasing hormone activates noradrenergic neurons of the locus coeruleus. Brain Res. 1983; 270:363-67.

70. Valentino RJ. Corticotropin releasing factor (CRF) and physiological stressors alter activity of rat noradrenergic locus ceruleus (LC) neurons in a similar manner. In: Abstracts of the UCLA Symposia on Molecular and Cellular Biology, Keystone, Colorado. J Cell Biochem, Suppl 12D, 1988;210Q.

71. Cole BJ, Koob GF. Propranalol antagonizes the enhanced conditioned fear produced by CRF. In: Abstracts of the UCLA Symposia on Molecular and Cellular Biology, Keystone, Colorado. J. Cell Biochem, Suppl 12D, 1988;102Q.

72. Munck AS, Guyre P, and Holbrook NJ. Physiological functions of glucocorticoids in stress and their relations to pharmacological actions. Endocr Rev 1984; 5:25-44.

73. Dunn AJ, Gilderslane NB, Gary HL. Mouse brain tyrosine hydroxylase and glutamic acid decarboxylase following treatment with adrenocorticotropic hormone, vasopressin, and corticosterone. J Neurochem 1978; 31:977-982.

74. Gold PW, Goodwin FK, Chrousos GP. Clinical and biochemical manifestations of depression: Relationship to the Neurobiology of Stress. Part I. N Engl J Med, in press.

75. Gold PW,. Goodwin FK, Chrousos GP. Clinical and biochemical manifestations of depression: Relationship to the Neurobiology of Stress. Part II. N Engl J Med, in press.

76. Egeland JA, Gerhard DS, Davls DL. Bipolar affective disorders linked to DNA markers on chronosome II. Nature 1987; 325:783-87.

77. Aghajanian GK,Cedarbaum MJ, Wong RY. Evidence for norepinephrine mediated-collateral inhibition of locus coeruleus neurons. Brain Res 1977; 136:570-77.

78. Kling MA, Chrousos GP, Doran AR, Calabrese JR, Gold PW. CSF CRH in depression and Cushing's disease. Presented at the 139th Annual Meeting of the American Psychiatric Association. Syllabus and Proceedings in Summary Form, American Psychiatric Association 1986; 75, abstract.

79. Muller OA, Stalla GK, v Werder K. Corticotropin releasing factor: a new tool for the differential diagnosis of Cushing's syndrome. J Clin Endocrinol Metab 1983; 57:227-9.

80. Pieters GFFM, Hermus ARMM, Smals AGH, Bartelink AKM, Benraad TJ, Kloppenborg PWC. Responsiveness of the hypophyseal-adrenocortical axis to corticotropin releasing factor in pituitary-dependent Cushing's disease. J Clin Endocrinol Metab 1983; 57:963-8.

81. Nakahara M, Shibasaki T, Shizume K, et al. Corticotropin-releasing factor test in normal subjects and patients with hypothalamic-pituitary-adrenal disorders. J Clin Endocrinol 1983; 57-963-8.

82. Lytras N, Grossman A, Perry L, et al. Corticotropin releasing factor: responses in normal subjects and patients with disorders of the hypothalamus and pituitary. Clin Endocrinol (Oxf) 1984; 20:71-84.

INDEX